T0338274

NATURAL AND ARTIFICIAL PHOTOSYNTHESIS

NATURAL AND ARTIFICIAL PHOTOSYNTHESIS

SOLAR POWER AS AN ENERGY SOURCE

Edited by

Reza Razeghifard

Copyright © 2013 by John Wiley & Sons, Inc. All rights reserved.

Published by John Wiley & Sons, Inc., Hoboken, New Jersey.
Published simultaneously in Canada.

For general information on our other products and services or for technical support, please contact our Customer Care Department within the United States at (800) 762-2974, outside the United States at (317) 572-3993 or fax (317) 572-4002.

Wiley also publishes its books in a variety of electronic formats. Some content that appears in print may not be available in electronic formats. For more information about Wiley products, visit our web site at www.wiley.com.

Library of Congress Cataloging-in-Publication Data:

Razeghifard, Reza, 1964- editor of compilation.
  Natural and artificial photosynthesis : solar power as an energy source / editor, Reza Razeghifard.
     pages   cm
  Includes bibliographical references and index.
  ISBN 978-1-118-16006-0 (cloth)
 1. Renewable energy sources.   2. Photosynthesis–Industrial applications.   I. Title.
  TJ811.8.N37 2013
  621.042–dc23

                                                                                    2013002851

Printed in the United States of America

10 9 8 7 6 5 4 3 2 1

CONTENTS

The world is facing environmental issues due to increasing CO_2 concentration in the atmosphere and an energy crisis from the depletion of fossil fuels. Photosynthesis offers a promising solution to these global problems since it is the process that has made our planet habitable by releasing oxygen while capturing CO_2 by converting it into reduced carbon sources. The biochemical compounds are produced from sunlight, CO_2, and water, which are all unlimited resources. The process relies on several steps including the conversion of light energy into an electron flow, extracting electrons from water molecules, and generating energy rich and reducing compounds for CO_2 fixation. The purpose of this book is to describe what is known about this complex process and how it can be applied to produce clean renewable energy.

Chapter 1 gives an overview of solar energy and introduces four ways for its direct use, including photosynthesis. Chapter 2 explains all aspects of light reactions in detail including light absorption, excitation energy transfer, photochemistry, and electron transfer proteins. CO_2 assimilation pathways in photoautotrophic and chemoautotrophic organisms are discussed, including properties of the RuBisCO enzyme, evolution of photosynthesis, and CO_2 concentrating mechanisms.

Learning from natural photosynthesis, artificial systems can be constructed for harvesting solar energy to produce photocurrents and drive chemical reactions. Photosystems evolved over billions of years as perfect nanomachines capable of generating an electron flow linked to chemical reactions using light energy. Chapter 3 describes principles of solar energy utilization by photosystem II, an optimized nanoscale machine for efficient light-induced water splitting. Current knowledge on the structure of photosystem II and the mechanism of water oxidation are provided. Learning from properties of the photosystem II complex, three approaches are proposed for constructing suitable technical devices to convert solar energy into fuels. These approaches are: the use of genetically modified photosynthetic organisms, development of hybrid artificial photosynthetic systems consisting of units isolated from biological materials and functional synthetic entities, and synthesis of biomimetic systems. Chapter 4 continues the artificial photosynthesis topic by reviewing systems capable of converting solar energy into an electron flow for generating photocurrents using natural or engineered photosystem proteins assembled on conductive surfaces or producing fuels such as hydrogen gas by connecting photosystems to hydrogenase enzymes. Recent advancements in mimicking water oxidation using manganese complexes and introducing manganese chemistry in proteins through protein design are covered. Constructing photoactive proteins using chlorophyll derivatives to initiate electron transfer between redox cofactors is also mentioned. Chapter 5 introduces the photochemistry of ruthenium and photoinduced transfer of electrons in ruthenium complexes in the presence of electron donors and acceptors. Particular attention is given to Ru(II) complexes synthesized as photoactive models for mimicking the Mn-cluster of PSII or active sites of hydrogenases.

Growing microalgae offers a practical solution through photosynthesis for the sequestration of CO_2 into biomass and the production of biohydrogen gas as a CO_2 neutral fuel. Chapter 6 describes the mechanisms of CO_2 fixation into algal biomass and hydrogen production by hydrogenase and nitrogenase enzymes in green algae, vegetative cells of non-nitrogen-fixing and heterocysts of nitrogen-fixing cyanobacteria. Furthermore, different modes of hydrogen production such as direct and indirect biophotolysis, light and dark fermentations, and use of chemicals are explained. Chapter 7 discusses the CO_2 uptake efficiency of cyanobacteria and mitigation of costs through captured-carbon products from engineered cyanobacteria. Several useful and direct products of captured-carbon including isobutyraldehyde, isobutanol, fatty acids, hydrocarbons, 1-butanol, isoprene, and poly-3-hydroxybutyrate are described. Chapter 8 presents genetic and metabolic engineering of microalgae for biohydrogen production and provides the rationale for creating microalgae mutants with improved hydrogen production. The chapter examines hydrogen production by microalgae engineered to have O_2-tolerant hydrogenases, increased specificity of electron flow toward H_2 production, impaired processes competing for NADPH such as CO_2 fixation, and less active PSII or truncated light-harvesting complexes. Hydrogen production and its relation to nitrogen and sulfur metabolisms, and cyclic electron flow around photosystem I are also discussed.

Microalgae are well-adapted microorganisms and can grow in open ponds, raceways, or closed photobioreactors under varieties of growth conditions. Chapter 9 discusses the advantages of using algae for biofuel production in the form of biodiesel, hydrocarbons, hydrogen, and ethanol. Cultivation conditions for biofuel production including light, nutrient removal, and biomass harvesting are also discussed. Recent achievements in algal biofuels including the design of advanced photobioreactors and cost-effective technologies for biomass harvesting and drying are mentioned. Strategies for enhancement of algal biofuel production including exploration of growth conditions, nutrients, and high-value coproducts are covered. Chapter 10 describes hydrogenase enzymes that are involved in diverse hydrogen-producing algae. Approaches for enhancing algal hydrogen production such as immobilization processes, increasing the resistance of algae to stress conditions, optimization of bioreactor conditions, and genetic engineering approaches to improve photosynthetic efficiency are discussed. Chapter 11 focuses on the design of photobioreactors for culturing microalgae and the factors that affect their productivity such as light distribution, gas exchange, and shear stress. Chapter 12 describes issues related to the design and operation of industrial cultivation systems for microalgal mass production. Kinetics, mass balance, and energy balance of microalgal growth, and gas–liquid mass transport in open systems and different photobioreactor configurations are discussed. At the end of this chapter, a case study for intensive production of bio-oil is described through defining maximum productivity and a rough cost estimation of the production plant.

Economic viability becomes the key factor for the production of biofuels from all available sources including plants or algae. Chapter 13 explores a range of biofuel feedstocks and summarizes the current advances and technological challenges for industrial scale production of biofuel, and associated production costs. The economic analyses for the production of biodiesel from microalgae and bioethanol from macroalgae are presented. Chapter 14 investigates the economic aspects of commercial production of bioethanol from microalgal biomass. The economics of the production process in terms of unit and operation costs are evaluated. Technical feasibility of the overall production process for various unit steps such as cultivation, dewatering, pretreatment, hydrolysis, fermentation, and product recovery are assessed. Chapter 15 highlights the strategy of process integration for production of multiple valuable chemicals as a partial solution to advance the economics and sustainability of algal biotechnology. Examples of chemicals derived from lipid, carbohydrate, and protein fractions of microalgae that appear promising for simultaneous production within integrated bioprocesses at the commercial scale or showed to be promising for scale-up are presented in this chapter. The relevant biotechnological processes and production systems including cultivation conditions that lead to the overproduction of these chemicals are discussed further. Chapter 16 explores the opportunities for the production of fuels and chemicals from lignocellulosic biomass. The production of fermentable sugars from biomass by pretreatment and enzymatic hydrolysis processes of cellulose and hemicellulose are discussed. Thermochemical conversion of biomass to fuels and chemicals at gasification, pyrolysis, and liquefaction steps are covered. A summary of assessments of the screening of potential candidates for the production of value-added chemicals from sugars, syngas, lignin, and biomass is also provided.

The technological advancements summarized in this book illustrate that biofuels can be produced by natural or artificial photosynthetic systems. These approaches are based on producing biofuel in an economically sustainable and feasible way using solar energy. It also highlights some technological challenges and their solutions for mass production of cost-effective biofuels. It is crucial that we continue worldwide research on photosynthesis as it can offer renewable sources of energies such as hydrogen gas, bioethanol, or bio-oil for making biodiesel or even jet fuels.

REZA RAZEGHIFARD

Helena M. Amaro, Centro Interdisciplinar de Investigação Marinha e Ambiental (CIMAR/CIIMAR)–Labotarório de Biodiversidade Costeira, Porto, Portugal.

Paul T. Anastas, Center for Green Chemistry and Green Engineering, Yale University, New Haven, CT, USA.

Lucy A. Arthur, School of Engineering and Science, Victoria University, Victoria, Australia.

Shota Atsumi, Department of Chemistry, University of California, Davis, CA, USA.

Lars Olof Björn, Department of Biology, Lund University, Lund, Sweden; Key Laboratory of Ecology and Environmental Science in Guangdong Higher Education, School of Life Science, South China Normal University, Guangzhou, China.

Diego Castano, Division of Math, Science & Technology, Farquhar College of Arts & Sciences, Nova Southeastern University, Fort Lauderdale, FL, USA.

Michael K. Danquah, Bio-Engineering Laboratory (BEL), Department of Chemical Engineering, Monash University, Victoria, Australia.

Debabrata Das, Department of Biotechnology, Indian Institute of Technology Kharagpur, West Bengal, India.

William O.S. Doherty, Centre for Tropical Crops and Biocommodities, Queensland University of Technology, Brisbane, Australia.

Niels Thomas Eriksen, Institute of Biotechnology, Chemistry and Environmental Engineering, Aalborg University, Aalborg, Denmark.

Ela Eroglu, Centre for Strategic Nano-Fabrication, School of Chemistry and Biochemistry, and ARC Centre of Excellence in Plant Energy Biology, The University of Western Australia, Crawley, Australia.

M. Glória Esquível, Instituto Superior de Agronomia (ISA)/Centro de Botânica Aplicada à Agricultura (CBAA), Lisboa, Portugal.

Dimitrios G. Giarikos, Division of Math, Science & Technology, Farquhar College of Arts & Sciences, Nova Southeastern University, Fort Lauderdale, FL, USA.

Govindjee, Department of Plant Biology, Department of Biochemistry and Center of Biophysics and Computational Biology, University of Illinois at Urbana-Champaign, Urbana, IL, USA.

Mark D. Harrison, Centre for Tropical Crops and Biocommodities, Queensland University of Technology, Brisbane, Australia.

Razif Harun, Bio-Engineering Laboratory (BEL), Department of Chemical Engineering, Monash University, Victoria, Australia and Department of Chemical and Environmental Engineering, Universiti Putra Malaysia, Serdang, Malaysia.

Philip A. Hobson, Centre for Tropical Crops and Biocommodities, Queensland University of Technology, Brisbane, Australia.

Hassan J, Bio-Engineering Laboratory (BEL), Department of Chemical Engineering, Monash University, Victoria, Australia.

Azadeh Kermanshahi-pour, Center for Green Chemistry and Green Engineering, Yale University, New Haven, CT, USA.

Kanhaiya Kumar, Department of Biotechnology, Indian Institute of Technology Kharagpur, West Bengal, India.

F. Xavier Malcata, Department of Chemical Engineering, University of Porto, Porto, Portugal.

Antonio Marzocchella, Chemical Engineering Department, Università degli Studi di Napoli Federico II, Napoli, Italy.

Mark P. McHenry, School of Engineering and Energy, Murdoch University, Murdoch, WA, Australia.

Pouria Mehrani, Orbital Australia Pty. Ltd., Balcatta, WA, Australia.

Navid R. Moheimani, Algae R&D Centre, School of Biological Sciences and Biotechnology, Murdoch University, Murdoch, WA, Australia.

Sagadevan G. Mundree, Centre for Tropical Crops and Biocommodities, Queensland University of Technology, Brisbane, Australia.

Ian M. O'Hara, Centre for Tropical Crops and Biocommodities, Queensland University of Technology, Brisbane, Australia.

John W. K. Oliver, Department of Chemistry, University of California, Davis, CA, USA.

Giuseppe Olivieri, Chemical Engineering Department, Università degli Studi di Napoli Federico II, Napoli, Italy.

Anjana Pandey, Nanotechnology and Molecular Biology Laboratory, Centre of Biotechnology, University of Allahabad, Allahabad, India, and Department of Biotechnology, Motilal Nehru National Institute of Technology (MNNIT), Allahabad, India.

Teresa S. Pinto, Instituto Superior de Agronomia (ISA)/Centro de Botânica Aplicada à Agricultura (CBAA), Lisboa, Portugal.

Reza Razeghifard, Division of Math, Science & Technology, Farquhar College of Arts & Sciences, Nova Southeastern University, Fort Lauderdale, FL, USA.

Gernot Renger, Technical University Berlin, Institute Chemistry, Max-Volmer-Laboratory of Biophysical Chemistry, Berlin, Germany.

Piero Salatino, Chemical Engineering Department, Università degli Studi di Napoli Federico II, Napoli, Italy.

Dmitriy Shevela, Centre for Organelle Research, Department of Mathematics and Natural Science, University of Stavanger, Stavanger, Norway.

Li J. S. Shu, Bio-Engineering Laboratory (BEL), Department of Chemical Engineering, Monash University, Victoria, Australia.

Steven M. Smith, ARC Centre of Excellence in Plant Energy Biology, and Centre for Metabolomics, Faculty of Science, The University of Western Australia, Crawley, Australia.

Matthew Timmins, ARC Centre of Excellence in Plant Energy Biology, and Centre for Metabolomics, Faculty of Science, The University of Western Australia, Crawley, Australia.

Archana Tiwari, Department of Biotechnology, Guru Nanak Girls College, Ludhiana, Punjab, India.

Zhanying Zhang, Centre for Tropical Crops and Biocommodities, Queensland University of Technology, Brisbane, Australia.

Julie B. Zimmerman, Center for Green Chemistry and Green Engineering, Yale University, New Haven, CT, USA.

ΔpH	pH difference across thylakoid membrane
$\Delta\psi$	Electric potential difference across thylakoid membrane
3PG	3-Phosphoglycerate
ACP	Acyl carrier protein
ARA	Arachidonic acid
BChl	BacterioChl
C3	A type of CO_2-assimilation system that involves 3-C intermediates
C4	A type of CO_2-assimilation system that involves 4-C intermediates
CA	Carbonic anhydrase
Car	Carotenoid
CCM	Carbon-concentrating mechanism
Ce6	Chlorin *e*6
CEF	Cyclic electron flow around PSI
Chl	Chlorophyll
CPP	Captured-carbon products
CSTR	Continuously stirred reactor
CTR	CO_2 transfer rate
Cyt	Cytochrome
DCMU	3-(3,4-Dichlorophenyl)-1,1-dimethylurea
DHA	Docosahexaenoic acid
DIC	Dissolved inorganic carbon
E	Photon energy (used with subscript indicating wavelength in nm)
EET	Excitation energy transfer
EPA	Eicosapentaenoic acid
ETC	Electron transfer chain
FAD	Fatty acid desaturase
FAME	Fatty acid methyl ester
Fd	Ferredoxin
Fe–S	Iron–sulfur cluster
FFAs	Free fatty acids
FNR	Ferredoxin $NADP^+$ oxidoreductase or reductase
FRET	Förster resonance energy transfer
Ga	Billion years
GAPDH	Glyceraldehyde 3-phosphate dehydrogenase
GS	Glutamate synthase
Hox	Bidirectional hydrogenase

HTR	Horizontal tubular photobioreactor
HTU	Hydrothermal upgrading
Hyd	Hydrogenase
ITO	Indium–tin oxide
LHC	Light-harvesting complex
MEC	Major equipment cost
Mn-cluster	Manganese cluster
NADHdh	NADH dehydrogenase
NADP$^+$	Oxidized form of nicotinamide adenine dinucleotide phosphate
NADPH	Reduced form of nicotinamide adenine dinucleotide phosphate
Ni-NTA	Ni-nitrilotriacetic acid
NiR	Nitrite reductase
NLP	Nanolipoprotein particle
n-side	Stromal side of thylakoid membrane
OEC	Oxygen-evolving complex
OTE	Optically transparent electrode
OTR	Oxygen transfer rate
P	Primary electron donor or photochemically active pigment
P/R	Photosynthesis/respiration ratio
PAR	Photosynthetically active radiation
PBP	Phycobiliprotein
PC	Phycocyanin or plastocyanin
PCP	Peridinin–Chl a-protein
PE	Phycoerythrin
PEP	Phosphoenolpyruvate
PFR1	Pyruvate–ferredoxin oxidoreductase
PG	Photogenerator
PHB	Poly-3-hydroxy butyrate
pmf	Proton-motive force
PNS	Purple non-sulfur
PQ	Plastoquinone
PRC	Carbonic anhydrase
PSI	Photosystem I
p-side	Lumen side of thylakoid membrane
PSII	Photosystem II
PT	Proton transfer
PUFA	Unsaturated fatty acid
RC	Reaction center
RET	Respiratory electron transport
ROS	Reactive oxygen species
RuBisCo	Ribulose bisphosphate carboxylase/oxygenase
S/V	Surface to volume ratio
SDH	Succinate dehydrogenase
SHF	Separate hydrolysis and fermentation
SiR	Sulfite reductase

SSF	Simultaneous saccharification and fermentation
SulP	Sulfate permease
SVO	Straight vegetable oil
TAG	Triacyglycerol
TAP	Tris-acetate-phosphate medium
TAP-S	(Sulfur-deprived) tris-acetate-phosphate medium
TCA cycle	Tricarboxylic acid cycle
TE	Thioesterase
TrxR	Ferredoxin–thioredoxin reductase
VTR	Vertical tubular photobioreactor
WOC	Wateroxidizing catalyst
WSCP	Water-soluble Chl binding protein
XR	Xanthinereductase
η	Light conversion efficiency

CHAPTER 1

Physics Overview of Solar Energy

DIEGO CASTANO

1.1 INTRODUCTION

Undoubtedly the most important factor in the study of solar energy is the sun, the local star and the gravitational stake to which the earth is tethered. All forms of energy on earth, except for nuclear, can ultimately be traced to the sun. There are on the order of a hundred billion stars in the Milky Way galaxy and about a hundred billion galaxies in the known universe. Stars began forming several hundred thousand years after the Big Bang, which, based on current theories of cosmology, happened 13.7 billion years ago. The sun itself formed about 5 billion years ago and should shine for another 5 billion. The earth consequently finds itself in a somewhat special place in space and time. The low entropy state of the solar system ensures vital change for billions of years to come. In terms of thermodynamics, the sun represents an effective high temperature reservoir of temperature 6000 kelvins that bathes the earth (the low temperature reservoir at an average 287 kelvins) with a radiant intensity of approximately 1400 W/m^2 above the atmosphere. Attenuation due to absorption and scattering in the atmosphere reduces this value. The exact value of solar irradiance at the earth's surface depends on the sunlight's path through the atmosphere and is generally no greater than 1000 W/m^2. The solar irradiance translates into a naive, maximal power output of approximately 10^{17} watts (compare this to the world's average energy consumption rate in 2010, of just over 10^{13} watts [1]). It would appear that the sunlight reaching just earth (the earth subtends a mere 4.6×10^{-8}% of the whole solid angle at the sun; see Fig. 1.1) can supply earth's needs 10,000 times over. In fact, Freeman Dyson considered the possibility that advanced civilizations would surround their suns with enough orbiting artificial satellites (this system is referred to as a Dyson sphere) to harness most of the star's solar energy. In the case of the sun, an optimal Dyson sphere would generate on the order of 10^{27} watts, a significant fraction of the luminosity of the sun ($L_\odot = 4.8 \times 10^{27}$ W).

Natural and Artificial Photosynthesis: Solar Power as an Energy Source, First Edition. Edited by Reza Razeghifard.
© 2013 John Wiley & Sons, Inc. Published 2013 by John Wiley & Sons, Inc.

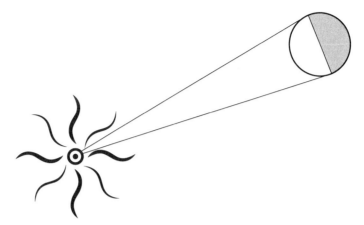

FIGURE 1.1 The earth subtends six nanosteradians at the sun.

In fact, about 30% of the sunlight reaching the earth is reflected back into space above the atmosphere. The albedo, or fraction of reflected sunlight, depends strongly on such factors as snow and cloud cover. The 70% of the incident sunlight not reflected is effectively transmitted through the atmosphere, which is mostly transparent to visible light, to the surface. The thermal, or infrared, radiation emitted by the surface of the earth is then absorbed by the atmosphere and reradiated. This atmospheric greenhouse effect is what leads to a global mean surface temperature of 287 kelvins as compared to a freezing 254 kelvins without an atmosphere.

1.2 THE SUN

The standard theories of particle physics and cosmology describe the Big Bang as the moment of creation of space, time, matter, and energy. At first, there was just one superforce to mediate the interactions between particles. The superforce eventually separated out into the four known fundamental forces of today, that is, gravity, electromagnetism, weak and strong nuclear. All four of these forces are important in the development and function of the sun.

Early on (about 20 minutes after the Big Bang) the universe is a cauldron containing, among other particles, electrons and various light nuclei. Recombination, which occurs around 300,000 years after the Big Bang, is the term used to describe the moment in the history of the universe when electromagnetically neutral bound states of matter were first possible in large numbers. In this era, the universe is a rarified gas consisting mostly of hydrogen (75%) and helium (25%). Gravity starts to make its presence known and leads to clumping due to inhomogeneities in the matter distribution. As a hydrogen/helium gas clump shrinks, the gravitational energy is turned into kinetic energy, resulting in temperatures at the core high enough to fuse the hydrogen and helium. At a temperature of ten million kelvins, protons can tunnel through the Coulomb repulsion barrier, resulting in the first nuclear reaction of the so-called proton–proton cycle, which is prevalent in stars like the sun,

$$^1H + {}^1H \rightarrow {}^2H + e^+ + \nu_e \tag{1.1}$$

Note that the fusing of the protons involves the strong nuclear force, and the appearance of the neutrino indicates that the weak nuclear force was also involved in this process. This reaction is followed by the reaction

$$^2H + {}^1H \rightarrow {}^3H + \gamma \tag{1.2}$$

The last reaction in the cycle can be any one of four with the most common in the sun being

$$^3He + {}^3He \rightarrow {}^4He + {}^1H + {}^1H \tag{1.3}$$

The net energy released along this sequence is 27 MeV. This is only one sequence of many possible ones, all of which lead to ^4He. In stars like the sun this eventually leads to an inert helium core with the hydrogen fusion occurring in the shell surrounding the core. This causes the star to grow and become a red giant. The helium core continues to collapse gravitationally until the temperature increases to the point at which helium–helium fusion can occur. The fusion of heavier and heavier nuclei continues in stars that are massive enough until ^{56}Fe is reached. This isotope of iron has the largest binding energy per nucleon, and fusion reactions that produce heavier nuclei are consequently endothermic. Although heavy elements can still be produced in stars, most heavy elements are created in supernovas.

One way astronomers classify stars is by their luminosities and spectral types or surface temperatures. The resulting scatter plot is called a Hertzsprung–Russell diagram (Fig. 1.2). Most stars, including the sun, fall in the region referred to as the main sequence.

1.3 LIGHT

The history of the developments in the theory of light is a long one. The Greeks, from as far back as Pythagoras, believed that light emanated from visible bodies. Some even philosophized that it traveled at a finite speed. However, most people before the 17th century believed that light was instantaneous. Galileo is credited with being the first well-known scientist to attempt to measure the speed of light. His experiment involved him and an assistant on distantly separated hills with lanterns and some sort of time measuring device, perhaps a water clock. Due to the inherent lack of precision in the design of the experiment, his result was ambiguous. Of the speed of light, he is alleged to have said, "If not instantaneous, it is extraordinarily rapid." He also concluded that the speed was at least ten times faster than sound. About a decade later, Ole Rømer used essentially the same idea as Galileo—that of measuring the

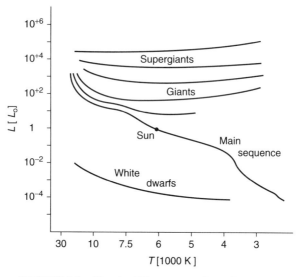

FIGURE 1.2   Sketch of Hertzsprung–Russell diagram.

time it takes a light signal to cover a spatial path—but Rømer did it with a longer path, the diameter of the earth's orbit around the sun. His results were considerably better ($c = 2 \times 10^8$ m/s). It took about another 200 years to perfect an earth bound experiment, but Leon Foucault performed an experiment involving rotating mirrors (based on one designed by Hippolyte Fizeau using rotating toothed wheels) that was able to measure the speed of light accurately to good precision and that agrees with modern measurement to four significant figures ($c = 2.998 \times 10^8$ m/s).

The nature of light, whether it be a particle or wave phenomenon, was a topic of debate in the 17th century. Isaac Newton was on the particle side and referred to the constituent particles as corpuscles. On the wave side was Christian Huygens. Newton's reputation helped the particle side, but the phenomena of interference and diffraction weighed heavily on the side of the wave theory. Today, at least in classical terms, light is recognized as a wave phenomenon, but it is interesting to note that quantum mechanics has forced reconsideration and Newton's corpuscles can be thought resurrected as photons.

The unification of electricity and magnetism made by James Clerk Maxwell in the mid-19th century put the wave nature of light on firm theoretical footing. Maxwell's correction to Ampere's law led him to wave equations for electric and magnetic fields with a wave speed relation that involved electric permittivity and magnetic permeability constants

$$c = \frac{1}{\sqrt{\varepsilon_0 \mu_0}} \tag{1.4}$$

This calculable wave speed coincided with the experimentally measured speed of light. The conclusion was inescapable: light is a manifestation of electromagnetic waves. In 1800 Frederick William Herschel had ascertained the existence of an invisible form of light by noting that the shadow region beyond the red end of a prism-induced spectrum of sunlight registered a higher temperature than the red lighted region. This invisible light is recognized today as the infrared.

Although it is far afield from the main thrust of this overview, it is perhaps interesting to note that questions immediately arose concerning the nature of the supporting medium for these waves. Whatever the medium was, it pervaded all of space and was referred to as the ether. The possibility of anisotropies in the speed of light due to the ether would occupy the minds of theorists and the efforts of experimentalists, in particular, Albert Michelson and Edward Morley, until the beginning of the 20th century. The null result of the late 19th century Michelson–Morley experiment to detect the so-called ether wind was, in retrospect, consistent with Albert Einstein's theory of relativity.

Along with the theory of relativity, the early 20th century saw the development of quantum mechanics. The understanding of various light phenomena was the catalyst for its inception. The story begins with the concept of a blackbody, an idealized body capable of absorbing all incident electromagnetic radiation. This is to be compared with a real body, which reflects and/or transmits some fraction of the incident radiation. A perfect absorber is also a perfect emitter by Kirchhoff's law of radiation, so a blackbody is also a perfect emitter. A kiln with a small opening is a good realization of a blackbody emitter. The radiation emitted was measured (using a bolometer), and two important facts were discovered. The intensity radiated (or total radiant emittance) at all wavelengths depends only on the absolute surface temperature of the blackbody according to the Stefan–Boltzmann law

$$I_b = \sigma T^4 \tag{1.5}$$

where $\sigma = 5.67 \times 10^{-8}$ W/(m^2·K^4) is the Stefan–Boltzmann constant. Also discovered was the fact that there was a peak to the intensity distribution, known as Wien's law, at

$$\lambda = \frac{0.003 \text{ m} \cdot \text{K}}{T} \tag{1.6}$$

Max Planck set out to theoretically derive the blackbody intensity distribution using the theory of electromagnetism and the laws of thermodynamics. He met with failure until he hypothesized the radiation quantum of energy

$$E = hf \tag{1.7}$$

where $h = 6.63 \times 10^{-34}$ J·s and f is the frequency of the radiation. With this assumption, he was able to derive Planck's law of blackbody radiation (see Fig. 1.3):

$$\frac{dI_b(\lambda, T)}{d\lambda} = \frac{2\pi hc^2}{\lambda^5} \frac{1}{e^{hc/\lambda k_B T} - 1} \tag{1.8}$$

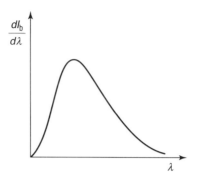

FIGURE 1.3 Characteristic blackbody curve.

By fitting the sun's radiation curve to this formula, its surface temperature can be deduced to be 6000 kelvins. The earth's own effective radiant temperature is around 240 kelvins.

Planck's idea was applied by Einstein to explain the photoelectric effect. In the late 19th century experiments conducted by Heinrich Hertz and Philipp Lenard showed that the energy of electrons ejected from a metal upon electromagnetic irradiation was independent of the radiation's intensity but depended on its frequency. By assuming that the electromagnetic radiation transferred only a quantum of energy to the electrons in the metal, Einstein was able to explain the effect. The predicted maximum energy of a free photoelectron was

$$KE_{max} = hf - \Phi \qquad (1.9)$$

where Φ is the work function, the minimum energy required to strip the electron from the metal. This is the basis of the photoemissive cell or phototube (see Fig. 1.4).

1.4 THERMODYNAMICS

It is often stated that thermodynamics was developed in response to the desire to make a better steam engine. As such it can be thought of as the study of thermal energy conversion. So what is energy? The term is pervasive in the modern world,

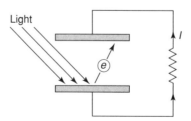

FIGURE 1.4 A current produced by the photoelectric effect.

and the concept is often reified, especially in scientifically informal settings. Energy is, in fact, an abstract concept, and its conservation (constancy in time) is a powerful organizational principle in physics. It comes in an apparent myriad of forms, such as solar, chemical, electrical, and even dark, but they can all be placed into two broad categories, kinetic and potential. Perhaps the most intuitive, if not the most perceptively apparent, is the kinetic type: If there is motion in a system, then there is energy. Conversely, and naively to be sure, energy is what makes things go.

There is (thermodynamic) energy associated with matter due to its atomic nature and resulting from the randomized (thermal) motion and from the interactions of the constituent particles. One of the most important results from the kinetic theory of gases is that temperature is a measure of the average kinetic energy of the constituents of matter. The finite size of bodies implies that the kinetic energies of their constituents must be changing due to collisions and oscillations. It is known from atomic theory that matter contains charged particles, such as electrons and protons. Maxwell's electromagnetic theory predicts that accelerated charged particles will radiate. Therefore all bodies at nonzero absolute temperature should radiate, and Eq. 1.8 predicts the spectrum of an ideal one.

In thermodynamics, the universe is divided into two parts, the system under consideration and its surroundings. All exchanges between system and surroundings are done across a boundary, the real or effective surface separating the two. The total energy of a thermodynamic system is referred to as its internal energy (U) and includes kinetic energies as well as the potential energies of particle interactions. In thermodynamics, heat (Q) is the process variable that represents the microscopically unobservable transfer of energy, whereas work (W) represents the macroscopically observable transfer of energy. Heat transfer is typically driven by diathermal, or unrestricted, contact between two bodies of differing temperatures. A boundary that does not permit the flow of heat is called adiabatic. The first law of thermodynamics is the statement of energy conservation. Any changes in the internal energy of a system must be the result of energy transfer across the boundary,

$$dU = d'Q - d'W \qquad (1.10)$$

where d' implies an inexact differential, and the work is considered done by the system.

Colloquially, the task of building an engine, for example, a steam engine, amounts to devising a way to turn internal into external energy, or how to use a source of heat to make something go. The first steam engine (the aeolipile; see Fig. 1.5) is credited to Hero of Alexandria, who lived in the 1st century A.D. A tropical hurricane is a natural example of a steam engine, effectively using the thermal energy in the surface water of the ocean to power wind. James Watt perfected the modern steam engine in the late 18th century. In all engines, some energy (in fact, generally quite a bit) is always wasted and in a form that is useless to the engine. The design details of the engine will affect the exact amount of wasted energy but there is always a nonzero amount, as surely as heat always flows spontaneously from higher to lower temperature. The second law of thermodynamics is the statement that the preceding empirical

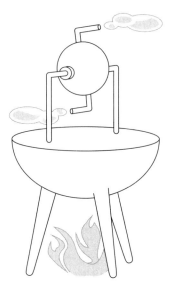

FIGURE 1.5   The aeolipile.

observation is indeed the case. The second law can be put into a mathematical form by noting that there exists a function of the extensive parameters of a system in a state of equilibrium that is maximized in any spontaneous process. The function is called the entropy (S) and for quasistatic processes

$$dS = \frac{d'Q}{T} \tag{1.11}$$

Engines exploit a temperature difference to extract useful work. The ideal cyclic engine, given two working temperatures, is one for which the entropy change of the universe is zero (i.e., its operation is reversible). Such an engine is called a Carnot engine, named after Sadi Carnot who first conceptualized it. The Carnot cycle consists of two isothermal processes and two adiabatic ones. The Carnot engine's efficiency depends on the two reservoir temperatures,

$$\varepsilon_{\text{Carnot}} = 1 - \frac{T_L}{T_H} \tag{1.12}$$

where $T_{L(H)}$ is the lower (higher) heat reservoir temperature. Unfortunately, the Carnot engine is an idealization. No heat transfer process is ever reversible in practice. Moreover, the sequence of processes associated with the Carnot cycle would be difficult to realize in a practical way. A more practical engine than Carnot's, with a theoretical efficiency that nevertheless matches Carnot's, is the Stirling engine (with regenerator). It is an external combustion engine that uses a single phase working substance (a gas, such as air). The Stirling cycle consists of two isothermal and two isochoric processes (see Fig. 1.6). Although they generally have low power outputs

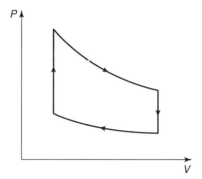

FIGURE 1.6   Stirling cycle.

for their size, Stirling engines are relatively easy to make and can exploit even small temperature differences.

There are at least four obvious ways to use direct energy from the sun. The first involves the direct absorption of sunlight. Architecturally, living spaces can be warmed by designing them to take advantage of sunlight. Water can also be heated directly by the sun for different uses such as space heating. By using devices, such as Fresnel lenses or parabolic mirrors, the sun's rays can be concentrated to increase input. Fresnel lenses have large aperture and dioptric power. Parabolic mirrors have the property that incident parallel light is focused without aberration. Both of these can collect light over a relatively large area and intensify it significantly.

The second way to exploit the thermal energy from the sun is to use an engine, perhaps in conjunction with some focusing system such as the ones discussed above. The Stirling engine can easily be adapted to exploit an external heat source and has consequently gained popularity in the solar energy business. Thermoelectric generators based on the Seebeck, or thermoelectric, effect can also be used to convert thermal energy into electricity, although these are usually less efficient than Stirling engines.

The third way is the direct conversion of sunlight into electricity and is the subject of Section 1.5.

1.5   PHOTOVOLTAICS

As discussed above, photoemission can be used to generate a current, but there is another related photovoltaic effect that involves semiconductors rather than metals. The discovery of the photovoltaic effect is credited to Alexandre-Edmond Becquerel in 1839. He discovered that illuminating one of the electrodes in an electrolytic cell caused the current to increase. The effect was also seen in solids, like selenium, in the late 19th century. In 1954, Bell Laboratories produced the first photovoltaic cell using a crystalline silicon semiconductor. In the photovoltaic effect, like in the photoelectric effect, light energy is absorbed by electrons, but, unlike in the photoelectric effect, the electrons are not ejected from the semiconductor.

Crystalline silicon, for example, has an extended, regular atomic pattern called a lattice. The basic unit of the lattice is a group of five silicon atoms. They form a geometrical array in which one is at the center of a tetrahedron and the other four at the centers of its faces. This structure is a consequence of the four outer valence electrons in silicon's third, or M, shell. When many identical atoms in close proximity are considered together, such as in a lattice, their electron's otherwise discrete energies (i.e., in isolation) are smeared into bands. In insulators, the highest band that is occupied by electrons at absolute zero is called the valence band. The next higher band, which is empty in this case, is called the conduction band. The energy gap between these bands in insulators is relatively large compared to characteristic thermal energies (0.025 eV at $T = 300$ K), so even above absolute zero an insulator has an effectively empty conduction band. In a conductor, the conduction band contains electrons even at absolute zero. In simplest terms, the electrons in the conduction band can be considered effectively free, that is, not bound to any one atom and therefore able to roam throughout the conductor. The electrons that are freed from covalent bonds leave a positively charged hole in the fixed lattice substratum. When a hole is filled by a neighboring electron, it in turn leaves behind a hole. The effective movement of the hole resembles that of a positively charged particle; the effect is similar to that of bubbles in a fluid. In a semiconductor, at absolute zero, the conduction band is empty like for an insulator. However, the energy gap is significantly smaller than in an insulator, and as the temperature rises thermal agitation is sufficient to lift electrons into the conduction band, making the material a conductor. Silicon has an energy gap of 1.1 eV at 300 K. In fact, every electron in the conduction band is accompanied by a hole in the valence band.

To change the balance of electrons in the conduction band and of holes in the valence band, a semiconducting material must be doped; that is, an impurity must be introduced into the otherwise homogeneous lattice. If phosphorus is introduced into the silicon lattice it will form the same four covalent bonds, but it will have an extra loosely bound electron. The energy level of these extra electrons lies in the energy gap just below the conduction band, so they readily become conduction electrons leaving behind a fixed positive ion. Impurities that have an extra valence electron, like phosphorus, are called donors. The semiconductor produced by donor doping is called n-type since the majority charge carriers are negative electrons. In an analogous fashion, silicon can be doped with an impurity that has one less valence electron, like boron. This type of impurity is called an acceptor. In this case the boron takes an electron from a neighboring silicon atom and creates a hole in the valence band. The energy level of these stolen (acceptor) electrons is just above the valence band. The holes are free to move as positive charge carriers, and the negative boron ions are fixed. The semiconductor produced by acceptor doping is called p-type since positive holes are the majority charge carriers.

Light incident on a semiconductor will generate electron–hole pairs if the photon energy is greater than the band gap. However, the pair will recombine unless the two carriers can be kept separated. Therefore to generate electrical power from incident

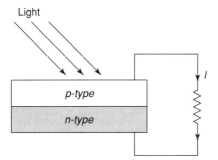

FIGURE 1.7 A solar cell in action.

light on these semiconductor materials requires a potential barrier. This barrier can be set up through the juxtaposition of p- and n-type materials. At the junction, both conduction and valance electrons move from the n-type side to the p-type side, thereby filling holes in the p-side and creating holes in the n-side. This process is finite and happens quickly. It creates a net fixed positive charge on the n-side of the junction and a net negative fixed charge on the p-side. This is known as the depletion zone and represents the potential barrier. It stops any further flow of electrons into the p-side and of holes into the n-side, but it does not prevent electrons from moving into the n-side and holes from moving into the p-side. Ideally a light-generated electron–hole pair will be separated by this mechanism and an emf generated.

If the two sides of the solar cell are connected through a load, a current will flow (see Fig. 1.7). Unfortunately, not all electron–hole pairs generated in this way will contribute to the current. The probability that a pair will contribute depends on many factors, one of which is the location within the cell where the pair forms. Those pairs created in the depletion zone will separate with certainty, then with high probability avoid combination and contribute to the current. Efficiency is typically a measure of output to input ratio. In the case of a solar cell, the input is solar energy and the output is electrical energy. The efficiency of a solar cell depends on many factors such as the one just described. Among other things, it also depends on the fraction of incident light absorbed and the nature (intensity and spectrum) of that light. The selenium cells of the late 19th century had 1% efficiency. The 1954 Bell Labs silicon cell had 4% efficiency. The maximum theoretical efficiency of a single p-n junction solar cell is given by the Shockley–Queisser limit of 31%. By using focusing devices and multijunction cells, that efficiency can be increased to 41%. A naive application of Eq. 1.12 would set an absolute upper limit of 95%. A more sophisticated calculation yields an upper limit of 86.6% [2].

1.6 PHOTOSYNTHESIS

This book is focused on the topic of photosynthesis. In this way, light energy generates an electron flow, which results in the production of energetic molecules that make reduced organic compounds. The photosynthetic mechanism absorbs light energy by using pigments, especially chlorophyll (Chl) molecules. Photosynthetic organisms only need 1% of the solar spectrum to provide enough biomass and oxygen to support life on earth. Two photosystems work in tandem to carry out oxygenic photosynthesis. They are very efficient systems, using most of their pigments as antennas to harvest the light energy and then to transfer it to a very few, special Chl molecules. Photochemistry begins when these special Chl molecules donate electrons, which travel through the electron transport chain, reaching nicotinamide adenine dinucleotide phosphate ($NADP^+$) at photosystem I. NADPH is used for CO_2 fixation into organic compounds. The ultimate electron donor in oxygenic photosynthesis is water, which is oxidized to oxygen by photosystem II.

For more detailed accounts of the subjects discussed in this overview, see [3–6], and [7].

REFERENCES

1. British Petroleum, Statistical Review of World Energy 2011, `http://www.bp.com/sectionbodycopy.do?categoryId=7500&contentId=7068481` (accessed September 1, 2011).

2. M. A. Green, *Third Generation Photovoltaics: Advanced Solar Energy Conversion*, Springer, New York, 2006.

3. J. W. Rohlf, *Modern Physics from α to Z_0*, 1st ed., Wiley, Hoboken, NJ, 1994.

4. M. L. Kutner, *Astronomy: A Physical Perspective*, 2nd ed., Cambridge University Press, Cambridge, U.K., 2003.

5. E. Hecht, *Optics*, 4th ed., Addison-Wesley, Reading, MA, 2002.

6. H. B. Callen, *Thermodynamics: An Introduction to the Physical Theories of Equilibrium Thermostatics and Irreversible Thermodynamics*, Wiley, Hoboken, NJ, 1960.

7. C. Kittel, *Introduction to Solid State Physics*, 8th ed., Wiley, Hoboken, NJ, 2005.

CHAPTER 2

Oxygenic Photosynthesis

DMITRIY SHEVELA, LARS OLOF BJÖRN, and GOVINDJEE

2.1 INTRODUCTION

2.1.1 Importance of Photosynthesis: Why Study Photosynthesis?

In a general sense the term *photosynthesis* is synthesis of chemical compounds by the use of light. In the more restricted sense, as we shall use it here, it stands for the process by which plants, algae, cyanobacteria, and phototrophic bacteria convert light energy to chemical forms of energy. Most photosynthesis is coupled to assimilation of carbon in the form of carbon dioxide or bicarbonate ions, but there exists also assimilation of CO_2 that is not coupled to photosynthesis, as well as photosynthesis that is not coupled to assimilation of carbon.

All life on Earth, with some exceptions, is completely dependent on photosynthesis. Most organisms that do not live directly by photosynthesis depend on the organic compounds formed by photosynthesis and, in many cases, also on the molecular oxygen formed by the most important type of photosynthesis, oxygenic photosynthesis. Even much of the energy fueling the ecosystems at deep-water hydrothermal vents depends on photosynthesis, since it is made available to organisms using molecular oxygen of photosynthetic origin. In addition, photosynthesis is biologically important in a number of more indirect ways. The stratospheric ozone layer protecting the biosphere from dangerous ultraviolet radiation from the sun is formed from photosynthesis-derived oxygen by a photochemical process. The photosynthetic assimilation of CO_2, and associated processes such as formation of carbonate shells by aquatic organisms, has (so far) helped to maintain the climate of our planet in a life-sustainable state. For basic descriptions of photosynthesis, see Rabinowitch [1] and Blankenship [2], and for reviews on all aspects of *Advances in Photosynthesis and Respiration Including Bioenergy and Other Processes*, see many volumes at the following web site: http://www.springer.com/series/5599.

Natural and Artificial Photosynthesis: Solar Power as an Energy Source, First Edition. Edited by Reza Razeghifard.
© 2013 John Wiley & Sons, Inc. Published 2013 by John Wiley & Sons, Inc.

Rosing et al. [3] speculate that photosynthesis has also caused the formation of granite and the emergence of continents. Granite is common, among bodies in the solar system, only on Earth. After oceans were first formed there were no continents, the surface of the Earth was completely aquatic. Granite with its lower density is, in contrast to the heavier basalt, able to "float high" on the Earth's liquid interior. Thus photosynthesis is very important for life on Earth, and worth a thorough study just for its biological and geological importance. In recent years it has also attracted much interest in connection with the search for a sustainable energy source that can replace nuclear power plants and systems that release greenhouse gases to the atmosphere. There is much interest in solar fuels today: see `http://blogs.rsc.org/cs/2012/09/25/a-centenary-for-solar-fuels/` for a special collection of articles and opinions to mark the centenary of Ciamician's paper "The Photochemistry of the Future" [4].

2.1.2 Oxygenic Versus Anoxygenic Photosynthesis

The form of photosynthesis that first comes to mind when the term is mentioned is that carried out by the plants we see around us. It is called *oxygenic photosynthesis* because one of its products is molecular oxygen, resulting from the oxidation of water [5]. This form of photosynthesis is also carried out by algae and by cyanobacteria (formerly called blue-green algae) (for a perspective on cyanobacteria, see Govindjee and Shevela [6]). Photosynthesis by bacteria other than cyanobacteria, on the other hand, does not involve evolution of O_2. Instead of water (H_2O), other electron donors, for example, hydrogen sulfide (H_2S), are oxidized. This latter type of photosynthesis is called *anoxygenic photosynthesis* [7,8]. In addition to these processes, some members of the "third domain of life," the Archaea, as well as some other organisms, carry out conversion of light into electric energy by carrying out light-dependent ion transport. Although this biological process, which strictly speaking is not photosynthesis, could also be a useful guide to technological applications, we shall not deal with it in this chapter (see, however, Oesterhelt et al. [9]).

The reactions of oxygenic photosynthesis in algae and plants take place within a special cell organelle, the *chloroplast* (see Fig. 2.1). The chloroplast has two outer membranes, which enclose the *stroma*. Inside the stroma is a closed membrane vesicle, the *thylakoid*, which contains the *lumen*. The stroma is the site where the CO_2 fixation reactions occur (the *dark reactions* of photosynthesis; described in Section 2.5); the thylakoid membrane is the site for the conversion of light energy into energy of the chemical bonds (the *light reactions*; discussed in Sections 2.2 and 2.3). In cyanobacteria, however, the thylakoid membrane is within the cytoplasm.

2.1.3 What Can We Learn from Natural Photosynthesis to Achieve Artificial Photosynthesis?

Natural photosynthesis is characterized by a number of features, which are useful to keep in mind when trying to construct useful and economically viable artificial systems [10]:

1. Use of antenna systems that concentrate the energy.
2. Regulation of antenna systems by light.

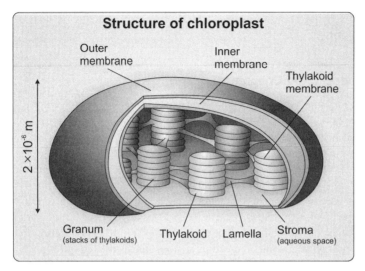

FIGURE 2.1   Three-dimensional diagrammatic view of a chloroplast.

3. Use of quantum coherence to increase efficiency.
4. Connection, in series, of two photochemical systems to boost electrochemical potential difference.
5. Protective systems and safety valves to prevent overload and breakdown.
6. Self-repair of damaged components.

We must consider constructing artificial systems from common and cheap materials that may be available everywhere. One should also bear in mind that perhaps plants do not optimize the process toward the same goal as we wish them to do. Maximizing energy conversion is not always the best strategy for an organism; they have evolved for survivability.

Attempts are being made on many fronts including mimicking the manganese–calcium cluster of PSII for energy storage (for more details see Chapters 3 and 4, and Refs. [11–18]).

2.1.4   Atomic Level Structures of Photosynthetic Systems

By means of X-ray diffraction studies of protein crystals and other methods, the detailed atomic structure of some photosynthetic systems are now available (for recent reviews on the structures of photosynthetic complexes, see Refs. [19–22]). They have revealed great similarity between the "cores" of the two photosynthetic systems (PSI and PSII) present in oxygenic organisms, and the "cores" of photosystems present in various groups of bacteria carrying out anoxygenic photosynthesis, indicating a common evolutionary origin of all photosynthesis (e.g., see Refs. [23–25]). Among many other structures of photosynthetic systems, we mention, at the very outset, that we now have available atomic level structure of the PSII at 1.9 Å resolution [26].

2.1.5   Scope of the Chapter

This chapter is intended as a background on natural photosynthesis for those interested in artificial photosynthesis. We start with a description of how light is used for creating positive and negative charges, and continue with how these charges are transferred through the molecular assemblies in the membranes. Next, we describe how the charge transport leads to creation of a pH difference across the photosynthetic membrane, and how charge and pH differences lead to the production of high-energy phosphate that can be used in chemical synthesis. Finally, we deal with the time dimension, how the type of photosynthesis present today has evolved over billions of years, and what can we expect of the future that we are ourselves able to influence. In addition, in the end, we consider some interesting photosynthesis-related questions relevant to whole land and aquatic plants.

2.2   PATH OF ENERGY: FROM PHOTONS TO CHARGE SEPARATION

2.2.1   Overview: Harvesting Sunlight for Redox Chemistry

The initial event in photosynthesis is the light absorption by pigments: chlorophylls (Chls), carotenoids (Cars), and phycobilins (in cyanobacteria and in some algae), contained in antenna protein complexes (for overviews of light-harvesting antenna, see Green and Parson [27], for Chls, see Scheer [28] for Cars, see Govindjee [29], and for phycobilins, see O'hEocha [30]). The absorbed energy is transferred from one antenna pigment molecule to another in the form of *excitation energy* until it reaches *reaction centers* (RCs), located in two large membrane-bound pigment–protein complexes named photosystem I (PSI) and photosystem II (PSII) (see Figs. 2.2 and 2.3). Due to the primary photochemistry, which takes place after trapping of the excitation energy by special photoactive Chl molecules in the RCs of these two photosystems, light energy is converted into chemical energy. This energy becomes available for driving the redox chemistry of the stepwise "extraction" of electrons from water and their transfer to $NADP^+$ (oxidized form of nicotinamide adenine dinucleotide phosphate) (for further details, see Section 2.3). In this section we briefly describe how photosynthetic organisms capture light energy and how this energy migrates toward the RC Chl molecules, where the primary photochemical reactions occur.

2.2.2   Light Absorption and Light-Harvesting Antennas

The function of all light-harvesting antennas in photosynthetic organisms is common to all, that is, capture of light energy through absorption of photons of different

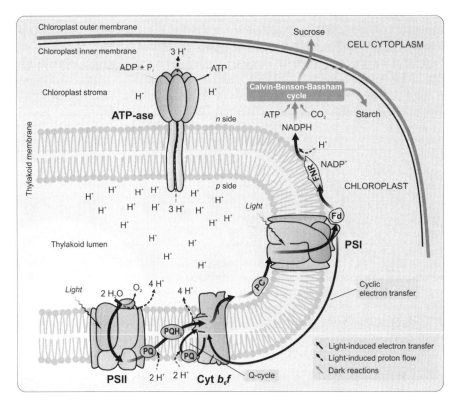

FIGURE 2.2   A schematic view of the photosynthetic thylakoid membrane and the protein complexes involved in the light-induced electron transfer (black solid arrows) and proton transfer (black dashed arrows) reactions in the thylakoid membrane of chloroplast in plants and algae. The end result of these light reactions is the production of NADPH and ATP. NADPH and ATP drive the "dark reactions" (grey arrows) of CO_2-fixation in stroma of the chloroplast via a cyclic metabolic pathway, the so-called the Calvin–Benson–Bassham cycle (also called by some as Calvin cycle, or Calvin-Benson cycle). This results in the reduction of CO_2 to energy-rich carbohydrates (e.g., sucrose and starch). See text for abbreviations and further details. Adapted from Messinger and Shevela [247].

wavelengths, and its transfer to RC complexes where photochemistry (the primary charge separation) takes place (see Fig. 2.3).

The process of photosynthesis starts in femtosecond time scale ($\sim 10^{-15}$ s) by light absorption in pigments, located in the light-harvesting antenna. Within less than a second, thylakoid membranes release O_2 and produce reducing power (reduced form of nicotinamide adenine dinucleotide phosphate or NADPH) and adenosine triphosphate (ATP). Kamen [31] used a pts (negative log of time) scale, analogous to the pH scale, to describe this process that spans pts of $+15$ to -1. The process of light absorption in any pigment molecule in the antenna, say, a Chl a molecule, implies that when a photon has the right energy ($E = hc/\lambda$, where h is Planck's constant, c is velocity of light, and λ is the wavelength of light), the molecule, which

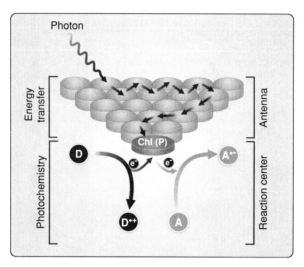

FIGURE 2.3   Excitation energy transfer in light-harvesting antenna that leads to primary photochemistry (charge separation) at the reaction center of photosynthetic organisms. Light energy is transferred through photosynthetic pigments of the light-harvesting antenna until it reaches reaction centers, where primary charge separation takes place. Abbreviations: P, reaction center Chl *a* molecule; e⁻, electron; A, an electron acceptor; D, an electron donor. For further description see text. Adapted from Messinger and Shevela [247].

is in the ground state, will go to its excited state (Chl*): one of the two outermost electrons, spinning in the opposite directions, is transferred to the higher excited states. An excited singlet state is produced (^{1}Chl a^*). This process is very fast: it occurs within a femtosecond, as mentioned above. Figure 2.4 shows the relation between the absorption spectrum of a Chl *a* molecule and its energy level diagram, the Jabłonski–Perrin diagram [32]. It shows that blue light (440 nm) will take the molecule to the *n*th excited state, whereas the red light (672 nm; or 678 nm, depending on the Chl *a* species) will take the molecule to its first excited state. The higher excited state is very unstable and within a pts of +14 to +13, the electron falls down to the lowest excited state; the extra energy is lost as heat. No matter what color of light is absorbed, the photochemical processes begin from this lowest excited state.

Plants and green algae have major and minor antenna complexes in both the photosystems (I and II). In PSII, there is a major complex, the LHCII (light-harvesting complex II), with many subcomplexes, and minor complexes that include CP43 (Chl–protein complex of 43 kDa mass) and CP47 (Chl–protein complex of 47 kDa mass). LHCII contains both Chl *a* and Chl *b* (the latter has absorption maxima at 480 nm and 650 nm), whereas CP43 and CP47 contain only Chl *a*. Chl *b* transfers energy to Chl *a* with 100% efficiency as has been known for a very long time (e.g., see Duysens [33, 34]). In addition, there are Cars that also transfer excitation energy to Chl *a*, with different efficiencies (see Govindjee [29]); the Cars, in general, are of two types: carotenes and xanthophylls (the mechanism of their energy transfer to

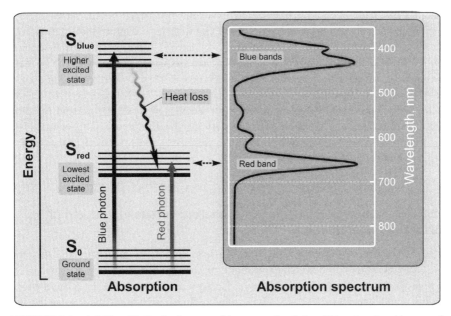

FIGURE 2.4   A Jabłonski–Perrin diagram of the energy levels in a Chl molecule with spectral transitions between them (vertical arrows) and absorption spectrum (turned 90° from the usual orientation) of Chl a corresponding to these levels. Diagram shows heat loss as radiationless energy dissipation (downward-pointing wiggly arrow): other radiationless energy dissipation processes such as fluorescence emission and intersystem crossing are not shown here. Note that the short- (blue) and long-wavelength (red) absorption bands of Chl absorption spectrum correspond to the absorption by this molecule of blue and red photons, respectively. Thus the red absorption band corresponds to the photon that has energy required for the transition from the ground state to the lowest excited state, while the blue absorption band reflects the transition to a higher excited state.

Chl a is, however, unique and different; e.g., see Zigmantas et al. [35] and Zuo et al. [36] for a discussion).

Brown algae, yellow-brown and golden-brown algae, and diatoms contain, in addition to Chl a, fucoxanthin as a xanthophyll, and various forms of Chl c, instead of Chl b [37]. Chls c_1 and c_2 have absorption maxima at ~630 nm; and Chl c_3 at 586 nm. Fucoxanthin absorbs in the green (535 nm) and gives the organisms brown color; cryptomonads and dinoflagellates contain peridinin (absorption peak at 440–480 nm [38]) instead of fucoxanthin. On the other hand, red algae have water-soluble red and blue pigment-proteins, the phycobilins: phycoerythrins (absorption peak at 570 nm), phycocyanins (at 630 nm), and allophycocyanins (at 650 nm), absorbing green to orange light [39,40].

The oldest oxygenic photosynthesizers are cyanobacteria (they were called blue-green algae before their prokaryotic nature was realized) (for a perspective, see Govindjee and Shevela [6]). Being prokaryotes they do not have chloroplasts. They contain Chl *a* and phycobilins like the red algae, and phycoerythrin, the red pigment, is also present in some cyanobacteria; these cyanobacteria capture light that is not absorbed by green algae [41, 42], and thus they have different ecological niches in nature [43, 44]. The major LHCs of cyanobacteria are the *phycobilisomes* (PBS) that are made of the *phycobiliproteins* attached to the cytoplasmic surface of thylakoid membrane (for further details on the cyanobacterial PBS, see Mimuro et al. [45] and Sidler [46]). Interestingly, Chl *b*, that is, with some exceptions, not present in wild type cyanobacteria, can be introduced by genetic engineering into cyanobacteria [47]. For recent overviews on the LHCs of photosynthetic organisms, see Collines et al. [48] and Neilson and Durnford [49].

2.2.3 Excitation Energy Transfer: Coherent Versus Incoherent or Wavelike Versus Hopping

2.2.3.1 A Bit of History In 1936, Gaffron and Wohl [50] were the first to discuss excitation energy transfer (or migration) among hundreds of Chl molecules, in what we now call "antennas" before it reaches what we now call "RCs", and what Emerson and Arnold in 1932 [51] had called a "unit" (that could be interpreted as a "photoenzyme"). The concept of the "photosynthetic unit" serving a photoenzyme (RC in today's language) was born in the experiments of Emerson and Arnold [51,52], who found that a maximum of only one oxygen molecule evolved per thousands of Chl molecules present (for a review, see Clegg et al. [53]). In 1943, Dutton et al. [54] and in 1946 Wassink and Kersten [55] were among the first to demonstrate efficient excitation energy transfer from fucoxanthin to Chl *a* in a diatom (*Nitzschia* sp.) using the method of sensitized fluorescence: excitation of fucoxanthin led to as much Chl *a* fluorescence as excitation of Chl *a* did (see Govindjee [29]). Using the same sensitized fluorescence method, Duysens (in 1952) [33] showed 100% excitation energy transfer from Chl *b* to Chl *a* in the green alga *Chlorella*, and about 80% transfer from phycocyanin to Chl *a* in the cyanobacterium *Oscillatoria*. In 1952, both French and Young [39] and Duysens [33] showed efficient excitation energy transfer, in red algae, from phycoerythrin to phycocyanin and from phycocyanin to Chl *a* (however, later, it was realized that a distinct kind of phycobiliprotein, allophycocyanin, carries energy from phycocyanin to Chl *a*). Such excitation energy transfers from one type of pigment to another may be dubbed "heterogeneous" excitation energy transfer. On the other hand, Arnold and Meek [56] showed for the first time that when Chl *a* molecules in *Chlorella* cells were excited with polarized light, an extensive depolarization of fluorescence was observed; this was evidence of excitation energy migration among Chl *a* molecules. Such energy migration can be dubbed as "homogeneous" energy transfer since it is between the same type of pigment molecules. One of the first measurements of the time of excitation energy transfer was performed by Brody in 1958 [57] when he observed a delay of about 500 ps for energy transfer from phycoerythrin to Chl *a,* using a home-built instrument for measuring lifetime of fluorescence. Furthermore, excitation energy transfer from various pigments to Chl *a*, and one spectral form of Chl *a* to another, was found to be temperature dependent down to 4 K (see Refs. [58–60]). For further historical details, see Govindjee [61].

2.2.3.2 *Mechanism of Excitation Energy Transfer*   There are two extreme cases [1]: (1) When there is a very strong coupling between the neighboring pigment molecules, the net result is that the exciton (excitation energy) formed from the absorbed photon belongs to all the pigment molecules, but not to one. There is thus quantum *coherence* in the system, and the motion of the exciton has a wavelike character. The exciton is *delocalized* [53, 62, 63]. (2) When there is a very weak (or even weak) coupling between the neighboring pigment molecules, the net result is that the excited pigment molecule formed from the absorbed photon belongs to that specific molecule only. There is thus quantum *incoherence* in the system, and the motion of the excitation energy has a *hopping* character. At one specific time, the excitation energy is said to be *localized* on a specific molecule [64–66]. Excitation energy transfer in this case is by the *Förster resonance energy transfer* (FRET), the magnitude of which depends (i) inversely on R^6, where R is the distance between the donor and the acceptor molecules; (ii) on the overlap integral of the absorption spectrum of the acceptor molecule and the emission spectrum of the donor molecule; and (iii) the so-called orientation factor, κ^2 [67, 68]. In photosynthetic systems, both the coherent (delocalized, wavelike) and incoherent (localized, hopping) mechanisms exist (for basics on fluorescence spectroscopy, see Colbow and Danyluk [69] and Lakowicz [70]).

Other sophisticated theories, besides the Förster theory, have evolved, which incorporate additional details and concepts and are applicable to several photosynthetic systems. They are the Redfield theory, the modified Redfield theory, and the generalized Förster theory [71–75].

When the pigment–pigment excitonic interaction coupling is weak, but the pigment–protein excitonic interaction–vibrational coupling is strong, and the excitation energy is localized, the classical FRET [66] mechanism applies (see Kleima et al. [76] as, for example, in the case of peridinin–Chl a system). However, when the pigment–pigment interaction coupling is strong, and the pigment–protein interaction coupling is weak, and the excitation is delocalized, a different mechanism called the *Redfield theory* applies (see Redfield [77] and Renger et al. [78]). On the other hand, when both the pigment–pigment and pigment–protein interaction couplings are strong, a *modified Redfield theory* applies [73]. For the case of energy transfer in LHCII, see Novoderezhkin et al. [79].

In PSII and PSI complexes, and in anoxygenic bacterial photosystems, we have pigment–protein domains that have strong coupling within them, but weak coupling between the domains. Thus interdomain excitation energy transfer would be by Förster theory, but the integrated mechanism would require extension of this theory to include coherence within individual domains (using Redfield or modified Redfield theory); the final description is called *generalized Förster theory* (see, e.g., a description of excitation energy transfer in PSII core complexes [80]).

In our opinion, the Förster theory must be applicable to excitation energy transfer within the phycobiliproteins, when excitation energy is transferred from phycocyanin to allophycocyanin and then from allophycocyanin to Chl a [41]. For examples and discussion of Förster energy transfer in other photosynthetic systems, see Şener et al. [81] and Jang et al. [82] and, for a historical perspective, see Clegg et al. [53]. On the other hand, Ishizaki and Fleming [62] and Ishizaki et al. [83] discuss the ramifications of coherent energy transfer in photosynthetic systems, whereas Collini et al. [84] show coherent energy transfer in marine algae at room temperature. Quantum coherence is inferred when oscillations of exciton state populations, lasting up to a few hundred femtoseconds, are observed. This happens when pigment–pigment interaction coupling is very strong. The reality is that both coherent and incoherent mechanisms occur in natural systems. However, there are many open questions that remain to be answered!

2.2.4 Concluding Remarks and Future Perspectives for Artificial Photosynthesis

Sunlight is a dilute form of energy traveling at a speed that is beyond our comprehension. Photosynthetic organisms have learned over billions of years of trial and error how to catch it, concentrate it, and convert it in an efficient way to electrical energy for further use. We have attempted here to describe what is known about this process. An urgent task for humanity is to explore this further and adapt the process for solving our present technological energy crisis by what we may call "artificial photosynthesis."

2.3 ELECTRON TRANSFER PATHWAYS

2.3.1 Overview of the Primary Photochemistry and the Electron Transfer Chain

The photosynthetic electron transfer chain (ETC) from water to $NADP^+$ is energized by two membrane-bound photosystems. Thus the end result of the steps of light harvesting by the antenna complexes of PSI and PSII (or by phycobilisomes in cyanobacteria) is the capture of excitation energy by the ensemble of unique photoactive Chl molecules, denoted P, in the RC of photosystems (see Fig. 2.3). In PSII, this special photoactive RC is composed of several Chl a molecules, dubbed P680, and in PSI, the special RC is a "heterodimeric" complex of Chl a and Chl a', dubbed P700 (the numbers 680 and 700 are based on the wavelengths of the absorption maxima of these Chls in the red region). The singlet excited states of these RC Chls ($^1P^*$) is where the primary photochemistry begins. This is followed by fast charge separation between $^1P^*$ and the neighboring primary electron acceptor (symbolized as A in Fig. 2.3) and thus the formation of the radical pair $P^{\bullet+} A^{\bullet-}$. The cation radical $P^{\bullet+}$ is reduced by the electron donor (denoted D in Fig. 2.3). An important point to realize here is that as soon as the cation radicals $P680^{\bullet+}$ (in PSII) and $P700^{\bullet+}$ (in PSI) are formed, the light energy has already been converted into chemical energy; this process has a very high quantum efficiency [85].

However, the steps involved in photosynthetic energy conversion are catalyzed not only by PSII and PSI, but also by two other protein complexes that are also located in the thylakoid membrane. These complexes are the cytochrome (Cyt) b_6f and the ATP synthase (ATP-asc) (see Fig. 2.2). PSII, PSI, and Cyt b_6f complexes contain almost all the redox active cofactors that allow light-induced transfer of electrons from H_2O to $NADP^+$ through the thylakoid membrane in the following sequence: $H_2O \rightarrow$ PSII $\rightarrow$ Cyt $b_6f \rightarrow$ PSI $\rightarrow NADP^+$. PSII is linked with the Cyt b_6f complex via the mobile lipophilic hydrogen atom carrier plastoquinone (PQ), in the membrane, while the Cyt b_6f is linked with PSI via a mobile water-soluble redox carrier plastocyanin (PC) in the thylakoid lumen (Fig. 2.2). This copper-containing protein PC can, in some cases, be substituted by the iron-containing carrier Cyt c_6 (also sometimes called Cyt c_{533}) [86, 87]. In addition to these mobile electron transfer carriers, there is a soluble [2Fe-2S]-containing protein ferredoxin (Fd), a one-electron carrier connected with PSI. Upon receiving an electron from PSI, Fd reduces $NADP^+$ to NADPH. This reduction is catalyzed by the membrane-associated flavoprotein called ferredoxin-$NADP^+$ reductase (FNR) [88, 89]. The scheme that depicts the ETC from water to $NADP^+$ via redox-active cofactors is called the *zig-zag* or the *Z-scheme* (Fig. 2.5). Its origin and development have been described recently by Govindjee and Björn [90].

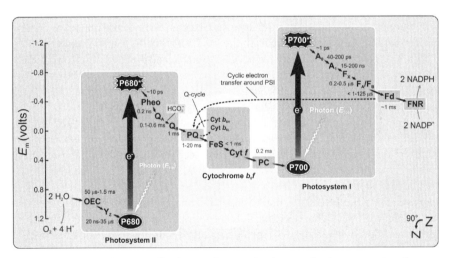

FIGURE 2.5   The zig-zag or Z-scheme of oxygenic photosynthesis representing the energetics of linear electron transfer from H_2O to $NADP^+$ plotted on redox midpoint potential (E_m, at pH 7) scale. The diagram also shows a cyclic electron transfer around PSI, Q-cycle, and half-times of several linear electron transfer steps. The two black vertical arrows symbolize the excitation of RC Chl *a* molecules (P680 and P700 in PSII and PSI, respectively); these lead to electrons in the ground state to be raised into a higher (singlet) excited state in response to the absorption of excitation energy from the light-harvesting antenna or by direct absorption of photons (wiggly white arrows). For further details and abbreviations of the components involved in the electron transfer see text. Adapted from Govindjee et al. [115] and Shevela [248].

The light-driven flow of electrons in the photosynthetic ETC also involves the flow of protons (see Fig. 2.2). There are two steps in the ETC from water to $NADP^+$, where protons are released into the lumen of the thylakoids: (1) during water oxidation (oxidation of two water molecules results in release of four protons), and (2) during oxidation of plastoquinol (PQH_2) (protons are released into the lumen that were initially taken up from the stromal side during the formation of PQH_2) (Fig. 2.2). The accumulation of these protons in the lumen of thylakoid generates the *proton motive force* (*pmf*) across the membrane. As a consequence of dissipation of the pmf (which includes membrane potential and proton gradient) through the ATP synthase, ATP is synthesized (see Section 2.4), according to the chemiosmotic hypothesis of Peter Mitchell (e.g., see Jagendorf [91]).

Thus the result of the light-induced electron transport in all oxygenic photosynthesizers is water splitting concomitant with O_2 evolution (see Chapter 3 by Gernot Renger for details), production of the reducing power (NADPH), and phosphorylation of ADP to ATP (see Section 2.4 and Fig. 2.2). With the energy stored in NADPH and ATP, carbon dioxide can be converted to carbohydrates through a complicated carbon-fixation cycle (the Calvin–Benson–Bassham cycle) (see Section 2.5).

2.3.2   Components Associated with P680 and P700 and the Entry into the Electron Transfer Chain

As mentioned above, the energy of light becomes converted into chemical energy upon formation of the cation radicals of the RC Chls, $P680^{\bullet+}$ (in PSII) and $P700^{\bullet+}$ (in PSI). These cation radicals are formed due to stable and directed charge separation that occurs upon absorption of photons or excitation energy transfer to P680 and P700 molecules. However, there are still debates as to the definition of P680 and P700 and the detailed steps involved in the primary photochemistry [92–96].

The term P680 was first used in 1965 by Rabinowitch and Govindjee [97]; Döring et al. [98] showed its existence experimentally, and the primary charge separation, within 3 ps, between it and a nearby pheophytin (Pheo) molecule (the primary electron acceptor) was first measured by Wasielewski et al. [99] (also see Greenfield et al. [100], and for historical overview on Pheo discovery, see Klimov [101]). Excitation energy reaching the PSII RC complex leads to the formation of the singlet excited state of P680, $^1P680^*$; what one calls P680 needs to be specified; some consider P_{D1} and P_{D2} as P680, whereas others include two other nearby Chl *a* molecules (Chl_{D1} and Chl_{D2}) (e.g., see Durrant et al. [102] and Fig. 2.6). The long-wavelength peak of P680 is at 680 nm ($E_{680} = 1.83$ eV), whereas the short-wavelength peak is at $\sim$440 nm [103]. It appears that we have two alternate primary steps: the primary charge separation is either from P_{D1} to $Pheo_{D1}$, or from Chl_{D1} to $Pheo_{D1}$. If the latter occurs first, then the oxidized Chl_{D1} oxidizes P_{D1}. However, there is now evidence that we have mixed electron transfer states [104]. The primary charge separation steps are over within a few picoseconds (3–7 ps), with the fastest steps being faster than 0.3–0.7 ps. As a result, the electron available on the $Pheo^{\bullet-}$ molecule enters into the ETC via a one-electron acceptor primary plastoquinone Q_A within PSII and then a two-electron acceptor Q_B, also within PSII.

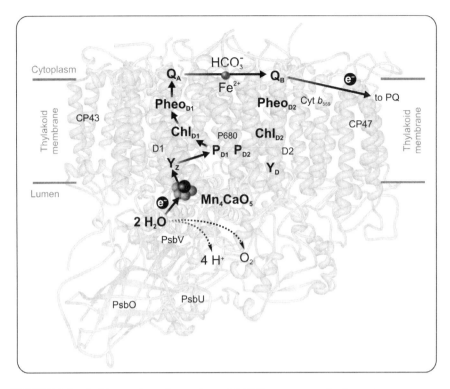

FIGURE 2.6   A side-on view of cyanobacterial PSII monomer and schematic arrangement of its central redox cofactors. Black solid arrows indicate the direction of electron transfer. Only a few proteins of PSII are symbolized. For further discussion and abbreviations see text. Adapted from Shevela et al. [249].

P700 was discovered by Kok in 1957 [105]. In contrast to PSII, PSI complex has not only the RC Chl a–Chl a' heterodimer P700, but also a large number of core PSI antenna molecules [106–108]. The long-wavelength absorption peak of P700 is at $\sim$700 nm ($E_{700} = 1.77$ eV). In contrast to P680, P700 has a much lower redox potential (ranging from $+400$ to $+470$ mV) (see chapters in Ke [103] and Golbeck [109]), but the primary charge separation in PSI is also over in a few picoseconds. Initially, it was thought that the primary charge separation is when P700 A_0 is converted to P700$^{\bullet +}$ $A_0^{\bullet -}$, where A_0 is a nearby Chl a molecule, previously known as the first electron acceptor (see Rutherford and Heathcote [110] and relevant chapters in Golbeck [109]; for a new proposed nomenclature of the redox active cofactors of the ETC in PSI, see Section 2.3.4 and Redding and van der Est [111]); however, recent data indicate that the primary charge separation may involve oxidation of another neighboring Chl a molecule, denoted as A and reduction of A_0 (see Fig. 2.7 and Refs. [94, 112, 113]). Thus these data suggest that the primary charge separation in PSI begins with the generation of a primary radical pair A$^{\bullet +}$ $A_0^{\bullet -}$, followed by fast reduction of A$^{\bullet +}$ by P700 and the formation of the secondary radical pair P700$^{\bullet +}$ $A_0^{\bullet -}$. The electron

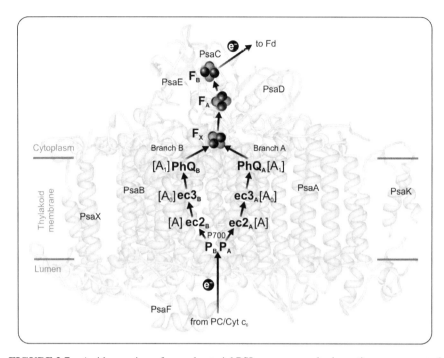

FIGURE 2.7 A side-on view of cyanobacterial PSI monomer and schematic arrangement of its central redox cofactors. Black solid arrows indicate the direction of electron transfer. Only a few proteins of PSI are shown. For further discussion and abbreviations see text. Adapted from Shevela et al. [249].

on $A_0^{\bullet-}$ that had been formed during charge separation in PSI enters the ETC, and through a series of intermediates, reaches mobile Fd. While oxidized $P680^{\bullet+}$ receives electrons from water, oxidized P700 receives electrons from PC (or alternately a Cyt c_6). For overview of the primary photochemistry in photosynthesis, see Renger [93].

2.3.3 Photosystem II: Function and Electron Transfer Pathway

All oxygenic photosynthetic organisms have PSII, a large membrane-integral pigment–protein complex, which exists as a dimer with a total mass of $\sim$700 kDa. In this unique enzyme, light-induced charge separation drives oxidation of water to molecular oxygen on its electron-donor side and the reduction of PQ to PQH_2 on its electron-acceptor side:

$$2\,H_2O + 2\,PQ + 4\,H^+_{stroma} \xrightarrow{h\nu} O_2 + 2\,PQH_2 + 4\,H^+_{lumen} \qquad (2.1)$$

Thus PSII complex functions as a light-driven *water:plastoquinone oxidoreductase* (for reviews on PSII, see Refs. [114–116]). The processes performed by PSII (charge separation, water oxidation, and PQ reduction) are fully discussed in

Chapter 3 by Renger. Therefore we only briefly describe the main structural features, redox-active cofactors, and the electron transfer pathway of this PSII complex.

According to recent crystallographic models of PSII with a resolution ranging from 2.9 to 1.9 Å [19, 26], each monomer of PSII contains 20 proteins (17 of those are transmembrane subunits and 3 are peripheral subunits on the lumenal side of the complex) and about 90 cofactors. Some of these proteins and major redox cofactors as well as their spatial arrangement are shown in Fig. 2.6. All electron transfer cofactors of PSII are bound to two central transmembrane proteins known as D1 (or PsbA) and D2 (or PsbD). These proteins exist as a D1/D2 heterodimer, which, together with a few other proteins, forms the core of PSII RC. As indicated in Fig. 2.6, the D1/D2 heterodimer binds two branches of redox cofactors related by a pseudo $C2$ symmetry. However, the electron transfer within PSII occurs via cofactors located mainly on the D1 side of the D1/D2 heterodimer. Thus this "active branch" contains two Chl a molecules assigned to P680 (P_{D1} and Chl_{D1}), the primary electron acceptor $Pheo_{D1}$, and the primary PQ electron acceptor Q_A; the latter is located on the D2 side (see Refs. [21, 26, 117], Chapter 3 by Renger, and Fig. 2.6). The "inactive branch" of PSII contains symmetrically related cofactors bound mainly on the D2 protein (P_{D2}, Chl_{D2}, and $Pheo_{D2}$), and only the secondary PQ electron acceptor, Q_B, located on the D1 protein, is a part of the ETC, and thus an exception here. The nonheme iron (Fe^{2+}) and an associated (bi)carbonate ion (HCO_3^-/CO_3^{2-}) are located between the quinones Q_A and Q_B.

The reactions of photosynthetic water oxidation in PSII are catalyzed by the so-called oxygen-evolving complex (OEC) located on the lumen side of the thylakoid membrane. The "heart" (inorganic core) of this catalytic complex of water splitting is a cluster of four Mn ions and one Ca ion with five bridging oxygen atoms (denoted as Mn_4CaO_5 cluster) [11, 26, 118]. Three extrinsic proteins, called PsbO (33 kDa), PsbP (23 kDa), and PsbQ (17 kDa), surround and stabilize the Mn_4CaO_5 cluster. In cyanobacteria, however, the two smaller subunits are substituted by PsbV (also known as Cyt c_{550}; 17 kDa) and PsbU (12 kDa) proteins (see Fig. 2.6; for recent reviews on the extrinsic proteins, see Bricker et al. [119] and Fagerlund and Eaton-Rye [120]).

As mentioned above, the end result of the primary charge separation in PSII is the formation of the radical pair $P_{D1}^{\bullet+} Pheo_{D1}^{\bullet-}$. The cation radical $P_{D1}^{\bullet+}$ (traditionally indicated as $P680^{\bullet+}$), which has a very high midpoint potential of $\sim$1.25 V [121,122], is able to sequentially withdraw electrons from the charge-accumulating Mn_4CaO_5 cluster, a catalyst that couples the slow (1–2 ms) four-electron oxidation chemistry of two water molecules with ultrafast (a few ps) one-electron photochemistry. This happens, in turn, via the redox active tyrosine residue of the D1 protein (generally labeled as Y_Z) located between the Mn_4CaO_5 cluster and $P_{D1}^{\bullet+}$. On the acceptor side of PSII, the formed $Pheo_{D1}^{\bullet-}$ very rapidly (within $\sim$200–300 ps) transfers the electron to a tightly bound Q_A molecule that acts as one-electron acceptor; this results in the formation of $P_{D1}^{\bullet+} Q_A^{\bullet-}$. Then the electron is further transferred from $Q_A^{\bullet-}$ to a loosely bound Q_B molecule that acts as a two-electron acceptor (sometimes called a "two-electron gate" of PSII). After a second light reaction followed by charge separation and the formation of $Pheo_{D1}^{\bullet-}$, and further of $Q_A^{\bullet-}$, the once-reduced Q_B^- accepts a second electron. The negative charge of the doubly reduced Q_B (Q_B^{2-}), on the electron-acceptor side of PSII, is further stabilized by protons derived from the stroma side of the thylakoid membrane, and PQH_2 is formed [115, 123]. Bicarbonate ion, tightly bound to the nonheme iron between Q_A and Q_B is known to play an important role in this protonation reaction (recently reviewed in Shevela et al. [124]). It takes $\sim$400 µs to form $Q_B{}^{2-}$ and almost 1 ms to form PQH_2. PQH_2 then leaves the complex and the empty Q_B binding site is filled by fresh PQ from the PQ pool in the thylakoid membrane. Meanwhile, PQH_2 diffuses in the membrane toward the Cyt $b_6 f$ complex: the electrons from PSII have thus entered the intersystem ETC (Figs. 2.2, 2.5, and 2.6).

2.3.4   Photosystem I: Function and the Electron Transfer Pathways

Just like PSII, PSI is a membrane-integral protein complex that is present in all oxygenic photosynthesizers. Similar to PSII, PSI is also a key player in the utilization of light energy for driving the redox reactions of the ETC. Thus PSI uses the energy of light to power the electron transfer from one mobile electron carrier PC (or alternatively Cyt c_6) to another carrier, Fd, and thereby provides the electrons for the reduction of $NADP^+$. It therefore acts as a *light-driven plastocyanin:ferredoxin oxidoreductase* (see Golbeck [109]). In cyanobacteria, an oligomeric from of PSI exists mainly as a trimer having a molecular weight of $\sim$1100 kDa. However, in plants, PSI always exists as a monomeric complex (for reviews on PSI, see Refs. [21, 125] and chapters in Golbeck [109]). Each monomer of PSI is known to contain 12 proteins and about 130 noncovalently bound cofactors [106, 126, 127].

The core of PSI complex is formed by two proteins, PsaA and PsaB. These two proteins coordinate the majority of the cofactors of the ETC within PSI, as well as most of the antenna Chls and the Cars of PSI. Moreover, PsaA/PsaB heterodimer is known to be involved in the docking of the mobile electron carrier (PC or Cyt c_6) to be near P700 on the electron donor (lumenal) side of PSI (see Fig. 2.7 and Grotjohann et al. [24]). On the other hand, three proteins (PsaC, PsaD, and PsaE) on the electron-acceptor side of PSI form the docking site of PSI for harboring the mobile electron acceptor Fd (see Fig. 2.7 and Refs. [106, 107]).

As in PSII, the cofactors of the ETC in PSI are arranged in two branches (A and B). These branches contain six Chls, which are P700 the "special pair" of Chl a and Chl a' heterodimer, also denoted as P_A and P_B according to Redding and van der Est [111], a pair of neighboring Chl a molecules ($ec2_A$ and $ec2_B$) also called A, and another pair of Chl a molecules ($ec3_A$ and $ec3_B$) denoted as A_0 in addition to two phylloquinone molecules A_1 (PhQ_A and PhQ_B) (Fig. 2.7). However, unlike in PSII, both branches in PSI are active for the electron transfer [128, 129]. Interestingly, the branch A in PSI of cyanobacteria was found to be more active than B, while in algae the branch B is known to be more active than A [130–133]. The components of two branches are followed by three iron–sulfur centers ([4Fe–4S] clusters) termed F_X, F_A, and F_B (see Fig. 2.7).

As described above, the primary and the secondary charge separation(s) lead to the formation of the secondary radical pair $P700^{\bullet+} A_0{}^{\bullet-}$ (also see Müller et al. [94] and Holzwarth et al. [112]). The generated $P700^{\bullet+}$ is then reduced by electrons obtained from the external electron donor PC (or Cyt c_6). On the other hand, the electron available on $A_0^{\bullet-}$ is transferred via a series of intermediates (including A_1, F_X, F_A, and F_B) toward the Fd docking site, where it finally reduces the external redox carrier Fd. Reduced Fd then provides electrons to the FNR, which catalyzes the production of NADPH, using $NADP^+$ and protons from the stroma (for more information, see Medina [88] and Aliverti et al. [89]). It takes about 200 ps for the electrons to reach A_1, and about a few ms to reach $NADP^+$ (see Fig. 2.5). The reduction of $NADP^+$ to NADPH completes the sequence of the linear ETC.

2.3.5   Intersystem Electron Transfer

2.3.5.1   *Cytochrome $b_6 f$ Complex and the "Q"-Cycle*   Light-induced electron transfer between the two photosystems is mediated by the membrane-integrated Cyt $b_6 f$ complex; here, this complex catalyzes the transfer of electrons from lipophilic PQH_2 to the soluble PC (see Figs. 2.2 and 2.5). For its property to oxidize PQH_2 and to reduce PC (or Cyt c_6), this complex is often also called the *plastoquinone:plastocyanin oxidoreductase* (for reviews, see Refs. [22, 134, 135]). Moreover, the intersystem electron transport network performed by the Cyt $b_6 f$ complex is coupled with transport of protons from the stroma to the lumen, thus contributing to the generation of a proton gradient, and an electrochemical potential across the thylakoid membrane that is further utilized by the ATP synthase for ATP synthesis.

In both cyanobacteria and plants, the Cyt $b_6 f$ complex exists as a dimer and contains several prosthetic groups [135, 136]. Each monomeric form of the Cyt $b_6 f$ consists of four large protein subunits: c-type cytochrome f (Cyt f), the Rieske iron–sulfur protein (FeS), cytochrome b_6 (Cyt b_6), and subunit IV (suIV). When PQH_2 is oxidized by the FeS protein, protons are released into the lumen. One of two available electrons delivered by PQH_2 passes along a linear ETC to the Cyt f subunit and further to a small mobile PC (or Cyt c_6), which then carries the electron toward PSI. The other electron is transferred through the two Cyt b_6 hemes, which then reduces a PQ to PQ^- (semiquinone). Upon receiving one of the two electrons from a new PQH_2 molecule (oxidized by the FeS), PQ^- is reduced to PQ^{2-} at the same time picking up two protons (and forming PQH_2), participating thereby in a cyclic process called the *Q-cycle*. Thus the Q-cycle increases the number of protons pumped across the membrane (Fig. 2.2). Overall, for every two electrons that reach PSI, four protons are translocated across the thylakoid membrane. For more detailed information on the Cyt $b_6 f$ complex and its electron transfer, see Refs. [22, 134–138].

2.3.5.2   *Linear Versus Cyclic Electron Transfer*   When the two photosystems, along with the Cyt $b_6 f$ complex, drive the electron transfer from water to $NADP^+$, we have the *linear electron transfer* (also referred to as *noncyclic* electron transfer) (see Figs. 2.2 and 2.5). The end result of such an electron transfer is the production of both NADPH and ATP required for the dark reactions of carbon fixation (see Section 2.5). However, under some conditions, electrons from the reducing side of PSI may cycle back toward the Cyt $b_6 f$ complex and/or PQ pool rather than to $NADP^+$ and then again back to the oxidizing side of PSI, performing the *cyclic*

electron transfer (see Bendall and Manasse [139] and Joliot et al. [140] and Figs. 2.2 and 2.5). As a result, no NADPH is produced, but the available energy may be used for ATP synthesis. This "extra" ATP can be utilized not only for carbon fixation reactions but also for other processes (e.g., for starch synthesis).

2.3.6 Water as a Source of Electrons for the Photosynthetic Electron Transfer Chain

An important issue here is to realize the unique role of the catalytic Mn_4CaO_5 cluster in the removal of electrons from water for the ETC. Water is a very poor electron donor and its oxidation into O_2 and protons requires a strong driving force. Although the oxidizing potential of cation radical $P_{D1}^{\bullet+}$ ($P680^{\bullet+}$) is very high (about 1.25 V) [121, 122], it is not strong enough to directly "extract" electrons from water. However, it has the ability to split water *involving* the Mn_4CaO_5 cluster. This is because this unique catalyst has the ability to first store four oxidizing equivalents and then to use them for a dovetailed four-electron water-splitting chemistry. Here, one-electron photochemistry is linked to the "extraction" of four electrons from two water molecules. See Chapter 3 by Renger for all aspects of photosynthetic water oxidation and O_2 evolution, its mechanisms and energetics.

2.3.7 Can the Rate Limitation of O_2 Production by Photosystem II Be Improved in Future Artificial Water-Splitting Systems?

The bottleneck reaction of the entire process of light-induced reactions of oxygenic photosynthesis is the diffusion and oxidation of PQH_2, and this could be as slow as 20 ms (see Fig. 2.5). Any improvement in the overall efficiency of electron transport would require that we engineer the system to make this reaction faster.

Interestingly, data obtained on PSII membrane fragments show that at high flashing rates the electron-acceptor side of PSII is rate limiting for O_2 evolution (water oxidation) [141, 142]. Such limitation of O_2 evolution by the reactions on the electron-acceptor side may indicate that there was no evolutionary pressure to make the relatively "slow" 1–2 ms turnover frequency of the OEC faster. Therefore one can assume that there is a chance to develop in future water-splitting artificial catalysts that have higher turnover rates than the "natural" Mn_4CaO_5 cluster of PSII.

2.4 PHOTOPHOSPHORYLATION

2.4.1 Overview

Photophosphorylation was discovered independently by Albert Frenkel in 1954 [143] in chromatophores of photosynthetic bacteria, and in the same year by Arnon et al. [144] in chloroplasts of higher plants. In addition, Strehler [145] had earlier observed photosynthetic ATP production, using the luciferase–luciferin system. The overall process is the production of ATP, on the ATP synthase, from ADP and inorganic phosphate (P_i), using the pmf, the sum of the ΔpH and $\Delta \psi$, the membrane potential, formed during light-induced electron flow in photosynthesis, on the thylakoid membrane [146] (see Section 2.3); for a historical perspective, see Jagendorf [91]. We note at the very outset that about 42 kJ of converted light energy is stored in each mole of ATP, and ATP is dubbed the *energy currency of life*.

In view of the earlier controversies between the chemical and the chemiosmotic mechanisms for ATP synthesis, we refer readers to a review of opinions that had, interestingly, all the parties as coauthors [147]. In 1978, Peter Mitchell received the 1978 Nobel Prize in Chemistry for the chemiosmosis hypothesis, and in 1997, Paul Boyer and John Walker shared the Nobel Prize, also in Chemistry, on how the ATP synthase converts this pmf to chemical energy via a rotary mechanism [148–150].

The pmf is formed during the noncyclic electron flow from water to $NADP^+$, involving PSII, Cyt b_6f, and PSI, as well as during the cyclic electron flow involving PSI and Cyt b_6f (see Figs. 2.2 and 2.5). As mentioned above (see Section 2.3.5.1), there is also a so-called Q-cycle around Cyt b_6f (for a perspective on the Q-cycle, see Crofts [151]). Oxidation of two water molecules, in PSII, is accompanied by the release of four protons into the lumen (the p-side); and consequent reduction and oxidation of two molecules of PQ leads to the transfer of another four protons from the stromal (the n-side) to the lumen. When PSI and the Q-cycle are involved, a much larger number of protons are available (Fig. 2.2).

2.4.2 Mechanism of ATP Synthesis

When Jagendorf and Uribe [152] suspended thylakoids in an acidic medium and then transferred them to an alkaline medium, and that too in darkness, but in the presence of ADP and P_i, ATP was produced, as if ΔpH was responsible for ATP synthesis! Furthermore, Witt et al. [153] found that an applied voltage in the medium was able to drive ATP synthesis. These and other early discoveries are thus the backbone of Mitchell's chemiosmotic hypothesis, which is simply that it is the energy available from the dissipation of the pmf, through the thylakoid membrane, from the p-side (lumenal) to the n-side (stromal) that makes ATP from ADP and P_i [146].

The ATP synthase, which has a molecular mass of $\sim$600 kDa, is $\sim$15 nm long and $\sim$12 nm wide. It is made up of F_0 (a hydrophobic part, embedded in the membrane) and F_1 (a hydrophilic part that protrudes into the stroma) (see Fig. 2.8). The F_0 has three subunits: a, b, and c (several copies, up to 15 in some cases), whereas F_1 has 5 subunits: α (3 copies); β (3 copies), γ (1 copy), δ (1 copy), and ε (1 copy). The ATP synthase is actually a rotary motor, with most of the F_0 units rotating, after protons are bound to its "c" subunits; the binding energy of protons, which are being translocated, is converted into mechanical energy leading to rotation (see Refs. [20, 150, 154]). Thus protons on the p-side of the thylakoid membrane begin the process of ATP synthesis [155]. The number of protons used for the synthesis of one molecule of ATP is not a constant number; in many cases four protons per ATP have been observed (see van Walraven and Bakels [156]).

In the stator (α, β, and δ) part of F_1, the mechanical energy in the above-mentioned rotary motion is converted into chemical energy needed to make the *high-energy*

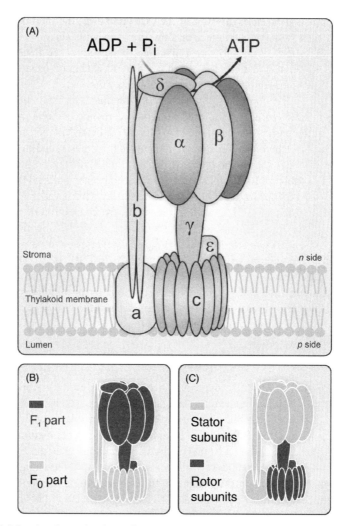

FIGURE 2.8 A schematic view of ATP synthase and its subunits. (A) Overall view of subunits of ATP synthase. (B) Two basic parts of ATP synthase: a membrane-integrated F_0 part and peripheral F_1 part (shaded individually as depicted in the figure). (C) Rotor and stator parts of ATP synthase (also shaded individually as depicted in the figure). For further details and abbreviations see text. Adapted from Shevela et al. [249].

phosphate bonds in ATP. On the α and β subunits, ADP and P_i are converted into ATP (for details, see Spetzler et al. [157]; also see a self-explanatory Fig. 2.8).

In conclusion, the pmf is used by the c subunits of F_0 to be converted into a mechanical rotation energy (a torque), which is then subsequently used by the α and β subunits of F_1, where this rotation energy is converted into chemical energy, and ATP is formed from ADP and P_i. Furthermore, ε and γ subunits are used to regulate this ATP synthase activity [158, 159].

2.4.3 Concluding Remarks

The ATP produced by the ATP synthase has many functions; it is used for the biosynthesis of many compounds, particularly in the steps leading to carbon fixation (see Section 2.5). It is also used, for instance, for protein synthesis in the chloroplast, while the ATP necessary for the synthesis of proteins and other compounds, outside of the chloroplast, is generated by respiration.

2.5 CARBON DIOXIDE TO ORGANIC COMPOUNDS

2.5.1 Overview of Carbon Dioxide Assimilation Systems in Oxygenic Organisms

Over the ages, many pathways have evolved both in photoautotrophic and chemoautotrophic organisms (Fig. 2.9). Figure 2.9A shows the photosynthetic carbon reductive pathway, known as the Calvin–Benson–Bassham cycle, which is the assimilation pathway in oxygenic organisms, as well as in some nonoxygenic ones. Figure 2.9B depicts a reductive tricarboxylic acid (TCA) cycle (known as the Arnon–Buchanan cycle, in some bacteria); it shares a common evolutionary origin with the well-known respiratory TCA cycle (the Krebs cycle), as it is essentially the same cycle running in reverse. The other two cycles occur in various prokaryotes: the reductive acetyl-CoA pathway (of methanogenic archaea) (Fig. 2.9C), and the 3-hydroxypropionate cycle (of some bacteria) (Fig. 2.9D).

During 2007–2008, two additional pathways for the assimilation of CO_2 were discovered in Archaea: the 3-hydroxypropionate/4-hydroxybutyrate cycle [160] and the dicarboxylate/4-hydroxybutyrate cycle [161]. Figure 2.10 represents the proposed reactions of the autotrophic 3-hydroxypropionate/4-hydroxybutyrate cycle in the thermoacidophilic archaeon *Metallosphaera sedula* (for further details, see Berg et al. [160]). For overviews of "classical" and "novel" metabolic pathways, see Refs. [162–165]. For thermodynamic constraints of carbon fixation pathways, see Bar-Even et al. [166].

In plants there exist two main pathways for CO_2 assimilation, the C3 (Calvin–Benson–Bassham cycle, mentioned above), and C4 (Hatch–Slack pathway) cycles. They are discussed below.

2.5.2 C3 Pathway Versus C4 Pathway

In C3 plants CO_2 is bound to the enzyme ribulose bisphosphate carboxylase oxygenase, often referred to as RuBisCO (or Rubisco). This enzyme converts it to the three-carbon compound phosphoglyceric acid, hence the term C3. Phosphoglyceric acid is then further processed in the Calvin–Benson–Bassham cycle, shown on the right side of Fig. 2.11. For a description of its discovery, see Benson [167] and Bassham [168].

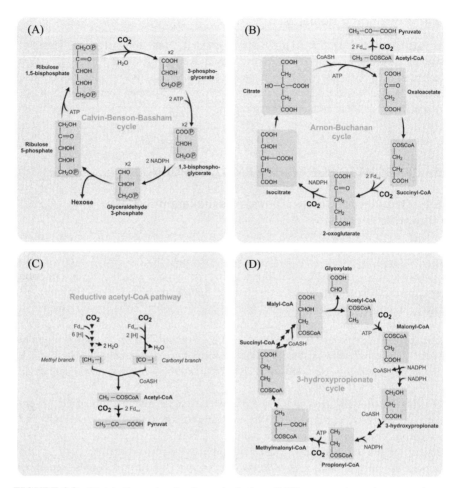

FIGURE 2.9   Metabolic cycles for the assimilation of CO_2 present in various organisms: (A) The Calvin–Benson–Bassham cycle. (B) The reductive TCA cycle (also called Arnon–Buchanan cycle). (C) The reductive acetyl-CoA pathway. (D) The 3-hydroxypropionate cycle. Of these, only the Calvin–Benson–Bassham cycle is present in oxygenic organisms (cyanobacteria, algae, and plants). Symbols in the cycles are: C, assimilated carbon; H, reduction equivalents; Fd_{red}, reduced ferredoxin; P or P within a circle, phosphate groups; CH_3-, enzyme-bound methyl group; $CO-$, enzyme-bound carbon monoxide. Reproduced from Sato and Atomi [163] and Strauss and Fuchs [164] with permission of John Wiley & Sons, Inc.

At low CO_2/O_2 ratios, C3 plants are at a disadvantage because Rubisco does not bind CO_2 tightly, and because O_2 competes with CO_2 at the Rubisco surface, and results in photorespiratory carbon loss. High temperature is also to the disadvantage of C3 plants, because photorespiration increases sharply with temperature. (Photorespiration is an oxygen-consuming process resulting from the fact that the CO_2-binding enzyme, Rubisco, also binds O_2, starting a complicated reaction sequence.) Finally, arid conditions disfavor C3 plants, because they have to keep their stomata more

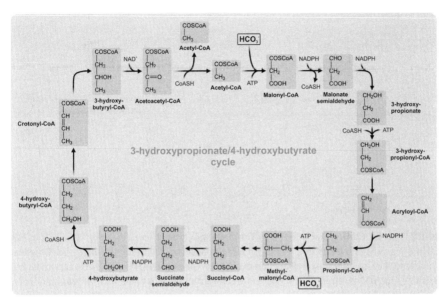

FIGURE 2.10   Proposed autotrophic 3-hydroxypropionate/4-hydroxybutyrate cycle in the archaeon *Metallosphaera sedula*. Reproduced from Sato and Atomi [163] with permission of John Wiley & Sons, Inc.

open, since they cannot maintain an equally steep diffusion gradient for CO_2 across the stomatal pores as do the C4 plants. Stomata are the tiny pores in plant leaf surfaces which let carbon dioxide in and oxygen and water vapor out. C4 plants are apparently at a disadvantage relative to C3 plants at high CO_2/O_2 ratios because of the additional energy expense needed to concentrate CO_2 in their bundle sheath cells (left side of Fig. 2.11). However, at low CO_2/O_2, C4 plants can achieve a high quantum yield of photosynthesis by suppressing photorespiration. For details on C4 photosynthesis and related CO_2 concentrating mechanisms, see chapters in Raghavendra and Sage [169].

2.5.3   C3 Versus C4 Plants During Glacial/Interglacial Periods

During the ice ages, the CO_2 content of the atmosphere was lower than during the interglacial periods. During the last glacial maximum, it was only half of the present content. The climate was also drier. This would favor C4 plants. But it was also colder. This would favor C3 plants. The outcome of these opposing tendencies is that C4 plants were more competitive against C3 plants during ice ages than they are now. But the total primary production of plants was only about 63% of that at the present [170].

The balance between C3 and C4 plants has shifted over time in a complex way and has varied from place to place. Greater C4 plant abundance only occurred when low CO_2 pressure coincided with increased aridity, as observed during the last glacial

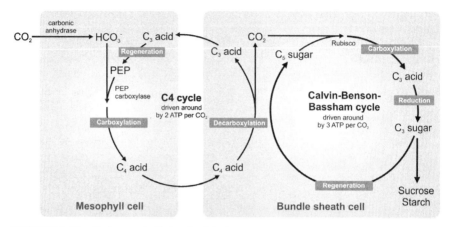

FIGURE 2.11 Carbon assimilation in C4 plants takes place in two cycles. The first cycle acts as a pump to deliver carbon dioxide at a high concentration to the second cycle, which is identical to the Calvin–Benson–Bassham cycle in C3 plants. The specific C4 cycle, driven by the energy in ATP, is slightly different in different groups of C4 plants, and the diagram here is much simplified.

maximum in Guatemala and in tropical Africa and India. Even at the minimum atmospheric CO_2 pressure during the last glacial maximum, the combined effect of high winter precipitation and low temperatures led to an expansion of C3 plants [171]. An investigation of the still arid Chinese loess plateau shows that the highest percentage of C4 plants did not occur there at the last glacial maximum (about 20,000 years ago), but much later, about 7000 years ago [172]. In the Ganges-Brahmaputra drainage area (including the Himalayas), the percentage of C4 plants, as deduced from marine sediments in the Bay of Bengal, declined almost continuously from the last glacial maximum to the present day [173, 174].

2.5.4 Concluding Remarks: Can the Natural Assimilation Pathways Be Improved to Help Solve the Energy Crisis?

The Calvin–Benson–Bassham cycle was the first assimilation pathway to be discovered, and it was first thought that it was the only one. Later, we learned that there were many other possibilities, each one adapted for a specific situation. A take-home lesson is that, since our needs are different from those of any other organism, we should not try to uncritically copy nature. There are indications that the natural fixation pathways can be "improved" in order to better comply with our specific needs. Bar-Even et al. [175] have proposed several artificial cycles. As an example we show the simplest one (Fig. 2.12).

We wish to aim at a future where less of our energy needs are met by burning fuel, but still there will probably always be a need for some fuel burning, and the fuel should be produced by us, and not from fossils. The ideal fuel from a climate-friendly aspect is hydrogen gas, which does not cause emission of CO_2 when burned, but transport issues must be solved. The kind of fuel we primarily want should be fluid (i.e., liquid

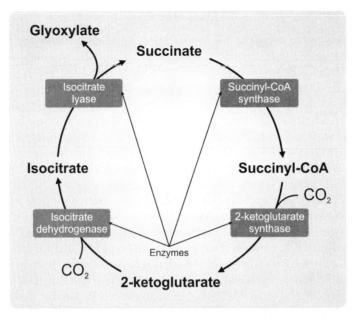

FIGURE 2.12   The shortest possible carbon-fixation cycle and the involved enzymes. Two CO_2 molecules are fixed to give glyoxylate, a two-carbon compound, as the cycle product. In fact, we are of the opinion that this cycle is not thermodynamically feasible and does not represent a viable alternative for carbon fixation. Adapted from Bar-Even et al. [175].

or gaseous), not carbohydrate, which plants mostly produce as primary assimilation products. To produce fluid fuels we need to handle carbon, but not necessarily in a process resembling the assimilation by photoautotrophic organisms.

2.6   EVOLUTION OF OXYGENIC PHOTOSYNTHESIS

2.6.1   Overview

Oxygenic photosynthesis utilizes two photosystems, PSI and PSII, coupled in series via many redox intermediates (see Section 2.3). Clearly this very complex machinery has evolved from simpler bacterial systems using a single photochemical reaction. Many bacteria that are not closely related are capable of anoxygenic photosynthesis; in fact, variants of this process take place in species belonging to most major groups of bacteria. For this reason it was for some time believed that the last common ancestor of bacteria was a photoautotroph. More recent insight showing that horizontal gene transfer is very common even between distantly related bacteria has changed this opinion. It is now believed that chemoautotrophy preceded photoautotrophy during biological evolution. Before molecular oxygen became common, other chemical species such as trivalent iron [176] and/or nitric oxide (NO) [177] would have served as electron sinks for respiration. Nevertheless, photosynthesis is a very ancient process, and its beginning is obscure.

2.6.2 Two Photosystems for Oxygenic Photosynthesis

As mentioned earlier, oxygenic photosynthesis depends on two photochemical reactions (by PSII and PSI), connected via the Cyt b_6f complex, and mobile electron carriers. PSI has a great similarity to a homodimeric type I RC that is present in modern green sulfur bacteria, and PSII has a great similarity to a type II RC that is present in purple bacteria. The differences between PSI and PSII are pronounced, and their amino acid sequences are very different. Yet the structural similarities are great enough to make it very plausible that they both have evolved from the same primitive type of RC [178] resembling a present-day type I RC of Chlorobiaceae [179, 180], and that in turn would have evolved from some component of a respiratory electron transport chain, probably a precursor of Cyt b [181, 182].

We do not know whether the present-day organisms, which have only one kind of RC, have evolved by gene loss from an early kind of prokaryote(s) with RCs of both type I and type II. The competing hypothesis is that organisms with both RCs have evolved from symbiosis between one organism with a type I RC and another one with a type II RC [25, 183, 184]. The current opinions seem to have been leaning more and more toward the former hypothesis. Pierson and Olson in 1987 [185], and, with new arguments, Nitschke et al. in 2010 [186] came to the conclusion that Heliobacteria (having only a type I RC) were derived from organisms with two RCs. In 1981 Olson [187] described how the two RCs evolved in a Chl a-containing organism carrying out anoxygenic photosynthesis. From this, three evolutionary paths diverged. One led to oxygenic photosynthesis. In the other two paths, one or the other RC was lost, and new RC Chls evolved. In 2007, Allen and Martin [188] gave a new twist to this and described how even an anoxygenic organism could profit from having two kinds of RC in order to be able to adapt to environments with varying redox potentials.

The very early appearance of cyanobacteria suggests that two photosystems must have evolved a long time ago. The last common ancestor for cyanobacteria and Chloroflexi is estimated to have lived about 2.6 billion years (Ga) ago [189, 190]. Whether this ancestor was oxygenic or not is not known, but the ability to evolve oxygen is thought to have originated at least 2.4 Ga ago. Nisbet et al. [191], based on carbon isotopic evidence for early appearance of type I Rubisco, conclude that "oxygenic photosynthesizers first appeared $\sim$2.9 Ga ago, and were abundant 2.7–2.65 Ga ago." Czaja et al. [192, 193], based on isotopic data for both iron and molybdenum, assume that free O_2 existed in the oceans 2.5–2.7 Ga ago; while the O_2 content of the atmosphere could still have been very low, they believe that their work "provides strong support for the development of oxygenic photosynthesis by at least 2.7 Ga, because anoxygenic photosynthesis will not produce the coupled variations in the measured Fe and Mo isotope compositions." Schwartzman et al. [194], based on atmospheric and climatic history, believe that oxygenic photosynthesis with bicarbonate, instead of water, as electron donor, and primitive cyanobacteria, existed already 2.8 Ga ago.

Thus the evidence is strong, and comes from various kinds of data that oxygenic photosynthesis took place already about 2.7 Ga ago, half a billion years before the "great oxygenation event." There are indications for the presence of primitive morphological forms of cyanobacteria (about 3.5 Ga ago) [195, 196]. However, there is no evidence that these first cyanobacteria were oxygenic. Other groups including photosynthetic bacteria, such as Chlorobia, must have split off from the cyanobacterial ancestors much earlier than 2.7 Ga ago, but their photosynthetic members could have acquired photosynthesis genes by horizontal gene transfer.

The next major step in the evolution of photosynthesis is the endosymbiotic event in which chloroplasts arose from cyanobacteria [197]. According to Falcón et al. [198], the clade leading to chloroplasts split off from extant cyanobacteria a little over 2 Ga ago.

2.6.2.1 Photosystem I: From Cyanobacteria to Angiosperm   Although the PSI core complexes are very similar in cyanobacteria and higher plants, this does not hold for the structure at the periphery [199, 200]. Another difference is that the cyanobacterial PSI core is trimeric heterodimer, while the plant PSI core appears as a monomeric heterodimer. The heterodimer has evolved from an ancestral homodimer. Trimerization of the cyanobacterial variant occurs via a polypeptide called PSaL [201]. The monomeric state of the plant PSI core allows for more peripheral proteins to be included in the system, and in particular for the constitutive LHCI antenna, and for the mobile LHCII antenna, which migrates from the grana and PSII during state transition [202]. The plant PSI–LHCI supercomplex contains 19 known protein subunits and approximately 200 noncovalently bound cofactors. The simpler PSI of cyanobacteria has no LHCI and only 12 subunits and 127 cofactors [106, 126]. In unicellular green algae, on the other hand, whose path of evolution separated from that of the land plants over 1 billion years ago, LHCI is about three times larger than that of higher plants [203, 204]. Both this larger antenna in aquatic green algae and the trimeric state in cyanobacteria may reflect the lower light levels in many aquatic environments.

On the reducing (electron-acceptor) side of PSI, changes have taken place during evolution from cyanobacteria to higher plant chloroplasts. Cyanobacterial PSI has only one docking site for Fd, which binds this cofactor loosely, and there is no docking site for FNR. The Fd thus has to diffuse away from the PSI complex to reduce $NADP^+$. Plant PSI has one loosely binding and one firmly binding Fd site, and also a site for the binding of the FNR. This allows a direct electron transfer in PSI, and thus there is less risk for competing reactions, such as reduction of O_2 to O_2^-. On the other hand, cyanobacteria have greater flexibility in case of nutrient deficiency. In the case of iron deficiency, they can use flavodoxin in place of the iron-containing Fd (Fig. 2.7). Flavodoxins occur in a wide range of bacteria and are thus not an invention of cyanobacteria. Although higher plants do not naturally contain flavodoxin, bioengineering can produce such plants, and they are more tolerant than natural plants to iron deficiency [205, 206]. The gene for flavodoxin may have been lost when plants first made their way onto land in an iron-rich environment [207].

In the case of copper deficiency, cyanobacteria (as well as red and brown algae) can use Cyt c_6 instead of PC. Plants and green algae (as well as some cyanobacteria) contain the related Cyt c_6 with unknown function, but this has a too low midpoint redox potential to be able to function in the same way. It is thought that Cyt c_6 preceded PC in evolution and was the electron carrier to PSI in the first cyanobacteria.

Geobacter sulfurreducens has a *c*-type Cyt closely resembling Cyt c_6 [208]. PC was probably introduced when oxygenic photosynthesis had already begun, and copper became more easily available than iron [209].

2.6.2.2 Photosystem II: Its Light-Harvesting Proteins and Water-Splitting Site

During evolution from bacterial type II RCs to plant PSII, several remarkable changes took place. The redox potential on the oxidizing side had to be increased to allow for the oxidation of water, and protection from the destructive action of molecular oxygen and its high-potential precursors had to be achieved [210,211]. Regulation mechanisms, the so-called state transitions, were developed to make PSI and PSII work "in step." Light-harvesting, pigment-carrying proteins and the remarkable water-oxidizing enzyme that collects the four charges for creation of a molecule of oxygen were added.

The primitive cyanobacterium *Gloeobacter violaceues* lacks thylakoids and does not have well-developed state transitions [212]. The absorption capacity (absorption cross section) of PSII can be varied by only a few percent [213]. In contrast, other cyanobacteria can acclimate to light and vary their distribution of energy to the photosystems both by movement of the phycobilisomes in the thylakoid membranes [214] and by chromatic acclimation ("chromatic adaptation"), that is, change of the pigment composition of the phycobilisomes. In higher plants the light-harvesting proteins of PSII have the major role of capturing light. The steps in the evolution of light-harvesting proteins of PSII have been described by Ballotari et al. [215].

As mentioned above, the OEC contains four atoms of manganese and one atom of calcium. It is natural to assume that it may have evolved from a precursor that is also a precursor to another extant manganese protein. Raymond and Blankenship [216] have explored this idea and found a structural similarity between the OEC and manganese catalase, but the latter enzyme contains only two manganese atoms. Dismukes et al. [217] envisage an early form of oxygenic photosynthesis in which bicarbonate (HCO_3^-), which is easier to oxidize to oxygen, rather than water served as the electron source. Bicarbonate under the early conditions with more reducing condition and higher concentration of CO_2 in the atmosphere and more bicarbonate in the ocean would have led to the formation of manganese bicarbonate complexes, $Mn_2(HCO_3)_4$. Could this have contributed the other two manganese atoms? For a review on the roles of bicarbonate in PSII, most of which have been shown to be on its electron-acceptor side, see Shevela et al. [124].

2.6.3 Evolutionary Acclimation to Decreasing CO_2 Availability

2.6.3.1 Is Rubisco a "Bad" Enzyme?

Rubisco has often been described as a "bad" enzyme, because it is relatively slow, it does not bind CO_2 very tightly, and O_2 competes with CO_2 at the binding site. Many pathways have evolved for the assimilation of CO_2, but Rubisco is the only CO_2-binding enzyme at which O_2 can compete with CO_2. Therefore it seems a bit surprising that it is Rubisco that has come to completely dominate the entrance to CO_2 assimilation in the oxygenic organisms. The explanations for this are the following:

1. The CO_2 concentration was much higher and the O_2 concentration negligible when water oxidation was "invented."
2. Various mechanisms evolved for concentrating CO_2 at the Rubisco surface.
3. Rubisco is perhaps not that bad after all.

2.6.3.2 Rubisco Properties We start the discussion with Rubisco being not that bad after all. Tcherkez et al. [218] explained that, for Rubisco, there exists an intrinsic and inescapable conflict between the rate constant for carboxylation, k_{cat}^c, and the CO_2/O_2 specificity, $S_{c/o}$ (see Fig. 2.13): the higher the specificity, the lower the rate. They argued that all Rubiscos are almost perfectly tuned to the requirement of different organisms in their environment. From Fig. 2.13 it seems as if the cyanobacterial Rubiscos are superior to the eukaryotic ones, but they are, in fact, inferior with respect to another important property: they bind CO_2 less tightly with a Michaelis–Menten constant of around 300 μM as compared to 10–80 μM for the

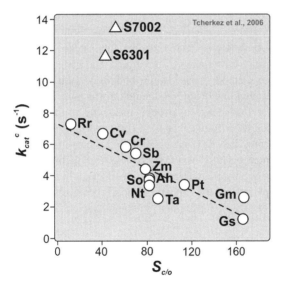

FIGURE 2.13 Rate constant for carboxylation as a function of the CO_2/O_2 specificity for Rubiscos from various species. The regression line is drawn neglecting the two cyanobacterial outliers (triangles). The Rubiscos were isolated from the following organisms: Ah, *Amaranthus hybridus* (C4 dicot); Cr, *Chlamydomonas reinhardtii* (green alga); Cv, *Chromatium vinosum* (bacterium); Gm, *Griffithsia monilis* (red alga); Gs, *Galdieria sulfuraria* (red alga); Nt, *Nicotiana tabacum* (C3 dicot); Pt, *Phaeodactylum tricornutum* (diatom); Rr, *Rhodospirillum rubrum* (bacterium); Sb, *Sorghum bicolor* (C4 monocot); So, *Spinacia oleracea* (C3 dicot); S6301, *Synechococcus* PCC 6301 (cyanobacterium); S7002, *Synechococcus* PCC 7002 (cyanobacterium); Ta, *Triticum aestivum* (C3 monocot); and Zm, *Zea mays* (C4 monocot). Adapted from Tcherkez et al. [218], copyright © 2006 National Academy of Sciences, U.S.A.

eukaryotes [219]. The relation between the properties of Rubisco is further discussed by André [220].

2.6.3.3 CO₂ Concentrating Mechanisms

Cyanobacteria compensate for the high Michaelis–Menten constant (the low affinity) for CO_2 by active uptake of bicarbonate into the cytoplasm in combination with carboxysomes. Carboxysomes are "Rubisco cages" surrounded by a barrier that is impermeable to CO_2. They also contain carbonic anhydrase that converts bicarbonate to carbon dioxide, resulting in a high CO_2 concentration at the Rubisco. The plasma membrane is permeable to carbon dioxide, so it is essential that the cytoplasm lacks carbonic anhydrase. In various groups of cyanobacteria several mechanisms for the active uptake of bicarbonate have evolved [221, 222]. Further, eukaryotic algae and some higher plants are equipped with various CO_2-concentrating mechanisms [223]. We have already mentioned the C4 mechanism above.

2.6.3.4 CO₂ Levels in the Atmosphere Over Time

We do not know what the availability of CO_2 was for the first photosynthetic organisms, but the concentration was much higher than now [224, 225]. From an initially very high level, it declined sharply around 4 Ga ago, when CO_2 was partly replaced by methane, resulting from the activity of methanogenic archaea [226].

Several methods have been used to estimate the past levels of atmospheric carbon dioxide. They do not show detailed agreement but give a reasonably consistent overall picture. One method is based on a budget of carbon, sulphur, and oxygen as the atoms circulate among different chemical forms and compartments over time. Several biological proxies have also been used, the most important is based on plant stomata. It has been found that both their size and their density on the leaf surface adapt to carbon dioxide concentration in the surrounding air, and they can be measured and counted on leaf fossils. Of course, this method can be used only for times since stomata evolved, in the late Silurian, about 420 million years ago. Figure 2.14 shows a comparison of the budget method (specifically the GEOCARBSULF model of Berner; for details, see Berner [227]) and various stomata-based proxies.

2.7 SOME INTERESTING QUESTIONS ABOUT WHOLE PLANTS

2.7.1 Overview

Here we first list some questions for the readers to ponder; then we present our views on these topics.

- Why are there grana in land plants but not in algae?
- Why are leaves darker on the upper side than on the lower side?
- How much do different layers in the leaf contribute to photosynthesis?
- How does photosynthesis interact with climate-atmosphere?

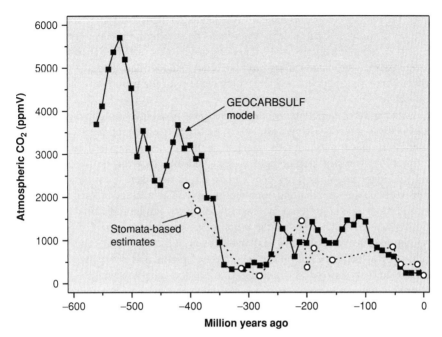

FIGURE 2.14 Atmospheric CO_2 over the Phanerozoic by the GEOCARBSULF model of Berner [227] (solid line, closed symbols), and stomata-based estimates [250–252] of van der Burgh et al. [252] (dashed line, open symbols).

- Is there photosynthesis without CO_2 assimilation (N_2 fixation in cyanobacteria, light-dependent NO_3^- assimilation in land plants)?
- How can animals carry out photosynthesis?

2.7.2 Why Are There Grana in Land Plants but Not in Algae?

A characteristic of land plants is that their chloroplasts contain grana, regions where the thylakoid membranes are appressed without any stroma regions between them (Fig. 2.1). Algae living in an aquatic environment do not have this structural arrangement.

The most likely explanation for this was advanced by Jan M. Anderson [228]. Her theory is based on the existence of different light environments in water and in air. The light environment of the red algae and the land plants is very different. Red algae live in water, often deeper than other algae. The light reaching them is filtered through a thick layer of water. This layer absorbs long-wavelength light more strongly than other visible (and photosynthetically active) light. Therefore an energy deficiency in light absorption by PSI relative to PSII could easily develop. However, extra energy can come to PSI by a process called *spillover*, from PSII [229]. Spillover of energy is possible since the light quanta absorbed by PSII are larger than needed to excite long-wavelength absorbing Chls in PSI. Furthermore, spillover is possible here because the two photosystems are intermingled in the algal chloroplasts. Because the spectrum of available light in deep water lacks far red light, there is never a risk that there will be too much energy for PSI, as compared to that in PSII. Furthermore, since light intensity is attenuated in deep water, it is advantageous for the red algae to retain the light-harvesting phycobilisomes, which came with the cyanobacteria giving rise to chloroplasts.

For land plants, the situation is different. The first plants colonizing land were small beech organisms living without competition from larger plants, exposed to full sunlight; their forerunners, the green algae, lived in very exposed habitats [230]. Their problem was not lack of light energy, and thus they did not have much use for phycobilisomes. With time, plants grew larger and were more numerous, and started to shade one another and compete for light. The average chloroplast became more and more shaded, filtered by other chloroplasts. For an individual chloroplast, it did not matter much whether the chloroplasts shading it were located in other plants, in other leaves on the same plant, or even in the same leaf. The light hitting the chloroplast became, during the evolution of plants and ecosystems, more and more depleted in short-wavelength light, while the long-wavelength light, on the long-wave edge of the chlorophyll absorption spectrum, was not attenuated to the same extent. The spectral situation contrasted sharply against that for chloroplasts in red algae. Now we speculate that the imbalance between the photosystems could not be adjusted by spillover, since the light quanta, absorbed by the most exposed PSI, were too small. Therefore PSI and PSII had to be separated to prevent spillover, or PSII would receive even less energy. Evolution has succeeded in this by development of grana in the chloroplasts of land plants (see Fig. 2.1). Grana are regions in the chloroplasts where the thylakoid membranes are closely stacked on top of one another and are enriched in PSII. The stacking of membranes and absence of PSI gave room for larger pigment antennas, not in the form of phycobilisomes, but in the form of protein-bound pigment (Chl a and Chl b) complexes. PSI is located in the more sparsely distributed membranes between the grana. There, it is in contact with stroma between the membranes, and this is advantageous because PSI delivers reducing equivalents via ferredoxin to $NADP^+$, which are then used for the reduction of carbon dioxide in the stroma region.

2.7.3   Why Are Leaves Darker on the Upper Side than on the Lower Side?

Almost all leaves, except those positioned almost vertically, appear darker on the upper (adaxial) side than on the lower (abaxial) side. We speculate that the reason for this is that they are adapted for receiving light from the upper side. The mesophyll on the upper side, a *palisade* with oblong cells, is arranged perpendicular to the leaf surface. Light entering through the transparent upper epidermis is partly absorbed by the chloroplasts in this layer, and partly penetrates deeper into the leaf through the vacuoles and cell walls [231]. Only little light is scattered back [232].

The mesophyll in the lower part of the leaf has a different structure, with large intercellulars and cells extending in different directions in a loose, spongy way. This

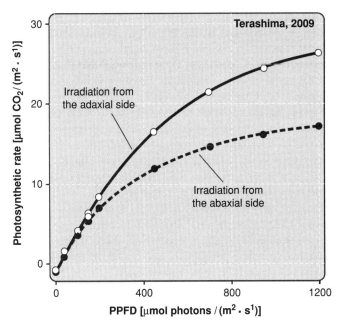

FIGURE 2.15   Net photosynthetic rate versus photosynthetic photon flux density for a leaf of *Helianthus annuus*, for the two sides of the leaf. Conditions are: 25°C, 390 μL/L CO_2, water vapor pressure deficit 0.7 kPa. Adapted from Terashima et al. [233], copyright © 2009, Oxford University Press.

layer scatters much of the light reaching it from above, back into the palisade layer (while some is absorbed in the spongy layer). Also, light hitting the lower side is to a large extent scattered back, giving the lower side of the leaf a lighter color. The greater reflectivity of the lower side of the leaf causes light hitting the lower side to have less effectiveness than light hitting the upper side (see Terashima et al. [233] and Fig. 2.15).

2.7.4   How Much Do Different Layers in the Leaf Contribute to Photosynthesis?

This question is not easy to answer. Although very sophisticated methods have been employed to answer this question, the results remain difficult to understand. In particular, there seems to be a difference between how CO_2 assimilation and how O_2 evolution are distributed throughout the thickness of the leaves. Experiments have been conducted on leaves of different species, but we limit ourselves here to the situation in a spinach leaf, which can be regarded as a leaf with rather ordinary structure, not adapted to any kind of extreme environment. Such a leaf is about 456 μm thick when shade-adapted (grown in weak light) and 631 μm thick when grown in strong light [234].

Light entering from the adaxial (upper) side of a leaf is gradually attenuated as one goes deeper into the leaf tissue. If the Chl molecules were uniformly distributed throughout the leaf, one would assume that the rate of light absorption and of photosynthesis would vary in proportion to the light available at various depths. Measurements of CO_2 assimilation in white light, however, indicate that the rate rises to a maximum, about 35% of the total thickness into the leaf, and then slowly decreases to about 40% of the maximum value near the abaxial (lower) side [232, 235]. The spongy mesophyll fixes about 40% of the total, a much higher proportion than what would be expected from a simple light-intensity model. The explanation may be that light in the spongy mesophyll is scattered back and forth simulating the situation in a "hall of mirrors" in an amusement park [232] so that the chloroplasts have enhanced chance of absorbing light coming from various directions.

Another strange result is that the evolution of O_2 does not show the same pattern as described above for CO_2 assimilation [236]. This conclusion was obtained by using a photoacoustic technique. Oxygen evolution was much more equally distributed across the whole leaf, nowhere going below about 60% of the maximum rate, which occurred at a depth of about 12% of the leaf thickness.

2.7.5 How Does Photosynthesis Interact with Climate-Atmosphere?

Photosynthetic organisms and their photosynthesis affect climate in several ways. What first comes to mind is that CO_2 is an important greenhouse gas, and its removal from the atmosphere keeps the Earth cool enough for life. Plants affect the CO_2 content of the atmosphere not only directly, but by converting CO_2 to organic carbon. Plants also increase weathering and conversion of silicate rock to carbonate. A drastic example of this is at the time (Ordovician) when vegetation first invaded land [237]. CO_2 was then removed from the atmosphere to such an extent that a series of glaciation periods followed.

There are, however, other ways in which plants and other oxygenic organisms affect the climate. The ozone produced from their "waste product" O_2 is also a greenhouse gas that affects the temperature of the Earth's surface, but it is also important in heating the stratosphere, and in fact creating it [238]. It is the heating by the absorption of ultraviolet (UV) radiation in ozone that makes the temperature increase with altitude in the higher parts of the atmosphere (in contrast to the lower atmosphere, the troposphere), and in this way, it makes the stratosphere stratified (without the strong vertical movement of the lower atmosphere) [238]. This is important for the total atmospheric circulation, and thus for climate.

Plants need to have openings, stomata, in their outer cell layer, the epidermis, in order to be able to absorb CO_2. This leads to transpiration, release of water vapor. Water vapor is an important greenhouse gas too, and land vegetation greatly increases the transfer of water from the surface to the atmosphere. This also affects cloudiness, an important climate determinant. Plants in various ways also contribute to condensation nuclei for the formation of water droplets in the atmosphere, which is important for cloud formation and precipitation [239]. Also, some unicellular photosynthetic algae in the oceans are thought to contribute to the formation of these nuclei by secreting sulfur compounds, which are converted, by bacteria, to dimethyl sulfide, $(CH_3)_2S$. This, in the atmosphere, is oxidized to sulfuric acid that attracts water and forms droplets [240].

On the other hand, climate affects photosynthesis. Photosynthesis is decreased by aridity and also by very low and very high temperatures, and in many cases also by cloudiness.

2.7.6   Is There Photosynthesis Without CO_2 Assimilation (N_2 Fixation in Cyanobacteria, Light-Dependent NO_3^- Assimilation in Land Plants)?

As mentioned earlier in this chapter, a type of photosynthesis exists without CO_2 assimilation. The most important case is the one tied to the assimilation of nitrogen.

Cyanobacteria have the ability to bind molecular nitrogen (N_2) by a special process. This involves the use of enzyme nitrogenase, but it is very sensitive to molecular O_2, which rapidly inactivates it if the two come together. Cyanobacteria have two strategies to keep the nitrogenase active, in spite of their own O_2. Many of them contain long cell chains with two types of cells: *vegetative cells* that carry out normal CO_2 assimilation and evolution of O_2 (using both PSI and PSII); and *heterocysts*, which assimilate N_2 using PSI, and organic compounds imported from the vegetative cells [241].

Other cyanobacteria, however, are unicellular or are filamentous but lack heterocysts and must use another strategy. In all cases of nitrogen fixation, PSI generates reduced Fd (see Section 2.3 and Fig. 2.7), which is the reductant for converting N_2 to ammonium ion, NH_4^+. One way of protecting nitrogenase from O_2 is to use some of this reduced Fd for reducing O_2 to O_2^- (superoxide ion), which can be handled further by superoxide dismutase and catalase. Another way is to separate the processes temporally and carry out the N_2 binding during night, when no oxygen evolution is taking place [242].

Higher plants too can use light for the reduction of nitrogen, but in this case only nitrogen in the form of nitrate is used. Nitrate is taken up by the roots, and in many plants, it is reduced right there, without the aid of light. But other plants transport the nitrate ions to the aboveground green parts and reduce it by NADH, which is produced indirectly from photosynthesis, via the organic photoassimilates [243], and the process is therefore stimulated by light.

2.7.7   How Can Animals Carry Out Photosynthesis?

Many animals harbor algae as endosymbionts [244]. The most important case is that of stone corals, which contain dinoflagellates of the genus *Symbiodinium*, which play an important role not only for the supply of organic compounds to supplement what the corals can catch as food, but also for the formation of the carbonate skeleton. Endosymbiotic algae that carry out photosynthesis are also very common among unicellular animals.

The most interesting examples of *photosynthetic animals* are perhaps a group of sea slugs that eat algae, but keep their chloroplasts and nucleic acid molecules; the chloroplasts are kept in a functional state for a long time [245, 246]. Some of these sea slugs use chloroplasts from red algae, others those from green algae to carry out photosynthesis. The most studied have been the slugs of the genus *Elysia*, for example, *E. chlorotica* that feeds on the green alga *Vaucheria litorea*. *Elysia chlorotica* can keep the chloroplasts active in its body for up to 10 months in the absence of any *Vaucheria* nucleus. This is remarkable since when the chloroplasts are in the alga, they must be constantly repaired and supplied with fresh protein molecules coded for by nuclear genes. Chloroplasts are not transferred to the slug offspring, but each generation must obtain them by feeding on *Vaucheria*. Only one nuclear gene for a chloroplast protein (i.e., the phosphoribulokinase gene) has been shown to have been transferred from *Vaucheria* to *Elysia*. It is thus far a mystery how *Elysia* can maintain chloroplasts active for so long!

2.8 PERSPECTIVES FOR THE FUTURE

The problem of supplying humanity with usable energy, which is tightly coupled to climate and other environmental issues, as well as to food supply and world peace, is an enormous challenge to human ingenuity. We have little time to solve it. Use of the "know-how" accumulated in photosynthetic organisms by evolution over billions of years may provide the shortcut we need to meet the time constraint. This gives us reason to continue the exploration of how the natural systems work, and how they have been able to sustain the biosphere of our, as far as we know, unique planet.

2.9 SUMMARY

While many bacteria carry out a simple form of photosynthesis using a single photo-system, cyanobacteria, algae, and plants perform oxygenic photosynthesis using two photochemical systems, connected in series. In the latter, water is oxidized to molecular oxygen, and organic compounds are produced from inorganic carbon compounds. It is a very complex process, which has several phases starting with absorption of light by pigments in antenna complexes, and channeling of the excitation energy to the reaction centers, where positive and negative charges are separated. The positive charges, obtained from PSII, are used to oxidize water in a four-step process involving manganese in several oxidation states, whereas the negative charges are transferred over a number of electron carriers, most of which are bound to membrane proteins, but there are also carriers in the membrane lipids or in water. PSI is involved in taking the electrons coming from PSII to the next level; the negative charges are finally used to reduce $NADP^+$. The electron transfer is coupled to transfer of protons across the membranes, which together with protons from the oxidation of water leads to energy stored in the form of a pH difference across the membranes. The energy, thus transiently stored, is used for the synthesis of ATP. The ATP is used, together with NADPH, for the synthesis of carbohydrate from carbon dioxide or bicarbonate.

Oxygenic photosynthesis is the main process that provides energy to the entire biosphere of our planet. It also gives rise to the ozone layer that protects us from solar ultraviolet radiation. Furthermore, by its consumption of carbon dioxide, it has for a long time kept the climate friendly to life, and it has been an indirect cause of various geological processes. There is now hope that by learning more about natural photosynthesis we shall be able to better address humanity's energy problems by harvesting solar energy in an economically feasible, environment-friendly, and sustainable way.

ACKNOWLEDGMENTS

DS gratefully acknowledges financial support by The Research Council of Norway (grant 192436) and the Artificial Leaf Project (K&A Wallenberg Foundation). LOB is grateful to South China Normal University for financial support. G is thankful to a Fulbright Specialist Award to India (October–December 2012) since the final version of this manuscript was completed while he was lecturing on this topic at the University of Indore and at the University of Hyderabad.

REFERENCES

1. E. Rabinowitch, Govindjee, *Photosynthesis*, Wiley, New York, 1969.
2. R. E. Blankenship, *Molecular Mechanisms of Photosynthesis*, Blackwell Publishing, Oxford, 2002.
3. M. T. Rosing, D. K. Bird, N. H. Sleep, W. Glassley, F. Albarede, The rise of continents—an essay on the geologic consequences of photosynthesis, Palaeogeography, Palaeoclimatology, Palaeoecology 232 (2006) 99–113.
4. G. Ciamician, The photochemistry of the future, Science 36 (1912) 385–394.
5. J. J. Eaton-Rye, B. C. Tripathy, T. D. Sharkey (Eds.), *Photosynthesis: Plastid Biology, Energy Conversion and Carbon Assimilation*, Springer, Dordrecht, 2012.
6. Govindjee, D. Shevela, Adventures with cyanobacteria: a personal perspective, Front. Plant Sci. 2:28 (2011) doi:10.3389/fpls.2011.00028.
7. C. N. Hunter, F. Daldal, M. C. Thurnauer, J. T. Beatty (Eds.), *The Purple Phototrophic Bacteria*, Springer, Dordrecht, 2009.
8. R. E. Blankenship, M. T. Madigan, C. E. Bauer (Eds.), *Anoxygenic Photosynthetic Bacteria*, Kluwer Academic Publishers, Dordrecht, 1995.
9. D. Oesterhelt, C. Brauchle, N. Hampp, Bacteriorhodopsin: a biological material for information processing, Q. Rev. Biophys. 24 (1991) 425–478.
10. G. D. Scholes, G. R. Fleming, A. Olaya-Castro, R. Van Grondelle, Lessons from nature about solar light harvesting, Nat. Chem. 3 (2011) 763–774.
11. M. M. Najafpour, Govindjee, Oxygen evolving complex in photosystem II: better than excellent, Dalton Trans. 40 (2011) 9076–9084.
12. P. D. Tran, L. H. Wong, J. Barber, J. S. C. Loo, Recent advances in hybrid photocatalysts for solar fuel production, Energy Environ. Sci. 5 (2012) 5902–5918.

13. D. Shevela, S. Koroidov, M. M. Najafpour, J. Messinger, P. Kurz, Calcium manganese oxides as oxygen evolution catalysts: O_2 formation pathways indicated by ^{18}O-labelling studies, Chem. Eur. J. 17 (2011) 5415–5423.

14. I. Zaharieva, P. Chernev, M. Risch, K. Klingan, M. Kohlhoff, A. Fischer, H. Dau, Electrosynthesis, functional, and structural characterization of a water-oxidizing manganese oxide, Energy Environ. Sci. 5 (2012) 7081–7089.

15. J. S. Kanady, E. Y. Tsui, M. W. Day, T. Agapie, A synthetic model of the Mn_3Ca subsite of the oxygen-evolving complex in photosystem II, Science 333 (2011) 733–736.

16. M. M. Najafpour, A. N. Moghaddam, S. I. Allakhverdiev, Govindjee, Biological water oxidation: lessons from Nature, Biochim. Biophys. Acta 1817 (2012) 1110–1121.

17. W. Lubitz, E. J. Reijerse, J. Messinger, Solar water-splitting into H_2 and O_2: design principles of photosystem II and hydrogenases, Energy Environ. Sci. 1 (2008) 15–31.

18. M. M. Najafpour, M. A. Tabrizi, B. Haghighi, Govindjee, A manganese oxide with phenol groups as a promising structural model for water oxidizing complex in photosystem II: a "golden fish," Dalton Trans. 41 (2012) 3906–3910.

19. A. Guskov, A. Gabdulkhakov, M. Broser, C. Glöckner, J. Hellmich, J. Kern, M. Frank, W. Saenger, A. Zouni, Recent progress in the crystallographic studies of photosystem II, ChemPhysChem 11 (2010) 1160–1171.

20. D. Bald, ATP synthase: structure, function and regulation of a complex machine. In: G. A. Peschek, C. Obinger, G. Renger (Eds.), *Bioenergetic Processes of Cyanobacteria: From Evolutionary Singularity to Ecological Diversity*, Springer, Dordrecht, 2011, pp. 239–261.

21. P. Fromme, I. Grotjohann, Structure of cyanobacterial photosystems I and II. In: G. A. Peschek, C. Obinger, G. Renger (Eds.), *Bioenergetic Processes of Cyanobacteria: From Evolutionary Singularity to Ecological Diversity*, Springer, Dordrecht, 2011, pp. 285–335.

22. G. Bernat, M. Rögner, Center of the cyanobacteria electron transport network: the cytochrome b_6f complex. In: G. A. Peschek, C. Obinger, G. Renger (Eds.), *Bioenergetic Processes of Cyanobacteria: From Evolutionary Singularity to Ecological Diversity*, Springer, Dordrecht, 2011, pp. 573–606.

23. J. Kargul, J. Barber, Structure and function of photosynthetic reaction centres. In: T. Wydrzynski, W. Hillier (Eds.), *Molecular Solar Fuels*, RSC Publishing, Cambridge, 2012, pp. 107–142.

24. I. Grotjohann, C. Jolley, P. Fromme, Evolution of photosynthesis and oxygen evolution: implications from the structural comparison of photosystems I and II, Phys. Chem. Chem. Phys. 6 (2004) 4743–4753.

25. M. F. Hohmann-Marriott, R. E. Blankenship, Evolution of photosynthesis, Annu. Rev. Plant Biol. 62 (2011) 515–548.

26. Y. Umena, K. Kawakami, J.-R. Shen, N. Kamiya, Crystal structure of oxygen-evolving photosystem II at a resolution of 1.9 Å, Nature 473 (2011) 55–60.

27. B. R. Green, W. W. Parson (Eds.), *Light-Harvesting Antennas in Photosynthesis*, Springer, Dordrecht, 2003.

28. H. Scheer, An overview of chlorophylls and bacteriochlorophylls: biochemistry, biophysics, functions and applications. In: B. Grimm, R. J. Porra, W. Rüdiger, H. Scheer (Eds.), *Chlorophylls and Bacteriochlorophylls*, Springer, Dordrecht, 2006, pp. 1–26.

29. Govindjee, Carotenoids in photosynthesis: a historical perspective. In: H. A. Frank, A. J. Young, G. Britton, R. J. Cogdell (Eds.), *The Photochemistry of Carotenoids: Applications in Biology*, Kluwer Academic Publishers, Dordrecht, 1999, pp. 1–19.

30. C. O'hEocha, Biliproteins of algae, Annu. Rev. Plant Physiol. 16 (1965) 415–434.

31. M. D. Kamen, *Primary Processes in Photosynthesis*, Academic Press, New York, 1963.

32. K. Sauer, Primary events and the trapping of energy. In: Govindjee (Ed.), *Bioenergetics of Photosynthesis*, Academic Press, New York, 1975, pp. 116–175.

33. L. N. M. Duysens, Transfer of excitation energy in photosynthesis, State University, Utrecht, 1952.

34. L. N. M. Duysens, Transfer of light energy within the pigment systems present in photosynthesizing cells, Nature 168 (1951) 548–550.

35. D. Zigmantas, R. G. Hiller, V. Sundström, T. Polívka, Carotenoid to chlorophyll energy transfer in the peridinin–chlorophyll-*a*–protein complex involves an intramolecular charge transfer state, Proc. Natl. Acad. Sci. U.S.A. 99 (2002) 16760–16765.

36. P. Zuo, B.-X. Li, X.-H. Zhao, Y.-S. Wu, X.-C. Ai, J.-P. Zhang, L.-B. Li, T.-Y. Kuang, Ultrafast carotenoid-to-chlorophyll singlet energy transfer in the cytochrome b_6f complex from *Bryopsis corticulans*, Biophys. J. 90 (2006) 4145–4154.

37. R. Fujii, M. Kita, M. Doe, Y. Iinuma, N. Oka, Y. Takaesu, T. Taira, M. Iha, T. Mizoguchi, R. Cogdell, H. Hashimoto, The pigment stoichiometry in a chlorophyll *a/c* type photosynthetic antenna, Photosynth. Res. 111 (2012) 165–172.

38. F. T. Haxo, J. H. Kycia, G. F. Somers, A. Bennett, H. W. Siegelman, Peridinin–chlorophyll *a* proteins of the dinoflagellate *Amphidinium carterae* (Plymouth 450), Plant Physiol. 57 (1976) 297–303.

39. C. S. French, V. K. Young, The fluorescence spectra of red algae and the transfer of energy from phycoerythrin to phycocyanin and chlorophyll, J. Gen. Physiol. 35 (1952) 873–890.

40. A. C. Ley, W. L. Butler, D. A. Bryant, A. N. Glazer, Isolation and function of allophycocyanin B of *Porphyridium cruentum*, Plant Physiol. 59 (1977) 974–980.

41. M. Mimuro, Photon capture, excitation migration and trapping and fluorescence emission in cyanobacteria and red algae. In: G. C. Papageorgiou, Govindjee (Eds.), *Chlorophyll a Fluorescence: A Signature of Photosynthesis*, Springer, Dordrecht, 2004, pp. 173-195.

42. R. Kana, O. Prasil, O. Komarek, G. C. Papageorgiou, Govindjee, Spectral characteristic of fluorescence induction in a model cyanobacterium, *Synechococcus* sp (PCC 7942), Biochim. Biophys. Acta 1787 (2009) 1170–1178.

43. M. Stomp, J. Huisman, L. J. Stal, H. C. P. Matthijs, Colorful niches of phototrophic microorganisms shaped by vibrations of the water molecule, ISME J. 1 (2007) 271–282.

44. M. Striebel, S. Behl, S. Diehl, H. Stibor, Spectral niche complementarity and carbon dynamics in Pelagic ecosystems, Am. Nat. 174 (2009) 141–147.

45. M. Mimuro, M. Kobayashi, A. Murakami, T. Tsuchiya, H. Miyashita, Oxygen-evolving cyanobacteria. In: G. Renger (Ed.), *Primary Processes of Photosynthesis, Part 1 Principles and Apparatus*, RSC Publishing, Cambridge, 2008, pp. 261–300.

46. W. A. Sidler, Phycobilisome and phycobiliprotein structures. In: D. A. Bryant (Ed.), *The Molecular Biology of Cyanobacteria*, Springer, Dordrecht, 1994, pp. 139–216.

47. D. Vavilin, H. Xu, S. Lin, W. Vermaas, Energy and electron transfer in photosystem II of a chlorophyll *b*-containing *Synechocystis* sp. PCC 6803 mutant, Biochemistry 42 (2003) 1731–1746.

48. A. M. Collins, J. Wen, R. E. Blankenship, Photosynthetic light-harvesting complexes. In: T. Wydrzynski, W. Hillier (Eds.), *Molecular Solar Fuels*, RSC Publishing, Cambridge, 2012, pp. 85–106.

49. J. Neilson, D. Durnford, Structural and functional diversification of the light-harvesting complexes in photosynthetic eukaryotes, Photosynth. Res. 106 (2010) 57–71.

50. H. Gaffron, K. Wohl, On the theory of assimilation, Naturwissenschaften 24 (1936) 81–90; 103–107.

51. R. Emerson, R. Arnold, A separation of the reactions in photosynthesis by means of intermittent light, J. Gen. Physiol. 15 (1932) 391–420.

52. R. Emerson, R. Arnold, The photochemical reaction in photosynthesis, J. Gen. Physiol. 16 (1932) 191–205.

53. R. M. Clegg, M. Sener, Govindjee, From Förster resonance energy transfer to coherent resonance energy transfer back. In: R. R. Alfano (Ed.), *Optical Biopsy VII*, SPIE Press, Bellingham, WA, 2010, p. 21.

54. H. J. Dutton, W. M. Manning, B. M. Duggar, Chlorophyll fluorescence and energy transfer in the diatom *Nitzschia closterium*, J. Phys. Chem. 47 (1943) 308–313.

55. E. C. Wassink, J. A. H. Kersten, Observations sur le spectre d'absorption et sur le role des carotenoides dans la photosynthèse des diatomes, Enzymologia 12 (1946) 3–32.

56. W. Arnold, E. S. Meek, The polarization of fluorescence and energy transfer in grana, Arch. Biochem. Biophys. 60 (1956) 82–90.

57. S. S. Brody, Transferts d'energie et spectres de fluorescence chez Porphyridium cruentum, J. Chimie Physique Physico-Chimie Biologique 55 (1958) 942–951.

58. F. Cho, J. Spencer, Govindjee, Emission spectra of *Chlorella* at very low temperatures (−269° to −196°), Biochim. Biophys. Acta 126 (1966) 174–176.

59. F. Cho, Govindjee, Low-temperature (4–77 K) spectroscopy of *Anacystis*: temperature dependence of energy transfer efficiency, Biochim. Biophys. Acta 216 (1970) 151–161.

60. F. Cho, Govindjee, Low-temperature (4–77 K) spectroscopy of *Chlorella*—temperature dependence of energy transfer efficiency, Biochim. Biophys. Acta 216 (1970) 139–150.

61. Govindjee, Chlorophyll *a* fluorescence: a bit of basics and history. In: G. C. Papageorgiou, Govindjee (Eds.), *Chlorophyll a Fluorescence: A Signature of Photosynthesis*, Kluwer Academic Publishers, Dordrecht, 2004, pp. 1–42.

62. A. Ishizaki, G. R. Fleming, Unified treatment of quantum coherent and incoherent hopping dynamics in electronic energy transfer: reduced hierarchy equation approach, J. Chem. Phys. 130 (2009) 234111.

63. A. Ishizaki, G. R. Fleming, Quantum coherence in photosynthetic light harvesting, Annu. Rev. Condensed Matter Phys. 3 (2012) 333–361.

64. T. Förster, Energiewanderung und Fluoreszenz, Naturwissenschaften 33 (1946) 166–175.

65. T. Förster, Zwischenmolekulare Energiewanderung und Fluoreszenz, Ann. Phys. 437 (1948) 55–75.

66. T. Förster, Delocalized excitation and excitation transfer. In: O. Sinanoglu (Ed.), *Modern Quantum Chemistry. Part III. Action of Light and Organic Crystals*, Academic Press, New York, 1965, pp. 93–137.

67. R. E. Dale, J. Eisinger, W. E. Blumberg, The orientational freedom of molecular probes. The orientation factor in intramolecular energy transfer, Biophys. J. 26 (1979) 161–193.

68. P. Wu, L. Brand, Orientation factor in steady-state and time-resolved resonance energy transfer measurements, Biochemistry 31 (1992) 7939–7947.

69. K. Colbow, R. P. Danyluk, Energy transfer in photosynthesis, Biochim. Biophys. Acta 440 (1976) 107–121.

70. J. R. Lakowicz, *Principles of Fluorescence Spectroscopy*, 2nd ed., Kluwer Academic/ Plenum Publishers, New York, 1999.

71. J. Voigt, T. Renger, R. Schödel, T. Schrötter, J. Pieper, H. Redlin, Excitonic effects in the light-harvesting Chl *a*/*b*–protein complex of higher plants, Phys. Status Solidi B 194 (1996) 333–350.

72. T. Renger, V. May, Multiple exciton effects in molecular aggregates: application to a photosynthetic antenna complex, Phys. Rev. Lett. 78 (1997) 3406–3409.

73. M. Yang, G. R. Fleming, Influence of phonons on exciton transfer dynamics: comparison of the Redfield, Förster, and modified Redfield equations, Chem. Phys. 275 (2002) 355–372.

74. T. Renger, Theory of excitation energy transfer: from structure to function, Photosynth. Res. 102 (2009) 471–485.

75. J. Strümpfer, M. Şener, K. Schulten, How quantum coherence assists photosynthetic light-harvesting, J. Phys. Chem. Lett. 3 (2012) 536–542.

76. F. J. Kleima, E. Hofmann, B. Gobets, I. H. M. van Stokkum, R. van Grondelle, K. Diederichs, H. van Amerongen, Förster excitation energy transfer in peridinin–chlorophyll–*a*–protein, Biophys. J. 78 (2000) 344–353.

77. A. G. Redfield, On the theory of relaxation processes, IBM J. Res. Dev. 1 (1957) 19–31.

78. T. Renger, V. May, O. Kühn, Ultrafast excitation energy transfer dynamics in photosynthetic pigment–protein complexes, Phys. Rep. 343 (2001) 137–254.

79. V. I. Novoderezhkin, M. A. Palacios, H. van Amerongen, R. van Grondelle, Excitation dynamics in the LHCII complex of higher plants: modeling based on the 2.72 Å crystal structure, J. Phys. Chem. B 109 (2005) 10493–10504.

80. G. Raszewski, B. A. Diner, E. Schlodder, T. Renger, Spectroscopic properties of reaction center pigments in photosystem II core complexes: revision of the multimer model, Biophys. J. 95 (2008) 105–119.

81. M. Şener, J. Strümpfer, J. Hsin, D. Chandler, S. Scheuring, C. N. Hunter, K. Schulten, Förster energy transfer theory as reflected in the structures of photosynthetic light-harvesting systems, ChemPhysChem 12 (2011) 518–531.

82. S. Jang, M. D. Newton, R. J. Silbey, Multichromophoric Förster resonance energy transfer, Phys. Rev. Lett. 92 (2004) 218301.

83. A. Ishizaki, T. R. Calhoun, G. S. Schlau-Cohen, G. R. Fleming, Quantum coherence and its interplay with protein environments in photosynthetic electronic energy transfer, Phys. Chem. Chem. Phys. 12 (2010) 7319–7337.

84. E. Collini, C. Y. Wong, K. E. Wilk, P. M. G. Curmi, P. Brumer, G. D. Scholes, Coherently wired light-harvesting in photosynthetic marine algae at ambient temperature, Nature 463 (2010) 644–647.

85. R. E. Blankenship, D. M. Tiede, J. Barber, G. W. Brudvig, G. R. Fleming, M. L. Ghirardi, M. R. Gunner, W. Junge, D. M. Kramer, A. Melis, T. A. Moore, C. C. Moser, D. G. Nocera,

A. J. Nozik, D. R. Ort, W. W. Parson, R. C. Prince, R. T. Sayre, Comparing photosynthetic and photovoltaic efficiencies and recognizing the potential for improvement, Science 332 (2011) 805–809.

86. D. S. Bendall, B. G. Schlarb-Ridley, C. J. Howe, Transient interactions between soluble electron transfer proteins. The case of plastocyanin and cytochrome c_6. In: G. A. Peschek, C. Obinger, G. Renger (Eds.), *Bioenergetic Processes of Cyanobacteria: From Evolutionary Singularity to Ecological Diversity*, Springer, Dordrecht, 2011, pp. 541–571.

87. G. Sandmann, Formation of plastocyanin and cytochrome c-553 in different species of blue-green algae, Arch. Microbiol. 145 (1986) 76–79.

88. M. Medina, Structural and mechanistic aspects of flavoproteins: photosynthetic electron transfer from photosystem I to $NADP^+$, FEBS J. 276 (2009) 3942–3958.

89. A. Aliverti, V. Pandini, A. Pennati, M. de Rosa, G. Zanetti, Structural and functional diversity of ferredoxin–$NADP^+$ reductases, Arch. Biochem. Biophys. 474 (2008) 283–291.

90. Govindjee, L. O. Björn, Dissecting oxygenic photosynthesis: the evolution of the "Z"-scheme for thylakoid reactions. In: S. Itoh, P. Mohanty, K. N. Guruprasad (Eds.), *Photosynthesis: Overviews on Recent Progress and Future Perspectives*, IK Publishers, New Delhi, 2012, pp. 1–27.

91. A. T. Jagendorf, Photophosphorylation and the chemiosmotic perspective, Photosynth. Res. 73 (2002) 233–241.

92. G. Renger, T. Renger, Photosystem II: the machinery of photosynthetic water splitting, Photosynth. Res. 98 (2008) 53–80.

93. G. Renger (Ed.), *Primary Processes of Photosynthesis: Principles and Apparatus*, RSC Publishing, Cambridge, 2008.

94. M. G. Müller, C. Slavov, R. Luthra, K. E. Redding, A. R. Holzwarth, Independent initiation of primary electron transfer in the two branches of the photosystem I reaction center, Proc. Natl. Acad. Sci. U.S.A. 107 (2010) 4123–4128.

95. V. I. Prokhorenko, A. R. Holzwarth, Primary processes and structure of the photosystem II reaction center: a photon echo study, J. Phys. Chem. B 104 (2000) 11563–11578.

96. G. Renger, A. R. Holzwarth, Primary electron transfer. In: T. J. Wydrzynski, K. Satoh (Eds.), *Photosystem II. The Light-Driven Water:Plastoquinone Oxidoreductase*, Springer, Dordrecht, 2005, pp. 139–175.

97. E. Rabinowitch, Govindjee, The role of chlorophyll in photosynthesis, Sci. Am. 213 (1965) 74–83.

98. G. Döring, H. Stiehl, H. T. Witt, A second chlorophyll reaction in the electron chain of photosynthesis, Z. Naturforsch. B 22 (1967) 639–641.

99. M. R. Wasielewski, D. G. Johnson, M. Seibert, Govindjee, Determination of the primary charge separation rate in isolated photosystem II reaction centers with 500-fs time resolution, Proc. Natl. Acad. Sci. U.S.A. 86 (1989) 524–528.

100. S. R. Greenfield, M. Seibert, Govindjee, M. R. Wasielewski, Direct measurement of the effective rate constant for primary charge separation in isolated photosystem II reaction centers, J. Phys. Chem. B 101 (1997) 2251–2255.

101. V. V. Klimov, Discovery of pheophytin function in the photosynthetic energy conversion as the primary electron acceptor of photosystem II, Photosynth. Res. 76 (2003) 247–253.

102. J. R. Durrant, D. R. Klug, S. L. S. Kwa, R. van Grndelle, G. Porter, J. P. Decker, A multimer model for P680, the primary electron-donor of photosystem II, Proc. Natl. Acad. Sci. U.S.A. 92 (1995) 4798–4802.

103. B. Ke, *Photosynthesis: Photochemistry and Photobiology*, Kluwer Academic Publishers, Dordrecht, 2001.

104. E. Romero, B. A. Diner, P. J. Nixon, W. J. Coleman, J. P. Dekker, R. van Grondelle, Mixed exciton charge-transfer states in photosystem II: Stark spectroscopy on site-directed mutants, Biophys. J. 103 (2012) 185–194.

105. B. Kok, Absorption changes induced by the photochemical reaction of photosynthesis, Nature 179 (1957) 583–584.

106. P. Jordan, P. Fromme, H. T. Witt, O. Klukas, W. Saenger, N. Krauss, Three-dimensional structure of cyanobacterial photosystem I at 2.5 Å resolution., Nature 411 (2001) 909–917.

107. A. Ben-Shem, F. Frolow, N. Nelson, Crystal structure of plant photosystem I, Nature 426 (2003) 630–635.

108. P. Fromme, P. Mathis, Unraveling the photosystem I reaction center: a history, or the sum of many efforts, Photosynth. Res. 80 (2004) 109–124.

109. J. H. Golbeck (Ed.), *Photosystem I. The Light-Driven Plastocyanin:Ferredoxin Oxidoreductase*, Springer, Dordrecht, 2006.

110. A. W. Rutherford, P. Heathcote, Primary photochemistry in photosystem I, Photosynth. Res. 6 (1985) 295–316.

111. K. Redding, A. van der Est, The directionality of electron transport in photosystem I. In: J. H. Golbeck (Ed.), *Photosystem I: The Light-Driven Plastocyanin:Ferredoxin Oxidoreductase*, Springer, Dordrecht, 2006, pp. 413–437.

112. A. R. Holzwarth, M. G. Muller, J. Niklas, W. Lubitz, Ultrafast transient absorption studies on photosystem I reaction centers from *Chlamydomonas reinhardtii*. 2: Mutations near the P700 reaction center chlorophylls provide new insight into the nature of the primary electron donor, Biophys. J. 90 (2006) 552–565.

113. W. Giera, V. M. Ramesh, A. N. Webber, I. van Stokkum, R. van Grondelle, K. Gibasiewicz, Effect of the P700 pre-oxidation and point mutations near A_0 on the reversibility of the primary charge separation in photosystem I from *Chlamydomonas reinhardtii*, Biochim. Biophys. Acta 1797 (2010) 106–112.

114. T. Wydrzynski, K. Satoh (Eds.), *Photosystem II. The Light-Driven Water:Plastoquinone Oxidoreductase*, Springer, Dordrecht, 2005.

115. Govindjee, J. Kern, J. Messinger, J. Whitmarsh, Photosystem II. In: *Encyclopedia of Life Sciences* (ELS), Wiley, Chichester, U.K., 2010. doi: 10.1002/9780470015902.a0000669.pub2, 15 pages.

116. J. Barber, Photosystem II: an enzyme of global significance, Biochem. Soc. Trans. 34 (2006) 619–631.

117. M. Broser, A. Gabdulkhakov, J. Kern, A. Guskov, F. Müh, W. Saenger, A. Zouni, Crystal structure of monomeric photosystem II from *Thermosynechococcus elongatus* at 3.6-Å resolution, J. Biol. Chem. 285 (2010) 26255–26262.

118. J. Yano, J. Kern, K. Sauer, M. J. Latimer, Y. Pushkar, J. Biesiadka, B. Loll, W. Saenger, J. Messinger, A. Zouni, V. K. Yachandra, Where water is oxidized to dioxygen: structure of the photosynthetic Mn_4Ca cluster, Science 314 (2006) 821–825.

119. T. M. Bricker, J. L. Roose, R. D. Fagerlund, L. K. Frankel, J. J. Eaton-Rye, The extrinsic proteins of photosystem II, Biochim. Biophys. Acta 1817 (2012) 121–142.

120. R. D. Fagerlund, J. J. Eaton-Rye, The lipoproteins of cyanobacterial photosystem II, J. Photochem. Photobiol. B Biol. 104 (2011) 191–203.

121. B. A. Diner, F. Rappaport, Structure, dynamics, and energetics of the primary photochemistry of photosystem II of oxygenic photosynthesis, Annu. Rev. Plant Biol. 53 (2002) 551–580.

122. H. Ishikita, B. Loll, J. Biesiadka, W. Saenger, E.-W. Knapp, Redox potentials of chlorophylls in the photosystem II reaction center, Biochemistry 44 (2005) 4118–4124.

123. A. Guskov, J. Kern, A. Gabdulkhakov, M. Broser, A. Zouni, W. Saenger, Cyanobacterial photosystem II at 2.9-Angstrom resolution and the role of quinones, lipids, channels and chloride, Nat. Struct. Mol. Biol. 16 (2009) 334–342.

124. D. Shevela, J. J. Eaton-Rye, J.-R. Shen, Govindjee, Photosystem II and the unique role of bicarbonate: a historical perspective, Biochim. Biophys. Acta 1817 (2012) 1134–1151.

125. N. Nelson, C. F. Yocum, Structure and function of photosystems I and II, Annu. Rev. Plant Biol. 57 (2006) 521–565.

126. P. Fromme, P. Jordan, N. Krauß, Structure of photosystem I, Biochim. Biophys. Acta 1507 (2001) 5–31.

127. P. Fromme, I. Grotjohann, Structural analysis of cyanobacterial photosystem I. In: J. H. Golbeck (Ed.), *Photosystem I. The Light-Driven Plastocyanin:Ferredoxin Oxidoreductase*, Springer, Dordrecht, 2006, pp. 47–69.

128. M. Guergova-Kuras, B. Boudreaux, A. Joliot, P. Joliot, K. Redding, Evidence for two active branches for electron transfer in photosystem I, Proc. Natl. Acad. Sci. U.S.A. 98 (2001) 4437–4442.

129. A. W. Rutherford, A. Osyczka, F. Rappaport, Back-reactions, short-circuits, leaks and other energy wasteful reactions in biological electron transfer: redox tuning to survive life in O_2, FEBS Lett. 586 (2012) 603–616.

130. R. O. Cohen, G. Shen, J. H. Golbeck, W. Xu, P. R. Chitnis, A. I. Valieva, A. van der Est, Y. Pushkar, D. Stehlik, Evidence for asymmetric electron transfer in cyanobacterial photosystem I: analysis of a methionine-to-leucine mutation of the ligand to the primary electron acceptor A_0, Biochemistry 43 (2004) 4741–4754.

131. N. Dashdorj, W. Xu, R. O. Cohen, J.H. Golbeck, S. Savikhin, Asymmetric electron transfer in cyanobacterial photosystem I: charge separation and secondary electron transfer dynamics of mutations near the primary electron acceptor A_0, Biophys. J. 88 (2005) 1238–1249.

132. W. V. Fairclough, A. Forsyth, M. C. W. Evans, S. E. J. Rigby, S. Purton, P. Heathcote, Bidirectional electron transfer in photosystem I: electron transfer on the PsaA side is not essential for phototrophic growth in *Chlamydomonas*, Biochim. Biophys. Acta 1606 (2003) 43–55.

133. T. Berthold, E. D. von Gromoff, S. Santabarbara, P. Stehle, G. Link, O. G. Poluektov, P. Heathcote, C. F. Beck, M. C. Thurnauer, G. Kothe, Exploring the electron transfer pathways in photosystem I by high-time-resolution electron paramagnetic resonance: observation of the B-side radical pair $P_{700}^{+} A_{1B}^{-}$ in whole cells of the deuterated green alga *Chlamydomonas reinhardtii* at cryogenic temperatures, J. Am. Chem. Soc. 134 (2012) 5563–5576.

134. W. A. Cramer, S. S. Hasan, E. Yamashita, The Q cycle of cytochrome *bc* complexes: a structure perspective, Biochim. Biophys. Acta 1807 (2011) 788–802.

135. D. Baniulis, E. Yamashita, H. Zhang, S. S. Hasan, W. A. Cramer, Structure–function of the cytochrome b_6f complex, Photochem. Photobiol. 84 (2008) 1349 1358.

136. E.A. Berry, M. Guergova-Kuras, L. S. Huang, A. R. Crofts, Structure and function of cytochrome *bc* complexes, Annu. Rev. Biochem. 69 (2000) 1005-1075.

137. J. F. Allen, Cytochrome b_6f: structure for signalling and vectorial metabolism, Trends Plant. Sci. 9 (2004) 130–137.

138. W. A. Cramer, H. Zhang, J. Yan, G. Kurisu, J. L. Smith, Transmembrane traffic in the cytochrome b_6f complex, Annu. Rev. Biochem. 75 (2006) 769–790.

139. D. S. Bendall, R. S. Manasse, Cyclic photophosphorylation and electron transport, Biochim. Biophys. Acta 1229 (1995) 23–38.

140. P. Joliot, A. Joliot, G. Johnson, Cyclic electron transfer around photosystem I. In: J. H. Golbeck (Ed.), *Photosystem I: The Light-Driven Plastocyanine:Ferredoxin Oxidoreductase*, Springer, Dordrech, 2006, pp. 639–656.

141. D. Shevela, J. Messinger, Probing the turnover efficiency of photosystem II membrane fragments with different electron acceptors, Biochim. Biophys. Acta 1817 (2012) 1208–1212.

142. R. Fromme, R. Hagemann, G. Renger, Comparative studies of electron transport and atrazine binding in thylakoids and PS II particles from spinach. In: J. Biggens (Ed.), *Progress in Photosynthesis Research*, Martinus Nijhoff Publishers, Dordrecht, 1987, pp. 783–786.

143. A. Frenkel, Light induced phosphorylation by cell-free preparations of photosynthetic bacteria, J. Am. Chem. Soc. 76 (1954) 5568–5569.

144. D. I. Arnon, F. R. Whatley, M. B. Allen, Photosynthesis by isolated chloroplasts. II. Photosynthetic phosphorylation, the conversion of light into phosphate bond energy, J. Am. Chem. Soc. 76 (1954) 6324–6329.

145. B. L. Strehler, Firefly luminescence in the study of energy transfer mechanisms. II. Adenosine triphosphate and photosynthesis, Arch. Biochem. Biophys. 43 (1953) 67–79.

146. P. Mitchell, Chemiosmotic coupling in oxidative and photosynthetic phosphorylation, Biol. Rev. 41 (1966) 445–501.

147. P. D. Boyer, B. Chance, L. Ernster, P. Mitchell, E. Racker, E. C. Slater, Oxidative phosphorylation and photophosphorylation, Annu. Rev. Biochem. 46 (1977) 955–966.

148. W. Junge, H. Lill, S. Engelbrecht, ATP synthase: an electrochemical ransducer with rotatory mechanics, Trends Biochem. Sci. 22 (1997) 420–423.

149. W. Junge, Protons, proteins and ATP, Photosynth. Res. 80 (2004) 197–221.

150. W. Junge, H. Sielaff, S. Engelbrecht, Torque generation and elastic power transmission in the rotary F_OF_1-ATPase., Nature 459 (2009) 364–370.

151. A. R. Crofts, The Q-cycle—a personal perspective, Photosynth. Res. 80 (2004) 223–243.

152. A. T. Jagendorf, E. Uribe, ATP formation caused by acid–base transition of spinach chloroplasts, Proc. Natl. Acad. Sci. U.S.A. 55 (1966) 170–177.

153. H. T. Witt, E. Schlodder, P. Gräber, Membrane-bound ATP synthesis generated by an external elecrical field, FEBS Lett. 69 (1976) 272–276.

154. R. E. McCarty, Y. Evron, E. A. Johnson, The chloroplast ATP synthase: a rotary enzyme?, Annu. Rev. Plant Physiol. Plant Mol. Biol. 51 (2000) 83–109.

155. D. Pogoryelov, C. Reichen, A. L. Klyszejko, R. Brunisholz, D. J. Muller, P. Dimroth, T. Meier, The oligomeric state of c rings from cyanobacterial F-ATP synthases varies from 13 to 15, J. Bacteriol. 189 (2007) 5895–5902.

156. H. S. van Walraven, R. H. A. Bakels, Function, structure and regulation of cyanobacterial and chloroplast ATP synthase, Physiol. Plant. 96 (1996) 526–532.

157. D. Spetzler, R. Ishmukhametov, T. Hornung, J. Martin, J. York, L. Jin-Day, W. Frasch, Energy transduction by the two molecular motors of the F_1F_o ATP synthase. In: J. J. Eaton-Rye, B. C. Tripathy, T. D. Sharkey (Eds.), *Photosynthesis: Plastid Biology, Energy Conversion and Carbon Assimilation*, Springer, Dordrecht, 2012, pp. 561–590.

158. B. E. Krenn, P. Aardewijn, H. S. VanWalraven, S. WernerGrune, H. Strotmann, R. Kraayenhof, ATP synthase from a cyanobacterial *Synechocystis* 6803 mutant containing the regulatory segment of the chloroplast gamma subunit shows thiol modulation, Biochem. Soc. Trans. 23 (1995) 757–760.

159. H. Konno, T. Murakami-Fuse, F. Fujii, F. Koyama, H. Ueoka-Nakanishi, C.-G. Pack, M. Kinjo, T. Hisabori, The regulator of the F1 motor: inhibition of rotation of cyanobacterial F1-ATPase by the e subunit, EMBO J. 25 (2006) 4596–4604.

160. I. A. Berg, D. Kockelkorn, W. Buckel, G. Fuchs, A 3-hydroxypropionate/4-hydroxybutyrate autotrophic carbon dioxide assimilation pathway in Archaea, Science 318 (2007) 1782–1786.

161. H. Huber, M. Gallenberger, U. Jahn, E. Eylert, I. A. Berg, D. Kockelkorn, W. Eisenreich, G. Fuchs, A dicarboxylate/4-hydroxybutyrate autotrophic carbon assimilation cycle in the hyperthermophilic Archaeum *Ignicoccus hospitalis*, Proc. Natl. Acad. Sci. U.S.A. 105 (2008) 7851–7856.

162. T. Sato, H. Atomi, Novel metabolic pathways in Archaea, Curr. Opin. Microbiol. 14 (2011) 307–314.

163. T. Sato, H. Atomi, Microbial inorganic carbon fixation. In: *Encyclopedia of Life Sciences* (ELS), Wiley, Chichester, U.K., 2010, pp. 1–12.

164. G. Strauss, G. Fuchs, Enzymes of a novel autotrophic CO_2 fixation pathway in the phototrophic bacterium *Chloroflexus aurantiacus*, the 3-hydroxypropionate cycle, Eur. J. Biochem. 215 (1993) 633–643.

165. S. Estelmann, W. H. Ramos-Vera, N. Gad'on, H. Huber, I. A. Berg, G. Fuchs, Carbon dioxide fixation in *Archaeoglobus lithotrophicus*: are there multiple autotrophic pathways?, FEMS Microbiol. Lett. 319 (2011) 65–72.

166. A. Bar-Even, A. Flamholz, E. Noor, R. Milo, Thermodynamic constraints shape the structure of carbon fixation pathways, Biochim. Biophys. Acta 1817 (2012) 1646–1659.

167. A. A. Benson, Following the path of carbon in photosynthesis: a personal story. In: Govindjee, J. T. Beatty, H. Gest, J. F. Allen (Eds.), *Discoveries in Photosynthesis. Advances in Photosynthesis and Respiration*, Springer, Dordrecht, 2005, pp. 793–813.

168. J. A. Bassham, Mapping the carbon reduction cycle: a personal perspective. In: Govindjee, J. T. Beatty, H. Gest, J. F. Allen (Eds.), *Discoveries in Photosynthesis. Advances in Photosynthesis and Respiration*, Springer, Dordrecht, 2005, pp. 815–832.

169. A. S. Raghavendra, R. Sage (Eds.), *C4 Photosynthesis and Concentrating Mechanisms*, Springer, Dordrecht, 2011.

170. I. C. Prentice, S. P. Harrison, P. J. Bartlein, Global vegetation and terrestrial carbon cycle changes after the last ice age, New Phytologist 189 (2011) 988–998.

171. Y. Huang, F. A. Street-Perrott, S. E. Metcalfe, M. Brenner, M. Moreland, K. H. Freeman, Climate change as the dominant control on glacial–interglacial variations in C3 and C4 plant abundance, Science 293 (2001) 1647–1651.

172. Z. Yao, H. Wu, M. Liang, X. Shi, Spatial and temporal variations in C3 and C4 plant abundance over the Chinese Loess Plateau since the last glacial maximum, J. Arid Environ. 75 (2011) 881–889.

173. V. Galy, L. François, C. France-Lanord, P. Faure, H. Kudrass, F. Palhol, S. K. Singh, C4 plants decline in the Himalayan basin since the Last Glacial Maximum, Quaternary Sci. Rev. 27 (2008) 1396–1409.

174. V. Galy, T. Eglinton, Protracted storage of biospheric carbon in the Ganges-Brahmaputra basin, Nature Geosci. 4 (2011) 843–847.

175. A. Bar-Even, E. Noor, N. E. Lewis, R. Milo, Design and analysis of synthetic carbon fixation pathways, Proc. Natl. Acad. Sci. U.S.A. 107 (2010) 889–8894.

176. P. R. Craddock, N. Dauphas, Iron isotopic compositions of geological reference materials and chondrites, Geostandards and Geoanalytical Research 35 (2011) 101–123.

177. A.-L. Ducluzeau, R. van Lis, S. Duval, B. Schoepp-Cothenet, M. J. Russell, W. Nitschke, Was nitric oxide the first deep electron sink?, Trends Biochem. Sci. 34 (2009) 9–15.

178. S. Sadekar, J. Raymond, R. E. Blankenship, Conservation of distantly related membrane proteins: photosynthetic reaction centers share a common structural core, Mol. Biol. Evol. 23 (2006) 2001–2007.

179. J. M. Olson, "Evolution of Photosynthesis" (1970), re-examined thirty years later, Photosynth. Res. 68 (2001) 95–112.

180. N. Nelson, A. Ben-Shem, The structure of photosystem I and evolution of photosynthesis, BioEssays 27 (2005) 914–922.

181. J. Xiong, C. E. Bauer, A cytochrome *b* origin of photosynthetic reaction centers: an evolutionary link between respiration and photosynthesis, J. Mol. Biol. 322 (2002) 1025–1037.

182. G. Giacometti, G. Giacometti, Evolution of photosynthesis and respiration: which came first?, Appl. Magn. Reson. 37 (2010) 13–25.

183. R. E. Blankenship, Early evolution of photosynthesis, Plant Physiol. 154 (2010) 434–438.

184. R. E. Blankenship, Origin and early evolution of photosynthesis, Photosynth. Res. 33 (1992) 91–111.

185. B. K. Pierson, J. M. Olson, Evolution of photosynthesis in anoxygenic photosynthetic prokaryotes. In: Y. Cochen, E. Rosenberg (Eds.), *Microbial Mats*, American Society for Microbiology, Washington, 1987, pp. 402–427.

186. W. Nitschke, R. van Lis, B. Schoepp-Cothenet, F. Baymann, The "green" phylogenetic clade of Rieske/cyt*b* complexes, Photosynth. Res. 104 (2010) 347–355.

187. J. M. Olson, Evolution of photosynthetic reaction centers, Biosystems 14 (1981) 89–94.

188. J. F. Allen, W. Martin, Evolutionary biology—out of thin air, Nature 445 (2007) 610–612.

189. F. Battistuzzi, A. Feijao, S. B. Hedges, A genomic timescale of prokaryote evolution: insights into the origin of methanogenesis, phototrophy, and the colonization of land, BMC Evol. Biol. 4 (2004) 44.

190. F. U. Battistuzzi, S. B. Hedges, A major clade of Prokaryotes with ancient adaptations to life on land, Mol. Biol. Evol. 26 (2009) 335–343.

191. E. G. Nisbet, N. V. Grassineau, C. J. Howe, P. I. Abell, M. Regelous, R. E. R. Nisbet, The age of Rubisco: the evolution of oxygenic photosynthesis, Geobiology 5 (2007) 311–335.

192. A. D. Czaja, C. M. Johnson, B. L. Beard, J. L. Eigenbrode, K. H. Freeman, K. E. Yamaguchi, Iron and carbon isotope evidence for ecosystem and environmental diversity in the ~2.7 to 2.5 Ga Hamersley Province, Western Australia, Earth Planet. Sci. Lett. 292 (2010) 170–180.

193. A. D. Czaja, C. M. Johnson, E. E. Roden, B. L. Beard, A. R. Voegelin, T. F. Nägler, N. J. Beukes, M. Wille, Evidence for free oxygen in the Neoarchean ocean based on coupled iron–molybdenum isotope fractionation, Geochim. Cosmochim. Acta 86 (2012) 118–137.

194. D. Schwartzman, K. Caldeira, A. Pavlov, Cyanobacterial emergence at 2.8 Gya and greenhouse feedbacks, Astrobiology 8 (2008) 187–203.

195. A. C. Allwood, M. R. Walter, B. S. Kamber, C. P. Marshall, I. W. Burch, Stromatolite reef from the Early Archaean era of Australia, Nature 441 (2006) 714–718.

196. T. Bosak, B. Liang, M. S. Sim, A. P. Petroff, Morphological record of oxygenic photosynthesis in conical stromatolites, Proc. Natl. Acad. Sci. U.S.A. 106 (2009) 10939–10943.

197. L. O. Björn, Govindjee, The evolution of photosynthesis and chloroplasts, Curr. Sci. 96 (2009) 1466–1474.

198. L. I. Falcón, S. Magallón, A. Castillo, Dating the cyanobacterial ancestor of the chloroplast, ISME J. 4 (2010) 777–783.

199. A. Amunts, N. Nelson, Plant photosystem I design in the light of evolution, Structure 17 (2009) 637–650.

200. A. Busch, M. Hippler, The structure and function of eukaryotic photosystem I, Biochim. Biophys. Acta 1807 (2011) 864–877.

201. C. Aspinwall, M. Sarcina, C. W. Mullineaux, Phycobilisome mobility in the cyanobacterium *Synechococcus* sp. PCC7942 is influenced by the trimerisation of photosystem I, Photosynth. Res. 79 (2004) 179–187.

202. S. Lemeille, J.-D. Rochaix, State transitions at the crossroad of thylakoid signalling pathways, Photosynth. Res. 106 (2010) 33–46.

203. M. Germano, A. E. Yakushevska, W. Keegstra, H. J. van Gorkom, J. P. Dekker, E. J. Boekema, Supramolecular organization of photosystem I and light-harvesting complex I in *Chlamydomonas reinhardtii*, FEBS Lett. 525 (2002) 121–125.

204. J. Kargul, J. Nield, J. Barber, Three-dimensional reconstruction of a light-harvesting complex I–photosystem I (LHCI-PSI) supercomplex from the green alga *Chlamydomonas reinhardtii*, J. Biol. Chem. 278 (2003) 16135–16141.

205. N. E. Blanco, R. D. Ceccoli, M. E. Segretin, H. O. Poli, I. Voss, M. Melzer, F. F. Bravo-Almonacid, R. Scheibe, M.-R. Hajirezaei, N. Carrillo, Cyanobacterial flavodoxin complements ferredoxin deficiency in knocked-down transgenic tobacco plants, Plant J. 65 (2011) 922–935.

206. M. D. Zurbriggen, V. B. Tognetti, M. F. Fillat, M.-R. Hajirezaei, E. M. Valle, N. Carrillo, Combating stress with flavodoxin: a promising route for crop improvement, Trends Biotechnol. 26 (2008) 531–537.

207. M. D. Zurbriggen, V. B. Tognetti, N. Carrillo, Stress-inducible flavodoxin from photosynthetic microorganisms. The mystery of flavodoxin loss from the plant genome, IUBMB Life 59 (2007) 355–360.

208. P. R. Pokkuluri, Y. Y. Londer, S. J. Wood, N. E. C. Duke, L. Morgado, C. A. Salgueiro, M. Schiffer, Outer membrane cytochrome c, OmcF, from *Geobacter sulfurreducens*: high structural similarity to an algal cytochrome c_6, Proteins. Struct. Funct. Genet. 74 (2009) 266–270.

209. M. A. De la Rosa, J. A. Navarro, M. Hervás, The convergent evolution of cytochrome c_6 and plastocyanin has been driven by geochemical changes. In: G. A. Peschek, C. Obinger, G. Renger (Eds.), *Bioenergetic Processes of Cyanobacteria: From Evolutionary Singularity to Ecological Diversity*, Springer, Dordrecht, 2011, pp. 607–630.

210. J. Allen, J. Williams, The evolutionary pathway from anoxygenic to oxygenic photosynthesis examined by comparison of the properties of photosystem II and bacterial reaction centers, Photosynth. Res. 107 (2011) 59–69.

211. A. Williamson, B. Conlan, W. Hillier, T. Wydrzynski, The evolution of photosystem II: insights into the past and future, Photosynth. Res. 107 (2011) 71–86.

212. R. Rippka, J. Waterbury, G. Cohen-Bazire, A cyanobacterium which lacks thylakoids, Arch. Microbiol. 100 (1974) 419–436.

213. G. Bernát, U. Schreiber, E. Sendtko, I. N. Stadnichuk, S. Rexroth, M. Rögner, F. Koenig, Unique properties vs. common themes: the atypical cyanobacterium *Gloeobacter violaceus* PCC 7421 is capable of state transitions and blue-light-induced fluorescence quenching, Plant Cell Physiol. 53 (2012) 528–542.

214. S. Yang, R. Zhang, C. Hu, J. Xie, J. Zhao, The dynamic behavior of phycobilisome movement during light state transitions in cyanobacterium *Synechocystis* PCC6803, Photosynth. Res. 99 (2009) 99–106.

215. M. Ballottari, J. Girardon, L. Dall'Osto, R. Bassi, Evolution and functional properties of photosystem II light harvesting complexes in eukaryotes, Biochim. Biophys. Acta 1817 (2012) 143–157.

216. J. Raymond, R. E. Blankenship, The origin of the oxygen-evolving complex, Coord. Chem. Rev. 252 (2008) 377–383.

217. G. C. Dismukes, V. V. Klimov, S. V. Baranov, Y. N. Kozlov, J. DasGupta, A. Tyryshkin, The origin of atmospheric oxygen on Earth: the innovation of oxygenic photosynthesis, Proc. Natl. Acad. Sci. U.S.A. 98 (2001) 2170–2175.

218. G. G. B. Tcherkez, G. D. Farquhar, T. J. Andrews, Despite slow catalysis and confused substrate specificity, all ribulose bisphosphate carboxylases may be nearly perfectly optimized, Proc. Natl. Acad. Sci. U.S.A. 103 (2006) 7246–7251.

219. Y. Savir, E. Noor, R. Milo, T. Tlusty, Cross-species analysis traces adaptation of Rubisco toward optimality in a low-dimensional landscape, Proc. Natl. Acad. Sci. U.S.A. 107 (2010) 3475–3480.

220. M. J. André, Modelling $^{18}O_2$ and $^{16}O_2$ unidirectional fluxes in plants: II. Analysis of Rubisco evolution, Biosystems 103 (2011) 252–264.

221. M. R. Badger, G. D. Price, CO_2 concentrating mechanisms in cyanobacteria: molecular components, their diversity and evolution, J. Exp. Bot. 54 (2003) 609–622.

222. M. R. Badger, D. Hanson, G. D. Price, Evolution and diversity of CO_2 concentrating mechanisms in cyanobacteria, Funct. Plant Biol. 29 (2002) 161–173.

223. J. A. Raven, M. Giordano, J. Beardall, S. C. Maberly, Algal evolution in relation to atmospheric CO_2: carboxylases, carbon-concentrating mechanisms and carbon oxidation cycles, Philos. Trans. R. Soc. Lond., B 367 (2012) 493–507.

224. A. M. Hessler, D. R. Lowe, R. L. Jones, D. K. Bird, A lower limit for atmospheric carbon dioxide levels 3.2 billion years ago, Nature 428 (2004) 736–738.

225. H. I. M. Lichtenegger, H. Lammer, J. M. Grießmeier, Y. N. Kulikov, P. von Paris, W. Hausleitner, S. Krauss, H. Rauer, Aeronomical evidence for higher CO_2 levels during Earth's Hadean epoch, Icarus 210 (2010) 1–7.

226. E. Nisbet, C. Fowler, The evolution of the atmosphere in the Archaean and early Proterozoic, Chin. Sci. Bull. 56 (2011) 4–13.

227. R. A. Berner, GEOCARBSULF: a combined model for Phanerozoic atmospheric O_2 and CO_2, Geochim. Cosmochim. Acta 70 (2006) 5653–5664.

228. J. M. Anderson, Insights into the consequences of grana stacking of thylakoid membranes in vascular plants: a personal perspective, Aust. J. Plant Physiol. 26 (1999) 625–639.

229. M. Yokono, A. Murakami, S. Akimoto, Excitation energy transfer between photosystem II and photosystem I in red algae: larger amounts of phycobilisome enhance spillover, Biochim. Biophys. Acta 1807 (2011) 847–853.

230. R. M. McCourt, C. F. Delwiche, K. G. Karol, Charophyte algae and land plant origins, Trends Ecol. Evol. 19 (2004) 661–666.

231. L. O. Björn, Interception of light by plants leaves. In: N. R. Baker, H. Thomas (Eds.), *Crop Photosynthesis: Spatial and Temporal Determinants*, Elsevier Science Publishers, New York, 1992.

232. T. C. Vogelman, J. N. Nishio, W. K. Smith, Leaves and light capture: light propagation and gradients of carbon fixation within leaves, Trends Plant. Sci. 1 (1996) 65–70.

233. I. Terashima, T. Fujita, T. Inoue, W. S. Chow, R. Oguchi, Green light drives leaf photosynthesis more efficiently than red light in strong white light: revisiting the enigmatic question of why leaves are green, Plant Cell Physiol. 50 (2009) 684–697.

234. M. Cui, T. C. Vogelman, W. K. Smith, Chlorophyll and light gradients in sun and shade leaves of *Spinacia oleracea*, Plant Cell Environ. 14 (1991) 493–500.

235. J. Sun, J. N. Nishio, T. C. Vogelmann, High-light effects on CO_2 fixation gradients across leaves, Plant Cell Environ. 19 (1996) 1261–1271.

236. T. Han, T. Vogelmann, J. Nishio, Profiles of photosynthetic oxygen-evolution within leaves of *Spinacia oleracea*, New Phytologist 143 (1999) 83–92.

237. T. M. Lenton, M. Crouch, M. Johnson, N. Pires, L. Dolan, First plants cooled the Ordovician, Nature Geosci. 5 (2012) 86–89.

238. K. Mohanakumar, *Stratosphere Troposphere Interactions: An Introduction*, Springer, Netherlands, 2008.

239. J. L. Lee, K. N. Liou, S. C. Ou, A three-dimensional large-scale cloud model: testing the role of radiative heating and ice phase processes, Tellus Dyn. Meteorol. Oceanogr. 44 (1992) 197–216.

240. D. D. Lucas, R. G. Prinn, Tropospheric distributions of sulfuric acid–water vapor aerosol nucleation rates from dimethylsulfide oxidation, Geophys. Res. Lett. 30 (2003) 2136.

241. R. Popa, P. K. Weber, J. Pett-Ridge, J. A. Finzi, S. J. Fallon, I. D. Hutcheon, K. H. Nealson, D. G. Capone, Carbon and nitrogen fixation and metabolite exchange in and between individual cells of *Anabaena oscillarioides*, ISME J. 1 (2007) 354–360.

242. B. Bergman, J. R. Gallon, A. N. Rai, L. J. Stal, N2 Fixation by non-heterocystous cyanobacteria, FEMS Microbiol. Rev. 19 (1997) 139–185.

243. L. Klepper, D. Flesher, R. H. Hageman, Generation of reduced nicotinamide adenine dinucleotide for nitrate reduction in green leaves, Plant Physiol. 48 (1971) 580–590.

244. R. K. Trench, The cell biology of plant–animal symbiosis, Annu. Rev. Plant Physiol. 30 (1979) 485–531.

245. B. J. Green, W.-Y. Li, J. R. Manhart, T. C. Fox, E. J. Summer, R. A. Kennedy, S. K. Pierce, M. E. Rumpho, Mollusc–algal chloroplast endosymbiosis. Photosynthesis, thylakoid protein maintenance, and chloroplast gene expression continue for many months in the absence of the algal nucleus, Plant Physiol. 124 (2000) 331–342.

246. M. E. Rumpho, K. N. Pelletreau, A. Moustafa, D. Bhattacharya, The making of a photosynthetic animal, J. Exp. Biol. 214 (2011) 303–311.

247. J. Messinger, D. Shevela, Principles of photosynthesis. In: D. Ginley, D. Cachen (Eds.), *Fundamentals of Materials and Energy and Environmental Sustainability*, Cambridge University Press, Cambridge, 2012, pp. 302–314.

248. D. Shevela, *Photosynthetic Water Oxidation. Role of Inorganic Cofactors and Species Differences*, VDM Verlag Dr. Müller, Saarbrücken, 2008.

249. D. Shevela, R. Y. Pishchalnikov, L. A. Eichacker, Govindjee, Oxygenic photosynthesis in cyanobacteria. In: A. K. Srivastava, A. N. Rai, B. A. Neilan (Eds.), *Stress Biology of Cyanobacteria: Molecular Mechanisms to Cellular Responses*, CRC Press/Taylor & Francis Group, Boca Raton, FL, pp. 3–40.

250. J. C. McElwain, W. G. Chaloner, Stomatal density and index of fossil plants track atmospheric carbon dioxide in the Palaeozoic, Ann. Bot. 76 (1995) 389–395.

251. J. C. McElwain, Do fossil plants signal palaeoatmospheric carbon dioxide concentration in the geological past?, Philos. Trans. R. Soc. Lond. B 353 (1998) 83–96.

252. J. Van Der Burgh, H. Visscher, D. L. Dilcher, W. M. Kurschner, Paleoatmospheric signatures in Neogene fossil leaves, Science 260 (1993) 1788–1790.

Apparatus and Mechanism of Photosynthetic Water Splitting as Nature's Blueprint for Efficient Solar Energy Exploitation

GERNOT RENGER

3.1 INTRODUCTION

All biological organisms represent open systems that are far from the thermodynamic equilibrium by an estimated average value of 20–25 kJ/mol [1]. As a consequence, fluxes of Gibbs free energy[1] are required as indispensable prerequisite for the development and sustenance of living matter [1, 2]. This energetic sine qua non condition is satisfied by the electromagnetic radiation in the near UV, visible, and near infrared region impinging the earth from the extraterrestrial nuclear fusion reactor, the sun, which acts as a unique Gibbs free energy source of the biosphere. It should be mentioned that some "exotic" organisms living in a special environment (e.g., in the "black smokers") use another form(s) of Gibbs free energy as driving force.

The annual energy input from the sun on our planet is estimated to be about 5×10^{21} kJ/mol [3]. Nearly half of this amount is either reflected by the atmosphere or absorbed by it (giving rise to storms and water movement in the oceans), the other half irradiates the earth surfaces, that is, oceans and land (at a ratio of 2–2.5). In the overall balance, the earth releases energy as blackbody radiation in the far infrared region (the maximum of the wavelength distribution is at about 10 μm), thus keeping the average temperature virtually constant. Only a very small fraction of less than 0.1% of these 2.5×10^{21} kJ/mol·yr is eventually transformed via photosynthesis

---

[1]The Gibbs free energy is defined as $G = H - TS$, where H and S are the enthalpy and entropy, respectively, and T is the absolute temperature (see Atkins and De Paula [14]).

Natural and Artificial Photosynthesis: Solar Power as an Energy Source, First Edition. Edited by Reza Razeghifard.
© 2013 John Wiley & Sons, Inc. Published 2013 by John Wiley & Sons, Inc.

into Gibbs free energy, thus providing the required driving force for the biosphere. This number clearly illustrates that more than enough energy supply is available for satisfying humankind's increasing demand provided that we are able to find the "key" for opening the door for large-scale technical exploitation of solar radiation. One way to achieve this goal is to learn from mother nature how to use solar radiation by analyzing the structural and functional principles that turned out to be most efficient during a long-lasting evolutionary process.

Among living matter two basically different types of organisms are to be distinguished with respect to the mode of exploitation of solar energy: (1) photoautotrophs, which can directly use solar radiation as the driving force for their living processes, and (2) photoheterotrophs, which are unable to perform this direct transformation and therefore need the uptake of energy-rich substances (food) that are eventually synthesized by photoautrotrophs and serve as the Gibbs free energy source through catabolic digestions.

The thermodynamics of these two types of direct and indirect solar energy exploitation can be summarized simply by a scheme that is shown in Fig. 3.1. In photoautotrophic organisms the light flux increases the Gibbs free energy content of a suitable storage system by providing the driving force for the transition from a lower level $G^\circ(A)$ of state A to a higher level $G^\circ(B)$ of state B. This process is referred to as photosynthesis and the specific amount of stored Gibbs free energy symbolized by $\Delta G^\circ_{store}(h\nu)$ depends on the properties of the components that are involved in the $A \xrightarrow{h\nu} B$ transition. The underlying thermodynamic principles of the exploitation of solar radiation through photosynthesis were outlined in 1845 by R. J. Mayer (first law of thermodynamics) and in 1886 by L. Boltzmann (second law of thermodynamics) [4].

In the heterotrophic organism an analogous type of transition takes place in the opposite direction, where food molecules with a higher Gibbs free energy content $G^\circ(B')$ are catabolized into a state A' with a lower level $G^\circ(A')$. The Gibbs free

BIOLOGICAL ORGANISMS

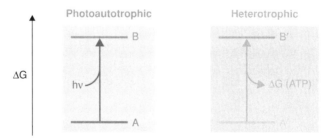

FIGURE 3.1 General scheme of the fundamental bioenergetic transformations of solar radiation as unique Gibbs energy: direct use in photoautotrophic organisms (left panel) and indirect exploitation via the uptake of food originating from photoautotrophic systems (right panel).

energy drop ΔG° is made available in the form of a universal "bioenergetic currency," the ATP molecule (for a review, see Renger and Ludwig [5]). This process is denoted as respiration (for the sake of simplicity no distinction will be made between anaerobic fermentation and aerobic respiration, see biochemistry textbooks for details).

3.2 OVERALL REACTION PATTERN OF PHOTOSYNTHESIS AND RESPIRATION

Based on the hypothesis of van Niel in 1941, the transition $A \to B$ in a photosynthetic organism can be described by the transfer of hydrogen from compounds with lower $G^\circ(A)$ to compounds with higher Gibbs free energy $G^\circ(B)$ [6]. The overall reaction of this transition is summarized by Eq. 3.1

$$H_2X + T \xrightarrow{h\mu} X + TH_2 \tag{3.1}$$

where H_2X is a suitable hydrogen donor and T a hydrogen acceptor substance (in most cases $NADP^+$) with the thermodynamic relation of $\Delta G^\circ(H_2X/X) < \Delta G^\circ(TH_2/T)$. This process is coupled to the formation of ATP via a mechanism referred to as chemiosmosis [7, 8]; that is, the directed (vectorial) proton flux through the enzyme ATP synthase provides the driving force for the endergonic reaction summarized by Eq. 3.2,

$$H^+ + ADP^{3-} + HPQ_4^{2-} \xrightarrow{\Delta G(\text{pmf})} ATP^{4-} + H_2O \tag{3.2}$$

where the protonation states of substrates and products refer to physiological pH and ΔG(pmf) symbolizes the proton motive force (pmf) of the transmembrane difference of the electrochemical potential of protons[2] (for a review on details of the coupling mechanism, see Junge [9]).

The products TH_2 (NADPH) and ATP give rise to assimilation of the primordial elements carbon, nitrogen, and sulfur. Of paramount importance among these processes is the CO_2 fixation into carbohydrates via the Calvin–Benson cycle ([10, 11] and references therein) which is by far the quantitatively largest process on the earth's surface with an estimated annual turnover rate of 200–300 billion tons of fixed CO_2 [3].

---

[2]The term "proton gradient" is often used in an incorrect manner. It should not be confused with the pH difference "ΔpH" across the coupling membrane because the gradient, by definition, is $\nabla[\vec{H^+}(\mathbf{r})]$, where ∇ is an operator (derivative to space coordinates) and $[H^+(\mathbf{r})]$ the scalar field of the local proton concentration at position $\vec{\mathbf{r}}$. Accordingly, the gradient is a space-dependent quantity and its actual value difficult to determine in biological membrane systems.

In heterotrophic organisms analogous hydrogen transfer reactions take place but in this case, in the opposite direction, that is, from state B' ($T'H_2$) to A' (H_2X') with $\Delta G°(T'H_2/T') > \Delta G°(H_2X'/X')$:

$$T'H_2 + X' \xrightarrow{\Delta G°(\text{pmf})} T' + H_2X' \tag{3.3}$$

where the Gibbs free energy available from this reaction is an intermediary "stored" as $\Delta G°(\text{pmf})$ and transformed into the "energy-rich" ATP according to Eq. 3.2.

3.3 BIOENERGETIC LIMIT OF SOLAR ENERGY EXPLOITATION: WATER SPLITTING

The efficiency of the fundamental bioenergetic processes (see Eqs. 3.1–3.3) depends on two parameters: (1) the specific storage capacity $\Delta G°_{\text{store}}(h\nu)$ and $\Delta G°(\text{pmf})$, respectively, and (2) the nature of the chemical compounds that enable a large-scale performance of transitions $A \xrightarrow{h\nu} B$ and $B' \xrightarrow{\Delta G°(\text{pmf})} A'$ through both availability in sufficient amounts and control of the reactivity in order to avoid dissipative side reactions.

From a Darwinian point of view it is not surprising that during a long-lasting evolutionary process the fundamental bioenergetics of life have been optimized to the highest possible level.

In the early stages of evolution the photoautotrophic organisms used substances like H_2, H_2S, and small organic molecules (formate, acetate, methanol) that are potent electron donors H_2X. However, this mode of solar energy exploitation, which still exists in present day "primitive" forms of life (for recent reviews, see Parson [12] and Vermeglio [13]), comprises limitations by both substrate abundance and specific storage capacity of Gibbs free energy, that is, comparatively small values of $\Delta G°_{\text{store}}(h\nu)$ and correspondingly $\Delta G°(\text{pmf})$.

Since living matter on earth is inevitably connected with bulk water phases (in fact the search for comparable forms of extraterrestrial life is always linked with the finding of water), the development toward more efficient systems was restricted per se by the thermodynamic limit of water splitting into the elements H_2 and O_2. Thus the maximum possible value of $\Delta G°_{\text{store,max}}(h\nu)$ is 237.17 kJ/mol [14]. In reality, the hydrogen is chemically bound to molecules T like $NADP^+$ and therefore $\Delta G°_{\text{store}}(h\nu)$ is lower by about 20 kJ/mol than $\Delta G°_{\text{store,max}}(h\nu)$.

This bioenergetic limit of photosynthetic water splitting into molecular dioxygen and bound hydrogen (NADPH) was reached about 2–3 billion years ago with the "invention" of a molecular machinery that enabled ancient prokaryotic cyanobacteria to perform this process ([15, 16, 17], for further reading, see Larkum [18], and Peschek et al. [19]). This evolutionary step was the Big Bang in the biosphere with two consequences of paramount importance: (1) the huge water pool of the earth's surface became available as a hydrogen source for living organisms and (2) the molecular dioxygen released as the "waste product" of photosynthetic water splitting led to the formation of an aerobic atmosphere [20, 21, 22], thus providing a very powerful oxidant that enabled an increase by more than a factor of 10 of the amount of Gibbs free energy that is extractable as $\Delta G°(\text{pmf})$ from the same food molecule (e.g., glucose) through aerobic respiration (for thermodynamic considerations, see Nicholls and Ferguson [23], and Renger [24]). Furthermore, the adventure of O_2 in the atmosphere led to the formation of the ozone layer in the stratosphere that established the protective "umbrella" against deleterious UV-B radiation [25], thus opening the road for global population of land by living organisms including human beings.

The ability of the biosphere in using the redox couple H_2O/O_2 for solar energy storage by "loading" the system through photosynthetic water splitting and "discharging" it via aerobic respiration was the energetic prerequisite for the development and sustenance of all higher forms of life on our planet with the wonderful facets of flora and fauna.

The global interplay between water splitting/formation and O_2 formation/ consumption in connection with the corresponding hydrogen metabolism/catabolism via CO_2 fixation/release leads to a stationary state of the content of both gases in the atmosphere. This situation, however, is prone to change due to humankind's increasing Gibbs free energy demand that is currently satisfied predominantly by the combustion of fossil fuels. As a consequence, the CO_2 of the atmosphere gradually increases with serious consequences on the global climate (the corresponding decrease of the oxygen content is negligible and without any effect).

3.4   HUMANKIND'S DREAM OF USING WATER AND SOLAR RADIATION AS "CLEAN FUEL"

The problem of exhausting fossil fuels and its possible solution by exploitation of solar radiation are old topics of scientific discussion as is illustrated by a general lecture presented by Ciamician 100 years ago (September 11, 1912) before the International Congress of Applied Chemistry in New York [26]. In this contribution he asked the question: "Is fossil solar energy the only one that may be used in modern life and civilization?" and denied it by concluding "If our black and nervous civilization, based on coal, shall be followed by a quieter civilization based on the utilization of solar energy, that will not be harmful to progress and to human happiness. The photochemistry of the future should not however be postponed to such distant times; I believe that industry will do well in using from this very day all the energies that nature puts at its disposal. So far, human civilization has made use almost exclusively of fossil solar energy. Would it not be advantageous to make better use of radiant energy?" Ciamician did not mention the environmental problems and the fossil fuels oil and gas of our modern civilization but he addressed the key point for the future.

Nature has taught us that water is the most suitable substance in exploiting solar radiation as the Gibbs free energy source of life. The use of water as the "coal of the

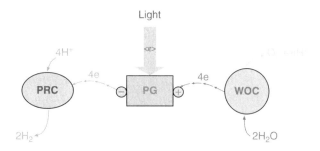

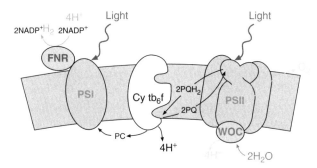

FIGURE 3.2   Scheme of light-induced water splitting into H_2 and O_2 (top panel) and into bound hydrogen and O_2 upon photosynthesis (bottom panel). Cyt b_6f = cytochrome b_6f complex, FNR = ferredoxin–NADP$^+$ oxidoreductase, PC = plastocyanin, PG = photogenerator, PQ = plastoquinone, PRC = proton reducing catalyst, PSI and PSII = photosystem I and photosystem II, respectively, WOC = water oxidizing catalyst (complex).

future" was already a dream of Jules Verne more than a century ago (1875) and the efforts in making this dream and the prediction of Ciamician a reality are topics of current worldwide research activities.

Regardless of any mechanistic details, the splitting of water by visible light into hydrogen (H_2) and dioxygen (O_2) requires the cooperation of three different operational units, as is schematically illustrated in the upper part of Fig. 3.2: (1) a photogenerator (PG) where electronically excited states (excitons symbolized by $<\varepsilon>$)[3] are trapped as a spatially separated electron–hole pair (exciton dissociation), (2) proton reducing complex (PRC) catalyzing the reduction of protons under formation of molecular hydrogen (H_2), and (3) water oxidizing complex (WOC) acting as the catalyst for oxidative water splitting into O_2 and four protons. Furthermore, an indispensable

---

[3] The term "exciton" is widely used for an excited electronic state, which resides on more than one pigment during its lifetime [27].

prerequisite for any practical application is the suitable separation of the spaces for formation of H$_2$ and O$_2$.

3.5 NATURE'S BLUEPRINT OF LIGHT-INDUCED WATER SPLITTING

For a discussion of possible options in achieving the goal of solar energy exploitation by water splitting it is worth considering at first the principles of the structural and functional organization scheme of this process in photosynthetic organisms.

Nature has perfectly solved the performance of light-driven water splitting at the microscopic level by developing during a long evolutionary process nanoscale devices that efficiently cooperate in generating O$_2$ and hydrogen chemically bound to NADP$^+$. The bottom part of Fig. 3.2 schematically summarizes the pattern of photosynthetic electron transport. Inspection of this scheme readily reveals three characteristics: (1) the overall process is energetically driven by two photogenerators designated as photosystem I (PSI) and photosystem II (PSII); (2) the operational units are anisotropically incorporated into a membrane (called the thylakoid membrane), thus performing reductive NADPH formation and oxidative water splitting in spatially separated compartments and simultaneously the concomitant coupling of these reactions with H$^+$ uptake and H$^+$ release, respectively, gives rise to generation of the pmf as the driving force for ATP synthesis (see Eq. 3.2); and (3) the catalyst for oxidative water splitting is an integral part of the PSII core (see the following description in this chapter) while the enzymatic formation of NADPH takes place at the water-soluble ferredoxin(Fd)–NADP$^+$:oxidoreductase often designated Fd-NADP$^+$ reductase (FNR) (see Hurley et al. [28] and Sétif and Leibl [29] and references therein).

3.6 TYPES OF APPROACHES IN PERFORMING LIGHT-DRIVEN H$_2$ AND O$_2$ FORMATION FROM WATER

When taking the photosynthetic apparatus which is nature's masterpiece for efficient light-induced water splitting as a blueprint for our considerations on the construction of technical devices, three basically different types of approaches can be distinguished: (1) use of photosynthetic organisms that are tuned by genetic engineering for H$_2$ production at the expense of bound hydrogen (biomass) formation, (2) building of hybrid systems consisting of synthetic components (e.g., metal complexes, nanoparticles, semiconductor electrodes) and isolated operational units of the photosynthetic apparatus (e.g., reaction centers, PSI and PSII complexes), and (3) construction of complete synthetic devices.

The design and generation of systems that enable light-driven water splitting is a rapidly growing field in current research activities and cannot be covered in this chapter. Therefore only a few representative examples for the different types will be selected from the wealth of approaches and briefly described; the main part of this chapter focuses on one particular and highly relevant topic–oxidative water splitting in photosynthesis.

3.6.1 Use of Photosynthetic Organisms

Several microorganisms (cyanobacteria) offer a most promising starting material that can be properly functionalized by genetic engineering. In this case the strongly reducing electrons generated by light-induced charge separation of PSI [29, 30] need to be funneled to PRC containing enzymes. The majority of these biological catalysts (hydrogenases, nitrogenases) containing strongly reducing iron–sulfur or iron–nickel clusters are highly sensitive to O_2 (see Refs. [31, 32, 33] and references therein) and therefore the reactions have to be performed under strictly anaerobic conditions. In order to satisfy this requirement some microorganisms have evolved specific cell structures like heterocysts in multicellular N_2 fixing cyanobacteria [34].

A very attractive proposal has been reported by Mazor et al. [35] for light-induced water splitting into H_2 and O_2 by using genetically modified cyanobacteria. In this approach H_2 and O_2 generation are not only performed in different reaction vessels but the overall process is also temporarily separated into two parts. In the first part of a cycle the cyanobacteria are tuned for light-driven water splitting into O_2 and metabolically bound hydrogen performed under aerobic conditions in reactor 1, followed by the transfer of the suspension into a second container (reactor 2) where in the second part of the cycle the bound hydrogen (biomass) formed in reactor 1 is transformed to H_2 under anaerobic conditions (established by thermal inactivating of PSII). This process is energetically driven by the light-induced reactions of PSI. After consumption of the bound hydrogen the cyanobacteria are transferred back to reactor 1 for starting the next cycle. In this way light-induced H_2 and O_2 production is performed in a periodic and alternating manner.

3.6.2 Hybrid Systems

Light-driven H_2 production can be achieved by using hybrid systems consisting of isolated genetically modified PSI complexes and functionally coupled catalysts for H^+ reduction. This process requires use of a sacrificial electron donor for feeding electrons into PSI. The catalysts (PRC) can be either inorganic constituents containing suitable transition metals or systems of biological origin like a hydrogenase [36]. A new road for using hydrogenases in hybrid systems was opened by the discovery of oxygen tolerant hydrogenases [37] and the resolution of the structure by X-ray diffraction crystallography (XRDC) [38].

Hybrid systems of isolated PSI complexes tethered with platinum nanoparticles were shown to be efficient in light-induced H_2 formation using ascorbate as the electron source [39, 40]. However, from a practical (economical) point of view, the use of the rare and rather expensive metal is counterproductive for large-scale applications and the search for catalysts containing "cheap" and widely available material is highly important. An attractive system could be the self-assembly of PSI complexes with a cobalt-containing catalyst (cobalamin). This system was reported to evolve H_2 under illumination at comparatively high rates with ascorbate as the electron donor [41]. Unfortunately, the currently available system is not stable so that further research is required to solve this problem.

Several attempts have been made to construct simple inorganic biomimetics of the catalytic site of hydrogenases containing binuclear iron centers connected via sulfur bridges. Most of these are only soluble in organic solvents [42] but water solubility could be achieved by suitable ligation [43, 44].

A complementary approach to a PSI-based hybrid system is the design of devices where a synthetic photogenerator is functionally coupled with a suitable hydrogenase. It was shown [45] that complexes consisting of CdS nanorods with a functionalized surface (3-mercaptopropionic acid) can be coupled successfully with [FeFe]-hydrogenase I from *Clostridium acetobutylicum*. This hybrid system performs at the expense of ascorbate as electron donor light-induced H_2 formation with a quantum efficiency of up to 20% at 405 nm. The system, however, is prone to inactivation after a total turnover number of about 10^6 due to photooxidation of the functionalized surface, which provides the linkage to the enzyme. Another approach is the linkage of an O_2 tolerant [NiFe] hydrogenase to Ru(bpy)$_3$ sensitized TiO_2 nanoparticles. This system achieved turnover numbers of 50 s^{-1} with TeOA as the sacrificial electron donor [46, 47].

The above hybrid systems can perform light-induced H_2 formation but are not able to use water as the electron source. In order to replace the sacrificial electron donor by H_2O, a functionally efficient coupling is required with a suitable device for light-driven water splitting into O_2 and bound hydrogen, which acts as the donor for PSI type or CdS nanorod hybrid systems. The development of such a system is a challenging and highly demanding task as is clearly illustrated by the long-lasting evolutionary process of photosynthetic organisms to achieve this goal by "constructing" the PSII core complex [15, 16, 17, 18, 19] with its unique properties that will be described in more detail in the following sections. Therefore it appears reasonable to consider the possibility of using isolated PSII complexes as the essential building block in hybrid systems. Reports on such systems are very rare. A successful functional connection of PSII core complexes from the thermophilic cyanobacterium *Thermosynechococcus* (*T.*) *elongatus* with gold nanoparticles (GNPs) has been reported by Noji et al. [48]. Likewise the same type of PSII preparation was successfully bound to a mesoporous indium–tin oxide (*meso*ITO) electrode. This system exhibited a turnover frequency of about 0.2 O_2/PSII•s under red light illumination [49]. This value is at least two orders of magnitude lower than the maximum turnover number of PSII core complexes, thus indicating limitations of the electron transfer to the electrode. The most serious problem, however, is the high sensitivity to photodegradation (the half-life time of this hybrid system under illumination was about 5 min). The susceptibility of the PSII complex to degradation is a general problem that has been solved in the natural system by the installation of a repair mechanism (see Section 3.8).

Apart from the development of suitable hybrid systems an efficient functional coupling between PSI- and PSII-type hybrid units is indispensable. Furthermore, their spatial array is a key factor in order to achieve the formation of H_2 and O_2 evolution in separate compartments.

3.6.3 Synthetic Systems

In these systems all parts result from employing modern material sciences. The key process in the transformation of solar radiation into Gibbs free energy is the dissociation of excitons into a sufficiently stabilized charge separation of electrons and holes. In photosynthesis these reactions take place in units referred to as reaction centers (RCs), which act as nanoscale single charge solar batteries. A large variety of macroscopic and mesoscopic synthetic devices have been developed in photovoltaics (solid state inorganic semiconductors, organic semiconductors, dye sensitized solar cells), which can be used for driving light-induced water splitting provided that connection with suitable PRC and WOC units is achieved. An alternative to these types of photovoltaic systems are biomimetic arrays at the molecular level, where a photochemically active chromophor P is connected with suitable redox partners. In the most simple case P is covalently linked with either a donor or acceptor molecule. These dyads perform charge separation; however, they do not permit a sufficient stability of the generated cation–anion radical pair. Therefore in analogy to the sequential reaction pattern of photosynthetic RCs [50] triads of the type D–P–A have been constructed (P = photochemically active pigment, A = electron acceptor, and C = electron donor and the dash symbolizes the covalent bond that could comprise a bridging molecule) which permit the formation of radical pairs $C^{+\bullet}$–P–$A^{-\bullet}$ with significantly longer lifetimes. More sophisticated systems are pentades consisting of five functional components (for reviews, see Refs. [51, 52, 53]). In these arrays the component P is often a porphyrin derivative and fullerenes were found to be useful electron acceptors, carotenoids can be used as electron donors [54, 55].

One essential point for efficient solar energy transformation is the harvesting of light. Photosynthetic organisms have developed highly efficient specialized pigment protein complexes (see chapters in Green and Parson [56] and Renger [57]) for optimal adaptation to strongly varying illumination conditions (diurnal rhythm, large differences of light intensities of photosynthetic organisms living in habitants like tropical rain forests and different water layers of oceans and lakes etc.). Numerous attempts have been reported during the last decades to synthesize pigment arrays for excitation energy transfer (EET) [58, 59, 60, 61, 62, 63] including the possibility of photoprotection by carotenoids [64] and the induction of nonphotochemical quenching [65]. The helical DNA structure was shown to offer a useful scaffold for pigment attachment in artificial antenna systems [66]. So far the high efficiency of photosynthetic antenna complexes has not been achieved. One promising approach seems to be the use of engineered proteins as matrices for binding of chromophores with tuned spectral properties [67]. This example nicely illustrates the huge potential of proteins for construction of molecular devices with special function(s). The great variability of proteins as biopolymers tailored to establish well controlled potential energy surfaces permits a precise tuning of the energetics of cofactors and their reaction coordinates. The availability of this "ideal" material has enabled biological systems to achieve, during a long-lasting process of evolutionary development, practically any degree of specificity, efficiency, and regulatory control that is required for a particular biological function. Therefore specifically designed synthetic polymers should be taken into consideration as a most promising tool in constructing efficient nanoscale devices for solar energy exploitation.

In order to obtain systems for light-induced water splitting into H$_2$ and O$_2$, the photogenerators with their attached antenna have to be coupled with PRC and WOC units. In this respect the construction of efficient and stable WOCs is the most challenging task (see Section 3.6.2). As a starting point of discussions on WOCs, some general features of the reaction pathway of oxidative water splitting into O$_2$ and protons in aqueous solution will be described.

3.6.4 Oxidative Water Splitting into O$_2$ and 4H$^+$

The top panel of Fig. 3.3 schematically illustrates the reaction pattern via one-electron redox steps leading from water to O$_2$ + 4H$^+$ and vice versa. Inspection of this scheme readily reveals two characteristics of the process: (1) the individual

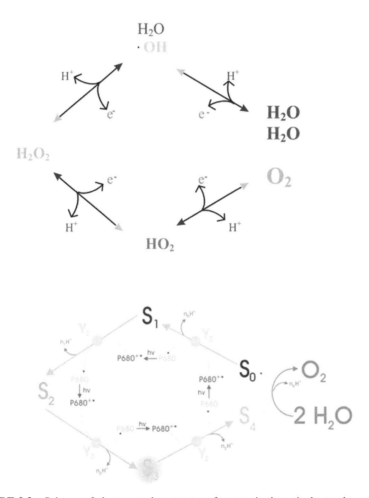

FIGURE 3.3   Scheme of electron and proton transfer steps in the univalent redox sequence between 2H$_2$O and O$_2$ + 4H$^+$ in aqueous solution (top panel) and extended Kok scheme of photosynthesis (bottom panel). See text for details.

redox steps are coupled with proton release/uptake and (2) the four-step sequence comprises the reactive oxygen species (ROS) $HO^\bullet$, H_2O_2, and $HO_2/O_2^{-\bullet}$, which are harmful to biological material (proteins, lipids) and therefore populations of these intermediates have to be suppressed when performing this process in sensitive biological systems. Furthermore, the Gibbs free energy gap ΔG° is quite different for the individual reactions [68]; in particular, the abstraction of the first electron from a water molecule leading to the $OH^\bullet$ radical under concomitant H^+ release is highly endergonic with a midpoint potential of $+2.3$ V at pH 7.0. Therefore this process cannot be driven by the energy of a photon in the red wavelength region.

Nature has solved these problems, that is, optimal coupling of electron and proton transfer steps, "taming" of the reaction sequence, and a proper tuning of its energetics, by "nanoscale encapsulation" of the process into a protein bound transition metal cluster. The protein matrix and bound water molecules form hydrogen bonds, which are of key relevance for a highly efficient performance of the overall process in both directions [5]. It is most important to note that nature has not only evolved two entirely different devices to catalyze the reaction in the forward (water oxidizing complex of PSII) and backward (cytochrome c oxidase of the respiratory chain) directions but also that these enzymes function in a unidirectional manner. Another striking feature is the invariance of the catalytic centers to evolutionary development from ancient primitive prokaryotic organisms to higher plants and animals [5].

Synthetic WOCs also contain transition metals either in the form of molecular complexes or in the form of solid state materials. A large variety of systems have been synthesized during the last three decades. This topic is currently a rapidly growing field that cannot be covered fully. Therefore, in the following, only a few representative examples are briefly described by focusing on the most recent reports.

3.6.5 Synthetic WOCs

Complexes of several transition metals catalyze oxidative water splitting in either homogeneous solution or heterogeneously as solid state metal oxides. The process can be driven energetically by using strong chemical oxidants like Ce^{4+} in solution or by connecting the catalyst with anodes of sufficient positive bias. During the last two decades numerous systems have been developed (for a review covering the time period 1979–2010, see Liu and Wang [69])

Most detailed analyses were performed on mono- and binuclear ruthenium complexes. Mechanistic studies on the binuclear "blue dimer" complex revealed that a peroxide linked to one Ru center is formed as an intermediate [70]. Likewise, a peroxidic intermediate is also formed on mononuclear Ru complexes [71]. These ruthenium complexes are efficient catalysts with either a strong oxidant (e.g., Ce^{4+}) in solution or when linked to doped semiconductor electrodes (e.g., ITO = indium-doped tin oxide) [72]. Complexes with iridium ions also efficiently catalyze oxidative water spitting [73, 74]. However, these metals are scarce and expensive and therefore not available in large-scale applications for economic reasons. Therefore current research activities are focused on the development of WOCs that are based on "cheap" transition metals like Mn, Fe, and Co. A variety of systems have been reported where mainly oxide units were used [75, 76, 77]. Ligation by polyoxymetalate (e. g., poly-tungstate) was shown to be successful in obtaining water-soluble self-assembling Co-based catalysts [78]. Another very useful form are nanoparticles or thin layers often attached to photoactive solid state semiconductor electrodes [79, 80, 81, 82, 83]. Most of these systems operate efficiently only in strongly acidic or alkaline solutions. For practical application, however, it is important to develop devices that are catalytically highly active under conditions that exist in natural water sources and tolerate contaminations in lakes, rivers, and oceans. An interesting Co-based WOC system has been described where the Co catalyst is electrochemically deposited on ITO electrodes. This system exhibits high activity of water oxidation to O_2 in rivers and seawater [84]. Another problem of artificial WOCs is the toxicity of some transition metals. In terms of availability and "biologically friendly" material, the development of suitable Mn-based WOCs appears to be a most attractive goal. Several manganese oxides often containing alkali and/or alkali earth metal ions were shown to exhibit catalytic activity [81, 85]. Nature offers the blueprint for our considerations on this topic because it uses exclusively this transition metal for the WOC in photosynthesis (see Section 3.11.1). The maximum rate of the photosynthetic WOC exceeds that of the majority of synthetic catalysts by at least three orders of magnitude [86]. This kinetic barrier has been surpassed by the synthesis of a mononuclear Ru complex where the ligands strongly support the transient formation of a binuclearly complexed püeroxide [87]. The turnover frequency of >300 s^{-1} matches that of the photosynthetic WOC but the total turnover number of about 8000 is two orders of magnitude lower compared to the natural system. Likewise, a Cu-containing catalyst was synthesized with a turnover number of about 100 s^{-1} [88].

3.6.6 Light-Induced Water Splitting in Photosystem II

The key steps in using water as the hydrogen source for solar energy exploitation in photosynthesis is the abstraction of four electrons from two substrate molecules under the release of O_2 and four protons. Two indispensable prerequisites have to be satisfied in order to perform this process: (1) tuning of the properties of the photoactive pigment to generate a species of sufficient oxidizing power as the driving force for electron abstraction from water and (2) "invention" and assembly of a catalytic device that enables the energetic tuning of a redox process involving the cooperation of four strongly oxidizing redox equivalents to split water into O_2 and four protons.

The result of a long evolutionary process ([15, 16, 17]; for further reading, see Larkum [18] and Peschek et al. [19]) was a multimeric complex designated photosystem II (PSII). This unique operational unit acts as water:plastoquinone-oxidoreductase:

$$2PQ + 2H_2O \xrightarrow{4h\nu} 2PQH_2 + O_2 \qquad (3.4)$$

The overall process of Eq. 3.4 comprises three types of reaction sequences (for a review, see Renger [89]):

Sequence (a): *Light- induced "stable" charge separation forming the strong oxidant* $P680^{+\bullet}$

$$\text{Ant} \bullet \text{P680PheoQ}_A \xrightarrow{\;h\nu\;} {}^1[\text{Ant} \bullet \text{P680}]^* \text{PheoQ}_A \rightarrow$$
$$\text{Ant} \bullet \text{P680}^{+\bullet}\text{Pheo}^{-\bullet}\text{Q}_A \rightarrow \text{Ant} \bullet \text{P680}^{+\bullet}\,\text{PheoQ}_A^{-\bullet} \qquad (3.5)$$

Sequence (b): *Oxidative water splitting*

$$4 \oplus + 2\text{H}_2\text{O} \rightarrow \text{O}_2 + 4\text{H}_{\text{Lumen}}^+ \qquad (3.6)$$

Sequence (c): *Reductive plastoquinol formation*

$$4\ominus + 4\text{H}_{\text{Stroma}}^+ + 2\text{PQ} \rightarrow 2\text{PQH}_2 \qquad (3.7)$$

where Ant symbolizes the antenna pigments, which transfer in a radiationless manner their electronic excitation generated by light absorption to the photoactive pigment denoted P680 in PSII, pheophytin *a* (Pheo) and Q_A are electron acceptors, the symbols $\oplus$ and $\ominus$ represent redox equivalents (including the cofactors $\text{P680}^{+\bullet}$ and Q_A involved) that give rise to oxidative water splitting (sequence (b)) and reductive plastoquinol formation (sequence (c)), respectively. $\text{H}_{\text{Lumen}}^+$ and $\text{H}_{\text{Stroma}}^+$ symbolize proton release into the thylakoid inner space (lumen) and uptake from the outer space (cytoplasm/stroma), respectively.

The top panels of Fig. 3.4 show on the left side the overall structure of the PSII core complex of the thermophilic cyanobacterium *Thermosynechcoccus* (*T.*) *elongatus* gathered from X-ray diffraction crystallography data and on the right side the array of the cofactors that perform the reaction sequences (a)–(b).

3.7 LIGHT-INDUCED "STABLE" CHARGE SEPARATION

Reaction sequence (a) and a simplified scheme of its energetics are presented in the bottom panel of Fig. 3.4. The formation of the radical pair $\text{P680}^{+\bullet}\text{Q}_A^{-\bullet}$ comprises three steps (for details, see Renger and Renger [90]): (i) primary charge separation leading to electron transport from the lowest excited singlet state ${}^1\text{Chl}_{\text{D1}}^*$ to Pheo_{D1}, thus forming the radical ion pair $\text{Chl}_{\text{D1}}^{+\bullet}\text{Pheo}_{\text{D1}}^{-\bullet}$ (for additional and/or alternative pathways, see Novoderezhkin et al. [91], and Shelaev et al. [92]); (ii) rapid reduction of $\text{Chl}_{\text{D1}}^{+\bullet}$ by $\text{P}_{\text{D1}} - \text{P}_{\text{D2}}$, where the spin in the radical $(\text{P}_{\text{D1}}\text{P}_{\text{D2}})^{+\bullet}$ is predominantly located on $\text{P}_{\text{D1}}^{+\bullet}$ in intact PSII with a functionally competent WOC [93, 94] but changes after a severe modification leading to significant spin distribution between P_{D1} and P_{D2}

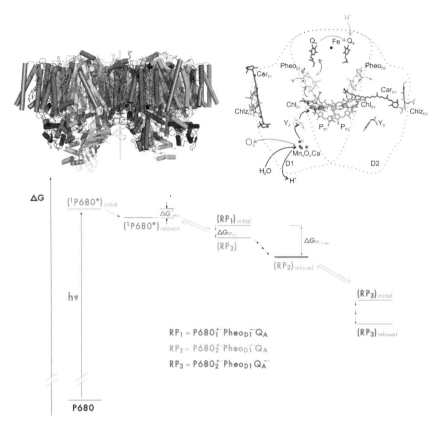

FIGURE 3.4   Top panel (provided by J. Kern), left side: View along the membrane plane of a dimer of PSII from *T. elongatus* as derived from the 3.0 Å resolution electron density. Right side: Schematic representation of the cofactor arrangement in the reaction center core. Helices are depicted as cylinders, cofactors are drawn in stick models. The C2 axis, relating the two monomers in the dimer, is indicated by a black dashed line. The large subunits D1 and D2, forming the reaction center, are surrounded by CP43 and CP47 as well as several membrane intrinsic low molecular weight subunits. At the lumenal side, the three extrinsic subunits PsbO, PsbV, and PsbU are known to interact with large loop regions of D1/D2/CP43/CP47. Organic cofactors are Chl, Pheo, PQ9, carotenoids, lipids, and heme, respectively; Ca^{2+}, Fe, and Mn are shown as spheres; the figure was generated using Pymol (Delano, 2003). On the right side cofactors in the PSII core are shown, the view is along the membrane plane, the coordinating protein subunits D1 and D2 are indicated by a dotted line. Arrows symbolize the direction of electron transfer leading to $P680^{+\bullet}QA^{-\bullet}$ formation (for further details see text), the oxidative water splitting under O_2 release, and formation of PQH_2 at the Q_B site. Bottom panel: Simplified scheme of the energetics of $P680^{+\bullet}QA^{-\bullet}$ formation. Symbols denote: $(^1P680^*)_{initial}$ and $(^1P680^*)_{relaxed}$ = excited singlet state of P680 in the initial state and in relaxed environment; RP1, RP2, and RP3 = radical pair states as indicated in the inset, indices "initial" and "relaxed" indicate the initial form and the species in the relaxed environment, respectively (electronic version prepared by S. Renger). See text for details. Reproduced from Kern and Renger [179] with permission of Springer.

concomitant with a drop of the E_m value [93], and (iii) stabilization of the radical pair formation by electron transfer from $Pheo_{D1}^{-\bullet}$ to Q_A [95].

The rate constants of the reactions (i) and (ii) giving rise to the radical pair $(P_{D1}P_{D2})^{+\bullet}Pheo_{D1}^{-\bullet}$ are still a matter of controversy because direct measurements are prevented due to spectral overlap of the six coupled pigments forming the [(Chl $a)_4$(Pheo $a)_2$] complex (for details, see Renger and Renger [90] and Renger and Holzwarth [95] and references therein). On the other hand, the reoxidation kinetics of $Pheo_{D1}^{-\bullet}$ by Q_A were measured directly by time resolved absorption changes reflecting $Pheo_{D1}^{-\bullet}$ reoxidation [96] and concomitant Q_A reduction [97, 98]. Values of 250–350 μs were obtained; that is, the stable radical pair $P680^{+\bullet}Q_A^{-\bullet}$ is formed within 1 ns and this state remains "stable" to recombination for more than 100 μs [99, 100].

The $P680^{+\bullet}$ cation radical of PSII is one of the most oxidizing species within biological systems. Its formation is the energetic prerequisite for oxidative water splitting. It must be emphasized that the unique redox property of $P680^{+\bullet}$, compared with the oxidizing species formed by light-induced charge separation in PSI and reaction centers of anoxygenic photosynthetic bacteria [50], was not achieved by cofactor replacement (BChl $a \rightarrow$ Chl a) but by evolutionary engineering of the protein environment [101, 102]. This finding is another striking example for illustrating the potential of proteins in tuning the properties of cofactors (for further discussion, see Renger [103]). In the case of PSII, the midpoint potential E_m is drastically increased from a value of about $+0.8$ V of Chl a/Chl $a^{+\bullet}$ in solution [104] to about $+1.25$ V [105] of the photoactive Chl a complex designated P680, whereas in PSI the modulation by the environment leads to an E_m shift in the opposite direction, that is, to values below $+0.5$ V [29] for $P700/P700^{+\bullet}$. A replacement of BChl by Chl, however, was required for another energetic reason, that is the population of a lowest excited singlet state with high enough electronic excitation energy to permit via the intermediate $P680^{+\bullet}Pheo_{D1}^{-\bullet}$ the formation of a "stable" radical pair $P680^{+\bullet}Q_A^{-\bullet}$ with sufficient redox free energy [50].

3.8   ENERGETICS OF LIGHT-INDUCED CHARGE SEPARATION

The energetics of light transformation into electrochemical Gibbs free energy can be calculated by the relation:

$$\eta_{max} = \eta_{max}^{\lambda(RC)} \bullet \frac{hc}{\lambda(RC)} \bullet \int_0^{\lambda(RC)} \rho(\lambda)d\lambda \left/ \int_0^{\infty} \frac{hc}{\lambda}\rho(\lambda)\,d\lambda \right. \tag{3.8}$$

where $\eta_{max}^{\lambda(RC)}$ is the maximum efficiency of the reaction center in PSII at threshold wavelength $\lambda(RC)$ of the photogenerator RC (reaction center) for photons that can be transformed into electrochemical Gibbs free energy while the excess energy of shorter wavelength photons is lost as heat, $\rho(\lambda)$, h, and c are the photon flux density of solar radiation impinging the earth's surface at wavelength λ, the Planck constant, and the velocity of light in vacuum, respectively.

The value of $\eta_{max}^{\lambda(RC)}$ can be calculated within the framework of a classical "Carnot type machine," where the flux of heat energy from a higher to lower temperature level is transformed into useful work (for details, see textbooks on physical chemistry, e.g., Atkins and De Paula [14]):

$$\eta_{max}^{\lambda(RC)} = (T_{solar} - T)/T_{solar} \qquad (3.9)$$

where T_{solar} is the "effective temperature" of the scattered nonpolarized solar radiation impinging on the RC and T its ambient temperature. The RC is assumed to be a "narrow band absorber," which uses nearly monochromatic red light at a wavelength $\lambda(RC)$ corresponding to the absorption maximum of the Chl a constituent(s). A value of about 73% was obtained [106] for $\eta_{max}^{\lambda(RC)}$ by using a calculated value of about 1100 K for T_{solar} [106, 107] and a wavelength of 700 nm for $\lambda(RC)$.

It must be emphasized that the maximum efficiency of a reversible system can only be achieved if no Gibbs free energy is used for driving any process at physiological rate. The efficiency of an energy converter acting as a thermodynamically open system that couples energy fluxes and forces is described by the rules of irreversible thermodynamics. Details of these calculations are beyond the scope of this chapter (for further reading, see Bell and Gudkov[108]). A value of about 20% was estimated for the maximum efficiency of solar energy conversion in the reaction centers of green plants when considering realistic rate constants for forward and back reactions [109]. However, in the photosynthetic apparatus the RCs are energetically driving the electron transport chain (Eq. 3.1) and ATP synthesis (see Eq. 3.2), which leads to further losses so that eventually a maximum efficiency of less than 5% is obtained for plant photosynthesis. When taking into account an additional Gibbs free energy demand for repair processes (vide infra) this number fits with experimental data on rapidly growing "cultures" like sugar cane under optimal conditions.

At first glance, the overall efficiency below 5% for photosynthesis seems to be surprisingly low when considering the long selection pressure during evolution and comparing it with devices like photovoltaic cells. However, a direct comparison with the latter systems is highly misleading because the energetics of photosynthetic organisms comprises all costs for synthesis of biomass and repair of the apparatus. Therefore the energy costs for driving all secondary reactions (e.g., electrolytic watersplitting into H_2 and O_2, formation of simple organic compound, e.g., glyceraldehyde) and for producing photovoltaic and other synthetic systems and dealing with their waste products must be taken into account in order to achieve meaningful numbers for comparison.

In photosynthetic organisms the repair is of particular relevance for PSII, which produces the strongly oxidizing cation radical P680$^{+\bullet}$ for driving oxidative water splitting (vide infra). When this latter reaction is disturbed or interrupted, the lifetime of P680$^{+\bullet}$ significantly increases (see Section 3.10) and deleterious oxidation reactions can take place. Likewise the recombination of the radical pair $P680^{+\bullet}Q_A^{-\bullet}$ populates $^3P680$ via transient population of the triplet state $^3\lfloor P680^{+\bullet}Pheo_{D1}^{-\bullet}\rfloor$. The triplet $^3P680$ acts as sensitizer for formation of the reactive oxygen species (ROS) singlet oxygen $(^1\Delta_g\ O_2)$ thus leading to deleterious processes (for details, see Vass and Aro[110]). The $^1\Delta_g\ O_2$ was shown to be the major ROS compound that gives rise to oxidative damage of plants [111]. In spite of different photoprotective mechanisms (see Jahns and Holzwarth [112] and Ruban et al. [113] and references therein), these deleterious effects cannot be prevented completely and therefore repair, including synthesis and replacement of the D1 protein [110,114, 115, 116], is required in order to maintain photosynthetic activity over long periods [116].

3.9 OXIDATIVE WATER SPLITTING: THE KOK CYCLE

A cornerstone in understanding the underlying reaction pattern of oxidative water splitting was the discovery of the period four oscillation of oxygen yield per flash. This pattern is observed when dark adapted algae cells or chloroplasts from higher plants are excited with a train of single turnover flashes [117]. Kok and co-workers [118] interpreted this characteristic feature by a scheme that is referred to as Kok cycle. The invariance of the flash-induced oxygen yield pattern to the fraction of competent PSII complexes unambiguously showed that each $P680^{+\bullet}$ is functionally coupled with virtually only one water oxidizing complex (WOC). Decades later this conclusion was confirmed by data on the structure of the PSII core complex (for a review, see Zouni [119]).

The bottom panel of Fig. 3.3 presents an extended version of the original Kok scheme in showing that the overall reaction sequence comprises two different types of reactions: (1) reduction of $P680^{+\bullet}$ by a component Y_Z and (2) stepwise oxidation of the WOC by Y_Z^{ox} until after the accumulation of four oxidizing redox equivalents molecular oxygen is formed and released via an exchange reaction with substrate water.

Inspection of this scheme also reveals that the oxidation steps are accompanied by proton release. It is important to note that the mode of coupling between electron transfer (ET) and proton transfer (PT) is of general relevance for many redox processes including photosynthetic water splitting and therefore the mechanisms of proton coupled electron transfer (PCET) reactions in different systems is a "hot" topic of current research activities [120, 121, 122, 123, 124].

So far the energetics of the redox steps could not be determined by direct measurements (for a discussion, see Renger [125]). The overall redox gap between $P680^{+\bullet}Y_z$ and the relaxed state of $P680\ Y_z^{ox}$ (see Section 3.10) was reported to be about 100 meV [126]. Data gathered from indirect lines of evidence revealed that the S_i state transitions are characterized by a redox gap $\Delta G^\circ(S_{i+1}/S_i)$ of 1.0–1.2 eV for $i =$ 1, 2, and 3, while $\Delta G^\circ(S_1/S_0)$ is significantly lower by 0.15–0.25 eV [105, 127]. These values illustrate the striking energetic tuning of the reaction coordinate of four step oxidative water splitting in the WOC of PSII compared to aqueous solution (for further discussion, see Renger [125]).

3.10 Y$_Z$ OXIDATION BY P680$^{+\bullet}$

Thermodynamic considerations reveal that the oxidation of Y$_Z$ by P680$^{+\bullet}$ requires deprotonation from the OH group of tyrosine residue Y$_Z$ [128, 129] and therefore the mechanism of this reaction implies questions on the mode of coupling between the proton transfer (PT) and electron transfer (ET) step. As the OH group of Y$_Z$ is able to form hydrogen bonds, this structural motif was anticipated to be of key relevance for the reaction [130]. Residue His 190 of polypeptide D1 was identified as the partner of the hydrogen bond to Y$_Z$ (see Hays et al. [131] and references therein) and this assignment was confirmed by XRDC structure analyses [132, 133, 134]. The top panel of Fig. 3.5 illustrates that the reaction comprises the electron transfer from the highest occupied molecular orbital (HOMO) of the tyrosine moiety to that of P680$^{+\bullet}$ and the concomitant proton transfer from the phenolic oxygen atom of Y$_Z$ to the

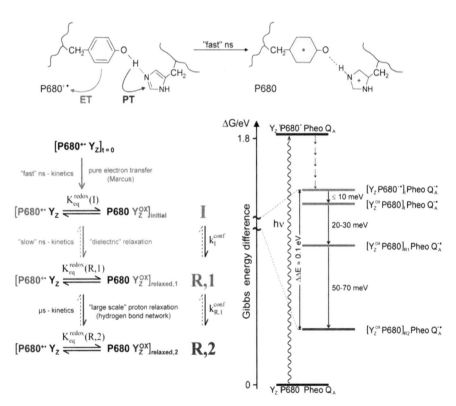

FIGURE 3.5 Top panel: Scheme of proton coupled electron transfer. Bottom panel: Simplified reaction sequence (left side) and corresponding energetics (right side) of P680$^{+\bullet}$ reduction by Y$_Z$ in PSII complexes with intact water oxidizing complex (WOC) in redox state S$_1$. The initial state I and the two relaxed states R,1 and R,2 are marked. See text for details. Reproduced from Renger and Renger [90] with permission of Springer.

nitrogen atom of His 190, which forms a strong hydrogen bond with Y_z. This type of reaction, referred to as multiple site electron and proton transfer (MS-PET), is not a specific feature of $P680^{+\bullet}$ reduction by Y_z but characteristic for a multitude of redox reactions [123]. The mode of hydrogen bonding of Y_z in PSII is of key relevance for the reaction pathway under physiological conditions and a special array is the indispensable prerequisite for Y_z oxidation at liquid helium temperature [135, 136].

The $P680^{+\bullet}$ reduction was found to exhibit a rather complex multiexponential kinetics in all different sample types of PSII with an intact WOC analyzed so far (for a review, see Renger [137] and references therein). At first glance this pattern appears to be very surprising because in these PSII complexes the overall reaction is dominated (more than 90%) by the electron donation from Y_z and both redox partners ($P680^{+\bullet}$ and Y_z) are bound to the D1/D2 heterodimeric protein matrix in a fixed geometry and mutual interaction (for details on the structure, see Refs. [132, 134, 138] and references therein). A consistent explanation of the kinetic pattern was offered by a model that is based on two assumptions [139, 140]: (1) the redox step is very fast and leads to an initial "quasiequilibrium" $[P680^{+\bullet}Y_z \leftrightarrow P680Y_z^{ox}]_{I}$,[4] with a constant $(K_{eq})_I$, and (2) at least two different types of relaxation processes including the rearrangement of a hydrogen bond network give rise to a shift toward markedly higher values $(K_{eq})_{relax}$, thus leading to a high percentage of transient Y_z^{ox} population and consequently to a low probability of misses in the S_i state transitions of each single turnover in PSII (for a detailed discussion of this effect, see Christen et al. [141]).

The bottom panel of Fig. 3.5 presents on the left side a scheme of the reaction pattern for the simplified model of a three-exponential kinetics. The first step is designated "fast ns" kinetics and the two relaxation steps as "slow ns kinetics" and "μs kinetics."

Estimations on the energetics of the different steps can be made when taking into account that the equilibration rate via the "fast ns" kinetics is sufficiently fast compared to the rate of subsequent relaxation processes. The bottom panel of Fig. 3.5 presents on the right side typical values obtained for PSII with the WOC in redox state S_1 (or S_0). In case of WOC attaining redox states S_2 or S_3, the first step is even slightly endergonic (for further details, see Kühn et al. [139]). This data reveals that the energetics of the overall process are dominated by the relaxation processes, thus illustrating again the important role of the protein matrix in tuning a reaction. Furthermore, the close similarity of the multiphasic kinetics in cyanobacteria and higher plants (vide supra) leads to the important conclusion that the reaction coordinate remained virtually invariant to evolutionary development.

As a consequence of the model of "dynamic" transitions only the "fast ns" kinetics reflect the properties of the redox step leading to $P680^{+\bullet}$ reduction by Y_z. The activation energy of this reaction was found to be about 10 kJ/mol in PSII membrane fragments from spinach with the WOC attaining the redox state S_1 [130]. Slightly

---

[4]In order to take into account the possible localization of the proton in the environment the oxidized Y_z will be symbolized by Y_z^{ox} rather than by the deprotonated radical $Y_z^{\bullet}$

higher values were obtained in studies on PSII core complexes from both cyanobacteria and higher plants [139, 142]. Evaluation of the experimental rate constant and activation energy within the framework of the Marcus–Levich–Hush theory of nonadiabatic electron transfer [143] leads to a reorganization energy of about 0.6±0.1 eV [140], thus confirming former calculations of about 0.5 eV [144]. Based on an empirical relationship between rate constant and distance [145], the edge to edge distance between P680$^{+\bullet}$ and Y$_z$ was calculated to be 10±1 Å [140], which is in perfect agreement with the value of 9.6 Å gathered from XRDC data [132, 134, 138] and also confirms our previously reported value of 9±2 Å [146].

The finding of a striking correspondence of the "calculated" values with the experimental distance is of high mechanistic relevance because it implies that the "fast ns" kinetics of P680$^{+\bullet}$ reduction by Y$_z$ can be described perfectly by the "theoretical" rate constant of a nonadiabatic electron transfer step. As a consequence, the rate of this reaction step is limited neither by any trigger reaction nor by the coupled proton transfer. This conclusion is in line with the lack of any isotope effect (KIE) on the "fast ns" kinetics due to replacement of exchangeable protons by deuterons, that is, $k_H/k_D < 1.05$ [147]. It must be emphasized that these characteristic features are exclusively observed in samples with an intact WOC (vide infra).

Inspection of the 1.9 Å XRDC structure reveals a close interaction between Y$_z$ and the WOC via hydrogen bonds. Therefore the mode of hydrogen bonding is expected to be different in redox states S$_0$ and S$_1$ versus S$_2$ and S$_3$ because the latter two redox states are characterized by the surplus of a positive charge [148, 149]. The altered hydrogen bond network and the localization of the surplus charge are inferred to be responsible for the different equilibrium constants K_j which determine the normalized amplitudes of the "fast ns," "slow ns," and "35 µs" kinetics (see Fig. 3.5, bottom part).

Figure 3.6 schematically illustrates the changes of the hydrogen bond network coupled with Y$_z$ oxidation and the differences due to the redox state of the WOC. It must be emphasized that no structure information is currently available on the details of hydrogen bonding in the individual redox states S$_i$ and the 1.9 Å XRDC structure reflects a superreduced S$_{-i}$ state, most likely S$_{-3}$ (see Section 3.11.1). Therefore this structure was used only as a frame and the presumed changes due to the Y$_z$ turnover during the S$_i$ state transitions are symbolized by differently colored ovals. It is important to note that the rearrangement of the hydrogen bond network due to the positive charge in S$_2$ and S$_3$ gives rise to an energetic destabilization of P680Y$_z^{ox}$ in the initial state compared to this state in S$_0$ and S$_1$ (vide supra). Values of 30–50 msV were obtained when the normalized amplitudes of the "fast ns" kinetics are used for the calculation of $(K_{eq})_I$ (for further details, see Kühn et al. [139]).

The reaction coordinate of P680$^{+\bullet}$ reduction by Y$_z$ is drastically changed in samples deprived of the WOC, as reflected by a marked increase of the activation energy (factor of about 3) and a significant kinetic H/D isotope effect of 2.7–3.3, which is completely absent for the "fast" and "slow" ns kinetics in intact PSII complexes (for a review, see Renger [150] and references therein). These findings convincingly illustrate that the wealth of interesting data obtained with this type of sample material does not reflect the situation of the intact PSII. A critical survey of the

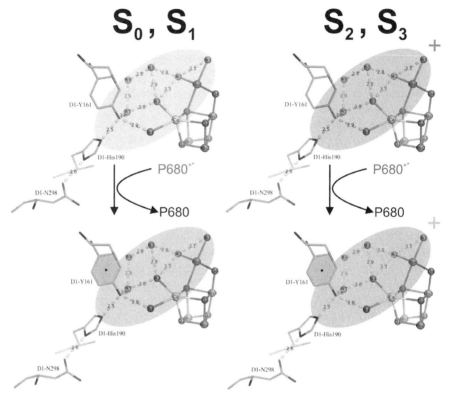

FIGURE 3.6 Schematic representation of the hydrogen bond network in the environment of Y_z (D1-Y161). The differently shaded ellipses symbolize different arrangements of individual hydrogen bonds within the network due to oxidation of Y_z (see bottom panels) and the surplus of a positive charge in S_2 and S_3 (right panels). For further details see text.

literature clearly shows that sound conclusions on the mechanism and in particular on hydrogen bonding of Y_z in vivo cannot be drawn from results gathered from samples with a destroyed WOC [150].

3.11 STRUCTURE AND FUNCTION OF THE WOC

Different lines of experimental evidence indicated that the catalytic site of the WOC is a Mn_4O_xCa cluster, where x denotes the number of μ-oxo-bridges (for reviews, see Messinger and Renger[127] and Yachandra [151]). Therefore the structure of the WOC is defined by (1) the spatial arrangement of the four manganese centers, the oxygen bridges, and the Ca^{2+} ion; (2) the first coordination sphere of the Mn_4O_xCa cluster, and (3) the surrounding protein matrix (second and third coordination sphere, hydrogen bond network(s)).

XRDC data revealed (vide infra) that the shortest van der Waals distance between Y_Z and the redox active manganese of the WOC is about 7 Å [132, 134, 138] and therefore Y_Z is not a constituent of the catalytic site itself although strong interactions exist via hydrogen bonds (vide supra)

One characteristic property of the WOC is its susceptibility to destruction under different stress conditions like heat, low and high pH, and UV-B irradiation (for a review, see Messinger and Renger [127]). The most striking feature, however, is the mode of Mn_4O_xCa cluster assembly. This process is highly endergonic, which requires a light-driven reaction sequence called photoactivation ([152, 153]; see also Salomon et al. [154]). As a consequence, the formation of a functional WOC is the last event in the ontogenetic development of the photosynthetic apparatus in oxygenic organisms [155, 156].

Detailed information on the following properties is required for understanding the unique nature of the photosynthetic WOC: (1) structure of the catalytic Mn_4O_xCa cluster and its coordination sphere, (2) electronic configuration and nuclear geometry of the catalytic site in the different redox states S_i of the Kok cycle, (3) kinetics of individual S_i state transitions in the WOC, (4) substrate/product pathways, and (5) mechanism of oxidative water splitting.

3.11.1 Structure of the Catalytic Mn–Ca Cluster and Its Coordination Sphere

X-ray diffraction crystallography (XRDC) and extended X-ray absorption fine structure (EXAFS) analyses are the currently most widely used methods to unravel the structure of the Mn_4O_xCa cluster. Site-directed mutagenesis of specific amino acid residues in combination with spectroscopic studies were used to gather information on the ligand sphere (for a review, see Debus [157]).

Several XRDC structures of PSII from thermophilic cyanobacteria have been published since 2001 [158, 159, 160, 161, 162]. A breakthrough in unraveling the WOC structure was achieved by obtaining a 1.9 Å XRDC resolution [134, 138], which offered for the first time a complete picture of the geometry of the Mn_4O_xCa cluster[5] including the oxygen atoms of the μ-oxo-bridges and of the water molecules that were not seen at the lower resolution of former studies (for a review, see Guskov et al. [132]). This structure shown in Fig. 3.7 most likely reflects a superreduced S_{-i} state because the manganese is expected to be reduced below the dark adapted state S_1 by electrons generated due to X-ray exposure of the crystals [163, 164, 165]. It was discussed that probably the S_{-3} state becomes populated [125] because this is a rather stable superredox state of the WOC that can be formed by chemical reduction

---

[5]The XRDC structure unambiguously shows that the cluster is identified as a Mn_4O_5Ca core, that is, $x = 5$, in a superreduced redox state, and therefore the symbol Mn_4O_5Ca is now widely used. However, Mn_4O_xCa appears to be more general because it comprises the possibility that the number of oxo-bridges could change during the Kok cycle in S_3 and S_4 due to formation of the O–O bond. Therefore Mn_4O_xCa will be used throughout this chapter keeping in mind that $x = 5$ for most of the S_i states.

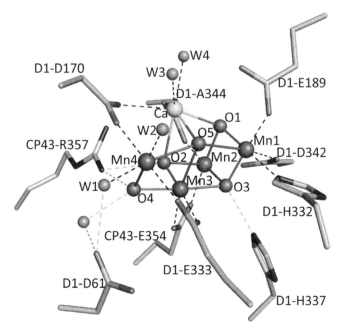

FIGURE 3.7 Structural arrangement of the Mn_4O_xCa cluster according to the X-ray diffraction crystallography data of Umena et al. [134]. For an assignment of amino acid residues and water molecules as ligands see text. Mn1, Mn2, Mn3, and Mn4 denote the different manganese ions of the WOC (see text for details). Figure prepared by Jian-Ren Shen and reproduced with permission from Umena et al. [134], copyright 2011.

with NH_2OH or NH_2NH_2 [166]. The idea of S_{-3} population in the irradiated crystals is supported by theoretical calculations [167].

Detailed information on distances between Mn and the nearest atoms of the co-ordination sphere were gathered from X-ray spectroscopy and the nuclear geometry of the Mn_4O_xCa cluster in the different S_i states was analyzed by EXAFS performed on samples where the WOC attains a definite S_i state. Progress in the detection technique led to a significant improvement of the distance resolution down to 0.09 Å with an accuracy of ± 0.02 Å [168]. The results obtained with this method unambiguously showed that the Mn_4O_xCa cluster in the dark stable S_1 state comprises three short Mn–Mn distances (two of about 2.7 Å and one of about 2.8 Å) and one long Mn–Mn distance of 3.3 Å. Likewise, two similar Ca–Mn distances (3.3–3.4 Å) were discovered [168].

A significant structural difference of the Mn_4O_xCa cluster was found between redox states S_0 and S_1 while only minor changes occur during the $S_1 \rightarrow S_2$ transition. This behavior contrasts with the marked rearrangement of the nuclei due to the $S_2 \rightarrow S_3$ transition as reflected by an elongation of the short (~2.7 Å) Mn–Mn distances to 2.82 and 2.95 Å [169]. Likewise, measurements on Sr^{2+}-containing PSII core complexes showed that also the Mn–Ca vectors (3.3–3.4 Å) decrease [170, 171]. The conclusion of significant structural changes coupled with redox transition $S_2 \rightarrow S_3$ is supported by other indirect lines of evidence: (1) the drastically different reactivity of S_2 and S_3 toward NH_2OH and NH_2NH_2 [172, 173], (2) differences in the binding of Ca^{2+} [174], and (3) the markedly higher activation energy compared to the other oxidation steps [175].

Optimization of the Mn_4O_xCa cluster geometry in different redox states has been performed by employing quantum mechanics/molecular mechanics (QM/MM) methods using XRDC and spectroscopic data as an experimental frame [176, 177, 178].

The ligation by amino acid residues has also been resolved by the 1.9 Å resolution XRDC structure. Figure 3.7 reveals that the Mn_4O_xCa cluster is coordinated by six carboxylate ligands (Asp170, Glu189, Glu333, Asp 342, and the C-terminus of Ala344 of polypeptide D1 and Glu354 of CP43) and one His (D1-H332) from the protein matrix and in addition by four water molecules. This mode of ligation by amino acid residues confirms former conclusions gathered from FTIR data on mutants of mesophilic cyanobacteria (for a review, see Debus [157]), except for Glu189 and His332 (both are ligands to Mn1) and His337, which coordinates to O3 rather than to Mn 2. One striking feature of the first coordination sphere is the bidentate ligation of the metal centers by carboxylate groups (D1-Asp170, D1-Glu333, D1-Asp342, D1-Ala344, and CP43-Glu354). This mode of bridging is inferred to stabilize the Mn_4O_xCa cluster and—even more important—to provide the indispensable scaffold for fixing and/or tuning functionally important mutual distances between the metal ions when the internal bridging mode is changed during the catalytic cycle, for example, by switching between mono- and bidentate ligation (see below) and formation or a change of μ-oxo-bridges between two metal ions [179].

Another very interesting feature of the 1.9 Å resolution XRDC structure is the hexa-coordination of all manganese centers. This finding is at variance with results of theoretical modeling by DFT where Mn1 (coordinated with His332) is inferred to be penta-coordinated [177, 180]. Analogous conclusions were derived for redox state S_2 of the WOC on the basis of analyses of data from EPR/ENDOR spectroscopy at low temperatures [176, 181, 182]. The key role of D1-His332 has been illustrated by X-ray spectroscopic studies on a D1-H332E mutant [183]. The mode of coordination of Mn1 is of mechanistic relevance and needs to be clarified by taking into account possible changes during the S_i state transitions.

In principle, ligand–metal interaction can be probed by using FTIR spectroscopy. The results of several studies [157, 184, 185, 186, 187] suggested that only two manganese of the Mn_4O_xCa cluster are redox active. This conclusion is in marked contrast to most of the published models (see Section 3.11.5). The unraveling of the origin of this serious discrepancy is of utmost relevance because it raises general questions on the interpretation of experimental results obtained by using different approaches under different conditions (low versus room temperature, hydration level, sample material, etc.).

Changes of the ligand sphere during the Kok cycle can be monitored by attenuated total reflection FTIR. It has been found that the orientation of the carboxylate group of CP43-Glu354 changes by $8°$ during the $S_1 \rightarrow S_2$ step due to transition from a bridging ligand of two metals to a bidentate coordination of one Mn [188].

The XDRC data unambiguously show that except for water molecules no inorganic ligand (Cl^-, HCO_3^-) is part of the first coordination sphere of the Mn_4O_xCa cluster [132, 134, 138]. Two Cl^- were found to be bound in the neighborhood which are part of a hydrogen bond network that was originally proposed by Olesen and Andréasson [189]. This network could facilitate $+H$ release from the catalytic site and might be involved in substrate water interaction (see also Section 3.11.4).

3.11.2 Electronic Configuration and Nuclear Geometry in the S_i States of the Catalytic Site

The electronic structure of each individual S_i state is determined by the valence of manganese and its first coordination sphere. For a precise description it appears useful to separate conceptually contributions due to manganese (M), ligands (L) except the substrate, and the substrate (W)[6] itself. Therefore the S_i states will be symbolized by $S_i = M_jL_kW_\ell$ with $i = j + k + l$, where $j = k = l = 0$ for S_0. This nomenclature permits a straightforward semantic distinction between metal- and ligand/substrate-centered oxidation steps in the WOC. In some cases it is of advantage to summarize the contributions of either the oxidation states of M and L, that is, $S_i = (ML)_mW_1$ with $m = j + k$, or the ligands and substrate, that is, $S_i = M_j(LW)_n$ with $n = k + l$ (for further details, see Messinger and Renger [127]). For the sake of simplicity integers are used for indices j, k, l, m, and n (e.g., $l = 0$, 1, and 2 correspond with substrate, oxo-radical, and peroxide-like species, respectively). In order to avoid misinterpretations, it must be emphasized that these indices represent only formal redox states—without giving any information on the protonation state—and even more important, they do not reflect the real charge at each atom in the Mn_4O_xCa cluster and its ligands including the substrate molecules. In fact, calculations employing density functional theory (DFT) and a charge population analysis revealed that actually Ca^{2+} carries the highest positive charge within the Mn_4O_xCa cluster although manganese attains the redox states III and IV (vide infra) [190, 191, 192]. More information on the electron density distribution within the WOC attaining the different S_i states are expected from future analyses on the basis of further spectroscopic studies in combination with advanced QM/MM calculations that include the protein environment and, in particular, the position of water molecules taking the refined 1.9 Å XRDC structure as a starting frame.

The electronic configuration of the Mn_4O_4Ca cluster can be probed by different spectroscopic methods like X-ray absorption near edge (XANES), EPR, and [55]Mn ENDOR spectroscopy (for recent reviews, see Refs. [127, 193, 194]).

In spite of numerous studies the most fundamental property—the overall oxidation state of the manganese in the Mn_4O_4Ca cluster—is still a matter of controversial discussions (see Gatt et al. [195] and references therein). Based on the interpretation

---

[6]Since at least one μ-oxo-bridge of the Mn_4O_4Ca cluster probably originates from a substrate water molecule (see Section 3.11.5) the symbol W also comprises this group (in general the μ-oxo groups will not be assigned to symbol L).

of EPR and ENDOR data the overall oxidation state of manganese in redox state S_2 is widely assumed to attain the configuration $Mn(III)Mn(IV)_3$ (see Kulik et al. [196] and references therein) with corresponding ramifications to the other S_i states (vide infra). This "overall high oxidation state" model has been questioned (see Gatt et al. [195]) and instead an "overall low oxidation state" model has been proposed where S_2 is characterized by $Mn(III)_3Mn(IV)$. Although a straightforward and unambiguous decision is still lacking so far, the vast majority of studies are in favor of the "overall high oxidation state" model and therefore this will be used exclusively in the present description.

Consensus exists on the $S_0 \rightarrow S_1$ and $S_1 \rightarrow S_2$ transitions of the WOC as to be metal (manganese) centered oxidation steps and therefore S_0 and S_1 are assigned to $Mn(III)_3Mn(IV)$ and $Mn(III)_2Mn(IV)_2$. A possible configuration $Mn(II)Mn(III)Mn(IV)_2$ for S_0 was reported to be highly unlikely [196]. In marked contrast, the nature of the redox step $S_2 \rightarrow S_3$ is not yet clarified: it could be either a metal- or a ligand-centered reaction (see Messinger and Renger [127] and references therein). The electronic configuration and nuclear geometry of S_3 are of key relevance for all considerations on the mechanism of O–O bond formation. Therefore this topic will be discussed in Section 3.11.5.

For a deeper understanding of the metal-centered reactions in the WOC it is essential to know whether or not the oxidation steps are localized on individual manganese ions or delocalized over the entire cluster. Interpretation of experimental ^{55}Mn ENDOR data monitored at 10 K on samples containing 3% methanol suggests that the oxidation states are localized. Analyses within the framework of a structural model of the Mn_4O_xCa cluster gathered from EXAFS studies on single crystals of *T. elongatus* [197] led to the following conclusions on the electronic structure of the S_i states [196]: $S_0 = Mn1(IV)Mn2(III)Mn3(IV)Mn4(III)$, $S_1 = Mn1(IV)Mn2(III)Mn3(IV)Mn4(III)$ and $S_2 = Mn1(IV)Mn2(IV)Mn3(IV)Mn4(III)$ (for numbering of manganese, see Fig. 3.7). On the other hand, a configuration of $S_1 = Mn1(III)Mn2(III)Mn3(IV)Mn4(IV)$ has been obtained by DFT calculations on the basis of the 1.9 Å XRDC structure [198]. As an alternative to this picture of localized oxidation states at individual Mn ions, X-ray resonant Raman scattering experiments suggest that the electron removed from the manganese during the $S_1 \rightarrow S_2$ transition probably originates from a delocalized orbital [199]. At present, extent and mode of delocalization and its possible variation with temperature are not known. In this respect it must be kept in mind that at low temperatures (10 K) an energetic difference of a few meV in the midpoint potential of the individual manganese ions is sufficient to give rise to localized oxidation states, whereas at room temperature this gap must exceed 50–100 meV for the same phenomenon. This problem is of high relevance for mechanistic consideration because the function of the WOC is completely blocked below characteristic threshold temperatures and intimately linked to protein dynamics (see Section 3.11.5). It is most important to clarify to what extent conclusions gathered from low temperature experiments really reflect the properties of a functionally competent WOC (for a discussion, see Renger [200]). A critical survey of the literature reveals that further detailed studies are required to find a satisfying answer.

3.11.3 Kinetics of Oxidative Water Splitting in the WOC

The kinetics of individual oxidation steps in the WOC and of the concomitant reduction of Y_z^{ox} have been resolved by measurements of flash-induced UV absorption changes and time resolved EPR studies, respectively (for a review, see Renger [137] and references therein). Table 3.1 compiles typical results obtained.

The oxidation steps $Y_z^{ox}S_i \rightarrow Y_zS_{i+1,i} + n_iH^+$ with $i = 0$, 1, and 2, respectively, exhibit virtually monoexponential kinetics with half-life times of 50–100 μs and about 200 μs for $i = 1$ and 2, respectively. Controversial values of about 50 μs[201] and 250 μs[202] were reported for the reaction with $i = 0$. A value of 250 μs appears to be more reliable [140]. This conclusion is confirmed by a new FTIR study [203].

In marked contrast to these reactions, the kinetics of the transition $Y_z^{ox}S_3 \rightarrow \rightarrow \rightarrow$ $Y_zS_0 + n_3H^+ + O_2$ appear to be more complex with a sigmoidal time course; comprising a lag phase that was interpreted as the reflection of a proton release that precedes the electron transfer from the Mn_4O_xCa cluster to Y_z^{ox} [202]. A precise determination of the lag phase, however, is difficult to gather from measurements of flash-induced absorption changes due to the inherent presence of misses and double hits in the S_i state transitions (see Kobayashi et al. [104]), which give rise to overlapping absorption changes (for an analysis, see Renger [137]). Comparative measurements of the kinetics of Y_z^{ox} reduction monitored via EPR spectroscopy and of O_2 release detected polarographically led to the conclusion that the lag phase should not be longer than 50 μs under physiological conditions but becomes extended when these change [204]. The extent of the lag phase was reported to be strongly dependent on pH [205]. Significant progress in unraveling the nature of the lag phase has been achieved by a detailed study on flash-induced transient IR changes, which clearly showed that this feature comprises both deprotonation and protein relaxation [206].

TABLE 3.1   Half-life Times $t_{1/2}$ (μs) of Y_z^{ox} Reduction at 8–10 °C and of S_i State Oxidation at 20 °C in *Spinacea oleracea* at 25 °C in *T. vulcanus* and Activation Energies E_A (kJ/mol) of the Reactions $Y_z^{ox}S_i \rightarrow Y_zS_{i+1-4\delta_{i3}} + \delta_{i3}O_2 + n_iH$ ($\delta_{i3} = 1$ for $i = 3$, otherwise zero)

Reaction	Y_z^{ox} Reduction		S_i State Oxidation					
Species	*Spinacea oleracea*		*Spinacela oleracea*				*T. vulcanus*	
Preparation	Thylakoids[a]	PSII m.f[a].	PSII m.f[b]		PSII core[c]		PSII core[d]	
Temperature	8.10 °C	8·10 °C	20 °C		20 °C		25 °C	
	$t_{1/2}$	$t_{1/2}$	$t_{1/2}$	E_A	$t_{1/2}$	E_A	$t_{1/2}$	E_A
$i = 0$	40–60	70	50	5	n.d.	n.d.	n.d.	n.d.
$i = 1$	85	110	100	12.0	75	14.8	70	9.6
$i = 2$	140	180	220	36.0	225	35.0	~130	26.8
$i = 3$	750	1400	1300	20/46	4100	21/67	1300	~16/60

[a]Razeghifard and Pace (1999).
[b]Renger and Hanssum (1992).
[c]Karge et al. (1997).
[d]Koike et al. (1987).

In addition to the peculiarity of a lag phase, the kinetics of Y_z^{ox} reduction by WOC in redox state S_3 are significantly slower (by factors of 5–15) and characterized by a markedly stronger dependence of the half-life times on different conditions compared to the other redox steps in the WOC (for details, see Messinger and Renger [127] and references therein).

A comparison of the kinetic parameters (rate constants, activation energies) led to the important conclusion that the reaction coordinate remained virtually invariant to evolutionary development from cyanobacteria up to the level of higher plants [207]. This finding shows that the WOC of PSII is a singular and unique entity in all oxygen evolving photosynthetic organisms.

3.11.4 Substrate/Product Pathways

The overall rate of the linear electron transport chain is limited by the PQH_2 re-oxidation at the Cyt b_6f complex with a rate constant of the order of $(10 \text{ ms})^{-1}$ [208]. Therefore a turnover number of oxidative water splitting of $>100 \text{ s}^{-1}$ under bright sunlight irradiation requires not only fast kinetics of the redox transitions at the catalytic Mn_4O_xCa cluster but also a sufficiently rapid uptake of substrate water and the correspondingly fast release of the two products O_2 and $4H^+$. The Mn_4O_xCa cluster is protected from dissipative reduction of the higher oxidation states (S_2, S_3) by a proteinaceous shield to the aqueous lumenal phase. In all oxygen-evolving organisms a barrier is formed by several extrinsic proteins [209, 210, 211] and references therein) that could also affect substrate/product pathways.

Rates of substrate water exchange with the aqueous bulk phase have been gathered from measurements of $^{16}H_2O/^{18}H_2O$ replacement. Biphasic kinetics with rate constants depending on the S_i states were obtained reflecting two different water molecules [212, 213]. The rate constants obtained for the slowly exchanging water are smaller by at least one order of magnitude than the rate limiting step of the electron transport chain [208]. Accordingly this $^{16}H_2O/^{18}H_2O$ turnover reflects neither the kinetics of substrate transport nor the rapid substrate/product replacement reaction $M_0L_0W_4 + 2H_2O \Leftrightarrow M_0L_0W_0 + O_2 + n_4H^+$. It is rather indicative of comparatively slow exchange reactions of water molecules around the Mn_4O_xCa cluster in the intermediary redox states S_0, S_1, S_2, and S_3. This "equilibration" is likely to be affected by the nature of hydrogen bond networks so that a retardation rather than acceleration of these exchange rates after removal of the extrinsic proteins PsbP and PsbQ [212, 213] can be rationalized.

Based on results obtained for the accessibility of the Mn_4O_xCa cluster to small hydrophilic reducing molecules (NH_2OH, NH_2NH_2) the substrate water transport to the catalytic site was estimated to occur within a few milliseconds [214], thus satisfying the conditions for high O_2 evolution rates under saturating continuous wave (CW) illumination. This estimation is in line with new theoretical model calculations (vide infra).

For further considerations it is important to note that in the case of oxidative water splitting the substrate concentration of the environment is very high (on the order of 55 M) and the product concentration is rather low (on the order of 250

μM in air saturated water at room temperature). On the contrary, in cytochrome c oxidase, which catalyzes the reaction in the opposite direction (O_2 reduction to H_2O coupled with a proton pump, see Renger and Ludwig [5]) at a binuclear transition metal (Fe–Cu) cluster consisting of heme a_3 and a Cu_B site, which is also buried in a protein matrix, the situation is reversed: that is, substrate concentration of the environment is rather low and the product concentration rather high. Therefore it appears worthwhile to compare the properties of the substrate/product transport in these enzymes.

A putative channel was found for the transport of the substrate O_2 to the catalytic binuclear heme a_3–Cu_B site in cytochrome c oxidase [215]. Therefore an analogous pathway was speculated to exist in the WOC for release of the product O_2 [89, 216]. Progress in resolution of the PSII structure has paved the way for an analysis of this idea (vide infra).

Crystallographic studies on Kr and Xe binding in cytochrome oxidase ba_3 from *Thermus thermophilus* revealed the existence of a continuous Y-shaped channel that is lined by hydrophobic amino acid residues, extending from the protein surface to the catalytic binuclear heme a_3– Cu_B center. Based on these data, Luna et al. [217] proposed a concerted sequence for the transfer steps, where the water molecule formed at the catalytic site is repelled by a hydrophobic surface of the O_2 channel. The channeling is different in cytochrome c oxidases aa_3 and probably requires some protein motions/rearrangements to permit access of O_2 to the active site. This mode of substrate channeling to the catalytic site may provide adaptation to environments with different saturation levels of O_2 [217]. On the contrary, in PSII the removal of the O_2 product from the catalytic site is probably of key relevance in order not only to permit rapid substrate/product exchange but also to diminish the probability of $^1\Delta_g O_2$ formation by hampering the contact of O_2 with the $^3P680$ state that is populated via $P680^{+\bullet}Q_A^{-\bullet}$ recombination in PSII (see Section 3.8).

The energetics of substrate/product exchange in the WOC has been concluded to be a highly exergonic process with an estimated Gibbs free energy of -100 to -200 meV [218]. Recent studies on the back pressure effect of molecular oxygen, however, led to the conclusion that the reaction sequence

$$Y_z^{ox}S_3 \Leftrightarrow [Y_zS_2(H_xO_2)^*] \Leftrightarrow Y_zS_0 + O_2 + n_3H^+ \qquad (3.10)$$

is only slightly exergonic, including the product/substrate exchange [219, 220]. This feature would imply that the population probability of the S_4 form $[S_2(H_xO_2)^*]$ at atmospheric oxygen pressure (index x symbolizes the unknown protonation state of the postulated complexed (*) peroxide $[S_2(H_xO_2)^*]$ in the S_4 state) is non-negligible and consequently a potential for H_2O_2 formation would exist. In this case the WOC could activate O_2 for inducing oxidative reactions. Furthermore, limitations for the extent of O_2 in the atmosphere could emerge as was discussed in a previous study [220]. However, the conclusion on a small Gibbs free energy gap between states $Y_z^{ox} S_3$ and $S_0 + O_2$ turned out to be wrong. It was questioned at first on the basis of fluorescence measurements [221]. New mass spectrometric data directly and unambiguously revealed that oxygen evolution is not suppressed by high O_2 pressure and that $\Delta G°$ is indeed rather large (on the order of -200 meV) [222] as originally proposed [218]. These energetics prevent an oxidase function of the WOC. The possible contribution of the O_2 transport pathway(s) to the unidirectional catalysis of oxidative water splitting in the WOC [89] remains to be clarified.

Four protons are the other product of oxidative water splitting to O_2. Analyses of the release pattern led to the conclusion that the oxidation steps of S_0 and S_2 are each coupled with the liberation of one proton while the oxidation of S_1 is not accompanied by a deprotonation and the sequence of the terminal oxidation of S_3 by Y_z^{ox} generates O_2 under release of two protons, that is, $n_i = 1, 0, 1,$ and 2 for $i = 0, 1, 2,$ and 3, respectively (see Suzuki et al. [223] and references therein). It must be emphasized that any electron transfer between redox groups that are embedded in a protein with protonatable groups gives rise to a Bohr type effect and therefore a noninteger stoichiometry is expected [200]. This was really found to be the case [223, 224].

The "chemical" protons released from the catalytic site are probably funneled via Asp61 of polypeptide D1 [132, 225, 226] into a pathway of the extrinsic PsbO protein. The unusual titration behavior of soluble PsbO was proposed to reflect the participation of Glu and Asp residues of this protein in an exit pathway of H^+ into the lumen [227]. A more refined analysis supported this idea [228].

Progress in the resolution of the XRDC structure opened the road for theoretical studies, which led to proposals for both substrate and product transport channels [229, 230, 231, 232]. The results obtained suggest that, in the WOC, substrate uptake and product release probably take place via different pathways. Figure 3.8 shows a model for possible channels of H_2O, O_2, and H^+.

Complementary information on O_2 transport channel(s) was gathered from Fourier transform ion cyclotron resonance mass spectrometry studies on the oxidation of amino acid residues in subunits D1, D2, CP43, and CP47. The results obtained were compared with the above modeling data of channels based on XRDC structure studies and the possible role of dynamic processes in the protein matrix for H_2O and O_2 transport has been emphasized [233].

Theoretical MD simulations using the method of multiple steered molecular dynamics led to information on the dynamics of the water transport to the Mn_4O_xCa cluster. It was concluded that H_2O access to the catalytic site occurs in a controlled way in order to avoid disadvantageous events that could lead to loss of Cl^- and/or Ca^{2+}. A transport rate of 5000 H_2O/s was obtained [234]. This finding shows that the substrate transport is sufficiently fast to avoid any kinetic limitations in the steady state turnover of the WOC under saturating light.

3.11.5   Mechanism of Oxidative Water Splitting

Important information on the mechanism of the redox steps in the WOC can be gathered from a comparison of experimental rate constants $k_{i+1,i}^{exp}$ with rate constants

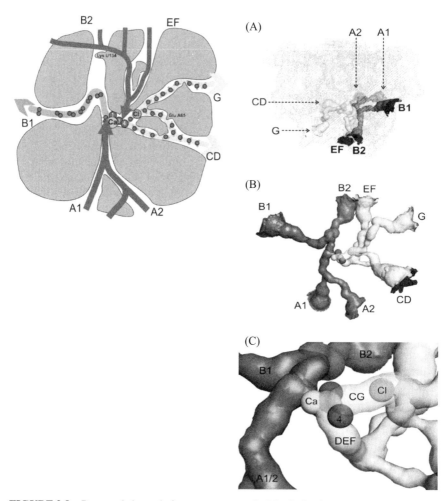

FIGURE 3.8 Proposed channels for water entry to the Mn_4O_xCa cluster (manganese marked by numbers) and for O_2 release. Figure provided by Athina Zouni and reproduced from Guskov et al. [132] with permission of John Wiley & Sons, Inc. and from F. Müh, A. Zouni, Light-induced water oxidation in photosystem II, Frontiers in Bioscience 16 (2011) 3072–3132, with permission of Frontiers in Bioscience.

$k_{i+1,i}^{NET}$ of nonadiabatic electron transfer (NET) calculated on the basis of the 1.9 Å XRDC structure by using the empirical rate constant–distance relationship [145]

$$\log k_{i+1,i}^{NET} = 13 - (1.2 - 0.8\rho)(R_{DA} - 3.6) - 3.1(\Delta G_{i+1,i}^0 + \lambda_{i+1,i})^2/\lambda_{i+1,i}$$

$$(3.11)$$

where R_{DA} = edge-to-edge distance between the reactants, ρ = packing density of protein atoms between the redox centers (ρ varies between 0 for vacuum and 1 for full package), $\Delta G_{i+1,i}^0 =$ Gibbs free energy gap between Y_z/Y_z^{ox} and S_i/S_{i+1}, and $\lambda_{i+1,i} =$ reorganization energy of the reaction.

The values of $(\Delta G_{i+1,i}^0 + \lambda_{i+1,i})^2/\lambda_{i+1,i}$ in Eq. 3.11 are gathered from the experimental activation energies $E_{A,i+1/i}$ (see Renger et al. [235]):

$$(\Delta G_{i+1,i}^0 + \lambda_{i+1,i})^2/\lambda_{i+1,i} = 4(E_{A;i+1,i} + 0.5\,RT) \qquad (3.12)$$

Rate constants on the order of 10^9 s^{-1} were obtained for $k_{2,1}^{NET}$ [5] when using a typical packing density of 0.76 in proteins containing redox active cofactors [145], the longest distance between Y_z and the redox active manganese $R_{DA} = 8.04$ Å [134, 138] and an activation energy of 10 kJ/mol for $Y_z^{ox}S_1 \to Y_zS_2 + n_1H^+$ [146, 236, 237]. It must be emphasized that 10^9 s^{-1} is the minimum number of $k_{2,1}^{NET}$ because the rate constants calculated for shorter distances R_{DA} are even larger. A number on the order of 10^9 s^{-1} for $k_{2,1}^{NET}$ exceeds the experimental value $k_{2,1}^{exp}$ by a factor of about 10^5 (see Table 3.1). Analogously, a rate constant $k_{3,2}^{NET}$ of about 10^6 s^{-1} for the reaction $Y_z^{ox}S_2 \to Y_zS_3 + n_2H^+$ results from calculation performed with $E_{A,3,2} = 36$ kJ/mol. This value is still larger by a factor of more than a hundred compared to the experimental $k_{3,2}^{exp}$ [125].[7]

The huge differences between $k_{i+1,i}^{exp}$ and $k_{i+1,i}^{NET}$ gathered from this analysis unambiguously show that the electron transfer is not the rate-limiting step of the S_i state transitions up to S_3 and therefore these processes have to be described by a sequential reaction of the type

$$Y_z^{ox}S_i \xrightarrow{k_{i+1,i}^{trigg}} (Y_z^{ox}S_i)^* \xrightarrow{k_{i+1,i}^{NET}} Y_zS_{i+1} + n_iH^+ \qquad (3.13)$$

for $i = 0$, 1, and 2 (the properties of the reaction $Y_z^{ox}S_3 \to\to Y_zS_0 + O_2 + n_3H^+$ will be discussed separately), where $k_{i+1,i}^{trigg}$ is the rate constant of a triggering process that precedes the electron transfer step and kinetically limits the overall reaction, that is, $k_{i+1,i}^{trigg} << k_{i+1,i}^{NET}$; $(Y_z^{ox}S_i)^*$ represents a "triggered" redox state S_i.

In contrast to the apparent monoexponential kinetics (vide supra) reflecting the rate-limiting triggering reactions for $i = 0$, 1, and 2, the time course of Y_z^{ox} reduction by the WOC in redox state S_3 is sigmoidal and the overall rate prone to variation by different conditions (vide supra). This behavior can basically be described by the following simplified reaction sequence:

$$Y_z^{ox}S_3 \xrightarrow{k_{3*,3}^{trigg}} \left(Y_z^{ox}S_3\right)^* \xrightarrow{k1,3} Y_z[S_2(H_xO_2)^*] \to \to \to Y_zS_0 + O_2 + n_3H^+ \qquad (3.14)$$

---

[7]No reliable information is available for $k_{1,0}^{NET}$ because the only value reported for $E_{A;1,0}$ [236] is most likely too small. However $E_{A;1,0}$ is expected to be $\leq E_{A,3,2}$ so that $k_{1,0}^{NET}$ is at least 10^6 s^{-1}, thus exceeding $k_{1,0}^{exp}$ by more than two orders of magnitude.

where $[S_2(H_xO_2)^*]$ reflects complexed peroxide in S_4 (vide infra) and n_3 the overall H^+ release without specifying the individual step(s) of deprotonation. Furthermore, for the sake of simplicity, the last steps of formation of complexed dioxygen ($[S_0(O_2)^*]$) and water/O_2 exchange (see Eq. 3.10) is omitted and back reactions are not explicitly shown.

Two different types of rate limitation of the sequence in Eq. 3.14 can be considered when taking into account the similar time course of Y_z^{ox} reduction by the WOC in S_3 and the O_2 release [204]: (1) the triggering process is slow compared to the redox reaction(s) and comprises a sequence of at least two steps, thus giving rise to the sigmoidal time course, or (2) a "fast" single triggering reaction with time constant similar to that of the other S_i state transitions is followed by an overall electron transfer, which is slower due to the participation of a fast uphill redox equilibrium characterized by a rather small constant. The available time resolved FTIR data [203] do not permit a straightforward distinction between both models. Intuitively, the second alternative appears to be more attractive and therefore this model has been analyzed by simulation of the kinetics of Y_z^{ox} reduction and O_2 release [140].

Regardless of mechanistic details of the sequences in Eq. 3.13 and Eq. 3.14, the conventional Kok scheme has to be extended as is shown in Fig. 3.9. This pattern implies that all redox transitions of the WOC are characterized by a sigmoidal

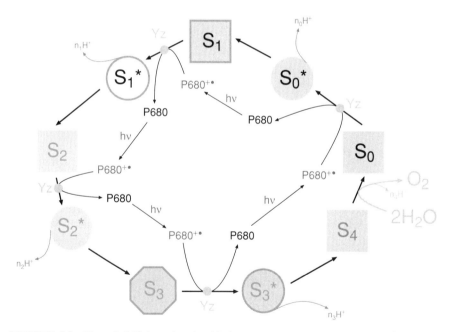

FIGURE 3.9 Extended Kok cycle of oxidative water splitting driven by P680$^{+\bullet}$ with Y_z acting as intermediate and the S_i states in their transient triggered conformation (symbolized by a star). For the sake of simplicity the "superreduced" S_{-i} states are omitted because they are not involved in the process of water oxidation. For further details see text. Reproduced from Kern and Renger [179] with permission of Springer.

time course. However, this feature cannot be resolved for the oxidation steps up to redox level S_3 because the effect is extremely small for $k_{i+1,i}^{trigg} << k_{i+1,i}^{NET}$ and therefore remains buried by the error margins originating from the signal/noise ratios. Likewise, the transient population probability of $(Y_Z^{ox}S_i)^*$ is negligibly small and therefore the triggered states $(Y_Z^{ox}S_i)^*$ escape from experimental characterization [140].

Another most important consequence emerges from the relation $k_{i+1,i}^{trigg} << k_{i+1,i}^{NET}$: the experimentally determined properties of $k_{i+1,i}^{exp}$ (activation energies, kinetic isotope effect, thermal blockage) reflect the characteristics of the trigger reaction(s) for $i = 0$, 1, and 2 rather than those of the redox step. The idea of triggered S_i state transitions is highly supported by the blockage of the reaction sequences in Eq. 3.13 and Eq. 3.14 below characteristic threshold temperatures for each S_i state [237, 238, 239]. Interestingly, an analogous blockage by freezing has also been observed for the electron transfer from $Q_A^{-\bullet}$ to Q_B at the acceptor side of PSII (and also in anoxygenic purple bacteria, see Refs. [240, 241, 242, 243]). The effect on the latter reaction was found to strictly correlate with protein dynamics as was shown by Mößbauer spectroscopy [243] and neutron scattering [244].

Figure 3.10 presents a comparison between the temperature dependence of protein dynamics and the thermal blockage of both the S_i state transitions and $Q_A^{-\bullet}$ reoxidation by Q_B in PSII membrane fragments. The data clearly show that all reactions except for the $S_1 \rightarrow S_2$ transition exhibit a strikingly similar dependence on protein dynamics [140]. The markedly lower threshold temperature for the $S_1 \rightarrow S_2$ transition compared to the redox steps of the other S_i state transitions probably originates from the different deprotonation pattern because the latter reactions are, but the former reaction is not,

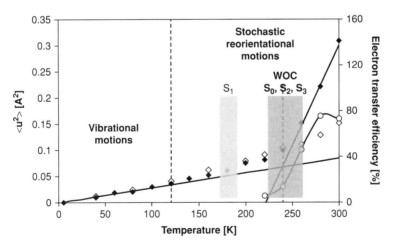

FIGURE 3.10   Temperature dependence of the average atomic mean square displacement $<u^2>$ for hydrated (full symbols) and dry PSII membrane fragments (open symbols). The temperature ranges of different mobility characteristics are separated by dashed lines. The thermal blockage of S_1 oxidation and S_0, S_2, and S_3 reactions are marked by shaded areas. For further details see text. Reproduced from Kühn et al., Photosynthesis Research 84 (2005) 317–323, with permission of Springer.

coupled with significant proton release into the lumen [223]. This feature suggests that the mode of proton shifts in the hydrogen bond network between Y_z and the Mn_4O_xCa cluster (see Fig. 3.6) is important for the trigger reactions of the WOC. In this respect it is important to note that a temperature-dependent nuclear rearrangement is still required for the $S_1 \rightarrow S_2$ transition while an analogous thermal threshold does not exist for the charge separation leading to $P680^{+\bullet}Q_A^{-\bullet}$ [245] and its dark recombination [246]. The rate of the latter reaction is limited by the nonadiabatic electron transfer (for a discussion, see Renger [103]).

So far the nature of the triggering reactions has been resolved neither for the acceptor nor for the donor side of PSII. It is proposed that rearrangements within the environment of the cofactors, in particular, of hydrogen bond network(s), enable the redox reactions of oxidative water splitting to take place at rather low overpotentials. The clarification of this essential point remains a challenging task of future research.

The cornerstone of the oxidation of two water molecules to O_2 is the covalent linkage of two oxygen atoms and therefore the mechanism of this process is a topic of high relevance and numerous research activities. The O–O bond was proposed to be formed by a population of binuclearly complexed peroxide as intermediate of oxidative water splitting in photosynthesis [217, 247]. This idea of a transient peroxidic state is supported by several theoretical studies and also by findings on ruthenium based model systems for oxidative water splitting (see Section 3.6.5) [248]. In the original model the complexed peroxide formation was assigned to redox state S_3. This hypothesis, however, is not accepted in the vast majority of currently discussed elaborated schemes, which all comprise the idea of a complexed peroxide as intermediate but assume that this state can be formed exclusively in S_4 ("S_4 dogma", see McEvoy and Brudvig [226] and references therein). Since the mechanistically most important question on the redox level (S_3 or S_4) of complexed peroxide formation has been outlined in detail in several reports [125, 249, 250], only a very brief discussion of this point will be presented here.

The "multiple S_3 state" model (for details, see Renger [125]) proposed for the transient population of a peroxidic form in S_3 is based on the postulate that redox state S_3 comprises an ensemble of Mn_4O_xCa clusters with different electronic configuration and nuclear geometry (symbolized by $S_3(W) = M_3L_0W_0$, $S_3(O) = M_2L_0W_1$, and $S_3(P) = M_1L_0W_2$; for nomenclature, see Section 3.11.2), which rapidly equilibrate (on the order of a few milliseconds or faster). The configuration $S_3(P) = M_1L_0W_2$ corresponding with a complexed peroxide is also postulated to be the "entatic state" for oxidation by Y_z^{ox}, giving rise to the reaction sequence in Eq. 3.14. As a consequence, the WOC is blocked in O_2 evolution when the population of $S_3(P)$ is prevented. A kinetic analysis revealed that the population probability of $S_3(P)$ is rather small ($<$ 5%) and therefore this state is expected to be extremely difficult to detect [140].

On the contrary, all "S_4 dogma" models assume that S_3 attains a single electronic configuration and nuclear geometry either in the form $S_3(W) = M_3L_0W_0$ ("Mn only" hypothesis) or in form of $S_3(O) = M_2(LW)_1$ ("Oxo radical" hypothesis). The Y_z^{ox} induced oxidation of either $S_3(W)$ or $S_3(O)$ is assumed to give rise to formation of the complexed peroxide $S_4(P) = Mn_2L_0W_2$ followed by the reaction(s) leading to O_2 release.

So far sound experimental evidence exists neither for the "S_4 dogma" models nor for the "multiple S_3 state" model. Progress in addressing this key question on the mechanism of O–O bond formation could be expected by using advanced theoretical analyses based on quantum mechanics/molecular mechanics (QM/MM) methodology. The studies reported so far led to different proposal on both redox level and mechanism of covalent linkage of two oxygen atoms at the Mn_4O_xCa cluster. Analyses on the basis of the "S_4 dogma" suggest that two oxo radicals in S_4, including a bridging oxo group, are linked together to a peroxide binuclearly complexed by two Mn [177, 251] or that a substrate water bound either to Mn or Ca reacts via a nucleophilic attack on a Mn(V)=O group (see McEvoy and Brudvig [226] and references therein). Alternatively, DFT calculations also suggest that the covalent O–O linkage can already take place in S_3 [252]. In this case the state Mn1(III)Mn2(IV)Mn3(IV)Mn4(IV)–O•H is assumed to be populated, which transfers to a peroxidic configuration Mn1(III)Mn2(III)Mn3(IV)Mn4(III)–O–OHCa, where the peroxide is also binuclearly complexed but in this case by Mn4 and Ca. The oxidation of this state by Y_Z^{ox} is inferred to lead eventually to ground state triplet O_2 via transient formation of complexed superoxide.

When discussing the mechanism of O–O bond formation another most important point has to be taken into account—the role of protons. The key role of proton accepting group(s) for the energetics of water oxidation has been outlined in former studies [253]. A wealth of information is available on the functional relevance of the mode of protonation for processes running in the opposite direction, that is, cleavage of the O–O bond in oxygen enzymology [254, 255, 256, 257, 258, 259] and in model systems [260, 261]. It is therefore expected that analogous effects exist also for the chemical linkage of two oxygen atoms [137]. In fact, theoretical calculations on the reaction pathway of water oxidation at a mononuclear Ru catalyst reveal that the coupling with a proton shift is essential for O–O bond formation [262].

At present no detailed information is available on the hydrogen bond network in redox state S_3 (see Section 3.10). Likewise, the mode of coupling with proton transfer steps is not specified. For this reason the essential role of protons is discussed only in a generalized manner by introducing two assumptions (for details see Renger [137]): (1) in redox state S_3 the local proton gradient $\nabla H^+(\vec{r},t,S_3)$ in the neighborhood of the two substrate oxygen atoms is the key player for linking them together to an O–O bond, and (2) the dependence on space vector $\vec{r}$ determines the population probability of the three S_3 state configurations ($M_3L_0W_0$, $M_2(LW)_1$, and $M_1(LW)_2$), and the time dependence due to the dynamics of the hydrogen bond network modulates the fast equilibration between these electronic configurations of redox state S_3. The coupling of the Mn_4O_xCa cluster with the protein matrix via interaction of μ-oxo groups with protons could also be of functional relevance. DFT calculations revealed that hydrogen bonding between D1-His 337 and the μ-oxo-bridge atom O5 (see Fig. 3.7) can give rise to an increase of the distance between Mn centers [263]. In this way the O–O bond formation could be tuned via protein dynamics.

In addition to effects on $\nabla H^+(\vec{r},t,S_3)$, structural changes of the ligand geometry can also markedly reorient the redox active orbital of the manganese, thus modulating the reactivity by orders of magnitude (for a general discussion on the modulation of the properties of metal–oxo complexes by ligand fields, see Betley et al. [264]). An illustrative example of this effect is the tuning of the reactivity of the binuclear nonheme iron center in the catalytic site of some oxygenases by reorientation of a Glu ligand due to the binding of a "regulator" protein [265].

In spite of lacking detailed information it is clear that protein dynamics and flexibility modulate the scalar proton field $H^+(\vec{r},t)$ around the Mn_4O_xCa cluster and Y_z^{ox} in all S_i states. This conclusion offers an explanation for the essential functional role of protein dynamics and flexibility for a functionally competent WOC as reflected by the thermal blockage of the S_i state transitions below characteristic threshold temperatures (vide supra). Likewise a decrease of sample hydration leads to suppression of these reactions at room temperature [266].

3.12 CONCLUDING REMARKS

This chapter has described principles of solar energy exploitation by water splitting, with special emphasis on the performance of this process in oxygen evolving photosynthetic organisms. During a long-lasting evolutionary process nature has "constructed" an optimized nanoscale machine. This fascinating unique device offers an excellent blueprint in stimulating our considerations on the development of artificial systems for large-scale and efficient use of solar radiation as a free energy source. The most delicate part of the overall process is the oxidation of two water molecules leading to O_2 formation and H^+ release. Therefore the discussions in this chapter focus on problems of this reaction and how it is performed in the multimeric photosystem II (PSII) core complex of the photosynthetic apparatus. It is emphasized that living organisms can use the enormous potential of highly specialized biopolymers (i.e., proteins)in tuning energetics and kinetics of the cofactors as the functional sites of biological catalysis. Of special relevance for oxidative water splitting, taking place at a Mn_4O_xCa cluster, is an optimized coupling between electron and proton transfer steps. The protein matrix of PSII is postulated to provide a dynamic scalar field of proton activity, which is of essential functional relevance. As a consequence of this conclusion it seems worth considering the possibility of developing functionalized synthetic polymer matrices for both binding the catalytic sites and simultaneously providing a spatial separation of oxidative and reductive pathways of light-induced water splitting.

ACKNOWLEDGMENTS

I am very grateful to A. Zouni for a critical reading of the chapter and for providing Fig. 3.8 and to J.-R. Shen for Fig. 3.7 and the structure data used for Fig. 3.6. I would also like to thank the following people for preparing electronic versions of the figures: Susanne Renger (Figs. 3.1, 3.2, 3.4 bottom, and 3.9), Franz Josef Schmitt (Fig. 3.3), Jan Kern (Fig. 3.4 top), Philipp Kühn (Fig. 3.5), Christoph Theiss (Fig. 3.6), and Jörg Pieper (Fig. 3.10).

REFERENCES

1. H. J. Morowitz, *Foundations of Bioenergetics*, Academic Press, New York, 1978.
2. G. Nicolis, I. Prigogine, *Self-Organization in Nonequilibrium Systems: From Dissipative Structures to Order Through Fluctuations*, Wiley, Hoboken, NJ, 1977.
3. K. I. Zamaraev, V. N. Parmon, Potential methods and perspectives of solar-energy conversion via photocatalytic processes, Catal. Rev. Sci. Eng. 22 (1980) 261–324.
4. L. Boltzmann, *Der zweite Hauptsatz der mechanischen Wärmetheorie Populäre Schriften*, J. A. Barth, Leipzig (in German), 1905.
5. G. Renger, B. Ludwig, Mechanism of photosynthetic production and respiratory reduction of molecular dioxygen: a biophysical and biochemical comparison. In: G. Peschek, C. Obinger, G. Renger (Eds.), *Bioenergetic Processes of Cyanobacteria: From Evolutionary Singularity to Ecological Diversity*, Springer, Dordrecht, The Netherlands, 2011, pp. 337–394.
6. C. B. van Niel, The bacterial photosynthesis and their importance of the general problem of photosynthesis, Acta. Enzymology 1 (1941) 263–328.
7. P. Mitchell, Coupling of photophosphorylation to electron and hydrogen transfer by a chemiosmotic type of mechanism, Nature 191 (1961) 144–148.
8. P. Mitchell, Chemiosmotic. coupling in oxidative and photosynthetic phosphorylation, Biol. Rev. 41 (1966) 445–501.
9. W. Junge, Evolution of photosynthesis. In: G. Renger (Ed.), *Primary Processes of Photosynthesis: Basic Principles and Apparatus, Part II Reaction Centers/Photosystems, Electron Transport Chains, Photophosphorylation and Evolution*, RSC Publishing, Cambridge, 2008, pp. 447–487.
10. A. A. Benson, Following the path of carbon in photosynthesis: a personal story, Photosynth. Res. 73 (2002) 31–49.
11. M. Calvin, 40 Years of photosynthesis and related activities, Photosynth. Res. 21 (1989) 3–16.
12. W. W. Parson, Functional patterns of reaction centers in anoxygenic photosynthetic bacteria. In: G. Renger (Ed.), *Primary Processes of Photosynthesis: Basic Principles and Apparatus, Part II Reaction Centers/Photosystems, Electron Transport Chains, Photophosphorylation and Evolution*, RSC Publishing, Cambridge, 2008, pp. 57–109.
13. A. Vermeglio, Electron transport chain and phosphorylation. In: G. Renger (Ed.), *Primary Processes of Photosynthesis: Basic Principles and Apparatus, Part II Reaction Centers/Photosystems, Electron Transport Chains, Photophosphorylation and Evolution*, RSC Publishing, Cambridge, 2008, pp. 353–382.
14. P. W. Atkins, J. De Paula, *Atkins' Physical Chemistry*, 7th ed., Oxford University Press, New York, 2002.
15. R. Buick, The antiquity of oxygenic photosynthesis—evidence from stromatolites in sulfate-deficient archean lakes, Science 255 (1992) 74–77.
16. D. J. Des Marais, Evolution—when did photosynthesis emerge on earth?, Science 289 (2000) 1703–1705.
17. J. Xiong, C. E. Bauer, Complex evolution of photosynthesis, Annu. Rev. Plant. Biol. 53 (2002) 503-521.
18. A. Larkum, The evolution of photosynthesis. In: G. Renger (Ed.), *Primary Processes of Photosynthesis: Basic Principles and Apparatus, Part II Reaction Centers/Photosystems,*

Electron Transport Chains, Photophosphorylation and Evolution, RSC Publishing, Cambridge, 2008, pp. 491–521.

19. G. Peschek, C. Obinger, G. Renger, *Bioenergetic Processes of Cyanobacteria: From Evolutionary Singularity to Ecological Diversity*, 1st ed., Springer, Dordrecht, The Netherlands, 2011.

20. A. Bekker, H. D. Holland, P. L. Wang, D. Rumble, H. J. Stein, J. L. Hannah, L. L. Coetzee, N. J. Beukes, Dating the rise of atmospheric oxygen, Nature 427 (2004) 117–120.

21. J. F. Kasting, J. L. Siefert, Life and the evolution of earth's atmosphere, Science 296 (2002) 1066–1068.

22. N. Lane, *Oxygen: The Molecule That Made the World*, Oxford University Press, New York, 2002.

23. D. G. Nicholls, S. J. Ferguson, *Bioenergetics 2*, Academic Press, London, 1992.

24. G. Renger, Biological energy conservation. In: W. Hoppe, W. Lohmann, H. Markl, H. Ziegler (Eds.), *Biophysics*, Springer-Verlag, Berlin, 1983, pp. 347–371.

25. R. C. Worrest, M. M. Caldwell (Eds.), *Stratospheric Ozone Reduction, Solar Ultraviolet Radiation, and Plant Life*, Springer-Verlag, New York, 1986.

26. G. Ciamician, The photochemistry of the future, Science XXXVI (1912) 385–394.

27. R.M. Pearlstein, Chlorophyll singlet exciton. In: Govindjee (Ed.), *Energy Conversion by Plants and Bacteria. Photosynthesis*, vol. 1, Academic Press, New York, 1982, pp. 293–330.

28. J. K. Hurley, G. Tollin, M. Medina, C. Gomez-Moreno, Electron transfer from ferredoxin and flavodoxin to ferrredoxin-NADP$^+$ reductase. In: J. H. Golbeck (Ed.), *Photosystem I: The Light-Driven Plastocyanin:Ferredoxin Oxidoreductase*, Springer, Dordrecht, 2006, pp. 456–476.

29. P. Sétif, W. Leibl, functional pattern of photosystem I in oxygen evolvingorganisms. In: G. Renger (Ed.), *Primary Processes of Photosynthesis: Basic Principles and Apparatus, Part I Photophysical Principles, Pigments and Light Harvesting/Adaptation/Stress*, RSC Publishing, Cambridge, 2008, pp. 147–191.

30. V. P. Shinkarev, I. R. Vassiliev, J. H. Golbeck, A kinetic assessment of the sequence of electron transfer from F_X to F_A and further to F_B in photosystem I: the value of the equilibrium constant between F_X and F_A, Biophys. J. 78 (2000) 363–372.

31. P. M. Vignais, B. Billoud, Occurrence, classification, and biological function of hydrogenases: an overview, Chem. Rev. 107 (2007) 4206–4272.

32. M. L. Ghirardi, M. C. Posewitz, P. C. Maness, A. Dubini, J. P. Yu, M. Seibert, Hydrogenases and hydrogen photoproduction in oxygenic photosynthetic organisms, Annu. Rev. Plant. Biol. 58 (2007) 71–91.

33. H. Bothe, O. Schmitz, M. G. Yates, W. E. Newton, Nitrogenases and hydrogenases in cyanobacteria. In: G. Peschek, C. Obinger, G. Renger (Eds.), *Bioenergetic Processes of Cyanobacteria: From Evolutionary Singularity to Ecological Diversity*, Springer, Dordrecht, The Netherlands, 2011, pp. 137–157.

34. P. Fay, Oxygen relations of nitrogen-fixation in cyanobacteria, Microbiol. Rev. 56 (1992) 340–373.

35. Y. Mazor, H. Toporik, N. Nelson, Temperature-sensitive PSII and promiscuous PSI as a possible solution for sustainable photosynthetic hydrogen production, Biochim. Biophys. Acta (2012). doi:10.1016/j.bbabio.2012.01.005.

36. C. E. Lubner, A. M. Applegate, P. Knorzer, A. Ganago, D. A. Bryant, T. Happe, J. H. Golbeck, Solar hydrogen-producing bionanodevice outperforms natural photosynthesis, Proc. Natl. Acad. Sci. U.S.A. 108 (2011) 20988–20991.

37. O. Lenz, M. Ludwig, T. Schubert, I. Bürstel, S. Ganskow, T. Goris, A. Schwarze, B. Friedrich, H_2 conversion in the presence of O_2 as performed by the membrane-bound [NiFe]-hydrogenase of *Ralstonia eutropha*, Chem. Phys. Chem. 11 (2010) 1107–1119.

38. J. Fritsch, P. Scheerer, S. Frielingsdorf, S. Kroschinsky, B. Friedrich, O. Lenz, C. M. T. Spahn, The crystal structure of an oxygen–tolerant hydrogenase uncovers a novel iron–sulphur centre, Nature 479 (2011) 249–252.

39. R. A. Grimme, C. E. Lubner, D. A. Bryant, J. H. Golbeck, Photosystem I/molecular wire/metal nanoparticle bioconjugates for the photocatalytic production of H_2, J. Am. Chem. Soc. 130 (2008) 6308–6309.

40. L. M. Utschig, N. M. Dimitrijevic, O. G. Poluektov, S. D. Chemerisov, K. L. Mulfort, D. M. Tiede, Photocatalytic hydrogen production from noncovalent biohybrid photosystem I/Pt nanoparticle complexes, J. Phys. Chem. Lett. 2 (2011) 236–241.

41. L. M. Utschig, S. C. Silver, K. L. Mulfort, D. M. Tiede, Nature-driven photochemistry for catalytic solar hydrogen production: a photosystem I–transition metal catalyst hybrid, J. Am. Chem. Soc. 133 (2011) 16334–16337.

42. P. D. Tran, V. Artero, M. Fontecave, Water electrolysis and photoelectrolysis on electrodes engineered using biological and bio-inspired molecular systems, Energy Environ. Sci. 3 (2010) 727–747.

43. M. L. Singleton, J. H. Reibenspies, M. Y. Darensbourg, A cyclodextrin host/guest approach to a hydrogenase active site biomimetic cavity, J. Am. Chem. Soc. 132 (2010) 8870–8871.

44. F. Quentel, G. Passard, F. Gloaguen, Electrochemical hydrogen production in aqueous micellar solution by a diiron benzenedithiolate complex relevant to [FeFe] hydrogenases, Energy Environ. Sci. 5 (2012) 7757–7761.

45. K. A. Brown, M. B. Wilker, M. Boehm, G. Dukovic, P. W. King, Characterization of photochemical processes for H_2 production by CdS nanorod–[FeFe] hydrogenase Complexes, J. Am. Chem. Soc. 134 (2012) 5627–5636.

46. E. Reisner, J. C. Fontecilla-Camps, F. A. Armstrong, Catalytic electrochemistry of a [NiFeSe]-hydrogenase on TiO_2 and demonstration of its suitability for visible-light driven H_2 production, Chem. Commun. 5 (2009) 550–552.

47. E. Reisner, D. J. Powell, C. Cavazza, J. C. Fontecilla-Camps, F. A. Armstrong, Visible light-driven H_2 production by hydrogenases attached to dye-sensitized TiO_2 nanoparticles, J. Am. Chem. Soc. 131 (2009)18457–18466.

48. T. Noji, H. Suzuki, T. Gotoh, M. Iwai, M. Ikeuchi, T. Tomo, T. Noguchi, Photosystem II–gold nanoparticle conjugate as a nanodevice for the development of artificial light-driven water-splitting systems, J. Phys. Chem. Lett. 2 (2011) 2448–2452.

49. M. Kato, T. Cardona, A. W. Rutherford, E. Reisner, Photoelectrochemical water oxidation with photosystem II integrated in a mesoporous indium–tin oxide electrode, J. Am. Chem. Soc. 134 (2012) 8332–8335.

50. G. Renger, The light reactions of photosynthesis, Current Science 98 (2010) 1305–1319.

51. D. Gust, T. A. Moore, A. L. Moore, Molecular mimicry of photosynthetic energy and electron-transfer, Acc. Chem. Res. 26 (1993) 198–205.

52. A. Osuka, S. Marumo, N. Mataga, S. Taniguchi, T. Okada, I. Yamazaki, Y. Nishimura, T. Ohno, K. Nozaki, A stepwise electron-transfer relay mimicking the primary charge separation in bacterial photosynthetic reaction center, J. Am. Chem. Soc. 118 (1996) 155–168.

53. M. R. Wasielewski, Photoinduced electron-transfer in supramolecular systems for artificial photosynthesis, Chem. Rev. 92 (1992) 435–461.

54. V. Garg, G. Kodis, M. Chachisvilis, M. Hambourger, A. L. Moore, T.A. Moore, D. Gust, Conformationally constrained macrocyclic diporphyrin-fullerene artificial photosynthetic reaction center, J. Am. Chem. Soc. 133 (2011) 2944–2954.

55. G. Kodis, Y. Terazono, P. A. Liddell, J. Andreasson, V. Garg, M. Hambourger, T. A. Moore, A. L. Moore, D. Gust, Energy and photoinduced electron transfer in a wheel-shaped artificial photosynthetic antenna-reaction center complex, J. Am. Chem. Soc. 128 (2006) 1818–1827.

56. B. R. Green, W. W. Parson, *Light-Harvesting Antennas in Photosynthesis*, Kluwer Academic, Dordrecht, The Netherlands, 2003.

57. G. Renger (Ed.), *Primary Processes of Photosynthesis: Basic Principles and Apparatus, Part I Photophysical Principles, Pigments and Light Harvesting/Adaptation/Stress*, RSC Publishing, Cambridge, 2008.

58. A. Harriman, Energy transfer in synthetic porphyrin arrays. In: V. Balzani (Ed.), *Supramolecular Photochemistry*, D. Reidel, Boston, 1987, pp. 207–223.

59. J. Iehl, J. F. Nierengarten, A. Harriman, T. Bura, R. Ziessel, Artificial light-harvesting arrays: electronic energy migration and trapping on a sphere and between spheres, J. Am. Chem. Soc. 134 (2012) 988–998.

60. J. Z. Li, J. R. Diers, J. Seth, S. I. Yang, D. F. Bocian, D. Holten, J. S. Lindsey, Synthesis and properties of star-shaped multiporphyrin-phthalocyanine light-harvesting arrays, J. Org. Chem. 64 (1999) 9090–9100.

61. A. Morandeira, E. Vauthey, A. Schuwey, A. Gossauer, Ultrafast excited state dynamics of tri- and hexaporphyrin arrays, J. Phys. Chem. A 108 (2004) 5741–5751.

62. J. K. Sprafke, D. V. Kondratuk, M. Wykes, A. L. Thompson, M. Hoffmann, R. Drevinskas, W. H. Chen, C. K. Yong, J. Karnbratt, J. E. Bullock, M. Malfois, M. R. Wasielewski, B. Albinsson, L. M. Herz, D. Zigmantas, D. Beljonne, H. L. Anderson, Belt-shaped pi-systems: relating geometry to electronic structure in a six-porphyrin nanoring, J. Am. Chem. Soc. 133 (2011) 17262–17273.

63. A. Uetomo, M. Kozaki, S. Suzuki, K. Yamanaka, O. Ito, K. Okada, Efficient light-harvesting antenna with a multi-porphyrin cascade, J. Am. Chem. Soc. 133 (2011) 13276–13279.

64. M. Kloz, S. Pillai, G. Kodis, D. Gust, T. A. Moore, A. L. Moore, R. van Grondelle, J. T. M. Kennis, Carotenoid photoprotection in artificial photosynthetic antennas, J. Am. Chem. Soc. 133 (2011) 7007–7015.

65. Y. Terazono, G. Kodis, K. Bhushan, J. Zaks, C. Madden, A. L. Moore, T. A. Moore, G. R. Fleming, D. Gust, Mimicking the role of the antenna in photosynthetic photoprotection, J. Am. Chem. Soc. 133 (2011) 2916–2922.

66. P. K. Dutta, R. Varghese, J. Nangreave, S. Lin, H. Yan, Y. Liu, DNA-directed artificial light-harvesting antenna, J. Am. Chem. Soc. 133 (2011) 11985–11993.

67. J. W. Springer, P. S. Parkes-Loach, K. R. Reddy, M. Krayer, J. Y. Jiao, G. M. Lee, D. M. Niedzwiedzki, M. A. Harris, C. Kirmaier, D. F. Bocian, J. S. Lindsey, D. Holten,

P. A. Loach, Biohybrid photosynthetic antenna complexes for enhanced light-harvesting, J. Am. Chem. Soc. 134 (2012) 4589–4599.

68. A. B. Anderson, T. V. Albu, Ab initio determination of reversible potentials and activation energies for outer-sphere oxygen reduction to water and the reverse oxidation reaction, J. Am. Chem. Soc. 121 (1999) 11855–11863.

69. X. Liu, F. Wang, Transition metal complexes that catalyze oxygen formation from water: 1979–2010, Coord. Chem. Rev. 256 (2012) 1115–1136.

70. D. Moonshiram, J. W. Jurss, J. J. Concepcion, T. Zakharova, I. Alperovich, T. J. Meyer, Y. Pushkar, Structure and electronic configurations of the intermediates of water oxidation in blue ruthenium dimer catalysis, J. Am. Chem. Soc. 134 (2012) 4625–4636.

71. J. J. Concepcion, M. K. Tsai, J. T. Muckerman, T. J. Meyer, Mechanism of water oxidation by single-site ruthenium complex catalysts, J. Am. Chem. Soc. 132 (2010) 1545–1557.

72. Z. F. Chen, A. K. Vannucci, J. J. Concepcion, J. W. Jurss, T. J. Meyer, Proton-coupled electron transfer at modified electrodes by multiple pathways, Proc. Natl. Acad. Sci. U.S.A. 108 (2011) E1461–E1469.

73. Y. Lee, J. Suntivich, K.J. May, E.E. Perry, Y. Shao-Horn, Synthesis and activities of rutile IrO_2 and RuO_2 nanoparticles for oxygen evolution in acid and alkaline solutions, J. Phys. Chem. Lett. 3 (2012) 399–404.

74. N. D. Schley, J. D. Blakemore, N. K. Subbaiyan, C. D. Incarvito, F. D'Souza, R. H. Crabtree, G. W. Brudvig, Distinguishing homogeneous from heterogeneous catalysis in electrode-driven water oxidation with molecular iridium complexes, J. Am. Chem. Soc. 133 (2011) 10473–10481.

75. C. Y. Cummings, F. Marken, L. M. Peter, K. G. U. Wijayantha, A. A. Tahir, New Insights into water splitting at mesoporous α-Fe_2O_3 films: a study by modulated transmittance and impedance spectroscopies, J. Am. Chem. Soc. 134 (2012) 1228–1234.

76. N. S. McCool, D. M. Robinson, J. E. Sheats, G. C. Dismukes, A Co_4O_4 "cubane" water oxidation catalyst inspired by photosynthesis, J. Am. Chem. Soc. 133 (2011) 11446–11449.

77. Y. Surendranath, M. Dinca, D. G. Nocera, Electrolyte-dependent electrosynthesis and activity of cobalt-based water oxidation catalysts, J. Am. Chem. Soc. 131 (2009) 2615–2620.

78. Q. S. Yin, J. M. Tan, C. Besson, Y. V. Geletii, D. G. Musaev, A. E. Kuznetsov, Z. Luo, K. I. Hardcastle, C. L. Hill, a fast soluble carbon-free molecular water oxidation catalyst based on abundant metals, Science. 328 (2010) 342–345.

79. D. B. Grotjahn, D. B. Brown, J. K. Martin, D. C. Marelius, M. C. Abadjian, H. N. Tran, G. Kalyuzhny, K. S. Vecchio, Z. G. Specht, S. A. Cortes-Llamas, V. Miranda-Soto, C. van Niekerk, C. E. Moore, A. L. Rheingold, Evolution of iridium-based molecular catalysts during water oxidation with ceric ammonium nitrate, J. Am. Chem. Soc. 133 (2011) 19024–19027.

80. B. Klahr, S. Gimenez, F. Fabregat-Santiago, J. Bisquert, T. W. Hamann, Electrochemical and photoelectrochemical investigation of water oxidation with hematite electrodes, Energy Environ. Sci. (2012). doi: 10.1039/C2EE21414H.

81. M. M. Najafpour, Calcium–manganese oxides as structural and functional models for active site in oxygen evolving complex in photosystem II: lessons from simple models, J. Photochem. Photobiol. B: Biol. 104 (2011) 111–117.

82. J. J. H. Pijpers, M. T. Winkler, Y. Surendranath, T. Buonassisi, D. G. Nocera, Light-induced water oxidation at silicon electrodes functionalized with a cobalt oxygen-evolving catalyst, Proc. Natl. Acad. Sci. U.S.A 108 (2011) 10056–10061.

83. T. Takashima, K. Hashimoto, R. Nakamura, Mechanisms of pH-dependent activity for water oxidation to molecular oxygen by MnO_2 electrocatalyst, J. Am. Chem. Soc. 134 (2012) 1519–1527.

84. A. J. Esswein, Y. Surendranath, S. Y. Reece, D. G. Nocera, Highly active cobalt phosphate and borate based oxygen evolving catalysts operating in neutral and natural waters, Energy Environ. Sci. 4 (2011) 499–504.

85. M. Wiechen, H. M. Berends, P. Kurz, Water oxidation catalysed by manganese compounds: from complexes to "biomimetic rocks," Dalton Trans. 41 (2012) 21–31.

86. H. Yamazaki, A. Shouji, M. Kajita, M. Yagi, Electrocatalytic and photocatalytic water oxidation to dioxygen based on metal complexes, Coord. Chem. Rev. 254 (2010) 2483–2491.

87. L. Duan, F. Bozoglian, S. Mandal, B. Stewart, T. Privalov, A. Llobet, L. Sun, A molecular ruthenium catalyst with water-oxidation activity comparable to that of photosystem II, Nature Chem. 4 (2012) 418–423.

88. S. M. Barnett, K. I. Goldberg, J. M. Mayer. A soluble copper–bipyridine water-oxidation electrocatalyst, Nature Chem. 4 (2012) 498–502.

89. G. Renger, Molecular mechanism of water oxidation. In: G. S. Singhal, G. Renger, Govindjee, K. D. Irrgang, S. K. Sopory (Eds.), *Concepts in Photobiology: Photosynthesis and Photomorphogenesis*, Kluwer Academic Publishers and Narosa Pub. House, Delhi, 1999, pp. 292–329.

90. G. Renger, T. Renger, Photosystem II: the machinery of photosynthetic water splitting, Photosynth. Res. 98 (2008) 53–80.

91. V. I. Novoderezhkin, E. Romero, J. P. Dekker, R. van Grondelle, Multiple charge-separation pathways in photosystem II: modeling of transient absorption kinetics, Chem. Phys. Chem. 12 (2011) 681–688.

92. I. V. Shelaev, F. E. Gostev, V. A. Nadtochenko, A. Y. Shkuropatov, A. A. Zabelin, M. D. Mamedov, A. Y. Semenov, O. M. Sarkisov, V. A. Shuvalov, Primary light-energy conversion in tetrameric chlorophyll structure of photosystem II and bacterial reaction centers: II. Femto-and picosecond charge separation in PSII D1/D2/Cyt b_{559} complex, Photosynth. Res. 98 (2008) 95–103.

93. T. Okubo, T. Tomo, M. Sugiura, T. Noguchi, Perturbation of the structure of P680 and the charge distribution on its radical cation in isolated reaction center complexes of photosystem II as revealed by Fourier transform infrared spectroscopy, Biochemistry 46 (2007) 4390–4397.

94. K. Saito, T. Ishida, M. Sugiura, K. Kawakami, Y. Umena, N. Kamiya, J. R. Shen, H. Ishikita, Distribution of the cationic state over the chlorophyll pair of the photosystem ii reaction center, J. Am. Chem. Soc. 133 (2011) 14379–14388.

95. G. Renger, A. R. Holzwarth, Primary electron transfer. In: T. J. Wydrzynski, K. Satoh (Eds.), *Photosystem II: The Light-Driven Water: Plastoquinone Oxidoreductase. Advances in Photosynthesis and Respiration*, Vol. 22, Springer, Dordrecht, The Netherlands, 2005, pp. 139–175.

96. A. M. Nuijs, H. J. Vangorkom, J. J. Plijter, L. N. M. Duysens, Primary-charge separation and excitation of chlorophyll a in photosystem ii particles from spinach as studied by

picosecond absorbency-difference spectroscopy, Biochim. Biophys. Acta. 848 (1986) 167–175.

97. J. Bernarding, H. J. Eckert, H. J. Eichler, A. Napiwotzki, G. Renger, Kinetic-studies on the stabilization of the primary radical pair P680$^+$ Pheo$^-$ in different photosystem-II preparations from higher-plants, Photochem. Photobiol. 59 (1994) 566–573.

98. G. Renger, H. J. Eckert, A. Bergmann, J. Bernarding, B. Liu, A. Napiwotzki, F. Reifarth, H. J. Eichler, Fluorescence and spectroscopic studies of exciton trapping and electron-transfer in photosystem-II of higher-plants, Aust. J. Plant Physiol. 22 (1995) 167–181.

99. J. Haveman, P. Mathis, Flash-induced absorption changes of primary donor of photosystem-II at 820 nm in chloroplasts inhibited by low pH or Tris-treatment, Biochim. Biophys. Acta. 440 (1976) 346–355.

100. G. Renger, C. Wolff, Existence of a high photochemical turnover rate at reaction centers of system II in Tris-washed chloroplasts, Biochim. Biophys. Acta 423 (1976) 610–614.

101. K. Hasegawa, T. Noguchi, Density functional theory calculations on the dielectric constant dependence of the oxidation potential of chlorophyll: implication for the high potential of P680 in photosystem II, Biochemistry 44 (2005) 8865–8872.

102. H. Ishikita, W. Saenger, J. Biesiadka, B. Loll, E. W. Knapp, How photosynthetic reaction centers control oxidation power in chlorophyll pairs P680, P700, and P870, Proc. Natl. Acad. Sci. U.S.A. 103 (2006) 9855–9860.

103. G. Renger, Functional pattern of photosystem II, In: G. Renger (Ed.), *Primary Processes of Photosynthesis: Basic Principles and Apparatus, Part I Photophysical Principles, Pigments and Light Harvesting/Adaptation/Stress*, RSC Publishing, Cambridge, 2008, pp. 237–290.

104. M. Kobayashi, S. Ohashi, K. Iwamoto, Y. Shiraiwa, Y. Kato, T. Watanabe, Redox potential of chlorophyll d in vitro, Biochim. Biophys. Acta 1767 (2007) 596–602.

105. F. Rappaport and B. A. Diner, Primary photochemistry and energetics leading to the oxidation of the 4 (Mn) Ca cluster and to the evolution of molecular oxygen in photosystem II, Coord. Chem. Rev. 252 (2008) 259–272.

106. L. M. N. Duysens, The path of light energy in photosynthesis, Brookhaven Symp. Biol. 11(1959) 10–25.

107. R. S. Knox, Photosynthetic efficiency and exciton transfer and trapping. In: J. Barber (Ed.), *Primary Processes of Photosynthesis*, Vol. 2, Elsevier Scientific Pub., Amsterdam 1977, pp. 55–97.

108. L. N. Bell, N. D. Gudkov, Thermodynamics of light energy conversion. In: J. Barber (Ed.), *The Photosystems: Structure, Function, and Molecular Biology. Topics in Photosynthesis*, Vol. 11, Elsevier, Amsterdam, 1992, pp. 17–43.

109. M. Almgren, Thermodynamic and kinetic limitations on conversion of solar-energy into storable chemical free-energy, Photochem. Photobiol. 27 (1978) 603–609.

110. I. Vass, E.-M. Aro, Photoinhibition of photosynthetic electron transport. In: G. Renger (Ed.), *Primary Processes of Photosynthesis: Basic Principles and Apparatus, Part II Reaction Centers/Photosystems, Electron Transport Chains, Photophosphorylation and Evolution*, RSC Publishing, Cambridge, 2008, pp. 393–425.

111. C. Triantaphylides, M. Krischke, F. A. Hoeberichts, B. Ksas, G. Gresser, M. Havaux, F. Van Breusegem, M.J. Mueller, Singlet oxygen is the major reactive oxygen species involved in photooxidative damage to plants, Plant Physiology 148 (2008) 960–968.

112. P. Jahns, A. R. Holzwarth, The role of the xanthophyll cycle and of lutein in photoprotection of photosystem II, Biochim. Biophys. Acta 1817 (2012) 182–193.

113. A. V. Ruban, M. P. Johnson, C. D. P. Duffy, The photoprotective molecular switch in the photosystem II antenna, Biochim. Biophys. Acta 1817 (2012) 167–181.

114. M. Edelman, A. K. Mattoo, The D1 protein: past and future perspectives. In: B. Demming Adams, W. W. Adams III, A. K. Mattoo (Eds.), *Photoinhibition, Gene Regulation, and Environment. Advances in Photosynthesis and Respiration*, Vol. 21, Springer, Dordrecht The Netherlands, 2006, pp. 23–38.

115. K. Yokthongwattana, A. Melis, Photoinhibition and recovery in oxygenic photosynthesis: mechanism of a photosystem II damage and repair cycle. In: B. Demming Adams, W. W. Adams III, A. K. Mattoo (Eds.), *Photoinhibition, Gene Regulation, and Environment. Advances in Photosynthesis and Respiration*, Vol. 21, Springer, 2006, pp 175–191.

116. P. Mulo, I. Sakurai, E.M. Aro, Strategies for psbA gene expression in cyanobacteria, green algae and higher plants: from transcription to PSII repair, Biochim. Biophys. Acta 1817 (2012) 247–257.

117. P. Joliot, G. Barbieri, R. Chabaud, Un nouveau modele des centres photochimiques du systeme II, Photochem. Photobiol. 10 (1969) 309–329.

118. B. Kok, B. Forbush, M. McGloin, Cooperation of charges in photosynthetic O_2 evolution, Photochem. Photobiol. 11 (1970) 457–476.

119. A. Zouni, Evolution of photosynthesis. In: G. Renger (Ed.), *Primary Processes of Photosynthesis: Basic Principles and Apparatus, Part II Reaction Centers/Photosystems, Electron Transport Chains, Photophosphorylation and Evolution*, RSC Publishing, Cambridge, 2008, pp. 193–236.

120. C. Costentin, Electrochemical approach to the mechanistic study of proton-coupled electron transfer, Chem. Rev. 108 (2008) 2145–2179.

121. R. I. Cukier, A theory that connects proton-coupled electron-transfer and hydrogen-atom transfer reactions, J. Phys. Chem. B 106 (2002) 1746–1757.

122. S. Hammes-Schiffer, Hydrogen tunneling and protein motion in enzyme reactions, Acc. Chem. Res. 39 (2006) 93–100.

123. M. H. V. Huynh, T. J. Meyer, Proton-coupled electron transfer, Chem. Rev. 107 (2007) 5004–5064.

124. S. Y. Reece, D. G. Nocera, proton-coupled electron transfer in biology: results from synergistic studies in natural and model systems, Annu. Rev. Biochem. 78 (2009) 673–699.

125. G. Renger, Photosynthetic watersplitting: apparatus and mechanism, In: J. Eaton-Rye, Tripathy, T. D. Sharkey (Eds.), *Photosynthesis: Plastid Biology, Energy Conversion and Carbon Assimilation*, Springer, New York, 2012, pp. 359–410.

126. J. G. Metz, P. J. Nixon, M. Rogner, G. W. Brudvig, B. A. Diner, Directed alteration of the D1 polypeptide of photosystem-II—evidence that tyrosine-161 is the redox component, Z, connecting the oxygen-evolving complex to the primary electron-donor, P680, Biochemistry 28 (1989) 6960–6969.

127. J. Messinger, G. Renger, Photosynthetic water splitting. In: G. Renger (Ed.), *Primary Processes of Photosynthesis: Basic Principles and Apparatus, Part II Reaction Centers/Photosystems, Electron Transport Chains, Photophosphorylation and Evolution*, RSC Publishing, Cambridge, 2008, pp. 291–349.

128. M. Sjodin, S. Styring, B. Akermark, L. C. Sun, The mechanism for proton-coupled electron transfer from tyrosine in a model complex and comparisons with Y-z oxidation in photosystem II, Philos. Trans. R. Soc. B 357 (2002) 1471–1478.

129. C. Tommos, J. J. Skalicky, D. L. Pilloud, A. J. Wand, P. L. Dutton, De novo proteins as models of radical enzymes, Biochemistry 38 (1999) 9495–9507.

130. H. J. Eckert, G. Renger, Temperature-dependence of $P680^+$ reduction in O_2-evolving PS-II membrane-fragments at different redox states S_I of the water oxidizing system, FEBS Lett. 236 (1988) 425–431.

131. A. M. A. Hays, I. R. Vassiliev, J. H. Golbeck, R. J. Debus, Role of D1-His190 in proton-coupled electron transfer reactions in photosystem II: a chemical complementation study, Biochemistry 37 (1998) 11352–11365.

132. A. Guskov, A. Gabdulkhakov, M. Broser, C. Glockner, J. Hellmich, J. Kern, J. Frank, F. Muh, W. Saenger, A. Zouni, Recent progress in the crystallographic studies of photosystem II, Chem. Phys. Chem. 11 (2010) 1160–1171.

133. K. Saito, J.R. Shen, T. Ishida, H. Ishikita, Short hydrogen bond between redox-active tyrosine Y_z and D1-His190 in the photosystem II crystal structure, Biochemistry 50 (2011) 9836–9844.

134. Y. Umena, K. Kawakami, J. R. Shen, N. Kamiya, Crystal structure of oxygen-evolving photosystem II at a resolution of 1.9 Å, Nature 473 (2011) 55–65.

135. N. Ioannidis, G. Zahariou, V. Petrouleas, Trapping of the S_2 to S_3 state intermediate of the oxygen-evolving complex of photosystem II, Biochemistry 45 (2006) 6252–6259.

136. J. Sjoholm, K. G. V. Havelius, F. Mamedov, S. Styring, Effects of pH on the S-3 state of the oxygen evolving complex in photosystem II probed by EPR split signal induction, Biochemistry 49 (2010) 9800–9808.

137. G. Renger, Oxidative photosynthetic water splitting: energetics, kinetics and mechanism, Photosynth. Res. 92 (2007) 407–425.

138. K. Kawakami, Y. Umena, N. Kamiya, J. R. Shen, Structure of the catalytic, inorganic core of oxygen-evolving photosystem II at 1.9 Å resolution, J Photochem. Photobiol. B: Biol 104 (2011) 9–18.

139. P. Kühn, H. Eckert, H. J. Eichler, G. Renger, Analysis of the $P680^{+\bullet}$ reduction pattern and its temperature dependence in oxygen-evolving PSII core complexes from a thermophilic cyanobacteria and higher plants, Phys. Chem. Chem. Phys. 6 (2004) 4838–4843.

140. G. Renger, Mechanism of light induced water splitting in photosystem II of oxygen evolving photosynthetic organisms, Biochim. Biophys. Acta 1817 (2012) 1164–1176.

141. G. Christen, A. Seeliger, G. Renger, $P680^{+\bullet}$ reduction kinetics and redox transition probability of the water oxidizing complex as a function of pH and H/D isotope exchange in spinach thylakoids, Biochemistry 38 (1999) 6082–6092.

142. C. Jeans, M. J. Schilstra, D. R. Klug, The temperature dependence of $P680^+$ reduction in oxygen-evolving photosystem, Biochemistry 41 (2002) 5015–5023.

143. R. A. Marcus, N. Sutin, Electron transfers in chemistry and biology, Biochim. Biophys. Acta 811 (1985) 265–322.

144. G. Renger, H. J. Eckert, M. Volker, Studies on the electron-transfer from Tyr-161 of polypeptide D-1 to P680+ in PS-II membrane-fragments from spinach, Photosynth. Res. 22 (1989) 247–256.

145. C. C. Page, C. C. Moser, X. X. Chen, P. L. Dutton, Natural engineering principles of electron tunnelling in biological oxidation–reduction, Nature 402 (1999) 47–52.

146. M. Karge, K. D. Irrgang, G. Renger, Analysis of the reaction coordinate of photosynthetic water oxidation by kinetic measurements of 355 nm absorption changes at different temperatures in photosystem II preparations suspended in either H_2O or D_2O, Biochemistry 36 (1997) 8904–8913.

147. M. Karge, K. D. Irrgang, S. Sellin, R. Feinaugle, B. Liu, H. J. Eckert, H. J. Eichler, G. Renger, Effects of hydrogen deuterium exchange on photosynthetic water cleavage in PS II core complexes from spinach, FEBS Lett. 378 (1996) 140–144.

148. J. Lavergne, Optical-difference spectra of the S-state transitions in the photosynthetic oxygen-evolving complex, Biochim. Biophys. Acta 894 (1987) 91–107.

149. O. Saygin, H. T. Witt, Evidence for the electrochromic identification of the change of charges in the 4 oxidation steps of the photoinduced water cleavage in photosynthesis, FEBS Lett. 187 (1985) 224–226.

150. G. Renger, Role of hydrogen bonds in photosynthetic water splitting. In: K.-L. Han, G.-J. Zhao (Eds.), *Hydrogen Bonding and Transfer in the Excited State*, Wiley, Hoboken, NJ, 2010, pp. 433–461.

151. V. K. Yachandra, The catalytic manganese cluster:organisation of the metal ions. In: T. J. Wydrzynski, K. Satoh (Eds.), *Photosystem II: The Light-Driven Water: Plastoquinone Oxidoreductase. Advances in Photosynthesis and Respiration*, Vol. 22 Springer, Dordrecht, The Netherlands, 2005, pp. 139–175.

152. J. Dasgupta, A. M. Tyryshkin, S. V. Baranov, G. C. Dismukes, Bicarbonate coordinates to Mn^{3+} during photo-assembly of the catalytic Mn_4Ca core of photosynthetic water oxidation: EPR characterization, Appl. Magnetic Resonance 37 (2010) 137–150.

153. N. Tamura, G. Cheniae, Photoactivation of the water-oxidizing complex in photosystem-II membranes depleted of Mn and extrinsic proteins .1. biochemical and kinetic characterization, Biochim. Biophys. Acta 890 (1987) 179–194.

154. E. Salomon, G. Renger, N. Keren, Mn transport and the assembly of photosystem II. In: G. Peschek, C. Obinger, G. Renger (Eds.), *Bioenergetic Processes of Cyanobacteria: from Evolutionary Singularity to Ecological Diversity*, Springer, Dordrecht, The Netherlands, 2011, pp. 423–442.

155. G. M. Cheniae, I. F. Martin, Absence of oxygen evolving capacity in dark-grown *Chlorella*: the photoactivation of oxygen-evolving centers, Photochem. Photobiol. 17 (1973) 441–459.

156. R. J. Strasser, C. Sironval, Induction of photosystem II activity in flashed leaves, FEBS Lett. 28 (1972) 56–60.

157. R. J. Debus, The catalytic manganese cluster: protein ligation. In: T. J. Wydrzynski, K. Satoh (Eds.), *Photosystem II: The Light-Driven Water: Plastoquinone Oxidoreductase. Advances in Photosynthesis and Respiration*, Vol. 22, Springer, Dordrecht, The Netherlands, 2005, pp. 261–284.

158. A. Zouni, H. T. Witt, J. Kern, P. Fromme, N. Krauss, W. Saenger, P. Orth, Crystal structure of photosystem II from *Synechococcus elongatus* at 3.8 Å resolution, Nature 409 (2001) 739–743.

159. N. Kamiya, J. R. Shen, Crystal structure of oxygen-evolving photosystem II from *Thermosynechococcus vulcanus at* 3.7 Å resolution, Proc. Natl. Acad. Sci. U.S.A. 100 (2003) 98–103.

160. K. N. Ferreira, T. M. Iverson, K. Maghlaoui, J. Barber, S. Iwata, Architecture of the photosynthetic oxygen-evolving center, Science 303 (2004) 1831–1838.

161. J. Biesiadka, B. Loll, J. Kern, K. D. Irrgang, A. Zouni, Crystal structure of cyanobacterial photosystem II at 3.2 Å resolution: a closer look at the Mn-cluster, Phys. Chem. Chem. Phys. 6 (2004) 4733–4736.

162. B. Loll, J. Kern, W. Saenger, A. Zouni, J. Biesiadka, Towards complete cofactor arrangement in the 3.0 Å resolution structure of photosystem II, Nature 438 (2005) 1040–1044.

163. O.Carugo, K. D. Carugo, When X-rays modify the protein structure: radiation damage at work. Trends Biochem. Sci. 30 (2005) 213–219.

164. J. Yano, J. Kern, K. D. Irrgang, M. J. Latimer, U. Bergmann, P. Glatzel, Y. Pushkar, J. Biesiadka, B. Loll, K. Sauer, J. Messinger, A. Zouni, V. K. Yachandra, X-ray damage to the Mn$_4$Ca complex in single crystals of photosystem II: a case study for metalloprotein crystallography, Proc. Natl. Acad. Sci. U.S.A. 102 (2005) 12047–12052.

165. M. Grabolle, M. Haumann, C. Müller, P. Liebisch, H. Dau, Rapid loss of structural motifs in the manganese complex of oxygenic photosynthesis by X-ray irradiation at 10–300 K, J. Biol. Chem. 281 (2006) 4580–4588.

166. J. Messinger, G. Seaton, T. Wydrzynski, U. Wacker, G. Renger, S-3 state of the water oxidase in photosystem II, Biochemistry 36 (1997) 6862–6873.

167. A. Galstyan, A. Robertazzi, E. W. Knapp, Oxygen-evolving Mn cluster in photosystem II: the protonation pattern and oxidation state in the high-resolution crystal structure, J. Am. Chem. Soc. 134 (2012) 7442–7449.

168. J. Yano, Y. Pushkar, P. Glatzel, A. Lewis, K. Sauer, J. Messinger, U. Bergmann, V. Yachandra, High-resolution Mn EXAFS of the oxygen-evolving complex in photosystem II: structural implications for the Mn$_4$Ca cluster, J. Am. Chem. Soc. 127 (2005) 14974–14975.

169. W. C. Liang, T. A. Roelofs, R. M. Cinco, A. Rompel, M. J. Latimer, W. O. Yu, K. Sauer, M. P. Klein, V. K. Yachandra, Structural change of the Mn cluster during the $S_2 \rightarrow S_3$ state transition of the oxygen-evolving complex of photosystem II. Does it reflect the onset of water/substrate oxidation? determination by Mn X-ray absorption spectroscopy, J. Am. Chem. Soc. 122 (2000) 3399–3412.

170. Y. L. Pushkar, J. Yano, K. Sauer, A. Boussac, V. K. Yachandra, Structural changes in the Mn$_4$Ca cluster and the mechanism of photosynthetic water splitting, Proc. Natl. Acad. Sci. U.S.A. 105 (2008) 1879–1884.

171. V. K. Yachandra, J. Yano, Calcium in the oxygen-evolving complex: structural and mechanistic role determined by X-ray spectroscopy, J. Photochem. Photobiol. B: Biol. 104 (2011) 51–59.

172. J. Messinger, G. Renger, The reactivity of hydrazine with photosystem-II strongly depends on the redox state of the water oxidizing system, FEBS Lett. 277 (1990) 141–146.

173. J. Messinger, U. Wacker, G. Renger, Unusual low reactivity of the water oxidase in the redox state S_3 toward exogenous reductants. analysis of the NH_2OH and NH_2NH_2 modifications of flash-induced oxygen evolution in isolated spinach thylakoids, Biochemistry 30 (1991) 7852–7862.

174. A. Boussac, A. W. Rutherford, Nature of the inhibition of the oxygen-evolving enzyme of photosystem-II induced by NaCl washing and reversed by the addition of Ca^{2+} or Sr^{2+}, Biochemistry 27 (1988) 3476–3483.

175. G. Renger, Mechanistic and structural aspects of photosynthetic water oxidation, Physiologia Plantarum 100 (1997) 828–841.

176. W. Ames, D. A. Pantazis, V. Krewald, N. Cox, J. Messinger, W. Lubitz, F. Neese, Theoretical evaluation of structural models of the S_2 state in the oxygen evolving complex of photosystem II: protonation states and magnetic interactions, J. Am. Chem. Soc. 133 (2011) 19743–19757.

177. P. E. M. Siegbahn, Recent theoretical studies of water oxidation in photosystem II, J. Photochem. Photobiol. B: Biol. 104 (2011) 94–99.

178. K. Yamaguchi, H. Isobe, S. Yamanaka, T. Saito, K. Kanda, M. Shoji, Y. Umena, K. Kawakami, J.-R. Shen, N. Kamiya, M. Okumura, Full geometry optimizations of the mixed-valence $CaMn_4O_4X(H_2O)_4$ (X = OH or O) cluster in OEC of PS II: degree of symmetry breaking of the labile Mn–X–Mn bond revealed by several hybrid DFT calculations, Int. J. Quant. Chem. (in press) (2012).

179. J. Kern, G. Renger, Photosystem II: structure and mechanism of the water : plastoquinone oxidoreductase, Photosynth. Res. 94 (2007) 183–202.

180. P. E. M. Siegbahn, Structures and energetics for O_2 formation in photosystem II, Acc. Chem. Res. 42 (2009) 1871–1880.

181. N. Cox, L. Rapatskiy, J. H. Su, D. A. Pantazis, M. Sugiura, L. Kulik, P. Dorlet, A. W. Rutherford, F. Neese, A. Boussac, W. Lubitz, J. Messinger, Effect of Ca^{2+}/Sr^{2+} substitution on the electronic structure of the oxygen-evolving complex of photosystem II: a combined multifrequency EPR, Mn-55-ENDOR, and DFT study of the S_2 state, J. Am. Chem. Soc. 133 (2011) 3635–3648.

182. S. Schinzel, J. Schraut, A. V. Arbuznikov, P. E. M. Siegbahn, M. Kaupp, Density functional calculations of Mn-55, N-14 and C-13 electron paramagnetic resonance parameters support an energetically feasible model system for the S_2 state of the oxygen-evolving complex of photosystem II, Chem. Eur. J. 16 (2010) 10424–10438.

183. J. Yano, L. M. Walker, M. A. Strickler, R. J. Service, V. K. Yachandra, R. J. Debus, Altered structure of the Mn_4Ca Cluster in the oxygen-evolving complex of photosystem II by a histidine ligand mutation, J. Biol. Chem. 286 (2011) 9257–9267.

184. H. A. Chu, W. Hillier, R. J. Debus, Evidence that the C-terminus of the D1 polypeptide of photosystem II is ligated to the manganese ion that undergoes oxidation during the S-1 to S-2 transition: an isotope-edited FTIR study, Biochemistry 43 (2004) 3152–3166.

185. Y. Kimura, N. Mizusawa, A. Ishii, T. Ono, FTIR detection of structural changes in a histidine ligand during S-state cycling of photosynthetic oxygen-evolving complex, Biochemistry 44 (2005) 16072–16078.

186. Y. Kimura, N. Mizusawa, T. Yamanari, A. Ishii, T. Ono, Structural changes of D1 C-terminal alpha-carboxylate during S-state cycling in photosynthetic oxygen evolution, J. Biol. Chem. 280 (2005) 2078–2083.

187. M. Sugiura, F. Rappaport, W. Hillier, P. Dorlet, Y. Ohno, H. Hayashi, A. Boussac, Evidence that D1-His332 in photosystem II from *Thermosynechococcus elongatus* interacts with the S_3 state and not with the S_2 state, Biochemistry 48 (2009) 7856–7866.

188. M. Iizasa, H. Suzuki, T. Noguchi, Orientations of carboxylate groups coupled to the Mn cluster in the photosynthetic oxygen-evolving center as studied by polarized ATR-FTIR spectroscopy, Biochemistry 49 (2010) 3074–3082.

189. K. Olesen, L.-E. Andréasson, The function of the chloride ion in photosynthetic oxygen evolution. Biochemistry 42 (2003) 2025–2035.

190. E. M. Sproviero, J. A. Gascon, J. P. McEvoy, G. W. Brudvig, V. S. Batista, QM/MM models of the O_2 evolving complex of photosystem II, J. Chem. Theory Comput. 2 (2006) 1119–1134.

191. E. M. Sproviero, J. A. Gascon, J. P. McEvoy, G. W. Brudvig, V. S. Batista, Quantum mechanics/molecular mechanics study of the catalytic cycle of water splitting in photosystem II, J. Am. Chem. Soc. 130 (2008) 3428–3442.

192. S. Zein, L. V. Kulik, J. Yano, J. Kern, Y. Pushkar, A. Zouni, V. K. Yachandra, W. Lubitz, F. Neese, J. Messinger, Focusing the view on nature's water-splitting catalyst, Philos. Trans. R. Soc. B 363 (2008) 1167–1177.

193. K. Sauer, J. Yano, V. K. Yachandra, X-ray spectroscopy of the photosynthetic oxygen-evolving complex, Coord. Chem. Rev. 252 (2008) 318–335.

194. J. Yano, V. K. Yachandra, Oxidation state changes of the Mn_4Ca cluster in photosystem II, Photosynth. Res. 92 (2007) 289–303.

195. P. Gatt, R. Stranger, R. J. Pace, Application of computational chemistry to understanding the structure and mechanism of the Mn catalytic site in photosystem II—review, J. Photochem. Photobiol. B: Biol. 104 (2011) 80–93.

196. L. V. Kulik, B. Epel, W. Lubitz, J. Messinger, Electronic structure of the Mn_4O_xCa cluster in the S_0 and S_2 states of the oxygen-evolving complex of photosystem II based on pulse Mn-55-ENDOR and EPR spectroscopy, J. Am. Chem. Soc. 129 (2007) 13421–13435.

197. J. Yano, J. Kern, K. Sauer, M.J. Latimer, Y. Pushkar, J. Biesiadka, B. Loll, W. Saenger, J. Messinger, A. Zouni, V.K. Yachandra, Where water is oxidized to dioxygen: structure of the photosynthetic Mn_4Ca cluster, Science 314 (2006) 821–825.

198. T. Ichino, K. Yamaguchi, Y. Yoshioka, Effectiveness of optimizing geometry for $CaMn_4O_5$ cluster at 1.9 Å resolved OEC and proposal for oxidation mechanism from S_0 to S_3 states, Chem. Lett. 41 (2012) 18–20.

199. P. Glatzel, U. Bergmann, J. Yano, H. Visser, J. H. Robblee, W. W. Gu, F. M. F. de Groot, G. Christou, V. L. Pecoraro, S. P. Cramer, V. K. Yachandra, The electronic structure of Mn in oxides, coordination complexes, and the oxygen-evolving complex of photosystem II studied by resonant inelastic X-ray scattering, J. Am. Chem. Soc. 126 (2004) 9946–9959.

200. G. Renger, Mechanistic aspects of photosynthetic water cleavage, Photosynthetica 21 (1987) 203–224.

201. P. J. van Leeuwen, C. Heimann, P. Gast, J. P. Dekker, H. J. Vangorkom, Flash-induced redox changes in oxygen-evolving spinach photosystem-II core particles, Photosynth. Res. 38 (1993) 169–176.

202. F. Rappaport, M. Blanchard-Desce, J. Lavergne, Kinetics of electron-transfer and electrochromic change during the redox transitions of the photosynthetic oxygen-evolving complex, Biochim. Biophys. Acta 1184 (1994) 178–192.

203. T. Noguchi, H. Suzuki, M. Tsuno, M. Sugiura, C. Kato, Time-resolved infrared detection of the proton and protein dynamics during photosynthetic oxygen evolution, Biochemistry 51 (2012) 3205–3214.

204. M. R. Razeghifard, R. J. Pace, EPR kinetic studies of oxygen release in thylakoids and PSII membranes: a kinetic intermediate in the S_3 to S_0 transition, Biochemistry 38 (1999) 1252–1257.

205. L. Gerencser, H. Dau, Water oxidation by photosystem II: H_2O-D_2O exchange and the influence of pH support formation of an intermediate by removal of a proton before dioxygen creation, Biochemistry 49 (2010) 10098–10106.

206. T. Noguchi, H. Suzuki, M. Tsuno, M. Sugiura, C. Kato, Time-resolved infrared detection of the proton and protein dynamics during photosynthetic oxygen evolution, Biochemistry 51 (2012) 3205–3214.

207. G. Renger, Studies on structure and mechanism of photosynthetic water oxidation. In: G. A. Peschek, W. Löffelhardt, G. Schmetterer (Eds.), *The Phototrophic Prokaryotes*, Kluwer Academic/Plenum, New York, 1999, pp. 35–50.

208. H. H. Stiehl, H. T. Witt, Quantitative treatment of the function of plastoquinone in photosynthesis, Z. Naturforsch. 24B (1969) 1588–1598.

209. I. Enami, A. Okumura, R. Nagao, T. Suzuki, M. Iwai, J.R. Shen, Structures and functions of the extrinsic proteins of photosystem II from different species, Photosynth. Res. 98(2008) 349–363.

210. T. M. Bricker, J. L. Roose, R. D. Fagerlund, L. K. Frankel, J. J. Eaton-Rye, The extrinsic proteins of photosystem II, Biochim. Biophys. Acta 1817 (2012) 121–142.

211. T. M. Bricker, L. K. Frankel, Auxiliary functions of the PsbO, PsbP and PsbQ proteins of higher plant Photosystem II: a critical analysis, J. Photochem. Photobiol. B: Biol. 104 (2011) 165–178.

212. W. Hillier, T. Wydrzynski, ^{18}O water exchange in photosystem II: substrate binding and intermediates of the water splitting cycle, Coord. Chem. Rev. 252 (2008) 306–317.

213. W. Hillier, J. Messinger, Mechanism of photosynthetic oxygen production. In: T. J. Wydrzynski, K. Satoh (Eds.), *Photosystem II: The Light-Driven Water: Plastoquinone Oxidoreductase. Advances in Photosynthesis and Respiration*, Vol. 22, Springer, Dordrecht, The Netherlands, 2005, pp. 567–608.

214. J. Messinger, U. Wacker, G. Renger, Unusual low reactivity of the water oxidase in the redox state S_3 toward exogenous reductants. Analysis of the NH_2OH and NH_2NH_2 induced modifications of flash induced oxygen evolution in isolated spinach thylakoids, Biochemistry 30 (1991) 7852–7860.

215. S. Riistama, A. Puustinen, A. GarciaHorsman, S. Iwata, H. Michel, M. Wikstrom, Channelling of dioxygen into the respiratory enzyme, Biochim. Biophys. Acta 1275 (1996) 1–4.

216. J. M. Anderson, Does functional photosystem II complex have an oxygen channel?, FEBS Lett. 488 (2001) 1–4.

217. V. M. Luna, Y. Chen, J. A. Fee, C. D. Stout, Crystallographic studies of Xe and Kr binding within the large internal cavity of cytochrome $ba(3)$ from *Thermus thermophilus*: structural analysis and role of oxygen transport channels in the heme-Cu oxidases, Biochemistry 47 (2008) 4657–4665.

218. G. Renger, Theoretical studies about the functional and structural organization of the photosynthetic oxygen evolution. In: H. Metzner (Ed.), *Photosynthetic Oxygen Evolution*, Academic Press, London, 1978, pp. 229–248.

219. J. Clausen, W. Junge, H. Dau, M. Haumann, Photosynthetic water oxidation at high O_2 backpressure monitored by delayed chlorophyll fluorescence, Biochemistry 44 (2005) 12775–12779.

220. J. Clausen, W. Junge, Detection of an intermediate of photosynthetic water oxidation, Nature 430 (2004) 480–483.

221. D. R. J. Kolling, T. S. Brown, G. Ananyev, G. C. Dismukes, Photosynthetic oxygen evolution is not reversed at high oxygen pressures: mechanistic consequences for the water-oxidizing complex, Biochemistry 48 (2009) 1381–1389.

222. D. Shevela, K. Beckmann, J. Clausen, W. Junge, J. Messinger, Membrane-inlet mass spectrometry reveals a high driving force for oxygen production by photosystem II, Proc. Natl. Acad. Sci. 108 (2011) 3602–3607.

223. H. Suzuki, M. Sugiura, T. Noguchi, Monitoring proton release during photosynthetic water oxidation in photosystem II by means of isotope-edited infrared spectroscopy, J. Am. Chem. Soc. 131 (2009) 7849–7857.

224. J. Lavergne, W. Junge, Proton release during the redox cycle of the water oxidase, Photosynth. Res. 38 (1993) 279–296.

225. T. J. Meyer, M. H. V. Huynh, H. H. Thorp, The possible role of proton-coupled electron transfer (PCET) in water oxidation by photosystem II, Angew. Chem. Int. Ed. 46 (2007) 5284–5304.

226. J. P. McEvoy, G. W. Brudvig, Water-splitting chemistry of photosystem II, Chem. Rev. 106 (2006) 4455–4483.

227. T. Shutova, K. D. Irrgang, V. Shubin, V. V. Klimov, G. Renger, Analysis of pH-induced structural changes of the isolated extrinsic 33 kilodalton protein of photosystem II, Biochemistry 36 (1997) 6350–6358.

228. T. Shutova, V. V. Klimov, B. Andersson, G. Samuelsson, A cluster of carboxylic groups in PsbO protein is involved in proton transfer from the water oxidizing complex of photosystem II, Biochim. Biophys. Acta 1767 (2007) 434–440.

229. J. W. Murray, J. Barber, Structural characteristics of channels and pathways in photosystem II including the identification of an oxygen channel, J. Struct. Biol. 159 (2007) 228–237.

230. H. Ishikita, W. Saenger, B. Loll, J. Biesiadka, E. W. Knapp, Energetics of a possible proton exit pathway for water oxidation in photosystem II, Biochemistry 45 (2006) 2063–2071.

231. F. M. Ho, S. Styring, Access channels and methanol binding site to the $CaMn_4$ cluster in photosystem II based on solvent accessibility simulations, with implications for substrate water access, Biochim. Biophys. Acta 1777 (2008) 140–153.

232. A. Gabdulkhakov, A. Guskov, M. Broser, J. Kern, F. Muh, W. Saenger, A. Zouni, Probing the accessibility of the Mn_4Ca cluster in photosystem II: channels calculation, noble gas derivatization, and cocrystallization with DMSO, Structure 17 (2009) 1223–1234.

233. L. K. Frankel, L. Sallans, P. A. Limbach, T. M. Bricker, Identification of oxidized amino acid residues in the vicinity of the Mn_4CaO_5 cluster of photosystem II: implications for the identification of oxygen channels within the photosystem, Biochemistry 51 (2012) 6371–6377.

234. S. Vassiliev, T. Zaraiskaya, D. Bruce, Exploring the energetics of water permeation in photosystem II by multiple steered molecular dynamics simulations, Biochim. Biophys. Acta 1817 (2012) 1671–1678.

235. G. Renger, G. Christen, M. Karge, H. J. Eckert, K. D. Irrgang, Application of the Marcus theory for analysis of the temperature dependence of the reactions leading to photosynthetic water oxidation: results and implications, J. Biol. Inorg. Chem. 3 (1998) 360–366.

236. G. Renger, B. Hanssum, Studies on the reaction coordinates of the water oxidase in Ps-II membrane-fragments from spinach, FEBS Lett. 299 (1992) 28–32.

237. H. Koike, B. Hanssum, Y. Inoue, G. Renger, Temperature-dependence of S-state transition in a thermophilic cyanobacterium, *Synechococcus vulcanus* copeland measured by absorption changes in the ultraviolet region, Biochim. Biophys. Acta 893 (1987) 524–533.

238. S. Styring, A. W. Rutherford, Deactivation kinetics and temperature-dependence of the S-state transitions in the oxygen-evolving system of photosystem-II measured by EPR, Biochim. Biophys. Acta 933 (1988) 378–387.

239. H. M. Gleiter, E. Haag, Y. Inoue, G. Renger, Functional-characterization of a purified homogeneous photosystem-II core complex with high oxygen evolution capacity from spinach, Photosynth. Res. 35 (1993) 41–53.

240. P. Joliot, A. Joliot, Different types of quenching involved in photosystem II centers, Biochim. Biophys. Acta 305 (1973) 302–316.

241. G. Renger, H. M. Gleiter, E. Haag, F. Reifarth, Photosystem II: thermodynamics and kinetics of electron transport from $Q_A^{-\bullet}$ to Q_B ($Q_B^{-\bullet}$) and deleterious effects of copper (II), Z. Naturforsch. 48c (1993) 234–240.

242. F. Reifarth, G. Renger, Indirect evidence for structural changes coupled with $QB^{-\bullet}$ formation in photosystem II, FEBS Lett. 428 (1998) 123–126.

243. A. Garbers, F. Reifarth, J. Kurreck, G. Renger, F. Parak, Correlation between protein flexibility and electron transfer from Q_A to Q_B in PSII membrane fragments from spinach, Biochemistry 37 (1998) 11399–11404.

244. J. Pieper, G. Renger, Protein dynamics investigated by neutron scattering, Photosynth. Res. 102 (2009) 281–293.

245. J. L. Hughes, P. Smith, R. Pace, E. Krausz, Charge separation in photosystem II core complexes induced by 690–730 nm excitation at 1.7 K, Biochim. Biophys. Acta 1757 (2006) 841–851.

246. S. Reinman, P. Mathis, Influence of temperature on photosystem-II electron-transfer reactions, Biochim. Biophys. Acta 635 (1981) 249–258.

247. G. Renger, The postulation of a model for the mechanism of photosynthetic oxygen evolution, FEBS Lett. 81 (1977) 223–228.

248. R.Bianco, P. J. Hay, J. T. Hynes, Proton relay and electron flow in the O–O single bond formation in water oxidation by the ruthenium blue dimer, Energy Environ. Sci. 5 (2012) 7741–7746.

249. G. Renger, Photosynthetic water oxidation to molecular oxygen: apparatus and mechanism, Biochim. Biophys. Acta 1503 (2001) 210–228.

250. G. Renger, Coupling of electron and proton transfer in oxidative water cleavage in photosynthesis, Biochim. Biophys. Acta 1655 (2004) 195–204.

251. P. E. M. Siegbahn, The effect of backbone constraints: the case of water oxidation by the oxygen-evolving complex in PSII, Chem. Phys. Chem. 12 (2011) 3274–3280.

252. T. Saito, S. Yamanaka, K. Kanda, H. Isobe, Y. Takano, Y. Shigeta, Y. Umena, K. Kawakami, J.-R. Shen, N. Kamiya, M. Okumura, M. Shoji, Y. Yoshioka, K. Yamaguchi, Possible mechanisms of water splitting reaction based on proton and electron release pathways revealed for $CaMn_4O_5$ cluster of PSII refined to 1.9 Å X-ray resolution, Int. J. Quant. Chem. 112 (2012) 253–276.

253. L. I. Krishtalik, Energetics of multielectron reactions—photosynthetic oxygen evolution, Biochim. Biophys. Acta 849 (1986) 162–171.

254. A. J. Augustine, L. Quintanar, C. S. Stoj, D. J. Kosman, E. I. Solomon, Spectroscopic and kinetic studies of perturbed trinuclear copper clusters: the role of protons in reductive cleavage of the O–O bond in the multicopper oxidase Fet3p, J. Am. Chem. Soc. 129 (2007) 13118–13126.

255. M. R. A. Blomberg, P. E. M. Siegbahn, Different types of biological proton transfer reactions studied by quantum chemical methods, Biochim. Biophys. Acta 1757 (2006) 969–980.

256. R. Davydov, R. L. Osborne, S. H. Kim, J. H. Dawson, B. M. Hoffman, EPR and ENDOR studies of cryoreduced compounds II of peroxidases and myoglobin. Proton-coupled electron transfer and protonation status of ferryl, Biochemistry 47 (2008) 5147–5155.

257. E. Santoni, C. Jakopitsch, C. Obinger, G. Smulevich, Comparison between catalase-peroxidase and cytochrome c peroxidase. The role of the hydrogen-bond networks for protein stability and catalysis, Biochemistry 43 (2004) 5792–5802.

258. D. Q. Wang, J. J. Zheng, S. Shaik, W. Thiel, Quantum and molecular mechanical study of the first proton transfer in the catalytic cycle of cytochrome P450cam and its mutant D251N, J. Phys. Chem. B 112 (2008) 5126–5138.

259. E. Yikilmaz, D. W. Rodgers, A. F. Miller, The crucial importance of chemistry in the structure–function link: manipulating hydrogen bonding in iron-containing superoxide dismutase, Biochemistry 45 (2006) 1151–1161.

260. J. D. Soper, S. V. Kryatov, E. V. Rybak-Akimova, D. G. Nocera, Proton-directed redox control of O–O bond activation by heme hydroperoxidase models, J. Am. Chem. Soc. 129 (2007) 5069–5075.

261. F. F. Li, J. England, L. Que, Near-stoichiometric conversion of H_2O_2 to Fe-IV $=$ O at a nonheme iron(II) center. insights into the O–O bond cleavage step, J. Am. Chem. Soc. 132 (2010) 2134–2135.

262. J. L. Vallés-Pardo, M. C. Guijt, M. Iannuzzi, K. S. Joya, H. J. M. de Groot, F, Buda, Ab initio molecular dynamics study of water oxidation reaction pathways in mono-Ru catalysts, Chem. Phys. Chem. 134 (2012) 140–146.

263. S. Petrie, P. Gatt, R. Stranger, R. J. Pace, The interaction of His337 with the Mn_4Ca cluster of photosystem II, Phys. Chem. Chem. Phys. 14 (2012) 4651–4657.

264. T. A. Betley, Q. Wu, T. Van Voorhis, D. G. Nocera, Electronic design criteria for O–O bond formation via metal-oxo complexes, Inorg. Chem. 47 (2008) 1849–1861.

265. J. K. Schwartz, P. P. Wei, K. H. Mitchell, B. G. Fox, E. I. Solomon, Geometric and electronic structure studies of the binuclear nonheme ferrous active site of toluene-4-monooxygenase: parallels with methane monooxygenase and insight into the role of the effector proteins in O_2 activation, J. Am. Chem. Soc. 130 (2008) 7098–7109.

266. T. Noguchi, M. Sugiura, Flash-induced FTIR difference spectra of the water oxidizing complex in moderately hydrated photosystem II core films: effect of hydration extent on S-state transitions, Biochemistry 41 (2002) 2322–2330.

CHAPTER 4

Artificial Photosynthesis

REZA RAZEGHIFARD

4.1 INTRODUCTION

Photosystem I (PSI) and photosystem II (PSII) are photoactive membrane proteins capable of converting light energy into electron flow. They are very efficient because they can absorb the entire visible portion of sunlight using their pigments including chlorophylls, xanthophylls, and carotenoids. The absorbed light is then used to excite special chlorophyll (Chl) molecules, called primary electron donors (P). The excitation causes an electron to be ejected from P and moved away by a few electron acceptors to avoid charge recombination. The oxidized P gets reduced by a donor replacing its lost electron. The electron acceptor in photosystems can be a quinone or an iron–sulfur (Fe–S) cluster while the electron donor can be a heme, metal, or redox-active residue. The potentials and distances of the redox cofactors are tuned by the protein to maximize the efficiency of electron transfer. A good example for this is the different redox midpoint potentials (E_m) of primary electron donors of the bacterial reaction center (P870), PSI (P700), and PSII (P680). While the E_m for a P^+/P couple of P870 and P700 is about $+0.5$ V, P680 can generate an extremely strong oxidant with an E_m of about $+1.2$ V [1].

Factors that contribute to the high oxidation potential of P680 are a change from bacterioChl (BChl) to Chl a, and the presence of unique cofactors and protein subunits in PSII. Another contributing factor is that the protein has arranged the Chl molecules forming these primary donors differently. This causes P870, P700, and P680 to have different electronic couplings. The electronic coupling is weaker in P680 since it is more monomeric than the other two. Unlike P870 and P700, two more Chls are believed to be involved in the excited state of P680 [2–6]. The protein environment surrounding these special Chls are also different. Chl molecules interact with amino acids as their Mg axial ligands, and through H bonds and van der Waals contacts [7]. Mutagenesis studies of bacterial reaction centers provide a very useful tool to investigate the effect of specific amino acids on the oxidation potential of its primary

Natural and Artificial Photosynthesis: Solar Power as an Energy Source, First Edition. Edited by Reza Razeghifard.
© 2013 John Wiley & Sons, Inc. Published 2013 by John Wiley & Sons, Inc.

electron donor. By adding three more hydrogen bonds to the BChl special pair, the oxidation potential was increased to about 0.8 V. The modified reaction centers had a sufficiently high potential capable of oxidizing a nearby engineered tyrosine residue [8] and a Mn ion [9, 10, 11]. The Mn ion in the modified reaction center was able to perform light-induced conversion of superoxide into molecular oxygen [12]. Satisfying this requirement is the first step for mimicking the PSII function of water oxidation by Mn ions in the bacterial reaction center. Oxygenic photosynthesis extracts four electrons from two water molecules as the ultimate electron donor using four Mn ions of PSII. The four Mn ions together with a Ca ion form a cluster, which serves as both the binding and catalytic sites for splitting water molecules into molecular oxygen. Three Mn ions with Ca are bridged by oxygen atoms forming a cubane-like structure. The fourth Mn ion is linked to two of the manganese ions in the cubane by oxo-bridges (see Fig. 3.7 in Chapter 3). The Mn and Ca ions are ligated to the side chains of glutamic and aspartic acids, histidine, and alanine amino acids [13, 14, 15]. With the new detailed structural information providing more clues on how these proteins work, many more successful attempts were made to build artificial photosynthetic systems as highlighted in the following sections.

4.2 ORGANIC PIGMENT ASSEMBLIES ON ELECTRODES

Covalently attached electron donors and acceptors are designed and made to achieve a charge-separated state with long lifetime. Two approaches can be taken to prolong the charge separation: (1) increasing the number of building blocks used in these systems from triad to tetrad or pentad and (2) changing the components to favor forward electron transfer over charge recombination. The ultimate goal is to assemble these molecules in monolayers on conductive surfaces capable of producing photocurrents. An important part of these systems is a photoactive porphyrin that acts as a sensitizer. In the simplest design, the sensitizer is sandwiched between a donor and an acceptor or is connected to two acceptors. One of the very first porphyrin triads was designed with a carotenoid as the donor and a quinone as the acceptor [16]. Fullerenes then became popular acceptors due to their excellent electron affinity and small reorganization energies associated with the reduction of C_{60} to C_{60}^{-} [17]. The electron in C_{60}^{-} is delocalized over many carbons. The mixed self-assembled monolayers of ferrocene–porphyrin–C_{60} and boron–dipyrrin dye on a gold electrode generated a photocurrent with 50% quantum yield [18]. The $C_{60}^{\bullet-}$ gives the electron to an electron carrier such as methyl viologen or oxygen, while the oxidized ferrocene receives an electron from the gold electrode.

Porphyrins were also used as sensitizers in dye-sensitized solar cells (Fig. 4.1). The applications of photoactive ruthenium complexes are discussed in Chapter 5. In most of these systems, porphyrins are chemically grafted to a nanocrystalline TiO_2 layer coated on an optically transparent electrode (OTE) [19, 20]. The cell efficiency depends on the quantity of adsorbed porphyrins on the electrode surface [21]. The photoexcited porphyrins will inject their electrons into the conduction band of the semiconductor and then get reduced by a redox mediator (I_3^-/I^-). The circuit is

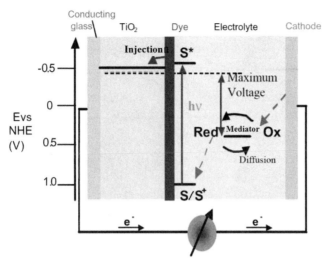

FIGURE 4.1   The design of dye-sensitized solar cells. Reproduced from Grätzel [119] with permission of Elsevier.

complete when the oxidized mediator reaches the counter electrode. An unprecedented conversion efficiency of 7% was achieved by these porphyrin-sensitized solar cells [22]. The conversion efficiency was varied by the porphyrin structure. By modifying substituent groups of porphyrins, spectral coverage and absorptivity were enhanced while their aggregation and orientation on the semiconductor were controlled. Porphyrin spectra are composed of two major absorption bands: Soret and Q bands. Extinction of the Q band was increased by elongating the π-conjugation and reducing the symmetry of a porphyrin monomer, and synthesizing porphyrin oligomers [23,24].The binding moiety also affected the efficiency of charge injection from the porphyrin into the semiconductor. Adding bulky substituents also improved the performance by increasing the electron lifetime [25]. The photocurrent generation was also improved by linking porphyrins together by enhancing the light-harvesting efficiency and electron injections of both porphyrins into TiO_2 [26]. Porphyrins bound to nanocrystalline TiO_2 films through carboxylate groups showed higher photon-to-current conversion efficiencies than those with a phosphonate anchoring group [27]. Also, the Zn-containing metalloporphyrins showed higher efficiency compared to the Cu-containing metalloporphyrins. Best performance was obtained when porphyrins with conjugated malonic acid groups as their anchoring groups were used [22]. Each porphyrin dye was adsorbed at an orientation of 35°–40° with respect to the TiO_2 surface plane [28]. The porphyrin layer was densely packed with intermolecular spacing of 3–4 Å between stacked porphyrins (Fig. 4.2).

The naturally abundant Chl molecules and their derivatives including chlorin $e6$ and copper Chl were also tried as sensitizers in solar cells [29, 30, 31]. A current efficiency of 5% was reported for a TiO_2 electrode coated with both fatty acid and Chl a molecules [32]. The hydrophobic fatty acids were needed since the natural

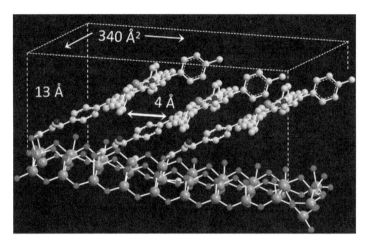

FIGURE 4.2 Orientation and molecular packing of organic dyes deposited on a TiO$_2$ surface. Reproduced with permission from Griffith et al. [28], copyright ©2011 American Chemical Society.

Chls do not have anchoring carboxylic groups. Electrodeposition and adsorption methods were applied to prepare and test Chl b-modified nanocrystalline SnO$_2$ semiconductor films [33]. Among five chlorin pigments synthesized with methyl, hexyl, dodecyl, 2-butyloctyl, or cholesteryl groups, the Chl a derivative with a dodecyl ester group performed the best with a TiO$_2$ electrode [34]. The chlorin pigment was further modified to increase the extinction coefficient of its Q bands by extending the π-conjugation by one additional double bond [35]. A similar strategy was used to synthesize a series of π-extended chlorin pigments by introducing various groups including phenyl and phenylethynyl [36]. A 20 nm red shift and significant increase in extinction coefficient of the Q$_y$ band was observed for a chlorin dimer with a rigid phenylbutadiynyl linker [36]. Introducing gold nanoparticles by electrodeposition on the OTE/TiO$_2$ electrode before its modification by Chl a molecules improved the performance of the cell [37]. This was mainly due to the ability of gold nanoparticles in accepting electrons from the excited dye and shuttling them to TiO$_2$ [38,39]. Chl c_1 and c_2 were also purified and tested for the current efficiencies in TiO$_2$-based solar cells [40].

4.3 PHOTOSYSTEM ASSEMBLIES ON ELECTRODES

Photosystem proteins were also immobilized on electrodes for the photocurrent generation. For stability purposes, thermophilic PSI proteins were mainly utilized for these applications. Compared to PSII, PSI is a much more stable protein. The idea was to reroute their internal electron flow by interfacing their cofactors with metal electrodes to create nanocircuits. Unlike pigments, photosystems are large protein complexes, which means there will be less coverage of the electrode surface with the photoactive complexes. Most protein-bound Chl molecules serve as light-harvesting pigments and only the primary electron donors are the photoactive species. The electrode coverage and orientation of photosystem proteins on the electrode surface then become main determining factors for current efficiency. The compact coverage also insulates the electrode surface and blocks access to reduced electron mediators. To bind PSI proteins to a nanoparticle gold electrode, a monolayer of 3-mercapto-1-propanesulfonate molecules was first self-assembled on the electrode surface. The binding was then promoted through the electrostatic interaction between positively charged ferredoxin-binding sites of PSI proteins and negatively charged 3-mercapto-1-propanesulfonate groups [41]. The photocurrent recorded from this electrode was maximized at around 670–680 nm, which corresponds to the absorption spectrum of PSI. L-Ascorbate was used as the sacrificial reagent in the electrolyte solution with 2,6-dichloroindophenol as a mediator.

Molecular wires were also designed to bind PSI proteins to a gold electrode within a self-assembled monolayer of 3-mercapto-1-propanesulfonate [42]. The molecular wire was introduced in the vitamin K_1 (phylloquinone) binding site. Organic solvent extraction in a water-saturated diethyl ether solution can remove the vitamin K_1 from PSI proteins. The molecular wire was designed with two parts: a naphthoquinone group for insertion into the quinone pocket and a viologen moiety for connecting to the sulfonate-covered gold electrode. The packing density of the molecular wire on the gold electrode was kept low to avoid steric hindrance. The length of the molecular wire was identical to vitamin K_1 to ensure that the viologen end of the wire was placed outside the PSI protein. The purpose of the molecular wire was to guide the electrons received by the naphthoquinone from the photoexcited P700 by sending them to the viologen moiety. The reduced viologen was detected by transient absorption measurements after the photoexcitation of P700. These trapped electrons were then transferred to the electrode. A similar approach was used for building a photosensor by binding the vitamin K_1-free PSI proteins to functionalized gold nanoparticles with molecular wires containing naphthoquinone and thiol groups [43]. The design was then adapted to an electronic imaging device (Fig. 4.3). The photosensing performance of the system was considerably increased by the addition of hexylamine and dodecylbenzenesulfonate surfactants [44]. These two neutral and anionic surfactants were added to form a hydrophobic layer around the gold nanoparticles on the gold electrode surface.

Solid mesoporous films made from TiO_2 nanoparticles were also shown to adsorb PSI proteins [45]. These films had large enough pores (23 nm diameter) to allow the PSI adsorption. The optical and photochemical properties of the PSI proteins were not significantly affected by the immobilization and they were able to generate a photocurrent. The photovoltaic performance of these systems was improved further by using bioengineered PSI proteins stabilized by surfactant peptides and taking advantage of the large surface areas of rough photoanodes made of TiO_2 nanocrystals or ZnO nanowires [46]. The nanocrystalline TiO_2 film used in this study had pores with 60 nm diameters to ensure a higher probability of the PSI adsorption. The bioengineered PSI had an amino acid segment with high affinity for ZnO incorporated into the PsaE subunit, thus promoting attachment to ZnO nanowires in an oriented

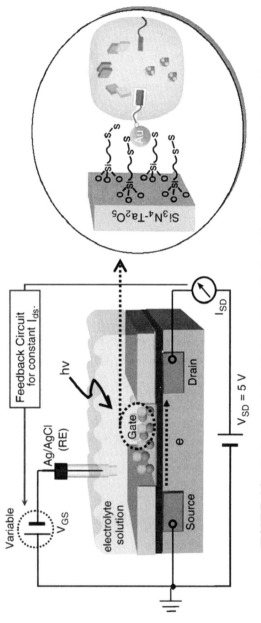

FIGURE 4.3 Electrodes made by binding vitamin K_1-free PSI proteins to functionalized gold nanoparticles. Reproduced from Terasaki et al. [43] with permission of Elsevier.

manner. The surfactant peptide used to increase the stability of dry PSI proteins on the nanostructured semiconducting electrodes was a cationic peptide surfactant [47, 48]. The surfactant peptide was designed to have a hydrophobic tail made of six amino acids (Ala, Val, or Ile) with an acetylated N terminus and a hydrophilic head made of lysine or arginine in the amidated C terminus. The nanocrystalline TiO_2 or ZnO nanowires played the role of ferredoxin as electron acceptors while the Co(II)/Co(III) ion containing electrolyte acted as the electron donor replacing plastocyanin. Covalent attachment of PSI proteins to gold electrodes was also achieved by modifying electrodes with aldehyde compounds capable of forming imine bonds with lysine residues of PSI proteins [49]. Thioaniline-functionalized PSI proteins were also prepared by reacting their lysine residues with a bifunctional reagent and *p*-aminothiophenol to form bisaniline crosslinks with a functionalized gold electrode [50]. By including Pt nanoclusters in the crosslinkers, a substantial 35-fold increase in the photocurrent was achieved. Some of these Pt nanoclusters were electrically attached to the last Fe–S cluster sites of the PSI proteins. The photocurrent was improved further by crosslinking ferredoxin proteins to the Pt nanoclusters. PSI mutants with surface-exposed cysteines were also prepared to allow direct binding of the protein complex to the electrode surface through sulfide bonds or linker molecules [51, 52]. To form efficient electric junctions, a total of four amino acids were mutated to cysteines in the extra membrane loops near P700. It was also shown that the dry PSI crystals generated very large photovoltages of 50 V, measured by Kelvin probe force microscopy, on solid surfaces such as gold, silicon carbide, and indium–tin oxide coated glass [53].

His tags were also engineered into PSI proteins to serve as binding sites to electrode surfaces functionalized with Ni-nitrilotriacetic acid (Ni-NTA) [48]. The same strategy was applied to immobilize PSII proteins on nanostructured gold electrodes [54, 55, 56, 57, 58]. Current densities and long-term stability were improved by immobilizing PSII proteins on gold electrode surfaces via osmium-containing redox polymers, which also acted as electron acceptors [59]. PSII proteins were integrated in a mesoporous indium–tin oxide electrode [60]. Upon photoexcitation, an electron flow to the electrode from quinones of PSII (Q_A and Q_B) was observed with concurrent water oxidation at the Mn_4Ca cluster. Water oxidation activity was also fully retained for PSII proteins bound to gold nanoparticles dispersed in solution [57]. Four to five PSII dimers were bound to a single particle and the oxygen activity was measured in the presence of exogenous electron acceptors. A photobiofuel cell was constructed with PSII-functionalized photoanodes and bilirubin oxidase/carbon nanotubes-modified cathode operating in aqueous solutions [61]. Upon illumination of the photoanodes, electricity was generated by electrons from water oxidation getting transferred to the cathode where O_2 is rereduced to water by bilirubin oxidase.

4.4   HYDROGEN PRODUCTION BY PHOTOSYSTEM I HYBRID SYSTEMS

Protons were reduced to H_2 gas by electrons delivered by PSI using metallic platinum (Pt) as the catalyst [62]. Photoprecipitation of Pt took place at the ferredoxin docking site of PSI where negatively charged $[PtCl_6]^{2-}$ were reduced. The electrons were supplied to the oxidized PSI proteins from sodium ascorbate by plastocyanin proteins, which are also needed for H_2 evolution. The covalent linkage of plastocyanin to platinized spinach PSI by a crosslinking reagent increased the hydrogen photoevolution by three-fold [63]. The system was improved further and a maximum yield of 5.5 μmol H_2/mg Chl/h was achieved using Cyt c_6 [64]. Cyt c_6 is the native electron donor to PSI in cyanobacteria.

In another approach, electrons from PSI were directly delivered to a [FeFe]-hydrogenase for making H_2 gas by connecting their electron transfer chains [65]. A molecular wire was used to connect the two proteins between their Fe–S clusters by Fe–S coordination bonds. Open coordination sites for the molecular wire (1,6-hexanedithiol) were introduced into these clusters by changing a Cys residue to a Gly through mutagenesis. The highest rates of light-induced H_2 production were obtained in the presence of Cyt c_6, ascorbate, and phenazine methosulfate as electron donors. The engineered PSI was a rebuilt protein made of PSI core and a recombinant PsaC subunit with the Cys to Gly mutation. The system was optimized further by crosslinking Cyt c_6 to PSI [66]. An average rate of 2.2 mmol H_2/mg Chl/h was reported for the light-induced H_2 evolution with octanedithiol as the molecular wire. The sample maintained the H_2-evolving ability over the course of 100 days at room temperature under anoxic conditions. The 1,6-hexanedithiol linker was used previously to conjugate the engineered PSI protein to Pt or Au nanoparticles for light-generated H_2 production [67]. 1,4-Benzenedithiol linker was also used to connect rebuilt PSI proteins to Pt nanoparticles [68]. Plastocyanin was again crosslinked to PSI to enhance the rate of electron donation to P700 and therefore increase H_2 generation. Membrane bound [NiFe]-hydrogenases (MBH) were also fused to PsaE subunits of PSI by genetic engineering at the electron-transferring subunits [69, 70, 71]. The fused hydrogenase enzyme was coupled to a PSI protein lacking the PsaE subunit. The hybrid protein complex with a His-tagged PSI protein was immobilized onto a gold electrode modified by Ni-NTA [69]. Electrons from the electrode were transferred to PSI by a soluble organic electron carrier. An oxygen tolerant [NiFe]-hydrogenase was used in this system and generated up to 3.0 mmol H_2/mg Chl/h (Fig. 4.4).

4.5 MIMICKING WATER OXIDATION WITH MANGANESE COMPLEXES

Photosystem II has evolved to catalyze the energetically demanding reaction of water oxidation. Inspired from its function and the discovery path to reveal the secret of this extraordinary enzyme, various manganese complexes were synthesized. In these model systems, an organic ligand framework is needed to bring the Mn ions together. Nitrogen and oxygen containing ligands are usually employed to obtain Mn clusters with μ-oxo bridges. These complexes were synthesized to achieve highly active water oxidation catalysts stable and robust enough for the fabrication of nanostructured oxygen-evolving anodes. For example, a dinuclear Mn(III) compound was

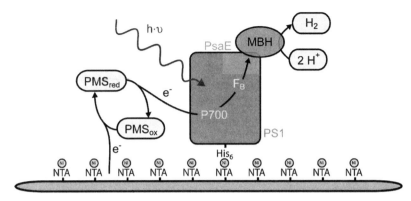

FIGURE 4.4 Making hydrogen gas by a fused PSI/hydrogenase enzyme system immobilized onto a gold electrode by Ni-NTA. Reproduced with permission from Krassen et al. [69], copyright ©2009 American Chemical Society.

synthesized capable of O_2 evolution in aqueous solution while illuminated by visible light in the presence of *p*-benzoquinone [72]. The proposed mechanism involved the formation of O_2 from two water molecules bridging the Mn ions. Another Mn(III/IV) dimer was shown to evolve oxygen in the presence of oxone, a highly oxidizing oxygen atom transfer reagent [73]. Initial rates of O_2 evolution for the Mn dimer ([Mn] = 125 µM) with terpyridine ligands was 61 µmol/h. This di-µ-oxo Mn dimer was also used to catalyze homogeneous water oxidation in the presence of sodium hypochlorite as the oxygen atom donor oxidant [74]. The oxidant was also responsible for changing the oxidation state of the dimer from Mn(III,III) to Mn(IV,IV). The mechanism proposed for oxygen formation included a nucleophilic attack of OH^- on the oxo group of Mn(V)=O. The active precatalyst dimer was then identified as Mn(III,IV) in the catalytic cycle of O_2 evolution with oxone [75]. The electrochemical oxidation of this dimer was later shown to induce the formation of a dimer-of-dimers [76]. The resulting tetranuclear Mn complex was then shown to catalyze water oxidation in aqueous solution when adsorbed on kaolin clay in the presence of Ce^{4+} as the oxidant [77]. The O_2 formation was proposed to occur by the reaction of water with a Mn-bound oxyl radical species.

Another Mn(III,III) dimer with imidazole and carboxylate groups as ligands oxidized water to O_2 in the presence of Ru complexes as single-electron oxidants or photosensitizers [78]. The complex was prepared as Mn(II,III) in organic solvents but got oxidized by atmospheric oxygen in aqueous solution shown by electron paramagnetic resonance spectroscopy. Mechanistically, water oxidation was then expected to occur by the complex cycling between the Mn(III,III) and Mn(V,V) oxidation states. Using Ru complexes as photosensitizers, Mn dimers adsorbed on a mica layer evolved O_2 under irradiation by visible light with persulfate as the electron acceptor [79]. The Mn oxidations and subsequent O_2 evolution were most likely initiated by electron transfer to the photoexcited Ru complexes. Successive photoinduced oxidations could be the result of electron hopping from one Ru complex to another until electrons were delivered to $S_2O_8^{2-}$ in the liquid phase.

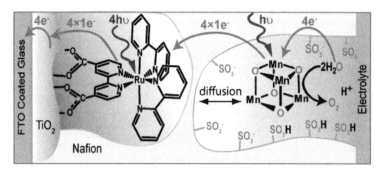

FIGURE 4.5 Artificial water oxidation by a manganese molecular catalyst coupled with Ru(II) photosensitizers coated on TiO_2 layers. Reproduced with permission from Brimblecombe et al. [81], copyright ©2010 American Chemical Society.

A manganese cubane cluster $[Mn_4O_4L_6]^+$ facilitated photocatalytic oxidation of water to O_2 when immobilized within a Nafion layer on a conducting electrode poised at 1 V [80]. The polymeric Nafion with hydrophobic pockets and anionic sulfonate groups provided binding sites for the hydrophobic and cationic Mn clusters and direct contact with the aqueous electrolyte. Nafion also assisted the water oxidation by transporting away the protons and stabilizing the Mn tetramer. It was postulated that O_2 was released upon the photodissociation of one phosphinate ligand from the Mn complex. A photoelectrochemical cell was also constructed by coupling these tetranuclear Mn–oxo clusters in Nafion films to Ru(II) sensitizers coated on TiO_2 layers [81,82]. This time only visible light was needed since the water oxidation was coupled to the reduction of photooxidized Ru dye with TiO_2 as the electron acceptor (Fig. 4.5). Another semiconductor/catalyst system made of nanocrystals of tungsten trioxide coated with a Mn catalyst was also built for the purpose of splitting water to O_2 under illumination [83]. It was postulated that the photogenerated holes from the semiconductor were responsible for oxidizing the Mn catalyst as the site of water oxidation. The system was operated without any sacrificial oxidant and its efficiency only dropped to 50% after 19 h.

4.6 PROTEIN DESIGN FOR INTRODUCING MANGANESE CHEMISTRY IN PROTEINS

In addition to PSII, Mn atoms are versatile cofactors found also in other enzymes such as arginase, aminopeptidase P, superoxide dismutase, and catalase. Manganese superoxide dismutase cycles between Mn(II) and Mn(III) to convert superoxide radical to oxygen and hydrogen peroxide. The manganese atom is bound in a trigonal bipyramidal geometry to the protein by three histidines and one aspartic acid [84]. Arginase uses a dimanganese active site to catalyze the hydrolysis of L-arginine to L-ornithine and urea [85]. Aminopeptidase P, which catalyzes the release of the N-terminal residue of a peptide with a proline as the second residue, also contains a dimanganese center [86]. It is suggested that Mn atoms can increase the acidity of the bridging water ligand to make a good nucleophile for attacking the substrate in both cases of arginase and aminopeptidase P. Nonheme Mn-containing catalases use their oxo-bridged Mn dimer to decompose hydrogen peroxide to O_2 and water by undergoing an oxidation/reduction cycle [87].

Protein design has offered a new approach to incorporate Mn ions into proteins. Metals can be ligated to proteins by amino acid side chains especially N of His imidazole, S of Cys thiolate, and O of Asp and Glu carboxylates. An artificial Mn binding site was created near the heme of cytochrome c peroxidase making it an active functional model for manganese peroxidase [88]. A dimeric protein formed from two helix–loop–helix peptides was shown to bind a di-Mn cluster by Glu and His residues in the hydrophobic core [89]. The two Mn(II) atoms were 3.6–3.9 Å apart and had vacant ligation sites to coordinate water molecules [90]. A helix–loop–helix polypeptide was specially designed with a binding site for a Mn ion made of three His and one Asp [91]. The designed peptide was specific to Mn(II) since it did not bind copper or iron and showed some superoxide dismutation activities.

Metaloprotein with prefabricated Mn complexes can also be constructed. For example, a Mn–salen complex with methyl thiosulfonate linking arms was introduced into the heme pocket of apo myoglobin by anchoring it to Cys residues [92]. Mn–salen complexed to albumin displayed efficient and selective oxidation of sulfide to sulfoxide [93]. Myoglobin was turned into an enzyme with some sulfoxidation activity, which depended on the anchoring position of the Mn–salen cofactor. Albumin was turned into an artificial superoxide dismutase by noncovalent conjugation of Mn(III) salophen [94]. A Mn(III)–porphyrin inserted into xylanase 10A enzyme showed enantioselectivity to p-methoxystyrene [95]. The anionic porphyrin had most likely inserted itself into the catalytic site of xylanase 10A which carries positively charged residues. By introducing Mn in place of Zn in carbonic anhydrase, the enzyme showed peroxidase activity oxidizing o-dianisidine and catalyzing the enantioselective epoxidation of styrene [96,97].

4.7 PROTEIN DESIGN AND PHOTOACTIVE PROTEINS WITH Chl DERIVATIVES

Chl and heme are the two main natural pigments that proteins use for their function. With some exception, they are both ligated to the protein through the central metal. The binding sites for Chls in natural proteins are tailored so that both the porphyrin head and the phytol tail are buried within the protein. Proteins usually use the side chain of an amino acid as an axial ligand to bind the Chl molecule through the metal center.

While most Chl-containing proteins are membrane proteins, nature has also made water-soluble Chl binding proteins (WSCP) and peridinin–Chl a-protein (PCP) [98, 99]. In addition to their natural Chl a groups these proteins can also bind BChl a, Chl b, or Chl d while maintaining a very similar fold [100, 101]. Chl b and d are structurally identical to Chl a except for the substitution of the methyl to a formyl group for Chl *b* and the vinly to a formyl group in Chl *d*. The PCP protein is a trimer with pigments organized into two clusters, each containing one Chl and four peridinin molecules positioned in the hydrophobic space between the two domains of each monomer. The Chl *a* molecule in PCP protein is coordinated from its central Mg atom to a water molecule, which is fixed by a His residue [102]. The distance between the Chl molecules in the two clusters is 17.4 Å. WSCP from cauliflower exists as a tetramer where it binds four tightly packed Chl molecules in the hydrophobic cavity formed in the center of the complex [103, 104] Chls are pentacoordinated through the central Mg ion to the backbone carbonyl group of proline residues [105]. Two Chl dimers form an open sandwich structure with the Chl planes inclined with respect to each other and separated by 9.0 Å at the center. This made WSCP an interesting model protein to study the spectroscopic properties of Chl in a single well-defined protein environment and its exciton interactions with nearby Chl molecules [106, 107]. The reconstitution of these proteins with Chl molecules happens almost instantly when the apo-protein is mixed with thylakoid membranes. However, they do not easily bind pure Chls in an aqueous solution possibly due to Chl aggregation. Similarly, binding studies of Chls to peptides require long incubation times and the presence of a detergent to increase the solubility of Chl molecules. For example, an artificial high-harvesting complex was created from synthetic α-helical peptides (14 residues long) containing a single His residue as the ligation site for BChl or Zn-BChl in the presence of *n*-octyl β-D-glycopyranoside detergent [108]. Using peptides as binding scaffold, a Chl *a* dimer was constructed by adding Chl dissolved in THF to a peptide containing a His residue [109]. The Chl *a* dimer bound to the peptide was red-shifted, giving rise to an absorption peak at 685 nm. In photosystems, the majority of Chl molecules are bound to membrane proteins, which have high proportion of hydrophobic amino acids. Membrane spanning peptides were then synthesized to make natural-like peptide/Chl complexes [110, 111, 112]. These hydrophobic peptides can partition readily into micelles especially when associated with Chls. A maquette was specially designed containing both lipophilic and hydrophilic domains to facilitate their incorporation into membranes [113, 114]. The binding sites for Chl molecules were included in the hydrophilic domains, which were capable of binding Ni-BChl. To avoid the need for detergent and hydrophobic domains, Chl derivatives can be prepared by removing the phytol tail. These derivatives are still photoactive just like their parent molecule of Chl because their extensive conjugated π molecular orbitals are unchanged. The central Mg can also be replaced by other metals. Some examples of photoactive Chl derivatives are chlorin *e*6 (C*e*6), pheophorbide, and tetracarboxyphenylchlorin. Water-soluble proteins and peptides can then bind these Chl derivatives through either metal ligation or covalent attachment. For example, Zn–pheophorbide *b* was covalently attached to a synthetic four-helix bundle protein through an aldehyde group and a modified lysine residue [115]. In another design, ligation through His residues allowed binding ZnC*e*6 to a synthetic four-helix bundle [116]. Photoactivity of the ZnC*e*6–peptide complex was tested in the presence of an external quinone electron acceptor with microsecond flashes as the excitation source. The electron transfer rate between the bound ZnCe6 and quinone was measured using time-resolved EPR spectroscopy. The light-induced oxidation of ZnCe6 was more efficient and its reduction through back reaction was accelerated compared to free ZnCe6 in solution. Synthetic quinones such as 2,6-dichloro-1,4-benzoquinone, *p*-benzoquinone, phenyl-*p*-benzoquinone, and 2,3 dimethoxy-5-methyl-1,4-benzoquinone are needed as efficient electron acceptors for multiple oxidations of the Chl derivative. Due to structural similarities between the Chl derivatives lacking the phytol tail and heme, hemoproteins become rational choices for binding Chl derivatives in their heme pocket. Apo-hemoproteins can readily be prepared by organic solvent extraction of heme. For example, apo-cyt b_{562} protein was successfully reconstituted with ZnCe6 [117]. The apo-cyt b_{562} protein was also modified carrying a covalently bound quinone as the electron acceptor for ZnCe6. Fluorescence lifetime measurements provided the ET rates between the ZnCe6 and the bound quinone inside the protein. A similar approach was applied to modify bacterioferritin protein by replacing its heme with ZnCe6 [118]. Upon illumination, the ZnCe6 was capable of oxidizing a dinuclear Mn center possibly through a tyrosine residue. The Mn ions were introduced in the natural di-iron binding site of bacterioferritin. These attempts for introducing the photoactive ZnCe6, efficient quinone as external electron acceptors, and Mn ions as electron donors in proteins are promising steps in creating semiartificial systems for mimicking water oxidation activity of PSII.

4.8 CONCLUSION

Some very good progress has been made in building artificial photosynthetic systems. Photosynthetic pigments are used as sensitizers in the construction of dye-sensitized solar cells. Many pigments are synthesized to improve their absorption spectra and abilities to anchor to conductive surfaces. Natural and engineered photosystem proteins are also immobilized on electrodes for photocurrent generation. Electrode surfaces are successfully functionalized to better adsorb the engineered photosystems in an oriented manner. Hydrogen gas is generated by modifying PSI proteins with metallic platinum or coupling them to hydroganese enzymes. Progress has also been made to extract electrons from water by preparing manganese complexes capable of performing water oxidation. Protein engineering is applied to introduce photoactivity and redox compounds into proteins to mimic electron transfer events of photosystems. These examples all indicate that artificial photosynthesis can become another promising future energy source.

ACKNOWLEDGMENT

This work was supported in part by Nova Southeastern University Faculty Research and Development Grant.

REFERENCES

1. H. Ishikita, W. Saenger, J. Biesiadka, B. Loll, E.-W. Knapp, How photosynthetic reaction centers control oxidation power in chlorophyll pairs P680, P700, and P870, Proc. Natl. Acad. Sci. U.S.A. 103 (2006) 9855–9860.

2. M. L. Groot, N. P. Pawlowicz, L. J. G. W. van Wilderen, J. Breton, I. H. M. van Stokkum, R. van Grondelle, Initial electron donor and acceptor in isolated photosystem II reaction centers identified with femtosecond mid-IR spectroscopy, Proc. Natl. Acad. Sci. U.S.A. 102 (2005) 13087–13092.

3. A. R. Holzwarth, M. G. Muller, M. Reus, M. Nowaczyk, J. Sander, M. Rogner, Kinetics and mechanism of electron transfer in intact photosystem II and in the isolated reaction center: pheophytin is the primary electron acceptor, Proc. Natl. Acad. Sci. U.S.A. 103 (2006) 6895–6900.

4. G. Raszewski, B. A. Diner, E. Schlodder, T. Renger, Spectroscopic properties of reaction center pigments in photosystem II core complexes: revision of the multimer model, Biophysical J. 95 (2008) 105–119.

5. B. Loll, J. Kern, W. Saenger, A. Zouni, J. Biesiadka, Towards complete cofactor arrangement in the 3.0 Å resolution structure of photosystem II, Nature 438 (2005) 1040–1044.

6. J. Barber, M. D. Archer, P680, the primary electron donor of photosystem II, J. Photochem. Photobiol. A: Chem. 142 (2001) 97–106.

7. L. Bjorn, G. C. Papageorgiou, R. E. Blankenship, Govindjee, A viewpoint: why chlorophyll *a*?, Photosynth. Res. 99 (2009) 85–98.

8. L. Kálmán, R. LoBrutto, J. P. Allen, J. C. Williams, Modified reaction centers oxidize tyrosine in reactions that mirror photosystem II, Nature 402 (1999) 696–699.

9. M. Thielges, G. Uyeda, A. Camara-Artigas, L. Kalman, J. C. Williams, J. P. Allen, Design of a redox-linked active metal site: manganese bound to bacterial reaction centers at a site resembling that of photosystem II, Biochemistry 44 (2005) 7389–7394.

10. L. Kalman, R. LoBrutto, J. P. Allen, J. C. Williams, Manganese oxidation by modified reaction centers from *Rhodobacter sphaeroides*, Biochemistry 42 (2003) 11016–11022.

11. L. Kalman, J. C. Williams, J. P. Allen, Energetics for oxidation of a bound manganese cofactor in modified bacterial reaction centers, Biochemistry 50 (2011) 3310–3320.

12. J. P. Allen, T. L. Olson, P. Oyala, W. J. Lee, A. A. Tufts, J. C. Williams, Light-driven oxygen production from superoxide by Mn-binding bacterial reaction centers, Proc. Natl. Acad. Sci. U.S.A. 109 (2012) 2314–2318.

13. Y. Umena, K. Kawakami, J. R. Shen, N. Kamiya, Crystal structure of oxygen-evolving photosystem II at a resolution of 1.9 Å, Nature 473 (2011) 55–60.

14. A. Zouni, H.-T. Witt, J. Kern, P. Fromme, N. Krauβ, W. Saenger, P. Orth, Crystal structure of photosystem II from *Synechococcus elongatus* at 3.8 Å resolution, Nature 409 (2001) 739–743.

15. K. N. Ferreira, T. M. Iverson, K. Maghlaoui, J. Barber, S. Iwata, Architecture of the photosynthetic oxygen-evolving centre, Science 303 (2004) 1831–1838.

16. T. A. Moore, D. Gust, P. Mathis, J. C. Mialocq, C. Chachaty, R. V. Bensasson, E. J. Land, D. Doizi, P. A. Liddell, W. R. Lehman, G. A. Nemeth, A. L. Moore, Photodriven charge separation in a carotenoporphyrin quinone triad, Nature 307 (1984) 630–632.

17. P. A. Liddell, J. P. Sumida, A. N. Macpherson, L. Noss, G. R. Seely, K. N. Clark, A. L. Moore, T. A. Moore, D. Gust, Preparation and photophysical studies of porphyrin C_{60} dyads, Photochemistry and Photobiology 60 (1994) 537–541.

18. H. Imahori, H. Norieda, H. Yamada, Y. Nishimura, I. Yamazaki, Y. Sakata, S. Fukuzumi, Light-harvesting and photocurrent generation by cold electrodes modified with mixed self-assembled monolayers of boron–dipyrrin and ferrocene–porphyrin–fullerene triad, J. Am. Chem. Soc. 123 (2001) 100–110.

19. M. Grätzel, Photoelectrochemical cells, Nature 414 (2001) 338–344.

20. M. Grätzel, K. Kalyanasundaram, Artificial photosynthesis—efficient dye-sensitized photoelectrochemical cells for direct conversion of visible-light to electricity, Current Science 66 (1994) 706–714.

21. H. Tsubomura, M. Matsumura, Y. Nomura, T. Amamiya, Dye sensitized zinc oxide: aqueous electrolyte: platinum photocell, Nature 261 (1976) 402–403.

22. W. M. Campbell, K. W. Jolley, P. Wagner, K. Wagner, P. J. Walsh, K. C. Gordon, L. Schmidt-Mende, M. K. Nazeeruddin, Q. Wang, M. Grätzel, D. L. Officer, Highly efficient porphyrin sensitizers for dye-sensitized solar cells, J. Phys. Chem. C. 111 (2007) 11760–11762.

23. M. Tanaka, S. Hayashi, S. Eu, T. Umeyama, Y. Matano, H. Imahori, Novel unsymmetrically π-elongated porphyrin for dye-sensitized TiO_2 cells, Chem. Commun. (2007) 2069–2071.

24. S. Hayashi, M. Tanaka, H. Hayashi, S. Eu, T. Umeyama, Y. Matano, Y. Araki, H. Imahori, Naphthyl-fused π-elongated porphyrins for dye-sensitized TiO_2 cells, J. Phys. Chem. C 112 (2008) 15576–15585.

25. M. Miyashita, K. Sunahara, T. Nishikawa, Y. Uemura, N. Koumura, K. Hara, A. Mori, T. Abe, E. Suzuki, S. Mori, Interfacial electron-transfer kinetics in metal-free organic dye-sensitized solar cells: combined effects of molecular structure of dyes and electrolytes, J. Am. Chem. Soc. 130 (2008) 17874–17881.

26. A. J. Mozer, M. J. Griffith, G. Tsekouras, P. Wagner, G. G. Wallace, S. Mori, K. Sunahara, M. Miyashita, J. C. Earles, K. C. Gordon, L. C. Du, R. Katoh, A. Furube, D. L. Officer, Zn–Zn porphyrin dimer-sensitized solar cells: toward 3-D light harvesting, J. Am. Chem. Soc. 131 (2009) 15621–15623.

27. M. K. Nazeeruddin, R. Humphry-Baker, D. L. Officer, W. M. Campbell, A. K. Burrell, M. Grätzel, Application of metalloporphyrins in nanocrystalline dye-sensitized solar cells for conversion of sunlight into electricity, Langmuir 20 (2004) 6514–6517.

28. M. J. Griffith, M. James, G. Triani, P. Wagner, G. G. Wallace, D. L. Officer, Determining the orientation and molecular packing of organic dyes on a TiO_2 surface using X-ray reflectometry, Langmuir 27 (2011) 12944–12950.

29. Y. Amao, Y. Yamada, K. Aoki, Preparation and properties of dye-sensitized solar cell using chlorophyll derivative immobilized TiO_2 film electrode, J. Photochem. Photobiol. A: Chem. 164 (2004) 47–51.

30. A. Kay, M. Grätzel, Artificial photosynthesis. 1. Photosensitization of TiO_2 solar-cells with chlorophyll derivatives and related natural porphyrins, J. Phys. Chem. 97 (1993) 6272–6277.

31. A. Kay, R. Humphrybaker, M. Grätzel, Artificial photosynthesis. 2. Investigations on the mechanism of photosensitization of nanocrystalline TiO_2 solar-cells by chlorophyll derivatives, J. Phys. Chem. 98 (1994) 952–959.

32. Y. Amao, K. Kato, Chlorophyll assembled electrode for photovoltaic conversion device, Electrochim. Acta 53 (2007) 42–45.

33. I. Bedja, S. Hotchandani, R. Carpentier, R. W. Fessenden, P. V. Kamat, Chlorophyll-b-modified nanocrystalline SnO_2 semiconductor thin-film as a photosensitive electrode, J. Appl. Phys. 75 (1994) 5444–5446.

34. X. F. Wang, H. Tamiaki, L. Wang, N. Tamai, O. Kitao, H. S. Zhou, S. Sasaki, Chlorophyll-a derivatives with various hydrocarbon ester groups for efficient dye-sensitized solar cells: static and ultrafast evaluations on electron injection and charge collection processes, Langmuir 26 (2010) 6320–6327.

35. X. F. Wang, O. Kitao, H. S. Zhou, H. Tamiaki, S. Sasaki, Extension of π-conjugation length along the Q_y axis of a chlorophyll a derivative for efficient dye-sensitized solar cells, Chem. Commun. (2009) 1523–1525.

36. S. Sasaki, K. Mizutani, M. Kunieda, H. Tamiaki, Synthesis and optical properties of C3-ethynylated chlorin and π-extended chlorophyll dyads, Tetrahedron 67 (2011) 6065–6072.

37. S. Barazzouk, S. Hotchandani, Enhanced charge separation in chlorophyll a solar cell by gold nanoparticles, J. Appl. Phys. 96 (2004) 7744–7746.

38. E. W. McFarland, J. Tang, A photovoltaic device structure based on internal electron emission, Nature 421 (2003) 616–618.

39. M. Grätzel, Applied physics—solar cells to dye for, Nature 421 (2003) 586–587.

40. X. F. Wang, C. H. Zhan, T. Maoka, Y. Wada, Y. Koyama, Fabrication of dye-sensitized solar cells using chlorophylls c'_1 and c'_2 and their oxidized forms c'_1 and c'_2 from *Undaria pinnatifida* (Wakame), Chem. Phys. Lett. 447 (2007) 79–85.

41. N. Terasaki, N. Yamamoto, T. Hiraga, I. Sato, Y. Inoue, S. Yamada, Fabrication of novel photosystem I–gold nanoparticle hybrids and their photocurrent enhancement, Thin Solid Films 499 (2006) 153–156.

42. Y. Yamanoi, N. Terasaki, M. Miyachi, Y. Inoue, H. Nishihara, Enhanced photocurrent production by photosystem I with modified viologen derivatives, Thin Solid Films 520 (2012) 5123–5127.

43. N. Terasaki, N. Yamamoto, K. Tamada, M. Hattori, T. Hiraga, A. Tohri, I. Sato, M. Iwai, M. Iwai, S. Taguchi, I. Enami, Y. Inoue, Y. Yamanoi, T. Yonezawa, K. Mizuno, M. Murata, H. Nishihara, S. Yoneyama, M. Minakata, T. Ohmori, M. Sakai, M. Fujii, Bio-photosensor: cyanobacterial photosystem I coupled with transistor via molecular wire, Biochim. Biophys. Acta-Bioenergetics 1767 (2007) 653–659.

44. M. Miyachi, Y. Yamanoi, Y. Shibata, H. Matsumoto, K. Nakazato, M. Konno, K. Ito, Y. Inoue, H. Nishihara, A photosensing system composed of photosystem I, molecular wire, gold nanoparticle, and double surfactants in water, Chem. Commun. 46 (2010) 2557–2559.

45. V. V. Nikandrov, Y. V. Borisova, E. A. Bocharov, M. A. Usachev, G. V. Nizova, V. A. Nadtochenko, E. P. Lukashev, B. V. Trubitsin, A. N. Tikhonov, V. N. Kurashov, M. D. Mamedov, A. Y. Semenov, Photochemical properties of photosystem I immobilized in a mesoporous semiconductor matrix, High Energy Chemistry 46 (2012) 200–205.

46. A. Mershin, K. Matsumoto, L. Kaiser, D. Y. Yu, M. Vaughn, M. K. Nazeeruddin, B. D. Bruce, M. Graetzel, S. G. Zhang, Self-assembled photosystem-I biophotovoltaics on nanostructured TiO_2 and ZnO, Scientific Reports 2 (2012) 1–7.

47. K. Matsumoto, M. Vaughn, B. D. Bruce, S. Koutsopoulos, S. G. Zhang, Designer peptide surfactants stabilize functional photosystem-I membrane complex in aqueous solution for extended time, J. Phys. Chem. B 113 (2009) 75–83.

48. R. Das, P. J. Kiley, M. Segal, J. Norville, A. A. Yu, L. Y. Wang, S. A. Trammell, L. E. Reddick, R. Kumar, F. Stellacci, N. Lebedev, J. Schnur, B. D. Bruce, S. G. Zhang, M. Baldo, Integration of photosynthetic protein molecular complexes in solid-state electronic devices, Nano Letters 4 (2004) 1079–1083.

49. C. J. Faulkner, S. Lees, P. N. Ciesielski, D. E. Cliffel, G. K. Jennings, Rapid assembly of photosystem I monolayers on gold electrodes, Langmuir 24 (2008) 8409–8412.

50. O. Yehezkeli, O. I. Wilner, R. Tel-Vered, D. Roizman-Sade, R. Nechushtai, I. Willner, Generation of photocurrents by bis-aniline-cross-linked Pt nanoparticle/photosystem I composites on electrodes, J. Phys. Chem. B 114 (2010) 14383–14388.

51. L. Frolov, Y. Rosenwaks, S. Richter, C. Carmeli, I. Carmeli, Photoelectric junctions between GaAs and photosynthetic reaction center protein, J. Phys. Chem. C 112 (2008) 13426–13430.

52. L. Frolov, Y. Rosenwaks, C. Carmeli, I. Carmeli, Fabrication of a photoelectronic device by direct chemical binding of the photosynthetic reaction center protein to metal surfaces, Advanced Materials 17 (2005) 2434–2437.

53. H. Toporik, I. Carmeli, I. Volotsenko, M. Molotskii, Y. Rosenwaks, C. Carmeli, N. Nelson, Large photovoltages generated by plant photosystem I crystals, Advanced Materials 24 (2012) 2988–2991.

54. A. Badura, B. Esper, K. Ataka, C. Grunwald, C. Woll, J. Kuhlmann, J. Heberle, M. Rogner, Light-driven water splitting for (bio-)hydrogen production: photosystem II as the central part of a bioelectrochemical device, Photochemistry and Photobiology 82 (2006) 1385–1390.

55. M. Vittadello, M. Y. Gorbunov, D. T. Mastrogiovanni, L. S. Wielunski, E. L. Garfunkel, F. Guerrero, D. Kirilovsky, M. Sugiura, A. W. Rutherford, A. Safari, P. G. Falkowski, Photoelectron generation by photosystem II core complexes tethered to gold surfaces, Chemsuschem 3 (2010) 471–475.

56. N. Terasaki, M. Iwai, N. Yamamoto, T. Hiraga, S. Yamada, Y. Inoue, Photocurrent generation properties of His tag-photosystem II immobilized on nanostructured gold electrode, Thin Solid Films 516 (2008) 2553–2557.

57. T. Noji, H. Suzuki, T. Gotoh, M. Iwai, M. Ikeuchi, T. Tomo, T. Noguchi, Photosystem II–gold nanoparticle conjugate as a nanodevice for the development of artificial light-driven water-splitting systems, J. Phys. Chem. Lett. 2 (2011) 2448–2452.

58. J. Maly, J. Krejci, M. Ilie, L. Jakubka, J. Masojidek, R. Pilloton, K. Sameh, P. Steffan, Z. Stryhal, M. Sugiura, Monolayers of photosystem II on gold electrodes with enhanced sensor response—effect of porosity and protein layer arrangement, Analytical and Bioanalytical Chemistry 381 (2005) 1558–1567.

59. A. Badura, D. Guschin, B. Esper, T. Kothe, S. Neugebauer, W. Schuhmann, M. Rogner, Photo-induced electron transfer between photosystem II via cross-linked redox hydrogels, Electroanalysis 20 (2008) 1043–1047.

60. M. Kato, T. Cardona, A. W. Rutherford, E. Reisner, Photoelectrochemical water oxidation with photosystem II integrated in a mesoporous, indium tin oxide electrode, J. Am. Chem. Soc. 134 (2012) 8332–8335.

61. O. Yehezkeli, R. Tel-Vered, J. Wasserman, A. Trifonov, D. Michaeli, R. Nechushtai, I. Willner, Integrated photosystem II-based photo-bioelectrochemical cells, Nature Commun. 3 (2012) 1–7.

62. J. F. Millsaps, B. D. Bruce, J. W. Lee, E. Greenbaum, Nanoscale photosynthesis: photocatalytic production of hydrogen by platinized photosystem I reaction centers, Photochemistry and Photobiology 73 (2001) 630–635.

63. B. R. Evans, H. M. O'Neill, S. A. Hutchens, B. D. Bruce, E. Greenbaum, Enhanced photocatalytic hydrogen evolution by covalent attachment of plastocyanin to photosystem I, Nano Letters 4 (2004) 1815–1819.

64. I. J. Iwuchukwu, M. Vaughn, N. Myers, H. O'Neill, P. Frymier, B. D. Bruce, Self-organized photosynthetic nanoparticle for cell-free hydrogen production, Nature Nanotechnology 5 (2010) 73–79.

65. C. E. Lubner, P. Knorzer, P. J. N. Silva, K. A. Vincent, T. Happe, D. A. Bryant, J. H. Golbeck, Wiring an [FeFe]-hydrogenase with photosystem I for light-induced hydrogen production, Biochemistry 49 (2010) 10264–10266.

66. C. E. Lubner, A. M. Applegate, P. Knorzer, A. Ganago, D. A. Bryant, T. Happe, J. H. Golbeck, Solar hydrogen-producing bionanodevice outperforms natural photosynthesis, Proc. Natl. Acad. Sci. U.S.A. 108 (2011) 20988–20991.

67. R. A. Grimme, C. E. Lubner, D. A. Bryant, J. H. Golbeck, Photosystem I/molecular wire/metal nanoparticle bioconjugates for the photocatalytic production of H-2, J. Am. Chem. Soc. 130 (2008) 6308–6309.

68. R. A. Grimme, C. E. Lubner, J. H. Golbeck, Maximizing H_2 production in photosystem I/dithiol molecular wire/platinum nanoparticle bioconjugates, Dalton Trans. (2009) 10106–10113.

69. H. Krassen, A. Schwarze, B. Friedrich, K. Ataka, O. Lenz, J. Heberle, Photosynthetic hydrogen production by a hybrid complex of photosystem I and [NiFe]-hydrogenase, ACS Nano 3 (2009) 4055–4061.

70. M. Ihara, H. Nishihara, K. S. Yoon, O. Lenz, B. Friedrich, H. Nakamoto, K. Kojima, D. Honma, T. Kamachi, I. Okura, Light-driven hydrogen production by a hybrid complex of a [NiFe]-hydrogenase and the cyanobacterial photosystem I, Photochemistry and Photobiology 82 (2006) 676–682.

71. M. Ihara, H. Nakamoto, T. Kamachi, I. Okura, M. Maeda, Photoinduced hydrogen production by direct electron transfer from photosystem I cross-linked with cytochrome c_3 to [NiFe]-hydrogenase, Photochemistry and Photobiology 82 (2006) 1677–1685.

72. M. Watkinson, A. Whiting, C. A. Mcauliffe, Synthesis of a bis-manganese water-splitting complex, J. Chem. Soc. Chem. Commun. (1994) 2141–2142.

73. J. Limburg, G. W. Brudvig, R. H. Crabtree, O-2 evolution and permanganate formation from high-valent manganese complexes, J. Am. Chem. Soc. 119 (1997) 2761–2762.

74. J. Limburg, J. S. Vrettos, L. M. Liable-Sands, A. L. Rheingold, R. H. Crabtree, G. W. Brudvig, A functional model for O–O bond formation by the O_2-evolving complex in photosystem II, Science 283 (1999) 1524–1527.

75. H. Y. Chen, R. Tagore, G. Olack, J. S. Vrettos, T. C. Weng, J. Penner-Hahn, R. H. Crabtree, G. W. Brudvig, Speciation of the catalytic oxygen evolution system: $Mn^{III/IV}_2(\mu\text{-}O)_2(terpy)_2(H_2O)_2](NO_3)_3 + HSO_5^-$, Inorg. Chem. 46 (2007) 34–43.

76. H. Y. Chen, J. W. Faller, R. H. Crabtree, G. W. Brudvig, Dimer-of-dimers model for the oxygen-evolving complex of photosystem II. Synthesis and properties of $[Mn^{IV}_4O_5(terpy)_4(H_2O)_2](ClO_4)_6$, J. Am. Chem. Soc. 126 (2004) 7345–7349.

77. Y. L. Gao, R. H. Crabtree, G. W. Brudvig, Water oxidation catalyzed by the tetranuclear Mn complex $[Mn^{IV}_4O_5(terpy)_4(H_2O)_2](ClO_4)_6$, Inorg. Chem. 51 (2012) 4043–4050.

78. E. A. Karlsson, B. L. Lee, T. Akermark, E. V. Johnston, M. D. Karkas, J. L. Sun, O. Hansson, J. E. Backvall, B. Akermark, Photosensitized water oxidation by use of a bioinspired manganese catalyst, Angew. Chem. Int. Ed. 50 (2011) 11715–11718.

79. M. Yagi, M. Toda, S. Yamada, H. Yamazaki, An artificial model of photosynthetic photosystem II: visible-light-derived O_2 production from water by a di-μ-oxo-bridged manganese dimer as an oxygen evolving center, Chem. Commun. 46 (2010) 8594–8596.

80. R. Brimblecombe, J. Chen, P. Wagner, T. Buchhorn, G. C. Dismukes, L. Spiccia, G. F. Swiegers, Photocatalytic oxygen evolution from non-potable water by a bioinspired molecular water oxidation catalyst, J. Mol. Catalysis a—Chemical 338 (2011) 1–6.

81. R. Brimblecombe, A. Koo, G. C. Dismukes, G. F. Swiegers, L. Spiccia, Solar driven water oxidation by a bioinspired manganese molecular catalyst, J. Am. Chem. Soc. 132 (2010) 2892–2894.

82. R. Brimblecombe, A. Koo, G. C. Dismukes, G. F. Swiegers, L. Spiccia, A tandem water-splitting device based on a bio-inspired manganese catalyst, Chemsuschem 3 (2010) 1146–1150.

83. R. Liu, Y. J. Lin, L. Y. Chou, S. W. Sheehan, W. S. He, F. Zhang, H. J. M. Hou, D. W. Wang, Water splitting by tungsten oxide prepared by atomic layer deposition and decorated with an oxygen-evolving catalyst, Angew. Chem. Int. Ed. 50 (2011) 499–502.

84. R. A. Edwards, H. M. Baker, M. M. Whittaker, J. W. Whittaker, G. B. Jameson, E. N. Baker, Crystal structure of *Escherichia coli* manganese superoxide dismutase at 2.1-angstrom resolution, J. Biol. Inorg. Chem. 3 (1998) 161–171.

85. M. C. J. Wilce, C. S. Bond, N. E. Dixon, H. C. Freeman, J. M. Guss, P. E. Lilley, J. A. Wilce, Structure and mechanism of a proline-specific aminopeptidase from *Escherichia coli*, Proc. Natl. Acad. Sci. U.S.A. 95 (1998) 3472–3477.

86. S. C. Graham, J. M. Guss, Complexes of mutants of *Escherichia coli* aminopeptidase P and the tripeptide substrate ValProLeu, Arch. Biochem. and Biophys. 469 (2008) 200–208.

87. V. V. Barynin, M. M. Whittaker, S. V. Antonyuk, V. S. Lamzin, P. M. Harrison, P. J. Artymiuk, J. W. Whittaker, Crystal structure of manganese catalase from *Lactobacillus plantarum*, Structure 9 (2001) 725–738.

88. S. K. Wilcox, C. D. Putnam, M. Sastry, J. Blankenship, W. J. Chazin, D. E. McRee, D. B. Goodin, Rational design of a functional metalloenzyme: introduction of a site for manganese binding and oxidation into a heme peroxidase, Biochemistry 37 (1998) 16853–16862.

89. L. D. Costanzo, H. Wade, S. Geremia, L. Randaccio, V. Pavone, W. F. DeGrado, A. Lombardi, Toward the *de novo* design of a catalytically active helix bundle: a substrate-accessible carboxylate-bridged dinuclear metal centre, J. Am. Chem. Soc. 123 (2001) 12749–12757.

90. W. F. DeGrado, L. D. Costanzo, S. Geremia, A. Lombardi, V. Pavone, L. Randaccio, Sliding helix and change of coordination geometry in a model di-MnII protein, Angew. Chem. Int. Ed. 42 (2003) 417–420.

91. U. P. Singh, R. K. Singh, Y. Isogai, Y. Shiro, Design and synthesis of de novo peptide for manganese binding, Int. J. Peptide Res. Therapeutics 12 (2006) 379–385.

92. D. K. Garner, L. Liang, D. A. Barrios, J. L. Zhang, Y. Lu, The important role of co-valent anchor positions in tuning catalytic properties of a rationally designed MnSalen-containing metalloenzyme, ACS Catalysis 1 (2011) 1083–1089.

93. P. Rousselot-Pailley, C. Bochot, C. Marchi-Delapierre, A. Jorge-Robin, L. Martin, J. C. Fontecilla-Camps, C. Cavazza, S. Menage, The protein environment drives selectivity for sulfide oxidation by an artificial metalloenzyme, Chembiochem 10 (2009) 545–552.

94. V. Oliveri, G. Vecchio, A novel artificial superoxide dismutase: non-covalent conjugation of albumin with a Mn-III salophen type complex, Eur. J. Med. Chem. 46 (2011) 961–965.

95. M. Allard, C. Dupont, V. M. Robles, N. Doucet, A. Lledos, J. D. Marechal, A. Urvoas, J. P. Mahy, R. Ricoux, Incorporation of manganese complexes into xylanase: new artificial metalloenzymes for enantioselective epoxidation, Chembiochem 13 (2012) 240–251.

96. A. Fernandez-Gacio, A. Codina, J. Fastrez, O. Riant, P. Soumillion, Transforming car-bonic anhydrase into epoxide synthase by metal exchange, Chembiochem 7 (2006) 1013–1016.

97. K. Okrasa, R. J. Kazlauskas, Manganese-substituted carbonic anhydrase as a new perox-idase, Chemistry—A European Journal 12 (2006) 1587–1596.

98. H. Satoh, A. Uchida, K. Nakayama, M. Okada, Water-soluble chlorophyll protein in Bras-sicaceae plants is a stress-induced chlorophyll-binding protein, Plant & Cell Physiology 42 (2001) 906–911.

99. E. Hofmann, P. M. Wrench, F. P. Sharples, R. G. Hiller, W. Welte, K. Diederichs, Structural basis of light harvesting by carotenoids: peridinin–chlorophyll–protein from *Amphidinium carterae*, Science 272 (1996) 1788–1791.

100. T. Schulte, R. G. Hiller, E. Hofmann, X-ray structures of the peridinin–chlorophyll–protein reconstituted with different chlorophylls, FEBS Lett. 584 (2010) 973–978.

101. J. L. Hughes, R. Razeghifard, M. Logue, A. Oakley, T. Wydrzynski, E. Krausz, Magneto-optic spectroscopy of a protein tetramer binding two exciton-coupled chlorophylls, J. Am. Chem. Soc. 128 (2006) 3649–3658.

102. T. Schulte, F. P. Sharples, R. G. Hiller, E. Hofmann, X-ray structure of the high-salt form of the peridinin–chlorophyll *a*–protein from the dinoflagellate *Amphidinium carterae*: modulation of the spectral properties of pigments by the protein environment, Biochem-istry 48 (2009) 4466–4475.

103. K. Schmidt, C. Fufezan, A. Krieger-Liszkay, H. Satoh, H. Paulsen, Recombinant water-soluble chlorophyll protein from *Brassica oleracea* Var. botrys binds various chlorophyll derivatives, Biochemistry 42 (2003) 7427–7433.

104. H. Satoh, K. Nakayama, M. Okada, Molecular cloning and functional expression of a water-soluble chlorophyll protein, a putative carrier of chlorophyll molecules in cauliflower, J. Biol. Chem. 273 (1998) 30568–30575.

105. D. Horigome, H. Satoh, N. Itoh, K. Mitsunaga, I. Oonishi, A. Nakagawa, A. Uchida, Structural mechanism and photoprotective function of water-soluble chlorophyll-binding protein, J. Biol. Chem. 282 (2007) 6525–6531.

106. J. L. Hughes, R. Razeghifard, M. Logue, A. Oakley, T. Wydrzynski, E. Krausz, Magneto-optic spectroscopy of a protein tetramer binding two exciton-coupled chlorophylls, J. Am. Chem. Soc. 128 (2006) 3649–3658.

107. C. Theiss, I. Trostmann, S. Andree, F. J. Schmitt, T. Renger, H. J. Eichler, H. Paulsen, G. Renger, Pigment–pigment and pigment–protein interactions in recombinant water-soluble chlorophyll proteins (WSCP) from cauliflower, J. Phys. Chem. B 111 (2007) 13325–13335.

108. A. Kashiwada, N. Nishino, Z.-Y. Wang, T. Nozawa, M. Kobayashi, M. Nango, Molecular assembly of bacteriochlorophyll a and its analogues by synthetic 4α-helix polypeptides, Chem. Lett. 2 (1999) 1301–1302.

109. A. Dudkowiak, T. Kusumi, C. Nakamura, J. Miyake, Chlorophyll a aggregates stabilized by a synthesized peptide, J. Photochem. Photobiol. B: Biol. 129 (1999) 51–55.

110. K. A. Meadows, P. S. Parkes-Loach, J. W. Kehoe, P. A. Loach, Reconstitution of core light-harvesting complexes of photosynthetic bacteria using chemically synthesized polypeptides. 1. Minimal requirements for subunit formation, Biochemistry 37 (1998) 3411–3417.

111. J. W. Kehoe, K. A. Meadows, P. S. Parkesloach, P. A. Loach, Reconstitution of core light-harvesting complexes of photosynthetic bacteria using chemically synthesized polypeptides. 2. Determination of structural features that stabilize complex formation and their implications for the structure of the subunit complex, Biochemistry 37 (1998) 3418–3428.

112. L. L. Eggink, J. K. Hoober, Chlorophyll binding to peptide maquettes containing a retention motif, J. Biol. Chem. 275 (2000) 9087–9090.

113. B. M. Discher, R. L. Koder, C. C. Moser, L. P. Dutton, Hydrophilic to amphiphilic design in redox protein maquettes, Curr. Opin. Chem. Biol. 7 (2003) 741–748.

114. D. Noy, P. L. Dutton, Design of a minimal polypeptide unit for bacteriochlorophyll binding and self-assembly based on photosynthetic bacterial light-harvesting proteins, Biochemistry 45 (2006) 2103–2113.

115. H. K. Rau, H. Snigula, A. Struck, B. Robert, H. Scheer, W. Haehnel, Design, synthesis and properties of synthetic chlorophyll proteins, Eur. J. Biochem. 268 (2001) 3284–3295.

116. A. R. Razeghifard, T. Wydrzynski, Binding of Zn-chlorin to a synthetic four-helix bundle peptide through histidine ligation, Biochemistry 42 (2003) 1024–1030.

117. S. Hay, B. B. Wallace, T. A. Smith, K. P. Ghiggino, T. Wydrzynski, Protein engineering of cytochrome b(562) for quinone binding and light-induced electrons transfer, Proc. Natl. Acad. Sci. U.S.A. 101 (2004) 17675–17680.

118. B. Conlan, N. Cox, J. H. Su, W. Hillier, J. Messinger, W. Lubitz, P. L. Dutton, T. Wydrzynski, Photo-catalytic oxidation of a di-nuclear manganese centre in an engineered bacterioferritin "reaction centre", Biochim. Biophys. Acta—Bioenergetics 1787 (2009) 1112–1121.

119. M. Grätzel, Dye-sensitized solar cells, J. Photochem. Photobiol. C—Photochem. Rev. 4 (2003) 145–153.

CHAPTER 5

Artificial Photosynthesis: Ruthenium Complexes

DIMITRIOS G. GIARIKOS

5.1 RUTHENIUM(II)

Ruthenium is a rare transition metal that belongs to the platinum group. Even though the metal is rare and its cost quite high, ruthenium complexes, and more specifically ruthenium(II) polypyridyl complexes, are ranked among the most researched compounds in inorganic chemistry mainly due to their unique photochemical properties. To better understand the photochemical properties of Ru(II) complexes, the electronic properties of ruthenium have to be examined, more specifically the metal ion Ru(II) but also the ligands associated with the complex by looking at both its crystal-field and ligand-field theory models. Both the central metal ion and the ligand interactions contribute to the complexes' photochemical properties.

Ru(II) complexes have a d^6 metal ion and typically a low spin octahedral configuration. The d orbitals in the octahedral crystal field split into a lower triple degenerate set (t_{2g}) and a higher energy double degenerate set (e_g) as shown in Fig. 5.1. The separation of the two sets of orbitals is the ligand-field splitting parameter for an octahedral structure, depicted as Δ_o. A complex has a low spin configuration when the value of Δ_o increases and a high spin configuration when the value of Δ_o decreases. The increase occurs for metal ions with higher oxidation states and also down a group (e.g., Fe, Ru, and Os). The higher oxidation states reflect the smaller size of the ions and consequently the shorter bond distance between the metal ion and the ligand which causes stronger interactions. Increasing down the group reflects the larger size of the 4d and 5d orbitals compared to the 3d orbitals, which leads to stronger metal ligand interactions by larger orbital overlap. The ligand-field splitting parameter Δ_o also varies with the identity of the ligand. Ligands are arranged in a spectrochemical

Natural and Artificial Photosynthesis: Solar Power as an Energy Source, First Edition. Edited by Reza Razeghifard.
© 2013 John Wiley & Sons, Inc. Published 2013 by John Wiley & Sons, Inc.

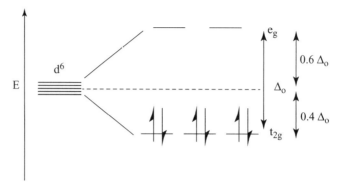

FIGURE 5.1   The energy diagram of the Ru(II) d orbitals in an octahedral crystal field.

series in order of increasing energy transitions that occur when they are part of a complex (the donor atom is underlined) [1]:

$$I^- < Br^- < S^{2-} < \underline{S}CN^- < Cl^- < N\underline{O}_2^- < N^{3-} < F^- < OH^- < C_2O_4^{2-}$$
$$< O^{2-} < H_2O < \underline{N}CS^- < CH_3CN < py(pyridine) < NH_3 < en(ethylenediamine)$$
$$< bpy(2, 2\text{-bipyridine}) < phen(1, 10\text{-phenanthroline}) < \underline{N}O_2^- < PPh_3 < \underline{C}N^- < CO$$

A ligand that increases the ligand-field splitting parameter Δ_o (such as CO, CN^-, or even phen and bpy) is considered a strong-field ligand, whereas one that decreases the parameter is a weak-field ligand. For the above reasons Ru(II) and especially Ru(II) polypyridyl complexes have a low spin configuration.

Looking at the metal ion alone, when Ru(II) absorbs visible light at a wavelength of approximately 450 nm, an electron from the t_{2g} orbital is promoted to the e_g orbital as shown in Fig. 5.2. More complex electronic transitions actually occur for photoactive Ru(II) complexes. These new transitions form from the interactions of the ligands' electronic transitions with Ru(II).

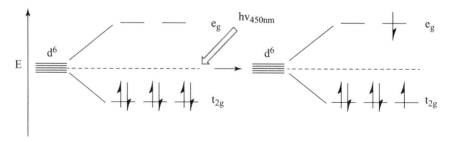

FIGURE 5.2   Promotion of the Ru(II) electron from the t_{2g} orbital to the e_g orbital.

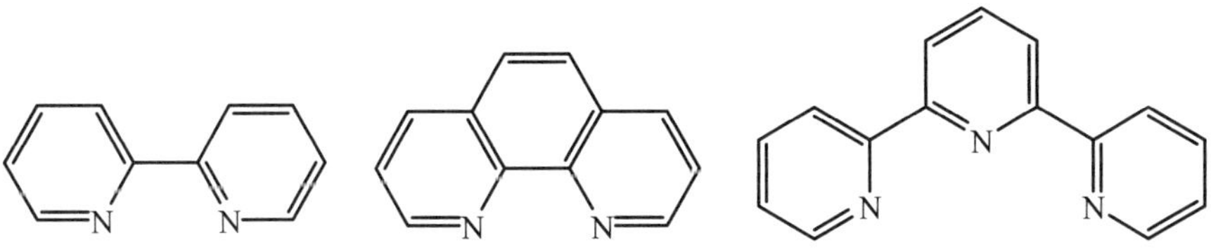

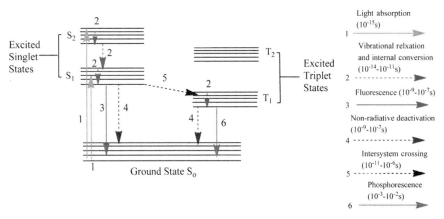

bpy   phen   tpy

FIGURE 5.3   The most common polypyridyl ligands used for transition metal complexes.

5.2   LIGAND INFLUENCE ON THE PHOTOCHEMISTRY OF Ru(II)

Due to their useful properties as sensitizers, polypyridyl ligands, which are all nitrogen-based heterocycles, have greatly been used to create transition metal complexes for use in various photonic and optoelectronic devices [2]. The most common polypyridyl ligands used are 2,2′-bipyridine (bpy), 1-10 phenanthroline (phen), and 2,2′:6′,2″-terpyridine (tpy), Fig. 5.3, which are considered to be strong-field ligands. These are stable, multidentate organic molecules that bind to Ru(II) through their nitrogen atoms.

When such molecules absorb light their electrons can populate upper excited states as shown in the Jablonski diagram in Fig. 5.4 [3,4]. These states are chemically different from their corresponding ground states. They are energetically unstable and have a very short life, in the range of nanoseconds. In the Jablonski diagram the molecule absorbs light in a vertical transition within a time frame ($\sim 10^{-15}$ s) that is too short for any nuclear displacement, according to the Frank–Cordon principle [5], and its electrons populate the upper excited singlet state S_2. The electrons then undergo vibrational relaxation (10^{-14}–10^{-11} s) and move to the lowest excited singlet level S_2. This high energy is dissipated very quickly by internal conversion, which

FIGURE 5.4   Jablonski diagram of organic molecules.

is a type of "horizontal energy" transition or a radiationless transition since it does not emit a photon. Vibrational relaxation occurs again and the electron moves to the lowest vibrational state of S_1, which may then undergo fluorescence (10^{-9}–10^{-7} s), giving off a photon or intersystem crossing to the triplet state T_1 (10^{11}–10^{-6} s). From the triplet state T_1, the electron can undergo an internal conversion (10^{-14}–10^{-11} s) and phosphorescence (10^{-3}–10^{-2} s), or deactivation. Once these ligands bind to a d^6 metal ion, there are three possible orbital disposition scenarios that can happen for a d^6 octahedral complex (Fig. 5.5).

In scenario 1, the intraligand π_L (HOMO) → π_L^* LUMO) electronic transitions are the lowest energy transitions. This occurs when the filled metal t_{2g} orbitals lie below the filled π_L orbitals of the ligand and the e_g orbitals lie above the ligand empty π_L^* orbitals. Eamples of scenario 1 are rhenium trispyridine and triphenanthroline complexes [6, 7, 8, 9]. In scenario 2, the filled metal t_{2g} orbital lies above the filled π_L ligand orbitals and the metal e_g orbitals lie below the empty ligand π_L^* orbitals. The lowest energy electronic transitions are the metal-centered Laporte forbidden [10] d→d* transitions t_{2g} (HOMO) →e_g (LUMO). Scenario 2 is typically exhibited by the hexacyanocobalt(III) ion. In scenario 3, the filled metal t_{2g} orbitals lie above the filled ligand orbitals and the empty ligand π_L^* orbital lies below the metal e_g orbital. The lowest energy electron transitions possible are d→ π_L^* t_{2g} (HOMO) → π_L^* (LUMO) transitions, which are also typical of weak-field ligands. Scenario 3 is found for Ru(II) and Os(II) complexes.

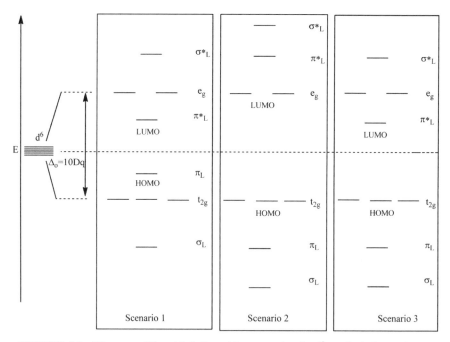

FIGURE 5.5   Three possible orbital disposition scenarios for d^6 octahedral complexes.

In the case of an octahedral Ru(II) complex with polypyridine molecules, there are new electronic transitions that occur due to the metal ion and ligand electronic transitions. The electronic transitions are a *metal centered transition* (MC), a *ligand to metal charge transfer* (LMCT), a *metal to ligand charge transfer* (MLCT), and a *ligand to ligand transition* (L-L), with their Jablonski diagram shown in Fig. 5.6 [11]. Since scenario 3 is the most common for Ru(II) complexes, the metal to ligand charge transfer Jablonski diagram is shown in Fig. 5.7.

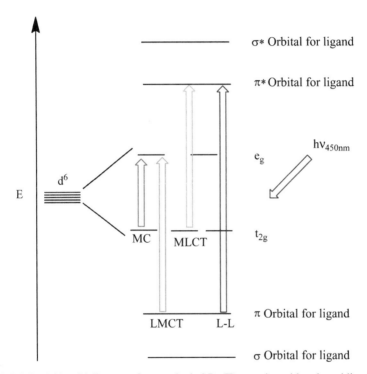

FIGURE 5.6   Jablonski diagram of an octahedral Ru(II) complex with polypyridine ligands.

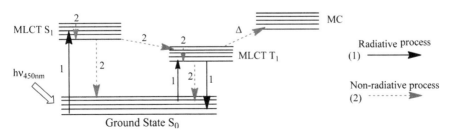

FIGURE 5.7   Jablonski diagram of MLCT transitions in transition metal complexes.

The Ru(II) polypyridine complexes are d^6 octahedral complexes so they are in the low spin t_{2g}^6 configuration. When they absorb light at 450 nm (2.8 eV) it promotes one of its electrons from the ground state S_0 to the singlet metal to ligand charge transfer (MLCT S_1) excited state. The electron then rapidly crosses to the triplet metal to ligand charge transfer (MLCT T_1) excited state via intersystem crossing. Internal heavy atom effects from the Ru(II) metal center promotes spin–orbit interactions that enhance intersystem crossing [4]. The MLCT T_1 excited state decays radiatively or nonradiatively to the ground state, emitting strongly at $\sim$620 nm ($\sim$2.0 eV) in water. The magnitude of the spin–orbit coupling can be high enough to enable spin-forbidden transitions [4], promoting electrons from the ground state directly to the triplet MLCT excited state [12, 13, 14, 15]. The electron in the MLCT T_1 excited state thermally converts to a higher energy MC state through a nonradiative process. The molecule in the MLCT T_1 state is a better oxidant and reductant than in the ground state and may initiate some redox reactions. In contrast, the MC state involves populating a higher energy d orbital of antibonding character, thereby weakening one or more Ru–N bonds, which may result in ligand substitution. The photoredox properties of Ru(II) polypyridyls can then be modulated by changing the nature or combination of ligands [16].

The absorption and emission spectra of $[Ru(bpy)_3]^{2+}$ in water are shown in Fig. 5.8. The high-energy bands at 185 nm and 285 nm are assigned to ligand-centered $\pi_L–\pi_L^*$ transitions (LC or L-L) [17]. The twin bands at 238 and 250 nm and the visible light absorption band at 452 nm are assigned to MLCT d–π_L^* transitions. The shoulder at 322 and 344 nm might be due to charge transfer or d–d (MC) transitions. Excitation in any of its absorption bands leads to an emission band with a maximum at approximately 602 nm [18].

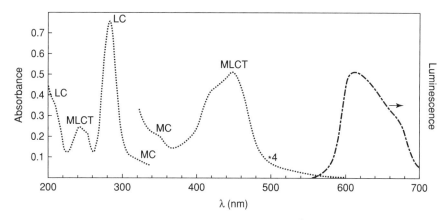

FIGURE 5.8 Absorption and emission spectra of $[Ru(bpy)_3]^{2+}$ in aqueous solution at room temperature. Reproduced from Kalyanasundaram [19, p. 166] with permission of Elsevier.

5.3   IMPORTANCE OF POLYPYRIDYL LIGANDS AND METAL ION FOR TUNING OF MLCT TRANSITIONS

Polypyridyl molecules have extended π systems and they possess nonbonding electrons (n) on their nitrogen atoms, which through light absorption can be promoted into the antibonding π^* orbitals on the aromatic ring (Fig. 5.3). The first electronic n$\rightarrow \pi^*$ transitions are usually the lowest energy transitions, followed closely by $\pi \rightarrow \pi^*$ transitions. Once the polypyridyl ligand is complexed with a metal ion, the n electrons of the nitrogen atom form a σ bond with the metal ion, which lowers the energy of the electrons so they can no longer be excited by low-energy light. On the other hand, the $\pi \rightarrow \pi^*$ transitions on the ring are not greatly shifted by the polypyridyl ligand being complexed with the metal ion and usually lie in the same region as in the free ligand [11]. This permits tuning of the excited state properties of d$\rightarrow \pi^*$ MLCT complexes through the use of ligands that have low-lying $\pi \rightarrow \pi^*$ transitions (π^* tuning) that can be exploited to extend d$\rightarrow \pi^*$ MLCT transitions to lower energies [15]. Research has focused on using long conjugated substituents as a way of lowering the π^* orbital energy to capture lower-energy photons [20].

The other approach to achieving lower-energy transitions by tuning excited state properties of Ru complexes involves the tuning on the metal t_{2g} orbital (t_{2g} tuning). The energy level of the metal t_{2g} orbital is largely determined by the electron density at the metal center and the types of substituents on the polypyridine ring. For example, electron-donating substituents increase this energy level while decreasing the energy associated with the MLCT transitions. Electron-withdrawing substituents and strong donor ligands lower this energy level while increasing the energy associated with the MLCT transitions. Figure 5.9 illustrates both strategies.

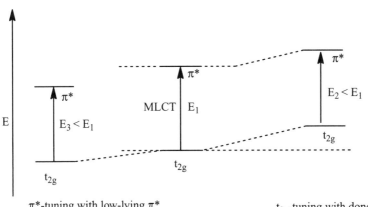

FIGURE 5.9   Schematic representation of t_{2g} and π^* tuning of MLCT transitions in polypyridine complexes.

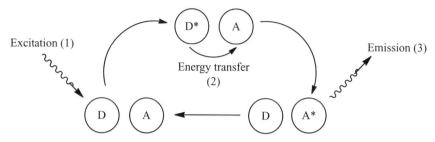

FIGURE 5.10   Photoinduced mechanism of the acceptor (A) by excitation of the donor (D). An asterisk(∗) is used to represent an excited state.

5.4   ELECTRON TRANSFER OF Ru(II) COMPLEXES

The photoinduced transfer of electrons in ruthenium complexes requires an electronic interaction between a donor and an acceptor. The general steps of the mechanism are described in Fig. 5.10. The donor (photosensitizer) is photoexcited (1) then, energy from the excited donor (D∗) is transferred to the acceptor (A) via a nonradiative process (2), which can be modeled using Förster–Dexter theory. Finally, the acceptor relaxes into a lower-energy state by emitting a photon (3). The process can reinitiate upon excitation of the donor. The transfer of electrons can occur either by the electron exchange (Dexter) mechanism or the coulombic (Förster) mechanism. In the Dexter mechanism, an excited donor (photosensitizer) group (D∗) and an acceptor group (A) exchange electrons via a bridging moiety or ligand L in a nonradiative process that requires direct orbital overlap. In the Förster mechanism a nonradiative dipole–dipole interaction occurs based on a resonance theory of energy transfer. As shown in Fig. 5.11, the deactivation of the donor species D∗→D generates an electric field

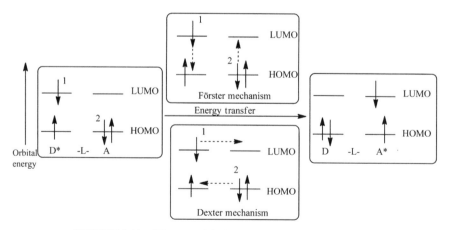

FIGURE 5.11   Diagram of the Dexter and Förster mechanisms.

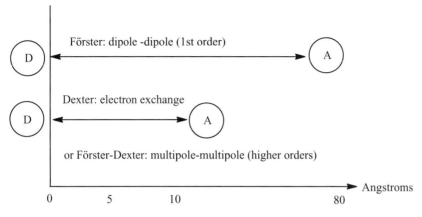

FIGURE 5.12   Ranges of interaction in Förster's and Dexter's models.

or transition dipole, which stimulates the formation of A* (induction of a dipole oscillation in A by D*) [21].

Both the Förster and Dexter models describe single-step photoinduced nonradiative energy transfers that occur between a donor D (photosensitizer) and an acceptor A. Förster first developed an approach based on electric dipole–dipole interaction [22], which was extended by Dexter [23] to include higher-order multipole interactions and exchange, hence the Förster–Dexter process. The main difference between these two models is the distance over which these interactions can occur. The Dexter mechanism operates efficiently over very short distances (<10 Å) since orbital overlap is needed, whereas in the Förster mechanism the transfer has been reported to occur over much longer distances, typically up to 10 nm away, and is strongly dependent on the spectral overlap of the emission spectrum of the donor and the absorption spectrum of the acceptor. In Dexter's model, the interaction strength decreases exponentially as the distance between these two parties increases, hence, the exchange mechanism is also called the short-range energy transfer. In Förster's model, it decreases to the inverse of the sixth power. This means that energy transfer rate for the electron exchange mechanism drops to negligibly small values (relative to the donor lifetime) as the separation distance between the donor and the acceptor increases, which makes the Dexter energy transfer insignificant beyond a separation distance of 10 Å between the sensitizer and the acceptor. Figure 5.12 depicts the ranges of interaction in the Förster and Dexter models.

5.5   LIGHT-HARVESTING COMPLEXES USING Ru(II) COMPLEXES

Photosynthesis is the most efficient way that nature converts solar radiation to other forms of usable energy. In this process, ordered assemblies of chlorophylls collect solar light and transfer that energy to a reaction center [24]. Antenna groups carry excitation to the reaction center, which produce multiple high-energy electrons to

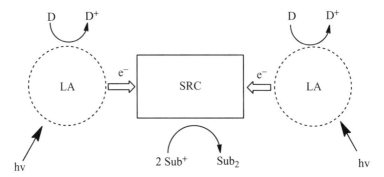

FIGURE 5.13 A schematic diagram of a photochemical system for photoinduced electron collection and multielectron catalysis. D = donor, LA = light absorber, SRC-electron storage and reaction center, Sub = substrate. Reproduced from Steed and Atwood [26] with permission of John Wiley & Sons, Inc.

produce O_2 from water in photosystem II or to reduce $NADP^+$ in photosystem I required for reducing CO_2. Researchers have been studying photosynthesis for decades and are now focusing on artificial photosynthesis to find alternative ways to keep up with the world's energy demands [25]. Artificial photosynthesis inspired the development of supramolecular photochemistry systems capable of multielectron transfer, storage, and catalysis. A simple schematic example of such a system is shown in Fig. 5.13. Light excitation of two light absorbers (LA) causes the transfer of the excited-state electrons to a central electron storage and reaction center (SRC). The SRC component receives the two electrons from the LA groups and carries out a two-electron reduction upon multiple substrates, therefore producing a covalently bound product (2 Sub^+ → Sub_2). Electron donor substrates (D), not a strong enough reductant to reduce Sub^+, are used to provide the electrons back to the LA groups.

Supramolecular photochemical systems such as these are being developed to convert the extremely inert and thermodynamically stable CO_2 molecule into reduced molecules that could be used as energy sources. Thermodynamically the one-electron reduction of CO_2 to $CO_2^{\bullet-}$ is an unfavorable process with a standard electrochemical potential of $E^o = -1.90$ V versus NHE in aqueous solution at pH 7 [27]. However, promoting CO_2 reduction via a proton-assisted multielectron transfer pathway is much more favorable. New transition metal complexes are being developed as homogeneous catalysts to drive these unfavorable reactions such as the two-electron two-proton reduction of CO_2 to CO and H_2O, which occurs at a potential of -0.53 V versus NHE, and the six-electron six-proton reduction of CO_2 to CH_3OH and H_2O, which occurs at a potential of -0.38 V versus NHE [27]. These transition metal catalysts can be applied to artificial photosynthesis systems since their absorption and reduction potential properties can be tuned to capture visible light and reduce CO_2. They also have multiple redox states, which is essential for multielectron transfer reactions, and they generate charge-separated excited states such as metal-to-ligand

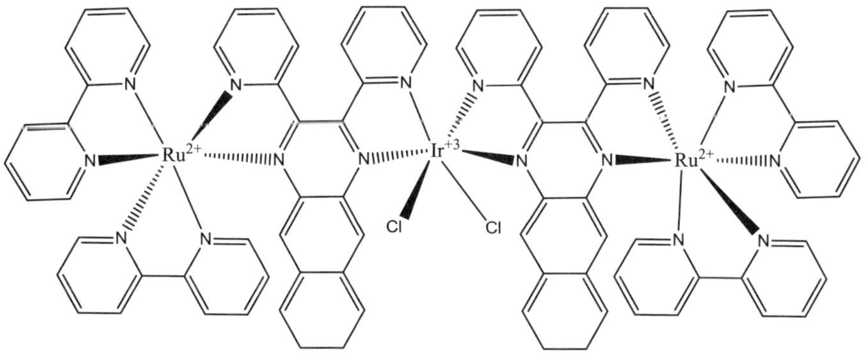

Compound 1

FIGURE 5.14 A photochemical molecular device compound 1, $[(bpy)_2Ru(dpb)]_2IrCl_2$ $(PF_6)_5$.

charge transfer (MLCT) that are long lived and energetic enough to couple with electron-transfer processes to reduce CO_2 via low-energy pathways. Ruthenium(II) polypyridyl complexes are used most often as the light absorbers because of their outstanding photophysical properties, more specifically the long lifetime of their triplet metal to ligand charge transfer (MLCT) excited state and the high emission quantum yield at ambient temperature even in solution [28, 29].

One of the first compounds reported to achieve photoinitiated electron collection using coupled ruthenium chromophores in a photochemical device is compound 1, $[(bpy)_2Ru(dpb)]_2IrCl_2(PF_6)_5$, where bpy = 2,2′bipyridine and dpq = 2,3-bis(2-pyridyl)benzoquinoxaline (Fig. 5.14) [30].

In compound 1, the LA groups are Ru(II) centers, the SRC group is a Ir(III) complex, and the electron donor D used is dimethylaniline. Photoabsorption and transmission by compound 1 was determined not to be very efficient but by photoexciting the Ru(II) ions, double electron transfer occurs to the Ir(III) center via the bridging ligand. The $[Ir(dbp)_2Cl_2]^+$ unit has been showed to deliver multiple electrons stored on the bridging ligand π^* orbitals to a substrate capable of catalyzing the reduction of CO_2 to HCO_2H in the presence of H^+ [31].

An example of a ruthenium-based photochemical complex that reduces CO_2 to CO more efficiently is compound 2, shown in Scheme 5.1. In compound 2 the LA groups are Ru(II) centers, the SRC group is a Re(I) complex, and the electron donor D used is dimethylaniline. Photoabsorption and transmission by compound 2 catalyzed the reduction of CO_2 to CO by the photoexcitation of the Ru(II) ions and double electron transfer to the dmb(4,4′-dimethyl-2,2′-bipyridine) ligand [32]. The Re(I) then reduces CO_2 to HCO_2H in the presence of H^+.

Although it would ultimately be desirable to reduce CO_2 all the way to CH_3OH or CH_4, this is an extremely difficult task. Most of the complexes known only perform two-electron reduction of CO_2 to CO or formate and a few to H_2. New transition

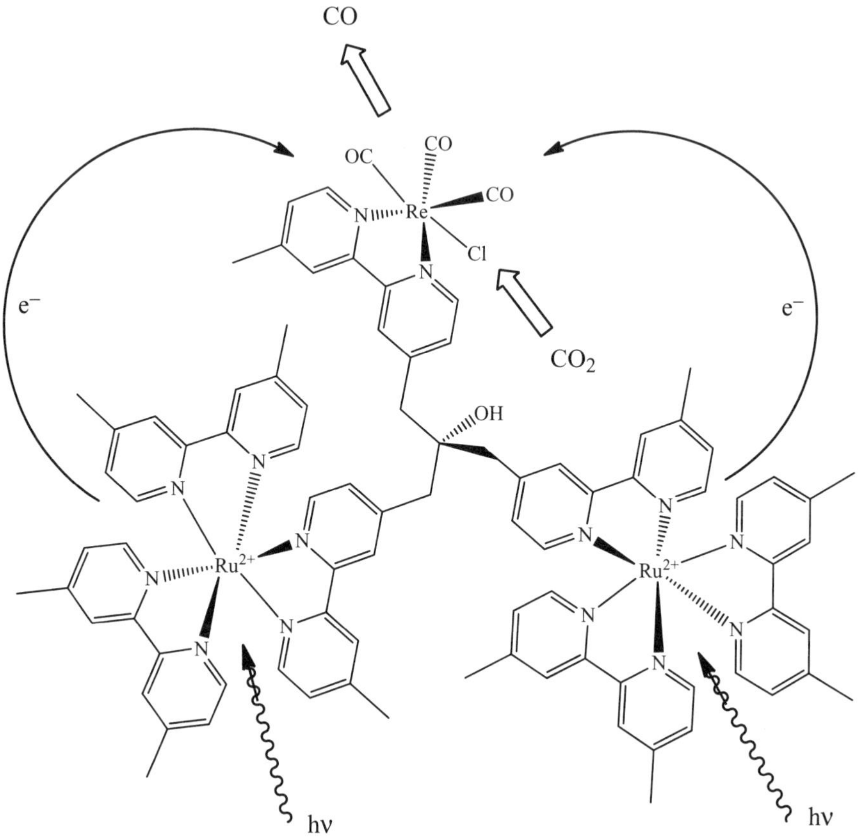

Compound 2

SCHEME 5.1 Photochemical reduction of CO_2 to CO by compound 2, $[(dmb_2Ru)_2(tb-carbinol)Re(CO)_3Cl](PF_6)_4(Ru_2Re)$, where dmb = 4,4'-dimethyl-2,2'-bipyridine and tb-carbinol = tris[(4'-methyl-2,2'-bipyridin-4-yl)methyl]carbinol.

metal catalysts and more specifically ruthenium complexes are being developed to reduce CO_2 beyond CO [33–35]. Hydrocarbons can then be produced from CO in the Fischer–Tropsch process.

In 2009 Geletii et al. [36] designed a photocatalytic system using ruthenium as the photoactive material to oxidize water. In this system $[Ru(bpy)_3]^{3+}$ is generated from $[Ru(bpy)_3]^{2+}$ by photooxidation using $S_2O_8^{2-}$ as a sacrificial electron acceptor [36]. The products $[Ru(bpy)_3]^{3+}$ and $SO_4^{\bullet-}$ are both strong oxidants which the later oxidize $[Ru(bpy)_3]^{2+}$ to form a second $[Ru(bpy)_3]^{3+}$. The absorption of two photons and the consumption of 2 equivalents of $S_2O_8^{2-}$ generates four $[Ru(bpy)_3]^{3+}$ (Eq. 5.1), which in turn oxidizes H_2O to O_2 (Eq. 5.2) and generates $[Ru(bpy)_3]^{2+}$ by

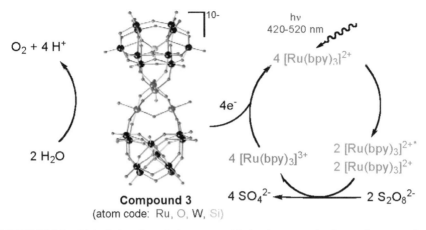

Compound 3
(atom code: Ru, O, W, Si)

SCHEME 5.2 Light-induced catalytic water oxidation by tetraruthenium polyoxometalate $[\{Ru_4O_4(OH)_2(H_2O)_4\}\text{-}(\gamma\text{-}SiW_{10}O_{36})_2]^{10-}$, compound 3, using $[Ru(bpy)_3]^{2+}$ as a photosensitizer and persulfate as a sacrificial electron acceptor. Reproduced with permission from Geletii et al. [36], copyright © 2009 American Chemical Society.

using the water oxidation catalyst (WOC) $[\{Ru_4O_4(OH)_2(H_2O)_4\}\text{-}(\gamma\text{-}SiW_{10}O_{36})_2]^{10-}$, compound 3 (net reaction in Eq. 5.3 via Scheme 5.2) [37].

$$4[Ru(bpy)_3]^{2+} + 2S_2O_8^{2-} + 2h\nu \rightarrow 4[Ru(bpy)_3]^{3+} + 4SO_4^{2-} \qquad (5.1)$$

$$4[Ru(bpy)_3]^{3+} + 2H_2O \rightarrow 4[Ru(bpy)_3]^{2+} + O_2 + 4H^+ \qquad (5.2)$$

$$2S_2O_8^{2-} + 2H_2O + 2h\nu \rightarrow 4SO_4^{2-} + O_2 + 4H^+ \qquad (5.3)$$

In 2010 Yin et al. [39] published the use of the photoactive $[Ru(bpy)_3]^{2+}$ in a similar manner for the oxidation of water by using a cobalt-based catalyst (WOC), which is a more abundant metal than ruthenium. The catalyst for the oxidation of water used was $[Co_4(H_2O)_2(PW_9O_{34})_2]^{10-}$, compound 4 (Fig. 5.15) (Eq. 5.4) [39].

$$4[Ru(bpy)_3]^{3+} + 2H_2O \xrightarrow{\ \ WOC\ \ } 4[Ru(bpy)_3]^{2+} + O_2 + 4H^+ \qquad (5.4)$$

Developing these water oxidation catalysts (WOCs) is a challenging task since an effective WOC must be fast, capable of water oxidation at a potential minimally above the thermodynamic value ($H_2O \rightarrow O_2 + 4H^+ + 4e^-$; 1.229 V × pH at 25 °C), and, critically, stable to air, water, and heat (oxidative, hydrolytic, and thermal stability) [40].

The ultimate goal would be to produce artificial photosynthesis by having a WOC to transfer electrons from water to photoactive materials that promote photoinitiated charge separation for reduction of protons to hydrogen or carbon dioxide to organic compounds, Scheme 5.3 [41].

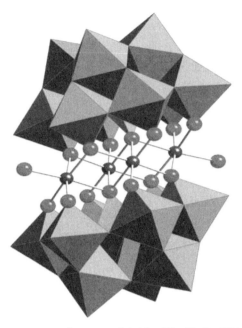

FIGURE 5.15 X-ray structure of compound 4, $Na_{10}[Co_4(H_2O)_2(PW_9O_{34})_2]$, in combined polyhedral ($[PW_9O_{34}]$ ligands) and ball-and-stick (Co_4O_{16} core) notation. Co central atoms; O/OH_2 are terminal; PO_4, tetrahedra; and WO_6, octahedra. Hydrogen atoms, water molecules, and sodium cations are omitted for clarity. Reproduced with permission from Hurst [38], copyright © 2010 American Association for the Advancement of Science.

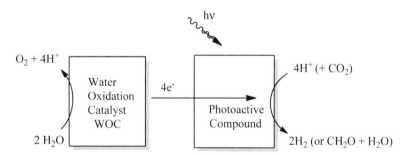

SCHEME 5.3 Generic scheme using a water oxidation catalyst and a photosensitizer capable of directing charge transport. Formaldehyde is shown as one of several possible CO_2 reduction products. Reproduced with permission from Hurst [38], copyright © 2010 American Association for the Advancement of Science.

5.6   Ru(II) ARTIFICIAL PHOTOSYSTEM MODELS FOR PHOTOSYSTEM II

A natural water oxidation catalyst exists in photosystem II (PSII) (Scheme 5.4 and see Chapter 3). There are two major components of PSII: the donor (D) side and the acceptor (A) side. The photochemistry in PSII begins when light hits the photooxidizable chlorophylls, called P680, in the reaction center of PSII. This causes the excitation of P680 forming the excited P680*, which undergoes charge separation by transferring an electron to quinone molecules Q_A and Q_B via a primary acceptor pheophytine (Pheo). The oxidized P680 replenishes its electron deficiency by taking electrons from a manganese cluster (Mn cluster) in the donor side of PSII via a redox-active tyrosine (Y_z). The Mn cluster replenishes its electron deficiency by taking electrons from water, generating molecular oxygen in the process and the Mn cluster returns to its most reduced state [42].

Efforts are being made to mimic some of the light-induced reactions in PSII by building supramolecular Ru–Mn complexes. In these artificial systems, a photoactive Ru(II) complex plays the role of the P680, to redox-active Mn donors and organic acceptors [44, 45]. Artificial photosynthetic systems are being built. The goal is to produce a sustainable and renewable fuel such as H_2 from H_2O, in a similar fashion as photosystem II produces H^+ from H_2O (Scheme 5.5).

In these supramolecular Ru–Mn complexes, the electron from the excited state of Ru(II) is transferred to an electron acceptor such as methyl viologen (MV^{2+}) or cobalt pentamine chloride $[Co(NH_3)_5Cl]^{2+}$. Then, the Mn complex demonstrates an intermolecular electron transfer to the photooxidized Ru(II) complex via the phenolic group of tyrosine. A review by Sun et al. [43] depicts synthesized supramolecular Ru–Mn complexes. Compound 5 as shown in Scheme 5.6 was designed to mimic

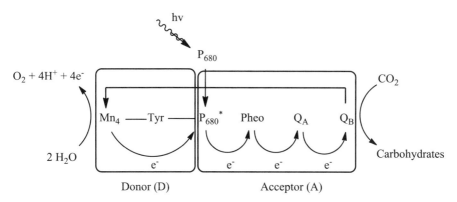

SCHEME 5.4   Schematic representation of the PSII reaction center. The arrows indicate the direction of the electron transfer that occurs resulting in the oxidation of water. Reproduced with permisson from Sun et al. [43], copyright © 2001 Royal Society of Chemistry.

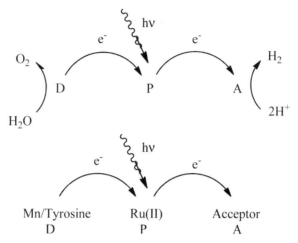

SCHEME 5.5 Schematic presentation of the electron transfer processes in a general artificial photosynthetic system and a system built on supramolecular Ru–Mn complexes. D is the electron donor, P the photosensitizer, and A the electron acceptor. Reproduced with permission from Sun et al. [43], copyright © 2001 Royal Society of Chemistry.

the redox components of natural PSII for a one-electron transfer sequence and for the redox potentials [42]. In flash photolysis experiments performed with compound 5, it was shown that the excited state of Ru(II) was quenched by intermolecular electron transfer to the electron acceptor MV^{2+} forming Ru(III). The Ru(III) was immediately reduced back to Ru(II) by fast intermolecular electron transfer from the Mn(II/II) cluster to provide Mn(II/III). Repeated flashes in the presence of the electron acceptor $[Co(NH_3)_5Cl]^{2+}$ showed that the manganese cluster of compound 5 can be oxidized from the Mn(II/II) state to the Mn(III/IV) state in a three-electron oxidation process. The Mn oxidation states were studied by EPR techniques [45]. This closely models the four-electron transfer cycle in PSII. The drawback to this system is that it does not extract electrons from water. Researchers are trying to find ways to improve this supramolecular artificial photosystem.

Another example of a supramolecular Ru–Mn complex that performs multielectron transfers is the Mn(II/II)–Ru(II)–NDI triad, compound 6 (Fig. 5.16). EPR and optical spectroscopic techniques were used to demonstrate the light-induced formation of the oxidized manganese dimer (Mn(II/II)) and the reduced naphthalenediimide ($NDI^{\bullet-}$) acceptor, and the kinetic evolution of the $[Ru(bpy)_3]^{2+}$ intermediate states and the ($NDI^{\bullet-}$) radical at different temperatures [31]. In a similar manner to compound 5 when sacrificial electron acceptors were used in an aqueous solution, a sequential three-step oxidation of the manganese dimer, bringing it from Mn(II/II) to Mn(III/IV) was observed. With the photoexcitation of Ru(II) in compound 6, the manganese dimer performs a three-electron transfer and the two covalently linked NDI moieties act as electron acceptors.

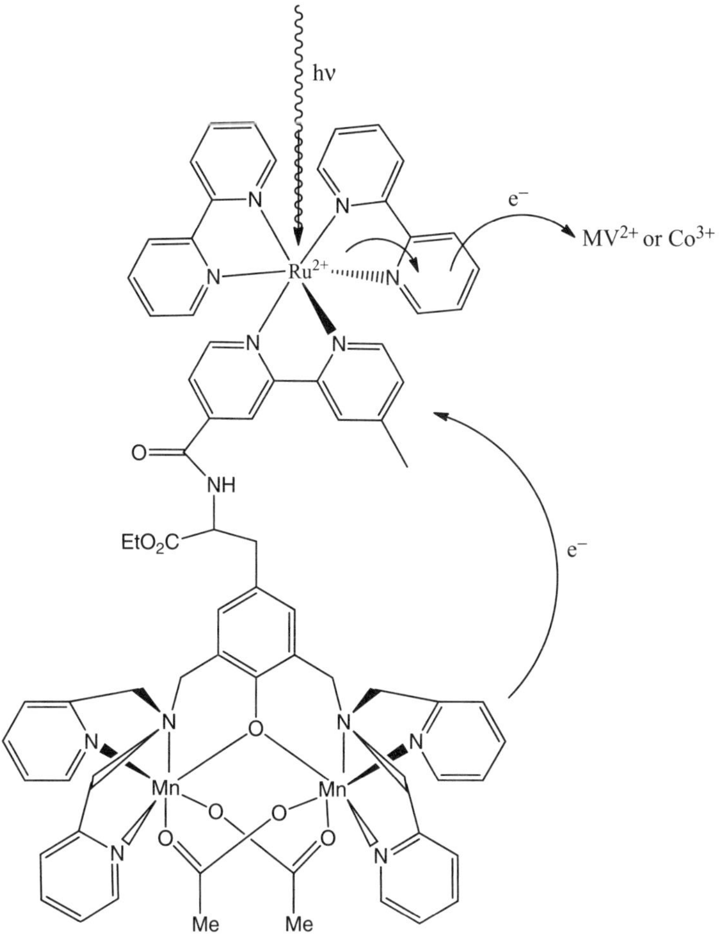

Compound 5

SCHEME 5.6 The intramolecular electron transfer process from a Mn(II,II) dimer to a photooxidized Ru(III) of compound 5 in the presence of methyl viologen or $[Co(NH_3)_5Cl]^{2+}$ as the electron acceptor.

In addition, Borgström et al. [44] estimated the reduction potentials for compound 6 as shown in Fig. 5.17. The reduction potentials are: $E_{1/2} = +0.88$ V($Ru^{3+/2+}$), $E_{1/2} = -0.97$ V(NDI^{0-}) (all vs. Fc$^{+/0}$) [46], the Mn(II/II)–Ru(II) dyad laking the NDI unit with $E_{1/2} = +0.10$ V Mn(II/III)–Mn(II/II), $E_{1/2} = +0.63$ V Mn(III/III)–Mn(II/III), $E_{1/2} = -1.6$ V ($[Ru(bpy)_3]^{2+/+}$) (all vs. Fc$^{+/0}$) [47, 48]. Based on this data and the tripled excited state energies of compound 4, $E_{00} = 2.05$ eV for ruthenium and $E_{00} = 2.03$ eV for NDI [49] neglecting coloumbic interactions.

The ultimate goal of the research is to mimic photosynthesis where electrons are extracted from water using energy from the sun and used to produce hydrogen gas as

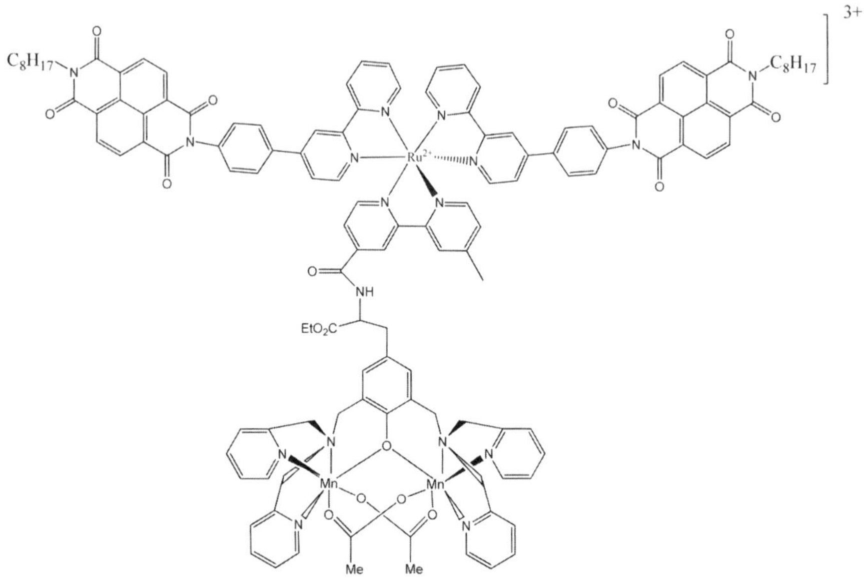

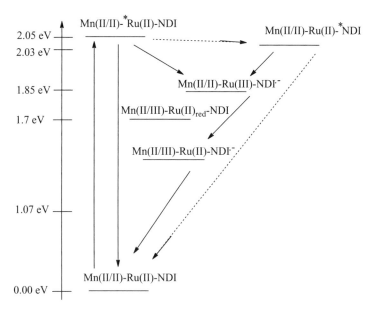

FIGURE 5.16 Compound 6, Mn(II/II)–Ru(II)–NDI triad. ([Mn$_2$(bpmp)(OAc)$_2$]$^+$, bpmp = 2,6-bis[bis(2-pyridylmethyl)-aminomethyl]-4-methylphenolate = Mn(II/II) acetate = OAc, tris-bipyridine ruthenium(II) = Ru(II), and naphthalenediimide = NDI.

FIGURE 5.17 Energy diagram for the different locally excited states (marked with an asterisk) and charge separated states for compound 6. The triplet NDI (*NDI) state was only observed in a PMMA matrix at $T < 120$ K. The state involving the reduced Ru(II) center, formally Ru^{II}(bpy$^{\bullet-}$)(bpy)$_2$ and here denoted Ru(II)$_{red}$, was not observed and probably not involved under any of the experimental conditions. Reproduced with permission from Borgström et al. [50], copyright © 2005 American Chemical Society.

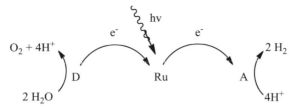

SCHEME 5.7   Schematic diagram of future solar energy system.

a means to renewable energy. Future research is being focused on the formation of a solar energy system using Ru(II) complexes where light will photoexcite Ru(II) and electrons will move from a donor species D to an acceptor A. These electrons will be taken from water and will be used to produce hydrogen as shown in Scheme 5.7.

5.7   Ru (II) ARTIFICIAL PHOTOSYSTEM MODELS FOR HYDROGENASE

In nature and more specifically in microorganisms, there are particular enzymes called hydrogenases that catalyze the reversible reduction of protons to hydrogen with remarkable activity [51, 52, 53, 54, 55]. There are two types of widely distributed hydrogenases, which contain different metals in their active sites: iron-only hydrogenases and nickel–iron (NiFe) hydrogenases [56].

Crystallographic data provided evidence that the active site of the iron-only hydrogenase enzymes contains two iron(I) ions linked to a bridging dithiolate ligand (Fig. 5.18A) [57, 58]. A biomimetic model of iron hydrogenase from a review by Sun et al. [59] is shown Fig. 5.18B as a catalyst for the electrochemical production of hydrogen.

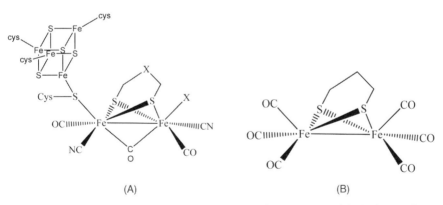

(A)   (B)

FIGURE 5.18   (A) X-ray crystallography has provided the structure of the hydrogen cluster in the iron-only hydrogenase active site where X is an unidentified ligand in which the resting state is proposed to be an oxygen ligand such as H_2O, OH^-, or O^{2-} [60]. (B) Model complex propyldithiolate (PDT)-bridged $[(\mu\text{-PDT})Fe_2(CO)_6]$ mimicking the structure as well as the function of the natural system.

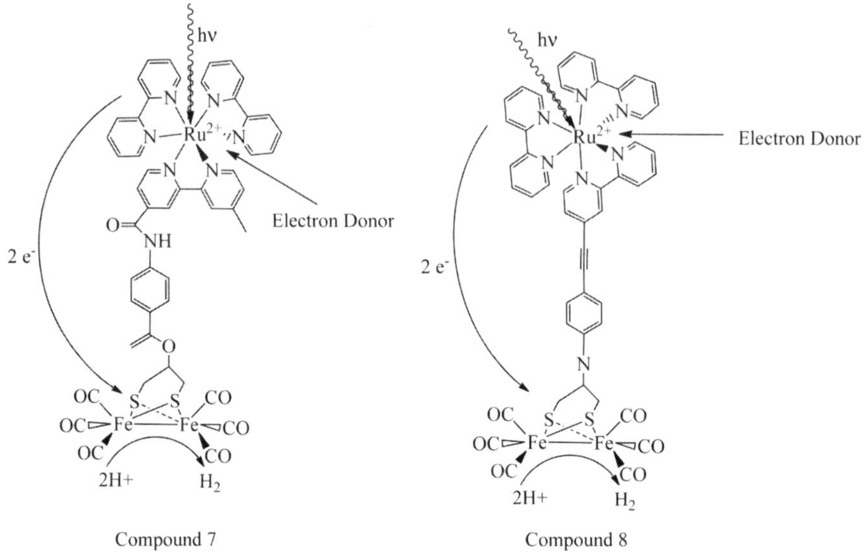

SCHEME 5.8 Schematic representation of compounds 7 and 8 for the projected photoinduced reduction of protons.

In an attempt to produce hydrogen gas by using light, researchers have covalently linked a biomimetic model of the iron hydrogenase active site to a ruthenium photosensitizer (Scheme 5.8, compounds 7 [59] and 8 [61]). The redox properties of both the ruthenium and diiron components in both dyads happened to be unfavorable for electron transfer and for the reduction of protons to molecular hydrogen.

However, a triad reaction system has been designed, as shown in Scheme 5.9, composed of $Ru(bpy)_3^{2+}$ as the photosensitizer, a diethyldithiocarbamate anion (dtc^-) as an electron donor, and $\{(i\text{-}SCH_2)_2X\}[Fe(CO)_3]_2$, where $X = CH_2$ or $NCH_2C_6H_5$,

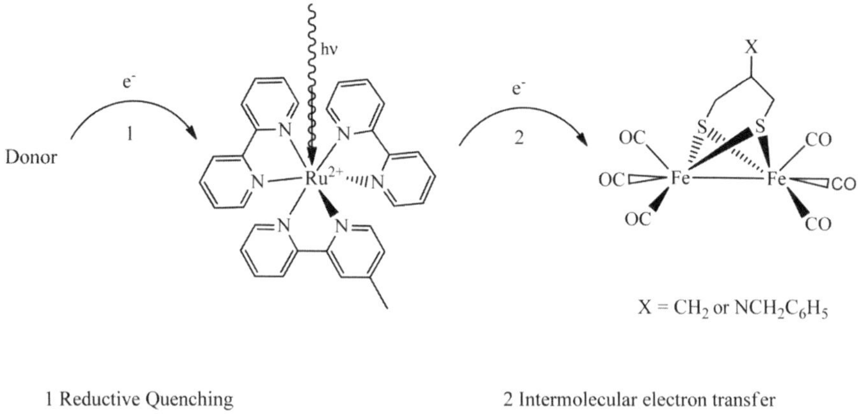

SCHEME 5.9 Bioinspired triad system for photoinduced electron transfer.

as a catalyst to mimic the diiron subsite [62]. The excited state of $Ru(bpy)_3^{2+}$ is quenched reductively by dtc$^-$, leading to the formation of $Ru(bpy)_3^+$. Consequently, one-electron transfers from $Ru(bpy)_3^+$ to the diiron dithiolate complex result in the formation of Fe(I)Fe(0) species, which is proposed to be a crucial intermediate for the dihydrogen evolution catalyzed electrochemically by the [2Fe2S] complexes, Scheme 5.10 [63,64]. The designed electron transfer processes were corroborated by laser flash photolysis, which gives promise in the light-driven hydrogen gas evolution using diiron complexes as surrogates for expensive noble platinum catalysts.

For the Ni–Fe hydrogenase from the enzyme *Desulfovibrio gigas*, crystallographic data provided evidence that the active site of this type of hydrogenase consists of one nickel atom and one iron atom, which are bridged by two cysteic thiolates and one unidentified ligand (depicted as X in Fig. 5.19) [65,66,67]. The bridging ligand X in the resting state is proposed to be an oxygen ligand such as H_2O, OH$^-$, or O^{2-} [59]. The role of metal atoms and bridging S and X ligands in the H_2 cleavage has so far been the subject of controversy.

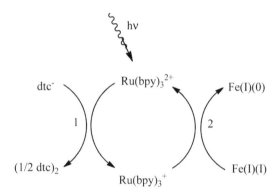

SCHEME 5.10   Proposed light-driven electron transfer process of the triad system.

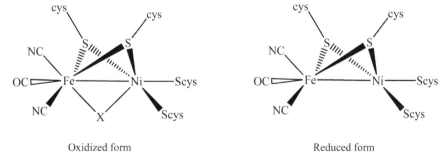

Oxidized form   Reduced form

FIGURE 5.19   Crystallographic evidence of the resting form of the core structure of [NiFe] hydrogenase.

Efforts have been made to elucidate the core structure of the active form of the [NiFe] hydrogenase by synthesizing model compounds, such as [NiRu] complexes by Rauchfuss and others [51, 68, 69, 70, 71, 72, 73]. A Ni(μ-S)$_2$(μ-H)Fe species is one of the candidates for the active form as shown in Fig. 5.20 [66, 74–76]. One of the first compounds synthesized as a model for the [NiFe] hydrogenase is the Ni(μ-H)Ru complex [(Ni(II)L)(H$_2$O)(μ -H)Ru(II)(η^6-C$_6$Me$_6$)](NO$_3$) (Fig. 5.21 and Scheme 5.11). Compound 9 was found to be catalytically active by catalyzing

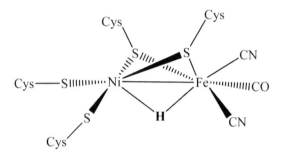

FIGURE 5.20   Candidate for the active form of [NiFe]hydrogenase, where Cys = cysteine.

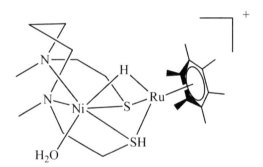

Compound 9

FIGURE 5.21   Structure of compound 9, [(Ni(II)L)(H$_2$O)(μ -H)Ru(II)(η^6-C$_6$Me$_6$)](NO$_3$), where L = N, N'-dimethyl-N, N'-bis(2-mercaptoethyl)-1,3-propanediamine.

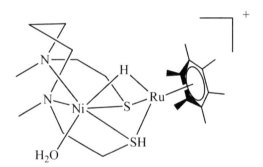

SCHEME 5.11   Synthesis of biomimetic catalyst, with Ru(C$_6$(CH$_3$)$_6$)$^{2+}$ playing the role of Fe(CN)$_2$(CO). The hydride ligand between the Ni and Ru centers is shown to transfer to unsaturated substrates.

the hydrogenation of benzaldehyde to the corresponding alcohol. Future efforts are being made to replace Ru with Fe, which is always used by the natural systems [77].

One of the most photocatalytically efficient and stable [FeFe] hydrogenase models obtained so far for H_2 production in pure water at room temperature is comprised of an artificial water-soluble [FeFe] hydrogenase mimic, compound 10, a photosensitizer of nanocrystal quantum dots (CdTe), and ascorbic acid (H_2A) (Scheme 5.12) [78]. The turnover numbers are competitive with those from current state-of-the-art catalytic

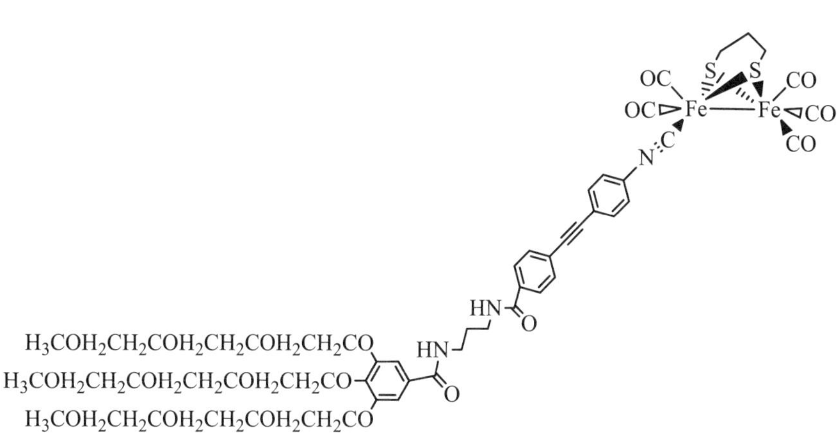

Compound 10

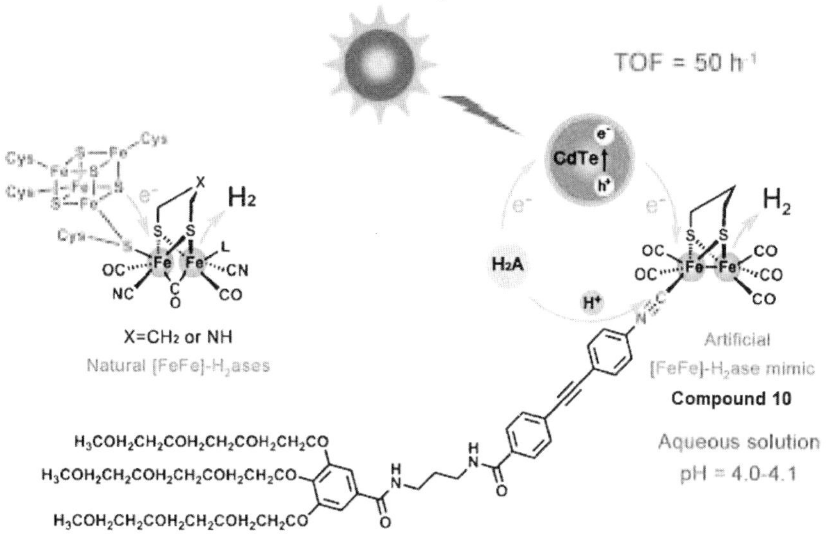

SCHEME 5.12 The structure of natural [FeFe] hydrogenase and the artificial [FeFe] hydrogenase mimic. Reproduced with permission from Wang et al. [80], copyright © 2011 John Wiley & Sons, Inc.

systems for H_2 production. This catalytic efficiency and stability indicate that both the catalyst and the photosensitizer can be regenerated effectively during the entire photocatalytic reaction. This is the first example of a [FeFe] hydrogenase mimic combined with nanocrystal quantum dots for light-driven H_2 evolution without any external manipulation [79].

5.8   CONCLUSION

The challenge for the future of energy and the answers will lie mainly in chemistry, with inorganic chemistry and especially transition metal chemistry playing a large role. Research studies will be foregoing especially using Ru-based photoactive complexes as sensitizers to create efficient and environmentally friendly photoactive model systems for photosystem II and hydrogenase in an attempt to produce light-driven energy sources such as hydrogen from water. It is not a simple task and there is still a long way to go to model nature and achieve its efficiency. Major efforts from scientists are needed before the goals can be reached.

REFERENCES

1. D. Shriver, P. Atkins, *Inorganic Chemistry*, 4th ed., W. H. Freeman, New York, 2006.
2. K. Kalyanasundaram, M. Gratzel, Applications of functionalized transition metal complexes in photonic and optoelectronic devices, Coord. Chem. Rev. 177 (1998) 347–414.
3. A. Gilbert, J. E. Baggott, *Essentials of Molecular Photochemistry*, Blackwell Scientific, London, 1991.
4. N. J. Turro, V. Ramamurthy, J. C. Scaiano, *Principles of Molecular Photochemistry: An Introduction*, University Science Books, Sausalito, 2009.
5. N. J. Turro, *Modern Molecular Photochemistry*, Science Books, Sausalito, CA, 1991.
6. P. J. Giordano, S. M. Fredericks, M. S. Wrighton, D. L. Morse, Simultaneous multiple emissions from Fac-Xre(CO)$_3$(3-Benzoylpyridine)$_2$-N-Pi*. Intraligand and charge-transfer emission at low-temperature, J. Am. Chem. Soc. 100 (1978) 2257–2259.
7. S. M. Fredericks, J. C. Luong, M. S. Wrighton, Multiple emissions from rhenium(I) complexes—intraligand and charge-transfer emission from substituted metal-carbonyl cations, J. Am. Chem. Soc. 101 (1979) 7415–7417.
8. G. Tapolsky, R. Duesing, T. J. Meyer, Synthetic control of excited-state properties in ligand-bridged complexes of rhenium(I)—intramolecular energy-transfer by an electron-transfer energy-transfer cascade, Inorg. Chem. 29 (1990) 2285–2297.
9. R. M. Leasure, L. Sacksteder, D. Nesselrodt, G. A. Reitz, J. N. Demas, B. A. Degraff, Excited-state acid-base chemistry of (alpha-diimine)cyanotricarbonylrhenium(I) complexes, Inorg. Chem. 30 (1991) 3722–3728.
10. C. E. Housecroft, A. G. Sharpe, *Inorganic Chemistry*, 2nd ed., Pearson Education Limited, England, 2005.
11. K. Kalyanasundaram, *Photochemistry of Polypyridine and Porphyrin Complexes*, Academic Press, London, 1991.

12. E. M. Kober, J. L. Marshall, W. J. Dressick, B. P. Sullivan, J. V. Caspar, T. J. Meyer, Synthetic control of excited-states—non-chromophoric ligand variations in polypyridyl complexes of osmium(II), Inorg. Chem. 24 (1985) 2755–2763.

13. E. M. Kober, B. P. Sullivan, W. J. Dressick, J. V. Caspar, T. J. Meyer, Highly luminescent polypyridyl complexes of osmium(II), J. Am. Chem. Soc. 102 (1980) 7383–7385.

14. R. H. Fabian, D. M. Klassen, R. W. Sonntag, Synthesis and spectroscopic characterization of ruthenium and osmium complexes with sterically hindering ligands. 3. Tris complexes with methyl-substituted and dimethyl-substituted 2,2′-bipyridine and 1,10-phenanthroline, Inorg. Chem. 19 (1980) 1977–1982.

15. A. P. K. Kalyanasundaram, *Polypyridyl Complexes of Ruthenium, Osmium, and Iron*, Academic Press, London, 1992, pp. 105–206.

16. K. Szacilowski, W. Macyk, A. Drzewiecka-Matuszek, M. Brindell, G. Stochel, Bioinorganic photochemistry: frontiers and mechanisms, Chem. Rev. 105 (2005) 2647–2694.

17. F. E. Lytle, D. M. Hercules, Luminescence of tris(2,2′-bipyridine)ruthenium(II) dichloride, J. Am. Chem. Soc. 91 (1969) 253–257.

18. K. Kalyanasundaram, *Photochemistry of Polypyridine and Porphyrin Complexes*, Academic Press, San Diego, CA, 1992.

19. K. Kalyanasundaram, Photophysics, photochemistry and solar energy conversion with tris(bipyridyl)ruthenium(II) and its analogues, Coord. Chem. Rev. 46 (1982), 159–244.

20. P. Wang, C. Klein, R. Humphry-Baker, S. M. Zakeeruddin, M. Grätzel, A high molar extinction coefficient sensitizer for stable dye-sensitized solar cells, J. Am. Chem. Soc. 127 (2005) 808–809.

21. J. W. Steed, J. L. Atwood, *Supramolecular Chemistry*, 2nd ed., Wiley, West Sussex, England, 2000.

22. T. Förster, *Comparative Effects of Radiation*, edited by M. Burton, Wiley, Hoboken, NJ, 1960.

23. D. L. Dexter, A theory of sensitized luminescence in solids, J. Chem. Phys. 21 (1953) 836–850.

24. H. Scheer, *Light Harvesting Antennas in Photosynthesis*, Kluwer Academic, Dordrecht, The Netherlands, 2003.

25. R. Eisenberg, D. G. Nocera, Forum on solar and renewable energy, Inorg. Chem. 44 (2005) 8642–8642.

26. J. W. Steed, J. L. Atwood, *Supramolecular Chemistry*, Wiley, Hoboken, NJ, 2000.

27. E. Fujita, Photochemical carbon dioxide reduction with metal complexes, Coord. Chem. Rev. 185-6 (1999) 373–384.

28. J.-L. Chen, Y. Chi, K. Chen, Y.-M. Cheng, M.-W. Chung, Y.-C. Yu, G.-H. Lee, P.-T. Chou, C.-F. Shu, New series of ruthenium(II) and osmium(II) complexes showing solid-state phosphorescence in far-visible and near-infrared, Inorg. Chem. 49 (2010) 823–832.

29. Q. X. Zhou, W. H. Lei, J. R. Chen, C. Li, Y. J. Hou, X. S. Wang, B. W. Zhang, A new heteroleptic ruthenium(II) polypyridyl complex with long-wavelength absorption and high singlet-oxygen quantum yield, Chemistry—A European Journal. 16 (2010) 3157–3165.

30. S. M. Molnar, G. Nallas, J. S. Bridgewater, K. J. Brewer, Photoinitiated electron collection in a mixed-metal trimetallic complex of the form {[(bpy)₂Ru(dpb)]₂IrCl₂}(PF₆)₅

(bpy = 2,2′-bipyridine and dpb = 2,3-bis(2-pyridyl)benzoquinoxaline), J. Am. Chem. Soc. 116 (1994) 5206–5210.

31. S. C. Rasmussen, M. M. Richter, E. Yi, H. Place, K. J. Brewer, Synthesis and characterization of a series of novel rhodium and iridium complexes containing polypyridyl bridging ligands—potential uses in the development of multimetal catalysts for carbon-dioxide reduction, Inorg. Chem. 29 (1990) 3926–3932.

32. Z.-Y. Bian, K. Sumi, M. Furue, S. Sato, K. Koike, O. Ishitani, Synthesis and properties of a novel tripodal bipyridyl ligand tb-carbinol and its Ru(II)–Re(I) trimetallic complexes: investigation of multimetallic artificial systems for photocatalytic CO(2) reduction, Dalton Trans. 6 (2009) 983–993.

33. C. Creutz, M. H. Chou, Rapid transfer of hydride ion from a ruthenium complex to C1 species in water, J. Am. Chem. Soc. 129 (2007) 10108–10109.

34. D. E. Polyansky, D. Cabelli, J. T. Muckerman, T. Fukushima, K. Tanaka, E. Fujita, Mechanism of hydride donor generation using a Ru(II) complex containing an NAD(+) model ligand: pulse and steady-state radiolysis studies, Inorg. Chem. 47 (2008) 3958–3968.

35. M. D. Doherty, D. C. Grills, J. T. Muckerman, D. E. Polyansky, E. Fujita, Toward more efficient photochemical CO(2) reduction: use of scCO(2) or photogenerated hydrides, Coord. Chem. Rev. 254 (2010) 2472–2482.

36. Y. V. Geletii, Z. Q. Huang, Y. Hou, D. G. Musaev, T. Q. Lian, C. L. Hill, Homogeneous light-driven water oxidation catalyzed by a tetraruthenium complex with all inorganic ligands, J. Am. Chem. Soc. 131 (2009) 7522–7523.

37. Y. V. Geletii, B. Botar, P. Koegerler, D. A. Hillesheim, D. G. Musaev, C. L. Hill, An all-inorganic, stable, and highly active tetraruthenium homogeneous catalyst for water oxidation, Angew. Chem. Int. Ed. 47 (2008) 3896–3899.

38. J. K. Hurst, In pursuit of water oxidation catalysts for solar fuel production, Science 328 (2010) 315.

39. Q. S. Yin, J. M. Tan, C. Besson, Y. V. Geletii, D. G. Musaev, A. E. Kuznetsov, Z. Luo, K. I. Hardcastle, C. L. Hill, A fast soluble carbon-free molecular water oxidation catalyst based on abundant metals, Science 328 (2010) 342–345.

40. R. Eisenberg, H. B. Gray, Preface on making oxygen, Inorg. Chem. 47 (2008) 1697–1699.

41. J. K. Hurst, In pursuit of water oxidation catalysts for solar fuel production, Science 328 (2010) 315–316.

42. V. K. Yachandra, K. Sauer, M. P. Klein, Manganese cluster in photosynthesis: where plants oxidize water to dioxygen, Chem. Rev. 96 (1996) 2927–2950.

43. L. Sun, L. Hammarström, B. Åkermark, S. Styring, Towards artificial photosynthesis: ruthenium–manganese chemistry for energy production, Chem. Soc. Rev. 30 (2001) 36–49.

44. M. Borgström, N. Shaikh, O. Johansson, M. F. Anderlund, S. Styring, B. Akermark, A. Magnuson, L. Hammarstrom, Light induced manganese oxidation and long-lived charge separation in a Mn-2(II,II)-Ru-II (bpy)(3)-acceptor triad, J. Am. Chem. Soc. 127 (2005) 17504–17515.

45. L. C. Sun, L. Hammarstrom, B. Akermark, S. Styring, Towards artificial photosynthesis: ruthenium–manganese chemistry for energy production, Chem. Soc. Rev. 30 (2001) 36–49.

46. O. Johansson, H. Wolpher, M. Borgstrom, L. Hammarstrom, J. Bergquist, L. C. Sun, B. Akermark, Intramolecular charge separation in a hydrogen bonded tyrosine–ruthenium(II)–naphthalene diimide triad, Chem. Commun. (2004) 194–195.

47. P. Huang, A. Magnuson, R. Lomoth, M. Abrahamsson, M. Tamm, L. Sun, B. van Rotterdam, J. Park, L. Hammarstrom, B. Akermark, S. Styring, Photo-induced oxidation of a dinuclear Mn-2(II,II) complex to the Mn-2(III,IV) state by inter- and intramolecular electron transfer to Ru-III tris-bipyridine, J. Inorg. Biochem. 91 (2002) 159–172.

48. G. Eilers, C. Zettersten, L. Nyholm, L. Hammarstrom, R. Lomoth, Ligand exchange upon oxidation of a dinuclear Mn complex—detection of structural changes by FT-IR spectroscopy and ESI-MS, Dalton Trans. (2005) 1033–1041.

49. J. E. Roger, L. A. Kelly, Nucleic acid oxidation mediated by naphthalene and benzophenone imide and diimide derivatives: consequences for DNA redox chemistry, J. Am. Chem. Soc. 121 (1999) 3854–3861.

50. M. Borgström, N. Shaikh, O. Johansson, M. F. Anderlund, S. Styring, B. A. k. A. Magnuson, L. Hammarström, Light induced manganese oxidation and long-lived charge separation in a Mn2II,II−RuII(bpy)$_3$−acceptor triad, J. Am. Chem. Soc. 127 (2005), 17504–17515.

51. M. Y. Darensbourg, E. J. Lyon, J. J. Smee, The bio-organometallic chemistry of active site iron in hydrogenases, Coord. Chem. Rev. 206 (2000) 533–561.

52. F. Gloaguen, J. D. Lawrence, M. Schmidt, S. R. Wilson, T. B. Rauchfuss, Synthetic and structural studies on $[Fe_2(SR)_2(CN)_x(CO)_{6-x}]^{x-}$ as active site models for Fe-only hydrogenases, J. Am. Chem. Soc. 123 (2001) 12518–12527.

53. S. J. George, Z. Cui, M. Razavet, C. J. Pickett, The di-iron subsite of all-iron hydrogenase: mechanism of cyanation of a synthetic {2Fe3S}–carbonyl assembly, Chemistry—A European Journal. 8 (2002) 4037–4046.

54. E. J. Lyon, I. P. Georgakaki, J. H. Reibenspies, M. Y. Darensbourg, Coordination sphere flexibility of active-site models for Fe-only hydrogenase: studies in intra- and intermolecular diatomic ligand exchange, J. Am. Chem. Soc. 123 (2001) 3268–3278.

55. M. Frey, Hydrogenases: hydrogen-activating enzymes, Chembiochem 3 (2002) 153–160.

56. G. Jaouen, *Bioorganometallics: Biomolecules, Labeling, Medicine*, Wiley-VCH, Weinheim, Germany, 2006.

57. J. W. Peters, W. N. Lanzilotta, B. J. Lemon, L. C. Seefeldt, X-ray crystal structure of the Fe-only hydrogenase (CpI) from *Clostridium pasteurianum* to 1.8 Å resolution, Science 282 (1998) 1853–1858.

58. Y. Nicolet, C. Piras, P. Legrand, C. E. Hatchikian, J. C. Fontecilla-Camps, *Desulfovibrio desulfuricans* iron hydrogenase: the structure shows unusual coordination to an active site Fe binuclear center, Structure with Folding & Design 7 (1999) 13–23.

59. L. C. Sun, B. Akermark, S. Ott, Iron hydrogenase active site mimics in supramolecular systems aiming for light-driven hydrogen production, Coord. Chem. Rev. 249 (2005) 1653–1663.

60. M. Carepo, D. L. Tierney, C. D. Brondino, T. C. Yang, A. Pamplona, J. Telser, I. Moura, J. J. G. Moura, B. M. Hoffman, 17O ENDOR detection of a solvent-derived Ni−(OH$_x$)−Fe bridge that is lost upon activation of the hydrogenase from *Desulfovibrio gigas*, J. Am. Chem. Soc. 124 (2002) 281–286.

61. S. Ott, M. Kritikos, B. Akermark, L. C. Sun, Synthesis and structure of a biomimetic model of the iron hydrogenase active site covalently linked to a ruthenium photosensitizer, Angew. Chem. Int. Ed. 42 (2003) 3285–3288.

62. Y. Na, J. X. Pan, M. Wang, L. C. Sun, Intermolecular electron transfer from photogenerated $Ru(bpy)_3^+$ to [2Fe$_2$S] model complexes of the iron-only hydrogenase active site, Inorg. Chem. 46 (2007) 3813–3815.

63. R. Mejia-Rodriguez, D. Chong, J. H. Reibenspies, M. P. Soriaga, M. Y. Darensbourg, The hydrophilic phosphatriazaadamantane ligand in the development of H_2 production electrocatalysts: iron hydrogenase model complexes, J. Am. Chem. Soc. 126 (2004) 12004–12014.

64. J. F. Capon, F. Gloaguen, P. Schollhammer, J. Talarmin, Catalysis of the electrochemical H-2 evolution by di-iron sub-site models, Coord. Chem. Rev. 249 (2005) 1664–1676.

65. M. Frey, *Structure and Bonding*, Springer, Berlin, 1998.

66. A. Volbeda, M. H. Charon, C. Piras, E. C. Hatchikian, M. Frey, J. C. Fontecillacamps, Crystal-structure of the nickel–iron hydrogenase from *Desulfovibrio gigas*, Nature 373 (1995) 580–587.

67. A. Volbeda, E. Garcin, C. Piras, A. L. deLacey, V. M. Fernandez, E. C. Hatchikian, M. Frey, J. C. FontecillaCamps, Structure of the [NiFe] hydrogenase active site: evidence for biologically uncommon Fe ligands, J. Am. Chem. Soc. 118 (1996) 12989–12996.

68. M. A. Reynolds, T. B. Rauchfuss, S. R. Wilson, Ruthenium derivatives of NiS_2N_2 complexes as analogues of bioorganometallic reaction centers, Organometallics 22 (2003) 4620–4620.

69. E. Bouwman, J. Reedijk, Structural and functional models related to the nickel hydrogenases, Coord. Chem. Rev. 249 (2005) 1555–1581.

70. D. J. Evans, C. J. Pickett, Chemistry and the hydrogenases, Chem. Soc. Rev. 32 (2003) 268–275.

71. A. C. Marr, D. J. E. Spencer, M. Schroder, Structural mimics for the active site of [NiFe] hydrogenase, Coord. Chem. Rev. 219 (2001) 1055–1074.

72. Z. L. Li, Y. Ohki, K. Tatsumi, Dithiolato-bridged dinuclear iron–nickel complexes $[Fe(CO)_2(CN)_2(mu\text{-}SCH_2CH_2CH_2S)Ni(S_2CNR_2)](-)$—modeling the active site of [NiFe] hydrogenase, J. Am. Chem. Soc. 127 (2005) 8950–8951.

73. W. F. Zhu, A. C. Marr, Q. Wang, F. Neese, D. J. E. Spencer, A. J. Blake, P. A. Cooke, C. Wilson, M. Schroder, Modulation of the electronic structure and the Ni–Fe distance in heterobimetallic models for the active site in [NiFe]hydrogenase, Proc. Nat. Acad. Sci. U.S.A. 102 (2005) 18280–18285.

74. R. Cammack, M. Frey, R. Robson, *Hydrogen as a Fuel: Learning from Nature*, Taylor & Francis, London, 2001.

75. S. Foerster, M. Stein, M. Brecht, H. Ogata, Y. Higuchi, W. Lubitz, Single crystal EPR studies of the reduced active site of [NiFe] hydrogenase from *Desulfiovibrio vulgaris* Miyazaki F, J. Am. Chem. Soc. 125 (2003) 83–93.

76. M. Brecht, M. van Gastel, T. Buhrke, B. Friedrich, W. Lubitz, Direct detection of a hydrogen ligand in the [NiFe] center of the regulatory H_2-sensing hydrogenase from *Ralstonia eutropha* in its reduced state by HYSCORE and ENDOR spectroscopy, J. Am. Chem. Soc. 125 (2003) 13075–13083.

77. T. B. Rauchfuss, A promising mimic of hydrogenase activity, Science 317 (2007) 43.

78. F. Wang, W.-G. Wang, X.-J. Wang, H.-Y. Wang, C.-H. Tung, A highly efficient photo-catalytic system for hydrogen production by a robust hydrogenase mimic in an aqueous solution, Angew. Chem. 123 (2011) 3251–3255.

79. F. Wang, W. G. Wen-Guang Wang, H. Y. Wang, G. Si, C. H. Tung, L. Z. Wu, Artificial photosynthetic systems based on [FeFe]-hydrogenase mimics: the road to high efficiency for light-driven hydrogen evolution, ACS Catal. (2012).

80. F. Wang, W.-G. Wang, X.-J. Wang, H.-Y. Wang, C.-H. Tung, L.-Z. Wu, A highly efficient photocatalytic system for hydrogen production by a robust hydrogenase mimic in an aqueous solution, Angew. Chem. 123 (2011) 3251–3255.

CHAPTER 6

CO_2 Sequestration and Hydrogen Production Using Cyanobacteria and Green Algae

KANHAIYA KUMAR and DEBABRATA DAS

6.1 INTRODUCTION

The presence of CO_2 influences the global environment to a great extent. The cyclic occurrence of ice ages on earth is now a well-established fact. Even though there exist contradictory theories on the long-term role of CO_2 emissions in bringing about ice ages in the last few hundred years, the impact of increasing CO_2 concentration in our atmosphere is quite visible. Rising global temperature, melting of glaciers, and increasing sea level are some interconnected effects of the increase in CO_2 level in the atmosphere. Fossil fuel is the major contributor (nearly 75%) of total CO_2 emissions [1]. Industries related to electricity generation, natural gas processing, cement, iron and steel manufacturing, fertilizer, production and ethanol and sugar production are the major fossil fuel-based industries dependent on carbon sources like coal, oil, and natural gas for fulfilling their energy requirements [2]. According to a report by the Carbon Dioxide Information Analysis Center (CDIAC), CO_2 emissions have increased 2750 times in the first decade of the 21st century compared to the mid-18th century. Moreover, based on the Keeling curve, the pace of the rise in the concentration of CO_2 is becoming progressively faster with time [3, 144]. This can be understood from the rate of increase in atmospheric CO_2, which was 1.94 ppm/year in 2011, more than twice the estimated value in 1959 [3]. In view of this, it is imperative to identify and explore ways in which CO_2 can be sequestered from the environment. Physical and chemical methods are widely used for CO_2 sequestration. Physical methods involve residual trapping, mineral trapping, deep saline aquifers, solubility trapping, and ocean sequestration [4, 143]. However, there are high costs associated with capturing, transporting, and storing CO_2. A biological method of CO_2

Natural and Artificial Photosynthesis: Solar Power as an Energy Source, First Edition. Edited by Reza Razeghifard.
© 2013 John Wiley & Sons, Inc. Published 2013 by John Wiley & Sons, Inc.

sequestration using green algae and cyanobacteria is another alternative available. Its advantages include mitigating the CO_2 and the use of biomass for the production of biofertilizers, industrially important biomolecules, and biofuels like biohydrogen, bioethanol, and biodiesel [5, 6, 7].

The world is also facing an energy crisis along with the global warming threat. According to one report, fossil fuel demand will exceed production demand after 2020 [8]. In this context, harvesting the abundant energy of the sun in an efficient manner is today's global demand. Hydrogen is considered an efficient and sustainable energy carrier of the sun is energy. After burning, it produces only water on reaction with oxygen [9]. It is considered an environmentally safe, renewable energy resource and the only CO_2 neutral fuel that doesn't contribute to the greenhouse effect. Biohydrogen has the potential to act as a sustainable alternative to conventional methods for H_2 production [10]. Biohydrogen production by green algae and cyanobacteria has its own importance. They produce hydrogen from the splitting of water, which can be sustained for longer periods of time.

The history of hydrogen production in cyanobacteria began in 1970, when the first evidence of hydrogen production was found in nitrogen-fixing heterocystous *Anabaena cylindrica* B-629 [11]. Later, hydrogen production potential was detected in both filamentous nitrogen-fixing and unicellular non-nitrogen-fixing cyanobacteria. Today, there are at least 14 hydrogen production genera of cyanobacteria [12]. The merits of a nitrogen-fixing system are that H_2 production is simultaneous with photosynthetic O_2 evolution, and no anaerobic treatment of the cells is required [13].

In the case of green algae, it was Gaffron and Rubin who first observed that the green alga *Scenedesmus obliquus* is related to molecular hydrogen in its metabolic pathway either by consumption of H_2 as an electron donor in CO_2 fixation or production of H_2 under anaerobic conditions in light and dark cycles [14]. Later, it was found that only after a period of dark followed by illumination do microalgae cells acquire the ability to produce H_2 because the dark period creates an anaerobic environment [15, 16, 17]. Thereafter, the focus of research shifted to creating anaerobic conditions inside the culture in a quick and efficient manner as the hydrogenase enzyme responsible for producing hydrogen is sensitive to oxygen. Sulfur deprivation on the green alga *Chlamydomonas reinhardtii* creates anaerobic conditions and causes the production of hydrogen by biophotolysis under light using the photosynthetic pathway [18]. It was envisaged by Melis and Happe that hydrogen can be produced in just one stage by careful supply of sulfur into the medium so that it can be consumed entirely after producing sufficient biomass [19]. An attempt was also made to produce hydrogen from *Chlamydomonas reinhardtii*, strain cc124, in a cyclic way by readdition of sulfur in the culture [20].

6.2 MICROBIOLOGY

Green algae and cyanobacteria are the vast group of both facultative photoautotrophic and photoheterotrophic microorganisms. They have a wide range of genetic diversity and can exist as unicells, colonies, and extended filaments. They are ubiquitously distributed throughout the biosphere and can grow under the widest possible variety of conditions from aquatic to terrestrial places. Microalgae can be cultivated under aqueous conditions ranging from freshwater to situations of extreme salinity and from open systems to highly controlled closed systems. Their uniqueness that separates them from other microorganisms is due to the presence of chlorophyll and having photosynthetic ability in a single algal cell, therefore allowing easy operation for biomass generation and effective genetic and metabolic research in a much shorter time period than with conventional plants.

Algae are generally classified into several taxonomic groups on the basis of their nature and properties, type, number and morphology of flagella, reverse food products chemistry and product of photosynthesis, the chemical and physical properties of their cell walls, the morphological characteristics of cells and thalli, the mode of reproduction, and so on. Among taxonomic groups, the Chlorophycophyta (green algae) division is one of the two subjects of interest for this chapter. Its diversity can be understood by its abundance in nature (Chlorophyta: $\sim$500 genera, $\sim$15,000 species, and 9 genomes being sequenced) [21]. However, only 30 genera have been investigated for their potential in hydrogen production, leaving a large scope for further exploration [22, 23]. A well-defined nucleus, a cell wall, chloroplast containing chlorophyll and other pigments, pyrenoid, (a dense region containing starch granules on its surface) stigma, and flagella are the major components of green algae used in H_2 production. Green algae contain both chlorophyll a and b types, mainly α and/or β carotene and rarely γ-carotene of carotenoids, and starch and oils as storage products. Green algae are similar to plants since both possess chlorophyll and other light-harvesting pigments (carotenoids) to capture energy from sunlight and to do photosynthesis [24, 25]. However, green algae differ from plants in having hydrogenase enzyme in some of the strains. The role of the hydrogenase in green algae growing under photosynthetic conditions in the natural environment was unclear for a long time [26]. *Chlamydomonas reinhardtii* is considered as a model green alga that reproduces by meiosis. Under ambient conditions, the cell density increases exponentially with a doubling time of approximately 24 h [27].

Cyanobacteria are older and more extensively studied microorganisms compared to green algae. There are 2654 classified species of cyanobacteria with 55 draft genomes sequenced [22]. They form a more diverse group in hydrogen metabolism than green algae. To date, nearly 100 strains have been found having the potential for hydrogen metabolism. Moreover, hydrogenase encoding enzymes were found in nearly 50 genera of all five taxonomic groups [22, 23, 28]. Filamentous colonies of cyanobacteria have the ability to differentiate into different cell types like vegetative cells, akinetes, and heterocysts. General functions of vegetative cells, akinetes, and heterocysts include the ability to carry out complete oxygenic photosynthesis, a resistance to climate, and having a potential to fix nitrogen, respectively. Heterocysts contain the enzyme complex nitrogenase, which converts atmospheric nitrogen into ammonium, a unique capacity among photosynthetic oxygenic organisms. Cyanobacteria have thick gelatinous cell walls. They lack flagella and quorum sensing molecules but have motile filaments called hormogenia. These are the only known prokaryotes having oxygenic photosynthesis for fixation of CO_2 like eukaryotic algae and plants.

6.3 BIOCHEMISTRY OF CO_2 FIXATION

6.3.1 CO_2 Assimilation and Concentrating Mechanisms in Algae

Microalgae and cyanobacteria, just like plants, sequester CO_2 and prepare food for their survival and growth through a biological process called photosynthesis. Photosynthesis is a combination of two processes. The first one consists of light-dependent reactions where light energy is conserved in the form of ATP and NADPH with the help of chlorophyll and other light-harvesting pigments. The second process includes carbon assimilation reactions where ATP and NADPH are used to reduce CO_2 to form triose phosphates, starch, sucrose, and other derived products. In green algae, photosynthesis takes place in specialized organelles called chloroplasts, where the photosynthetic apparatus is located in vesicles called thylakoids arranged in stacks. The location of the light-harvesting complex in cyanobacteria is different from green algae. These are found in the cytoplasm as isolated and freely lying photosynthetic lamellae. The photosynthetic apparatus consists of photosystem II (PSII) and photosystem I (PSI) connected by electron carriers in order of increasing redox potential forming a Z scheme of photosynthesis. The increasing redox potential of electron carriers facilitates the flow of electrons to PSI from PSII.

The presence of different pigments enables PSII and PSI to harvest light at different wavelengths. Each photosystem is equipped with a light-harvesting complex (LHC) made of many photosynthetic pigments like chlorophylls and other accessory molecules to help in harvesting light energy. In cyanobacteria, chlorophyll a is the major pigment apart from other accessory pigments like phycoerythrin and phycocyanin. Phycoerythrin gives the red-brown color to cyanobacteria and its formation become greater in green light. Phycocyanin is a blue pigment giving a blue-green color to the cyanobacteria. These pigments are embedded in photosynthetic lamellae. Cyanobacterial Chl a molecules are housed in a specialized proteinous structure called the phycobilisome. In total, there are four types of phycobilisomes present in the cyanobacteria, each absorbing light at a different wavelength. Phycocyanin and allophycocyanin of phycobilisomes are universal to cyanobacteria absorbing light at wavelengths of 620 and 650 nm, respectively [29]. Carotenoids and phycobilins are other light-absorbing pigments [30]. These microorganisms absorb light preferably in the range of 665–680 nm, where light is mainly absorbed by chlorophyll a. This wide range of light absorption may be the main reason for the high photosynthetic activity of these microorganisms [31].

The electrons are transferred from PSII to PSI, passing through plastoquinone, cytochrome b_6f complex and plastocyanine. PSI and PSII absorb visible light of 700 nm and 680 nm, respectively, and hence PSI requires lower energy (longer wavelength corresponds to lower energy) than PSII. Electrons at PSII come ultimately from the splitting of 2 moles of H_2O into 4 moles of H^+ and 4 moles of e^- with the release of 1 mole of O_2 [29]. The energy for splitting the water comes from photons so it is called biophotolysis (Eq. 6.1).

$$2H_2O \rightarrow 4H^+ + 4e^- + O_2 \qquad (6.1)$$

Electrons are delivered to cytochrome b_6f complexes by mobile plastoquinones (PQs) and from there to PSI complexes by plastocyanine proteins. These electrons are used to reduce photooxidized PSI complexes, which gave their electrons to ferredoxin (Fd) proteins [32]. Fd is loosely attached to the thylakoid membrane from outside and located in stroma where its reduced form is specific to ferredoxin NADP$^+$ reductase (FNR). The Fd transfers electrons via FNR to NADP$^+$ for the production of NADPH by acquiring protons from stroma (Eq. 6.2). The electron transport chain also results in proton (H$^+$) accumulation in the lumen, generating a proton gradient across the thylakoid membrane. Protons are released from lumen to stroma via ATP synthase, which generates energy in the form of ATP. In this way, electrons get utilized in a noncyclic manner, called noncyclic photophosphorylation. Both ATP and NADPH are generated in noncyclic photophosphorylation, in contrast to cyclic photophosphorylation (details are given in Section 6.5) where only ATP forms and electrons are not used to reduce NADP$^+$ into NAD(P)H.

$$2Fd_{red} + 2H^+ + NADP^+ \rightarrow 2Fd_{ox} + NAD(P)H + H^+ \tag{6.2}$$

The photoexcitation of each photosystem requires one photon, hence a total of two photons are required to transfer one electron produced at PSII to NADP$^+$. Two moles of water are oxidized to produce two moles of NAD(P)H (Eq. 6.3).

$$2H_2O + 2NADP^+ + 8 \text{ photons} \rightarrow O_2 + 2NAD(P)H + 2H^+ \tag{6.3}$$

ATP and NADPH generated during the photosynthesis process are used for the synthesis of carbohydrates and other organic compounds (Eq. 6.4). A step of photosynthesis in which CO$_2$ is converted into sugar with the help of ATP (adenosine-5′-triphosphate) is called the Calvin–Benson cycle [33] or simply the Calvin cycle. This reaction is catalyzed by the carboxylase activity of ribulose 1,5-bisphosphate carboxylase/oxygenase (RuBisCO) [34]. The Calvin cycle undergoes three major steps—carboxylation, reduction, and regeneration (Fig. 6.1). In the carboxylation step, three molecules of CO$_2$ are fixed to three molecules of ribulose 1,5-bisphosphate, a five-carbon compound using carboxylase activity of RuBisCO enzyme to form six molecules of glycerate 3-phosphate. It is followed by the reduction step where glycerate 3-phosphate gets reduced to glyceraldehyde 3-phosphate (Eq. 6.5). Out of six molecules of glycerate 3-phosphate only one molecule is used for synthesizing biosynthetic material and the remaining five molecules are used to regenerate the starting material (ribulose 1,5-bisphosphate) of the cyclic Calvin cycle.

$$CO_2 + H_2O + light \rightarrow (CH_2O)_n + O_2 \tag{6.4}$$

$$\text{Ribulose 1,5 bisphosphate} + CO_2 + H_2O \rightarrow 2 \text{ glyceraldehyde 3-phosphate} \tag{6.5}$$

A small fraction of glyceraldehyde 3-phosphate used in synthesizing biosynthetic material is utilized as an immediate energy source and the remaining portion is converted to sucrose for transport to the cytosol or stored in the chloroplast as starch

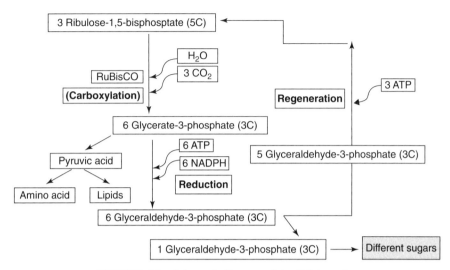

FIGURE 6.1 Schematic diagram of the Calvin cycle.

and glycogen in green algae and cyanobacteria, respectively. In green algae, these starch granules get stored in the pyrenoid of the chloroplast.

6.3.2 Carbon-Concentrating Mechanisms (CCMs)

RuBisCO has low affinity for CO_2 and at the normal atmospheric, it is only half saturated with the CO_2. In addition, it expresses oxygenase activity that produces glycolate 2-phosphate as the end product (Eq. 6.6):

$$Ribulose\ 1,5\ bisphosphate + O_2 \rightarrow glycerate\ 3\text{-phosphate} + glycolate\ 2\text{-phosphate}$$

$$(6.6)$$

Glycolate 2-phosphate produced by the oxygenase activity of RuBisCO consumes a significant amount of cellular energy and it is of no use to cell. Moreover, it also releases previously fixed CO_2 by the carboxylase activity of RuBisCO. It was found that the oxygenase activity of RuBisCO reduces biomass formation by nearly 50% [35]. Most of the microalgae and cyanobacteria have developed different CO_2-concentrating mechanisms (CCMs) to overcome the low affinity of RuBisCO for CO_2. CO_2 affinity constant $K_m(CO_2)$ in cyanobacteria was found to be nearly 200 μM compared to its nearly 10 μM dissolved concentration in water in equilibrium with air [5, 36]. This helps them to grow efficiently at low inorganic carbon (Ci) concentrations by inducing the expression of CCMs [37] leading to the accumulation of a large intracellular inorganic carbon pool from which CO_2 can be obtained efficiently for fixation by RuBisCO. The onset of CCMs in these microorganisms leads to expression of the enzyme carbonic anhydrase (CA), which plays a vital role in its success. It acts as a catalyst during the inter-conversion of CO_2 and HCO_3^- (Eq. 6.7) [5]. It ensures the availability of CO_2 by intracellular mobilization of the HCO_3^- pool for the carboxylase activity of RuBisCO [5,38]. CA is one of the most efficient enzymes known, and increases the intracellular CO_2 concentration 1000 times higher than in the external fluid. The carbon-concentrating mechanisms can be either biochemical or biophysical:

$$CO_2 + H_2O \leftrightarrow HCO_3^- + H^+ \tag{6.7}$$

6.3.2.1   Biochemical CO$_2$ Pump

C4 Mechanism   Phosphoenolpyruvate (PEP) carboxylase has more affinity for HCO_3^- and temporarily stores CO_2 in a four-carbon compound (Eq. 6.4). In a mesophyll cell, one molecule of HCO_3^- combines with one molecule of PEP, acting as the inorganic carbon carrier to form one molecule of oxaloacetate using PEP carboxylase enzyme, which further converts it into malate by malate dehydrogenase. This malate is then decarboxylated in plastid and acts as a local source of CO_2 at the site of RuBisCO, where it is fixed via the Calvin cycle. Examples of algae having evidence of C4 mechanism are *Udotea flabellum* and *Thalassiosira weissflogii* [35].

CAM Mechanism   Temporary storage of inorganic carbon in malate and its release and fixation by RuBisCO occur separately over time rather than spatial separation as in C4 mechanism. At night, CO_2 is stored in vacuoles as a malate, which makes CO_2 available to RuBisCO for its fixation in the daytime [35].

$$\text{Inorganic carbon} + C_3 \rightarrow C_4 \tag{6.8}$$

6.3.2.2   Biophysical CO$_2$ Pump

Active Transport of HCO$_3^-$/CO$_2$ Cyanobacteria and green algae mostly follow active transport of HCO_3^-/CO_2 across the barrier plasma membrane from outside dissolved inorganic carbon (DIC) to the closest membrane of chloroplast enclosing RuBisCO. Various transporters help in delivering HCO_3^- to the cytosol. DIC species transported into the cytosol then diffuse into the carboxysome in cyanobacteria and pyrenoid in microalgae where CA enzyme generates higher steady state concentration of CO_2, which in turn favors carboxylase activity over oxygenase activity [35]. In *Chlamydomonas reinhardtii* active transport of both HCO_3^- and CO_2 occurs across the plasmalemma but the CO_2 gets preference over HCO_3^- [39].

CO$_2$ Passive Influx   In this category of CCMs, CO_2 is converted to HCO_3^- by NADH dehydrogenase (NADHdh) at the plasmalemma of cyanobacteria and delivered into the cytosol. From the cytosol, it diffuses into the carboxysome/pyrenoid to be acted upon by the CA enzyme to release CO_2 [35].

Direct Access of HCO$_3^-$ In this carbon-concentrating mechanism, an acidic compartment gets HCO_3^- from an alkaline medium/compartment due its high concentration, where it is acted upon by either a relatively acid-stable CA or a proton-driven catalysis of HCO_3^- to CO_2 (Eq. 6.9). The CO_2 produced in this compartment then diffuses into the more alkaline compartment, the chloroplast containing RuBisCO [35]. It was found that the rate of direct transport of HCO_3^- across the plasma membrane is less than the active CO_2 influx [39].

$$\text{Alkaline}(HCO_3^-) \rightarrow \text{Acidic}(HCO_3^- \rightarrow CO_2) \rightarrow \text{Alkaline}(CO_2) \quad (6.9)$$

6.4 PARAMETERS AFFECTING THE CO₂ SEQUESTRATION PROCESS

6.4.1 Selection of Algal Species

Algae with high specific growth rates, less sensitive to temperature, pH, and shear stress, are considered ideal species for CO_2 sequestration. As added advantages, algae have a low risk of contamination and the potential to produce valuable metabolites [40]. Some of the green algae and cyanobacteria used in the CO_2 sequestration process are listed in Table 6.1. In the photobioreactor, shear stress to algal cells varies from place to place. For example, shear stress to algal cells in the riser of the airlift

TABLE 6.1 Some of the Cyanobacterial and Green Algal Strains in CO₂ Sequestration at Different Percentages of CO₂

Microorganisms	CO₂ Percentage (v/v)	Photobioreactor	Maximum Biomass Produced (g/L)	Biomass Productivity (mg dwt/ L/day)	References
Chlorella sorokiniana	5	Airlift	4.4	—	[46]
Chlorella sp.	—	Bubble column	1.445	—	[44]
Chlorella vulgaris	10	—	—	104.76	[41]
Scenedesmus obliquus	10	—	3.13	217.50 ± 11.2	[41]
Scenedesmus obliquus	5.5 (Flue gas)	—	—	203	[41]
Botryococcus braunii	10	—	—	26.55	[41]
Botryococcus braunii	5.5 (Flue gas)	—	—	77	[41]
Spirulina sp.	6	Three–stage serial tubular	—	220	[132]
Synechocystis aquatilis SI-2	10	Vertical flat plate	—	30 g/m²/day	[48]
Anabaena sp. ATCC 33047	Air	Bubble column	—	310	[40]
Phaeodactylum tricornutum	60	Airlift	6.2	—	[43]

TABLE 6.2   Pollutants Present in Flue Gas of Different Fossil Fuels [147, 148]

Types of Fuel	CO_2 (%, v/v)	CO (ppm)	NO_x (ppm)	SO_x (ppm)	Mercury (ppm)	Fly Ash (ppm)
Natural gas	10–12	70–110	50–70	1–3	0.0	6–8
Fuel oil	12–14	70–160	50–110	180–250	0.007	80–90
Coal	14–17	200–300	400–500	>2000	0.016	>2700

reactor is much higher due to the high velocity of air bubbles. However, cultures in the down-comer are usually in laminar motion and experience almost no shear stress. Similarly, in stirred tank and raceway ponds, cells undergo huge stress due to the mechanical means of agitation. In view of this, a strain having higher resistance to shear stress minimizes cell death. Resistance to shear stress varies from species to species. Some species are very sensitive to shear stress while some are resistant to it. *Dunaliella* lacks the cell wall, making it an extremely fragile microorganism. Shear stress to the cells of *Haematococcus*, known for carotenoid astaxanthin production, depends on its cell cycle. It undergoes green to red phase during the cell cycle, which enhances its resistance to shear stress due to the presence of immobile aplanospores [5]. *Spirulina, Chlorella*, and *Scenedesmus* sp. are some of the species having high biomass productivity, CO_2 fixation ability, and tolerance to relatively higher levels of stress [5, 41].

6.4.2   Effect of Flue Gas Component

Combustion of fossil fuels such as coal, oil, and gas is the largest source of CO_2 emissions globally. Flue gas emitted from these sources mostly contains nitrogen, CO_2, and O_2. In addition, it also contains small amounts of CO, H_2S, NO_x, SO_x, and particulate matters. The composition and characteristics of flue gas depend on the type of fuel and its combustion process. Among the different types of fossil fuel sources like fuel gas, oil, and coal, coal-fired plants generally have a higher percentage of CO_2 emissions [42] (Table 6.2).

CO_2 is responsible for the acidification of the medium along with other acidic gases such as NO_x, SO_x, and H_2S. CO_2 percent varies usually from 10% to 20% depending on the type of industry [142, 146]. Green algae and cyanobacteria both have an optimum CO_2 concentration where they show the highest specific growth rate. CO_2 sequestration efficiency is enhanced on increasing the CO_2 concentration, as it is the only carbon source for their growth, but a further increase in CO_2 concentration reduces the algal growth because of environmental stress imposed by it [43]. The optimum value of CO_2 concentration for the process varies from species to species, which usually lies between 2% and 10% of CO_2 concentration. *Chlorella* was reported to have the most suitable CO_2 value of 2% for optimal growth, and 10% of tolerance limit [44], while a similar species *Chlorella* T-1 was found to show the highest productivity at 10% of CO_2 concentration [45]. Similarly, *Chlorella sorokiniana* was found to have an optimal growth at a CO_2 concentration of 5% (v/v), while the growth was observed even at a CO_2 concentration of 100% [46].

6.4.3 Effect of Physiochemical Parameters

Along with physiochemical parameters like temperature and pH in the dark fermentation process, optimizing the incident light intensity is an additional component that needs to be taken into consideration in the case of algae. Being photosynthetic microorganisms, they require a suitable light intensity for the CO_2 sequestration and biomass production. Exposure of these cells to high light intensities for long periods causes photoinhibition due to overloading of pigments present in the photosystems and thus interfering in the light-harvesting complex synthesis and degradation cycle [47]. Some of the methods that may be found effective in improving interaction of light and cells are proper mixing, design of a suitable photobioreactor, and biological pigment reduction [5]. Uniform mixing ensures uniform distribution of nutrients as well as distribution of cells in light zones. It minimizes the total incident light intensity available to a particular cell with the added advantage of the flashing light effect [5]. Similarly, a photobioreactor can be designed to have a uniform distribution of light. Furthermore, reducing the size of the antenna complex can decrease the blockage of light penetration by pigmented algal cells [47].

6.4.4 Issues of Product Inhibition

Growth of pigmented cells interferes with light penetration inside the photobioreactor, leading to the self-shading effect [48]. However, at lower cell concentrations, all of the light energy is not captured by the cells and light is not completely utilized. A highly dense culture has an advantage in CO_2 sequestration as it makes cells more tolerable to a high concentration of inlet CO_2 [44]. In view of this, an optimum cell concentration completely utilizes the light without any self-shading. Usually optimum cell concentration lies in the range of 1–2 g/L [5, 48, 49]. To avoid the self-shading effect, running a fed-batch mode of operation involving periodic removal of cells [50] or using a continuous mode of operation are some alternative options.

O_2 production/accumulation in the photobioreactor is another factor inhibiting the CO_2 sequestration process by algae. Excess O_2 in the culture can reduce photosynthetic efficiency by the photobleaching process and RuBisCO becomes more an oxygenase rather than carboxylase as the surrounding O_2 concentration gets higher [5]. Taking this into account, it is necessary to take an efficient degassing system into consideration while designing the photobioreactor for CO_2 sequestration. However, photobioreactors with high surface to volume ratios, for example, horizontal tubular and flat panel photobioreactors, enhance the light penetration and trap more O_2 [51]. The problem of O_2 accumulation becomes more severe while scaling up the photobioreactor, which is usually done either by enhancing the length of the tubes, for example, in a horizontal tubular photobioreactor, or increasing the light-harvesting unit, for example, in a helical photobioreactor. In order to overcome the problem of oxygen accumulation, it is necessary to have provision for an additional degassing unit into the photobioreactor in such a way that even the smallest bubbles can disengage.

6.5 HYDROGEN PRODUCTION BY CYANOBACTERIA

6.5.1 Mechanism of Hydrogen Production

6.5.1.1 *Vegetative Cell-Based Hydrogen Production in Cyanobacteria*
Fd is the hub for electrons coming from the splitting of water. It can channel electrons to four different directions (Fig. 6.2). First, FNR can draw electrons from the reduced Fd for the formation of NAD(P)H, making a unidirectional flow of electrons. Second, Fd can donate electrons directly to the bidirectional hydrogense. The excess electrons generated during photosynthesis can be used to generate hydrogen, acting as an alternative electron sink that helps in redox balancing [52]. However, this fact is not completely established. Third, it can donate electrons to the Mehler reaction. Fourth, it can return the electrons to the PQ, hence making a cyclic flow of electrons via PSI.

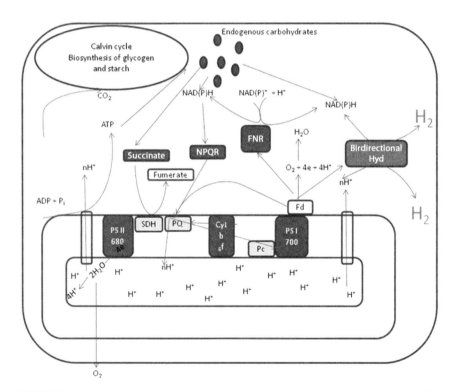

FIGURE 6.2   Schematic diagram of hydrogen production pathway in vegetative cells of cyanobacteria.

Bidirectional hydrogenase (Hox) can catalyze the evolution as well as the oxidation of molecular hydrogen [53]. The soluble or loosely membrane-associated cyanobacterial Hox is found in both vegetative cells of filamentous nitrogen-fixing cyanobacteria and unicellular non-nitrogen-fixing bacteria. The subcellular location of this enzyme even varies from place to place. In *Synechococcus* strain PCC 6301, it was found to be loosely associated with the cytoplasmic membrane, whereas in *A. variabilis* and *Synechocystis* strain PCC 6803, it was found in thylakoid membranes [54]. It is different from uptake hydrogenase, which is found predominantly in the heterocysts of filamentous cyanobacteria along with nitrogenase. Normally, hydrogenase and nitrogenase in cyanobacteria have their own importance and are not meant for hydrogen production. Hydrogenase protects cells from Mn^{2+} deficiency. It is considered important in the photoprotection of cyanobacteria from high light intensities. In addition, it makes chlorophyll *a* more stable in Mn^{2+} deficiency [55]. However, the bidirectional role of hydrogenase has not been explored completely and it is thought to control ion levels in cyanobacteria [12]. Hydrogen evolution takes place under anaerobic conditions, probably because of the presence of NAD(P) bidirectional hydrogenase used for the dissipation of excess reducing equivalents [32]. The advantage of the Hox enzyme is the production of hydrogen without any need for ATP [52]. Under dark fermentative conditions, it helps in accelerating the breakdown of endogenous carbohydrates by reoxidizing NAD(P)H [52]. In this condition, the cyanobacteria switch immediately to the constitutive fermentative pathway from the autotrophic pathway, leading to use of either reduced NAD(P)H or H_2 as a substrate for the Hox enzyme for hydrogen metabolism, while under light conditions reduced ferredoxin can act as a terminal electron donor to hydrogenase for hydrogen production [57].

Bidirectional hydrogenase mainly interacts with NAD(P)H and oxidizes it to produce hydrogen. Accumulated NAD(P)H in the vegetative cells has three fates. First, it can act as a feed for the Calvin cycle. Second, it can interact with the Hox enxyme for the production of hydrogen. Third, it can donate its electrons back to the PQ pool using NAD(P)–plastoquinone oxidoreductase (NPQR) [145]. However, in respiration electrons are fed into PQ via succinate dehydrogenase (SDH) [71]. PQ is a pool that can receive electrons from different electron carriers such as NAD(P)H, succinate, or reduced ferredoxin. Thus these electron carriers have been proposed to mediate the nonphotochemical reduction of the plastoquinone pool, and therefore to be the potential electron sources for PSII-independent H_2 production in chloroplasts [56]. The in vivo activity of the bidirectional enzyme in heterocystous strains increases considerably under anaerobic or microaerobic conditions, whereas in the unicellular non-nitrogen-fixing *Gloeocapsa alpicola* and *Chroococcidiopsis thermalis* the partial pressure of oxygen does not seem to have any significant influence [54]. These strains produce a little less hydrogen compared to the filamentous nitrogen-fixing cyanobacteria (Table 6.3). Moreover, it requires a highly reducing environment to protect their hydrogenase [57].

Examples of unicellular non-nitrogen-fixing cyanobacteria are *Synechocystis* PCC 6803 and *Gloeocapsa alpicola* CALU743 in which Hox gets activated in dark anaerobic conditions [32, 58]. *Gloeocapsa alpicola* CALU743 produces as high as 3.92

TABLE 6.3 Potential Hydrogen-Producing Cyanobacterial Strains

Cyanobacteria	Growth Conditions	Experimental Conditions	Rate of Hydrogen Production	References
Anabaena variabilis ATCC 29413	—	Pure argon	5.8 mL/L/h	[76]
Anabaena variabilis ATCC 29413	73% Ar, 25% N_2, 2% CO_2; 90 $\mu E/m^2/s$	Argon environment	45.16 $\mu mol/mg$ Chl *a*/h	[133]
Anabaena variabilis ATCC 29413	Argon enriched, with 25% N_2, 2% CO_2; 90 $\mu E/m^2/s$	95% Ar, 5% N_2; 90 $\mu E/m^2/s$	3.07 $\mu mol/mg$ Chl *a*/h	[133]
Anabaena variabilis ATCC 29413	2% CO_2–air gas mixture continuous tarbidostat; 113 $\mu E/m^2/s$	Argon environment	39.4 $\mu mol/mg$ Chl *a*/h	[134]
Anabaena variabilis PK17R	73% Ar, 25% N_2, 2% CO_2; 90 $\mu E/m^2/s$	93% Ar, 5% N_2, 2% CO_2; 90 $\mu E/m^2/s$	59.18 $\mu mol/mg$ Chl *a*/h	[133]
Anabaena variabilis 1403/4B	Air; 15 $\mu E/m^2/s$	No gas phase; cells immobilized in hollow fiber; 25 $\mu E/m^2/s$ on the top surface and 13 $\mu E/m^2/s$ on bottom surface of reactor	20 $\mu mol/mg$ Chl *a*/h	[135]
Anabaena variabilis IAM M-58	Air enriched; 20 $\mu E/m^2/s$	95% Ar, 5% CO_2; 100 $\mu E/m^2/s$	3.4 $\mu mol/mg$ Chl *a*/h	[66]
Anabaena variabilis mutant PK84	—	During growth CO_2 enriched air	10 mL/L/h	[76]
Anabaena variabilis PK84	2% CO_2–air gas mixture 113 $\mu E/m^2/s$, continuous mode	Ar	32.3 $\mu mol/mg$ Chl *a*/h	[134]
Anabaena variabilis PK84	25% N_2, 2% CO_2, 73% Ar; 90 $\mu E/m^2/s$	5% N_2, 2% CO_2, 93% Ar; 90 $\mu E/m^2/s$	167.6 mmol/mg Chl *a*/h	[133]

(continued)

TABLE 6.3 *(Continued)*

Cyanobacteria	Growth Conditions	Experimental Conditions	Rate of Hydrogen Production	References
Anabaena cylindrica B-629	0.3% CO_2–air gas mixture; 92 $\mu E/m^2/s$	97% Ar, 3% CO_2; 276 $\mu E/m^2/s$	0.74 $\mu mol/mg$ Chl *a*/h	[11]
Anabaena cylindrica	0.3% CO_2–air gas mixture; 20 W/m^2	97% Ar; 3% CO_2; 60 W/m^2	0.93 mmol/L/h	[78]
Synechocystis PCC 6803	—	Dark-light transition	0:3 mmol/mol Chl/s	[32]
Synechococcus PCC 602	Air; 20 $\mu E/m^2/s$	Ar with CO (13.4 μmol); 20–30 $\mu E/m^2/s$	0.66 $\mu mol/mg$ Chl *a*/h	[136]
Gloeobacter PCC 7421	Air; 20 $\mu E/m^2/s$	Ar with CO (13.4 μmol),C_2H_2 (1.34 μmol); 20–30 $\mu E/m^2/s$	1.38 $\mu mol/mg$ Chl *a*/h	[137]
Nostoc sp. PCC 7422	—	Ar + 5% CO_2, 200 $\mu E/m^2/s$	100 $\mu mole/mg$ Chl *a*/h	[66]
Nostoc muscorum IAM M-14	Air enriched; 20 $\mu E/m^2/s$	95% Ar, 5% CO_2; 100 $\mu E/m^2/s$	7.7 $\mu mol/mg$ Chl A/h	[66]
Oscillatoria sp. Miami BG7	Air enriched; 100 $\mu E/m^2/s$	Ar; 89 $\mu E/m^2/s$	0.15 $\mu mol/(dry\ weight)/h$	[138]
Gloeobacter violaceus PCC 142	Air enriched; 20 $\mu E/m^2/s$	Ar with CO (13.4 lmol) and C_2H_2 (1.34 mmol); dark	0.59 $\mu mol/mg$ Chl *a*/h	[136]

mol of hydrogen per mol of glucose, following acetate as the end fermentative product through dark fermentation. The maximum hydrogen production rate was similar to the heterocysts of nitrogen-fixing cyanobacteria but this value was much higher than the direct biophotolysis by non-nitrogen-fixing cyanobacteria. In *Synechocystis* PCC 6803 short burst of H_2 output was observed upon illumination in anaerobiosis, which was followed by a rapid H_2 uptake concomitant with CO_2 uptake because of reversible inactivation by photosynthetically produced O_2 [56].

6.5.1.2 Heterocyst-Based Hydrogen Production in Cyanobacteria

Overview of Heterocysts   Filamentous nitrogen-fixing cyanobacteria have two types of cells: the vegetative cells, and the thick-walled heterocysts. However, unicellular non-nitrogen-fixing bacteria have only vegetative cells. Vegetative cells differentiate into heterocysts, which are sites of intense reducing activity, having limited permeability to oxygen. Heterocysts have a specialized cell envelope, with the outer and inner layers of the envelope consisting of polysaccharides and glycolipids, respectively [59]. This prevents oxygen diffusion from outside [60] and separates the internal anaerobic environment from external aerobic environment. Formation of heterocysts is induced under some metabolic stress, such as few days of nitrogen starvation [54, 61]. Some amount of O_2 is also required in heterocysts for the oxidative metabolism of the carbohydrates. A high level expression of respiratory enzymes consumes and lowers the O_2 tension within the heterocysts [62]. The number of heterocysts can be as high as 5–10% of their vegetative cells [54]. Heterocysts are the source of nitrogen, which can be quantified from phycocyanin content as it may be the index of its nitrogen content [63]. It is important to know that heterocysts need to be surrounded by the vegetative cells to get a continuous supply of carbohydrates. It is assumed that there are 10 vegetative cells for every heterocyst. Vegetative cells carry out oxygenic photosynthesis, leading to the formation of carbohydrates in the form of glycogen as it has both PSII and PSI. However, anoxygenic photosynthesis takes place in heterocysts because they have only PSI and lack PSII. Apart from getting carbohydrates from vegetative cells, heterocysts also produce organic carbon and ATP for hydrogen production [64]. Heterocystous filamentous cyanobacteria produce hydrogen for 7–19 days in a nitrogen-free medium [65].

Machinery of Hydrogen Production   In some filamentous nitrogen fixing cyanobacteria, bidirectional hydrogenase has also been found in heterocysts along with vegetative cells [66]. However, in a filamentous dinitrogen-fixing cyanobacterium, *Nostoc punctiforme* ATCC 29133, nitrogenase and an uptake hydrogenase have been found but no bidirectional hydrogenase [53]. Both the nitrogenase and hydrogenase enzymes influence hydrogen production in the heterocysts.

Nitrogenase is found only in the heterocysts of cyanobacteria. The role of nitrogenase is mainly to fix gaseous nitrogen into ammonia and hydrogen is only the by-product of the nitrogen fixation step. Hydrogen is produced as a secondary metabolite to dissipate the excess energy and to balance the organism's redox potential

[29]. However, the turnover rate of nitrogenase is significantly slower in comparison to the bidirectional [NiFe] hydrogenases of vegetative cells (6.4 s^{-1} vs. 98 s^{-1}) [67], requiring a large quantity of nitrogenase to sustain hydrogen production for a longer period of time. Moreover, cells spend a large amount of their energy currency (ATP) for maintenance of a large quantity of nitrogenase, posing a potential physiological burden. Although both hydrogenase and nitrogenase are sensitive to oxygen, the latter is found to be more tolerant of oxygen than the former. However, nitrogenase can be converted from the active to the nonactive form after even a sudden and short-term exposure to high oxygen concentrations [68,69]. Despite the coproduction of oxygen in the biophotolysis of heterocystous cyanobacteria, the inhibitory effect of oxygen is a minimum compared to the hydrogenase-based system of green algae, which get inactivated at an O_2 partial pressure of 2% [65].

Uptake hydrogenase is found only in thylakoid membranes of heterocysts of filamentous nitrogen-fixing cyanobacteria while some reports mention its presence in both vegetative and heterocyst cells with a particular association with cytoplasmic membranes [54]. Uptake hydrogenase has been found in the heterocysts of all known filamentous nitrogen-fixing cyanobacteria and is proclaimed to be functionally connected with nitrogen fixation [54]. Uptake hydrogenase consumes hydrogen and conserves the lost energy by the cells in the form of hydrogen. By recycling hydrogen, uptake hydrogenase benefits cyanobacteria in the following three ways. First, hydrogen consumption in cyanobacteria is physiologically inevitable. The uptake hydrogenase oxidizes molecular hydrogen and reduces molecular oxygen for the generation of energy (ATP) [70]. In other words, it is a way to balance the redox potential of the cells by refilling the redox potential lost during nitrogen fixation. This necessitates the need of uptake and/or bidirectional [NiFe] hydrogenase in cells. It was found that upon mutation of uptake hydrogenase, the hydrogen consumption pathway is channeled via bidirectional [NiFe] hydrogenase [54] and further inactivation of the bidirectional hydrogenase drastically decreases the hydrogen productivity of cyanobacteria [13]. Second, the consumption of oxygen (Knallgas reaction) by uptake hydrogenase helps in creating an anoxic condition in the heterocysts and rescues the activity of hydrogenase as well as nitrogenase [71]. It is called the Mehler reaction or water–water cycle as consumed water at PSII gets regenerated through this reaction near PSI making a cycle of water [72]. Third, it provides reducing equivalents (electrons) to nitrogenase and other cells for their function [54].

Hydrogen Production Pathway In the heterocysts of cyanobacteria, the carbohydrates are imported mainly from the neighboring vegetative cells while the fixed nitrogen is siphoned outside the heterocysts into the vegetative cells. Fixed nitrogen may be in the form of nitrite, ammonia, and glutamine amino acid [59]. Both the transported carbohydrates and the carbohydrates present in the heterocysts are the main substrates for the hydrogen production. Carbohydrates oxidize via the pentose–phosphate pathway, yielding ATP, and reductants NADPH and CO_2 [59]. They donate their electrons to reduce ferredoxin, which in turn transfers them to the nitrogenase enzyme for the production of hydrogen [73]. Both uptake hydrogenase and nitrogenase enzymes have an affinity for reduced ferredoxin but they are also oxygen sensitive.

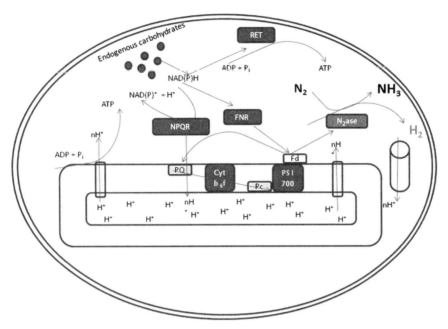

FIGURE 6.3   Schematic diagram of hydrogen production pathway in the heterocysts of cyanobacteria.

The ATP molecules required for nitrogen fixation reactions are also produced in the heterocysts by the cyclic electron flow around the PSI [62]. The formation of ATP and reductants are light dependent while the diffusion of reductants to heterocyst cells is slow and light independent [65]. As the rate of ATP generation in the heterocysts exceeds the rate of reductant supply, an improved hydrogen production takes place even in the intermittent light supply because the diffusion of reductants requires low light [65]. Electrons generated by catabolism of endogenous carbohydrates are channeled either to NPQR or FNR (Fig. 6.3). NPQR transfers electrons into the electron transport chain (ETC) via PQ [29], where, after following the electron transport chain, electrons reduce the Fd and interact with the nitrogenase system for ATP-dependent hydrogen production [60]. On the other hand, FNR transfers electrons to a heterocyst-specific ferredoxin, then to nitrogenase through reverse electron transport [59]. For the function of reverse electron transport, heterocysts has higher ratio of NADPH to $NADP^+$ and nearly tenfold enhanced rate of expression of FNR compared to vegetative cells [59]. In addition, NAD(P)H can also donate electrons to respiratory electron transport (RET), generating the necessary ATP for the nitrogenase reaction [59].

In comparison to green algae, cyanobacteria have developed an efficient system to make the environment for hydrogen producing enzymes anaerobic, but in order to achieve that it spends a significant amount of energy and its enzymes are intrinsically less efficient for hydrogen production. The most efficient hydrogen-producing enzymes are [FeFe] hydrogenases, which can have an activity 1000 times higher than nitrogenases and about 10–100 times better than [NiFe] hydrogenases [67]. The higher hydrogen production efficiency of the [FeFe] hydrogenase system over [NiFe] hydrogenases may be due to its high turnover rate and lower potential of Fd than NAD(P)H, making it a more efficient donor for H$^+$ reduction [32]. In contrast, [NiFe] hydrogenases work about 15 times better than nitrogenases for producing hydrogen and do not require ATP. However, the [NiFe] hydrogenase of cyanobacteria is more tolerant to oxygen than the [FeFe] hydrogenase of green algae [17]. Nitrogenase is also sensitive to oxygen, but less sensitive than hydrogenase [69]. The nitrogenase enzyme produces a small amount of hydrogen as a by-product of the nitrogen-fixing process, where ammonia is the main product. Under normal and optimal conditions of nitrogen fixation, hydrogen production takes place inevitably and irreversibly [74,75]. One molecule of H$_2$ is formed for every molecule of reduced N$_2$. The reduction of nitrogen to ammonia consumes electrons along with 16 ATP for the reduction of 1 mole of nitrogen (Eq. 6.10). This makes the process energy intensive and slow in nature.

$$N_2 + 8e^- + 8H^+ + 16ATP \rightarrow H_2 + 2NH_3 + 16(ADP + P_i) \qquad (6.10)$$

The ratio of electrons used for nitrogen fixation to the total electrons consumed in the nitrogenase-mediated reaction is called the electron allocation coefficient. The value of the electron allocation coefficient lies between 0.75 and 1. It can be derived from the fact that a large number of electrons can be channeled for hydrogen production even in the presence of nitrogen by decreasing the electron allocation coefficient [61]. In biophotolysis, there is a theoretical yield of O$_2$:H$_2$ in the ratio of 1:2 on a mole basis [65]. Nitrogen starvation has the following advantages for hydrogen production. First, it increases the frequency of heterocyst formation. Second, nitrogenase is found in cyanobacteria when they are grown under nitrogen-limiting condition [12] and can be activated by starving the cells with nitrogen. Third, it immediately channels all the electrons for hydrogen production in the absence of N$_2$ substrate for the nitrogenase reaction. In the absence of nitrogen, the ATP requirement per mole of hydrogen production gets reduced and it requires only 4 moles of ATP for the production of 1 mole of hydrogen (Eq. 6.11). Fourth, it conserves the ATP reserves for the extended period of hydrogen production [62].

$$2e^- + 2H^+ + 4ATP \rightarrow H_2 + 4(ADP + P_i) \qquad (6.11)$$

The activity of nitrogenase enzyme varies during the cell cycle. Its activity has the highest value during the exponential phase [76]. For this reason, nitrogen starvation at the end of the growth of cells is always beneficial for hydrogen production in the two-stage process of hydrogen production [77]. During growth, under ambient nitrogen and oxygen concentrations, recycling of hydrogen takes place by uptake and reversible hydrogenase enzymes [76]. For a similar reason, creating a nitrogen-free environment by sparging with argon and CO$_2$ gas mixture enhances the

TABLE 6.4   Characteristics of Direct and Indirect Biophotolysis in Cyanobacteria and Green Algae

Process	Characteristics
Direct biophotolysis	• Light-dependent process • Single-stage proces; simultaneous evolution of oxygen and hydrogen $$2H_2O + \text{light energy} \rightarrow 2H_2 + O_2$$ • Use of inherent metabolic pathway • Maximum light conversion efficiency • Direct conversion of light into hydrogen energy
Indirect biophotolysis	• Light-independent process • Two-stage process; separation of oxygen-evolving and hydrogen-producing processes $$6H_2O + 6CO_2 + \text{light energy} \rightarrow C_6H_{12}O_6 + 6O_2$$ $$C_6H_{12}O_6 + 6H_2O + \text{light energy} \rightarrow 12H_2 + 6CO_2$$

hydrogen production to a great extent [78]. In addition, replacement of air by argon in cyanobacteria inhibits hydrogen uptake owing to the absence of oxygen.

6.5.2   Mode of Hydrogen Production

6.5.2.1   *Direct Biophotolysis*   The process of hydrogen production can be grouped into direct and indirect biophotolysis. In direct biophotolysis, electrons required for hydrogen production come directly from the splitting of water, which is essentially a light-dependent process and uses the electron transport chain (Table 6.4). The energy level of electrons is boosted by light energy absorbed at PSII and PSI [57]. Direct biophotolysis occurs in both vegetative cells and heterocysts of cyanobacteria.

6.5.2.2   *Indirect Biophotolysis*   In indirect biophotolysis, reducing equivalents for hydrogenase or nitrogenase come from the endogenous carbohydrates such as starch in microalgae and glycogen in cyanobacteria [57]. Indirect biophotolysis is essentially a two-stage process. In the first stage, carbon is stored as endogenous carbohydrates in their respective form in microalgae and cyanobacteria; while in the second stage these stored reducing equivalents are used for hydrogen production through dark anaerobic fermentation similar to conventional anaerobic dark fermentation [57]. The two-stage process of indirect biophotolysis ensures temporal or spatial separation of hydrogen and oxygen [79]. The approach of a natural light and dark cycle was proposed by Miura et al. [80], where endogenous carbohydrates are stored in the light phase followed by anaerobic fermentation to produce hydrogen in the dark. Indirect biophotolysis is more common in cyanobacteria than green algae.

6.5.2.3   *Dark Fermentation*   Under dark anoxia, the catabolism of algal endogenous starch reservoirs leads to the generation of fermentative end products such as formate, acetate, or ethanol, to sustain a basal level of metabolism. Hydrogen is another fermentative metabolite produced during dark fermentation. It is believed that the use of hydrogen as an electron sink is preferred by the cells because hydrogen generated from proton reduction will diffuse out and away from cells more easily than organic acids or alcohols [52]. However, if the cell changes its reduced state faster than difffusion of hydrogen away from it, then Hox hydrogenase will reverse its activity. In that case hydrogen may serve only as a temporary reserve for electrons (and protons), and whose fate will depend on cellular conditions [52]. In the dark, the electron transport chain is inactive, hence the electrons for the reduction of Fd proteins are provided by pyruvates using pyruvate–ferredoxin oxidoreductase (PFR1) [81]. In this way, it is different from the algal pathway as it does not rely solely on reduced Fd as an electron donor. Here, electrons are believed to be shuttled directly to bidirectional [NiFe] hydrogenase at the level of NPQR [29].

6.5.3 Hydrogenase Versus Nitrogenase-Based Hydrogen Production

Hydrogenase-based hydrogen production is superior in theoretical maximal energy conversion efficiency, but it requires a daily anaerobic production-harvesting cycle as the bidirectional hydrogenase reversibly catalyzes the absorption of hydrogen during the night. Grown biomass is first kept under dark incubation followed by exposure of the cells to light for completion of the anaerobic hydrogen production [82]. Furthermore, even though nitrogenase-based hydrogen production is less attractive based on maximal energy efficiency, its merit lies in the irreversible nature of hydrogen production. In fact, it does not require the anaerobic production-harvesting cycle [66]. Most nitrogen-fixing cyanobacteria exhibit reversible hydrogenase activity, although it is less than nitrogenase. The affinity of [FeFe] hydrogenase for Fd makes a better system for hydrogen production, as reduced Fd enjoys the lower potential and hence interacts more efficiently with the hydrogenase enzyme than NAD(P)H does [56].

6.5.4 Factors Affecting Hydrogen Production in Cyanobacteria

- Traces of molybdenum are essential for nitrogen fixation. Its deficiency will increase the heterocysts' frequency and double the response to the nitrogen deficiency in *Anabaena cylindrical* while the nitrogenase activity gets reduced by one-fifth [63]. However, heterocysts' frequency and nitrogenase content are two independent variables.
- Nickel is involved in several biological processes and required for the synthesis of active hydrogenase at low concentrations [83]. However, higher concentrations increase the activity of uptake hydrogenase and lower the hydrogen production rate.
- Cellular nitrogen level (e.g., ammonia, nitrate, nitrite) as a key irreversible intermediate in the nitrogen fixation process suppresses the formation of heterocysts and the activity of enzymes involved in nitrogen fixation such as nitrogenase

and uptake [NiFe] hydrogenase [29]. In fact, all exogenously added nitrogen sources inhibit nitrogenase synthesis [84].

- Methane has been found to enhance hydrogen production in a few cyanobacteria [12].
- Carbon sources are also known to influence hydrogen production considerably by influencing the nitrogenase activity [85]. Glucose and CO_2 additions were found to be boosters of hydrogen production in some strains of cyanobacteria. It was hypothesized that they may become the external source of energy and/or reductants for the nitrogenase-catalyzed reaction [62]. However, at higher levels of CO_2 concentration, reduction in hydrogen production was observed probably because of the competition of the Calvin cycle pathway for ATP and reductants [83].
- The effect of an exogenous carbon source in hydrogen production is determined by light intensity. Some cyanobacteria produce hydrogen in the presence of light and some produce in the dark. In the absence of an exogeneous carbon source it was found that an increase in the light intensity resulted in enhanced hydrogen production. Moreover, hydrogen production rate and yield were found to be higher at high light intensities. However, the stimulatory effect of increased light intensity for hydrogen production was less pronounced for the culture lacking an exogenous carbon source [62].
- The optimum temperature for hydrogen production by cyanobacteria is found to be in the range of 30–40 °C [12].
- The effect of salinity for hydrogen production is different for marine and freshwater cyanobacteria. Hydrogen production by freshwater cyanobacteria gets reduced as the reductant energy gets channeled for the extrusion of sodium ions and prevention of sodium ion influx [12].
- N_2 was found to be a more potent inhibitor of hydrogen production than O_2. In fact, low levels of oxygen in the head space (1–5%, v/v) were generally found to be helpful for optimum hydrogen production. This was exploited in acetylene reduction assay, which was very useful for screening hydrogen-producing cyanobacteria [62].

6.5.5 Recent Advances in the Field of Hydrogen Production Using Cyanobacteria

6.5.5.1 Immobilization Immobilization of cyanobacterial cells in a suitable matrix has a number of advantages, such as protecting the cells from mechanical disruption and deactivation, providing constant light intensity, improving volumetric productivity, and stabilizing the cells' activity. Hydrogen production by cyanobacteria using the immobilization process has been done in two stages. In the first stage, cyanobacterial cells are immobilized in nitrate-limiting conditions during growth in proper light for the accumulation of glycogen and activation of hydrogenase enzymes. In the second stage, hydrogen is produced in the dark by the fermentation of glycogen [86]. An anaerobic condition was created by continuously sparging with argon gas.

Matrix can be made up of a number of different materials such as glass fiber, alginate, or polyurethane. Glass fiber, a cheap and translucent material, was used as a matrix for the immobilization of cells since they naturally adhere to it. Alginate matrix is another choice for immobilization as it is cheap, nontoxic, and can be produced in large quantities from renewable sources. In addition, immobilization of cyanobacteria in polyurethane foam was found to stabilize the PSI and PSII photoactivities in *N. muscorum*, resulting in enhanced hydrogen evolution [83].

6.5.5.2 *Reverse Micelles* Reverse micelles are the self-assembling surfactant molecules whose polar heads are directed toward the interior of a water-containing sphere while the aliphatic and nonpolar tails are directed outward in the nonpolar organic phase. Reverse micelles possess many unique properties that make them suitable for hydrogen production. Their ability to encapsulate whole cells and to create a thermostable anaerobic environment is highly useful for hydrogen production as the hydrogen-producing enzymes are highly susceptible to an aerobic environment [87]. In fact, a reverse micelle having a translucent surface makes a suitable microvolume photobioreactor for hydrogen production with its compartmentalized cells. The anionic surfactants like single chain sodium dodecyl sulfate (SDS) and *N*-ethyl hexyl sodium sulfosuccinate (AOT), and the cationic surfactants such as cetyl trimethyl ammonium chloride (CTAC) or bromide (CTAB) are the most commonly used surfactants for this purpose [12].

6.5.5.3 *Nanotechnology* Nanotechnology is another area that can be utilized for hydrogen production using cyanobacteria. It can be used for ex vivo hydrogen production employing the enzymes responsible for hydrogen generation. Nanolipoprotein particles (NLPs) help in the efficient isolation and stabilization of hydrogen-producing enzymes. NLPs can provide suitable environments for low-temperature conditions, higher selectivity to hydrogen production, and potential immobilization in their nanopore membranes [12]. Recently, membrane-bound [NiFe] type hydrogenase of thermophilic *Pyrococcus furiosus* was isolated in NLPs and used for the production of hydrogen [88]. NLPs successfully solubilized the membrane-bound hydrogenase. The enzyme kept its activity in NLPs and showed less sensitivity to oxygen. It was envisaged that hydrogen production can reach nearly theoretical yields using this strategy [12].

6.6 MECHANISMS OF H₂ PRODUCTION IN GREEN ALGAE

There are many green algae known for their hydrogen production capabilities linked to photosynthetic products (Table 6.5). In photosynthesis, water is the direct source of electrons, where reduced Fd proteins channel electrons for the production of NAD(P)H using FNR. However, under special conditions the reduced Fd is oxidized by giving electrons to protons, transferred from lumen to stroma during ATP synthesis (Eq. 6.12), to be turned into molecular hydrogen with the help of hydrogenase enzymes [89]. A special condition arises when an anaerobic environment is established

TABLE 6.5 Potential Hydrogen-Producing Green Algal Strains

Green Algae	Photobioreactor	Growth Conditions	H_2 Production Conditions	Rate of Hydrogen Production	References
Chlamydomonas reinhardtii strain C137 (mt$^+$)	Flat bottles	—	TAP (−S)	2 mL/L/h	[104]
Chlamydomonas reinhardtii	Vial	—	Dark–light transition	1.34 mL/(mg Chl)/h^a	[32]
Chlamydomonas reinhardtii cc124	—	97% air 3% CO_2; acetate (17 mM); 43 W/m^2	Argon; S-free acetate (17 mM); 65 W/m^2	2.1 mL/L/h	[109]
Chlamydomonas reinhardtii 137C+	—	Air; acetate (17 mM); 22 W/m^2	Argon; S-free acetate (17 mM); 26 W/m^2	4.3 mL/L/h	[139]
Chlamydomonas MGA 161	—	—	5% CO_2; 25 W/m^2	4.48 mL/L/h	[140]
Chlorella sorokiniana Ce	—	—	Acetate; 120 µE/m^2/s	1.35 mL/L/h	[141]
Chlamydomonas reinhardtii	—	—	Acetate (17 mM); 100 µE/m^2/s	2.1 mL/L/h	[139]

^{a}Chlorophyll concentration of culture was 40 µg /mL

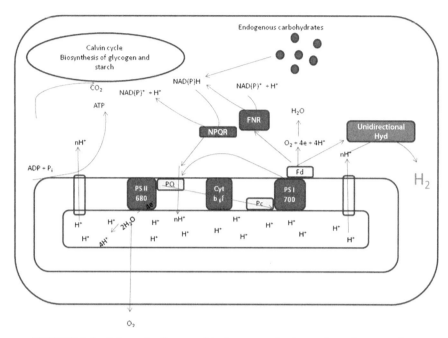

FIGURE 6.4 Schematic diagram of hydrogen production pathway in green algae.

in the culture [90]. Under this situation, there is a lack of final electron acceptor (O_2) and hence green algae need some other pathways to dispose of its excess electrons. Thus reduced Fd channels electrons toward hydrogenase enzymes for the hydrogen production instead of FNR (Fig. 6.4).

$$2Fd_{red} + 2H^+ \rightarrow H_2 + 2Fd_{ox} \qquad (6.12)$$

During photosynthetic H_2 production, two protons and two electrons combine to yield one H_2 molecule. Similar to cyanobacteria, direct biophotolysis in green algae uses water as the direct source of electrons and the process is essentially a light-dependent process, where light helps the transfer of electrons produced after the splitting of water.

Contrary to direct biophotolysis, in indirect biophotolysis water is not split directly by light [91, 92]. However, in this case starch (endogenous carbon reserve) formed during photosynthesis gets degraded producing reducing equivalents (electrons). Similar to cyanobacteria, indirect biophotolysis in green algae has two distinct stages in series. Starch is accumulated first as a result of photosynthesis producing molecular oxygen followed by the fermentation of this carbon reserve to produce molecular hydrogen. It is to be noted that electrons necessary for hydrogen production come indirectly from the first stage. Separate stages have separate gaseous products (O_2 and H_2), which has many advantages such as preventing hydrogenase inactivation by oxygen, easy purification of hydrogen from O_2/H_2 mixture, and preventing a possible explosion due to O_2 and H_2 reaction [93, 94]. This can be further divided into two categories: anaerobic dark fermentation and light-dependent anaerobic fermentation of endogenous substrate. Starch can be degraded through different metabolic pathways, such as glycolytic (EMP), Entner–Doudoroff (ED), and hexose monophosphate shunt (HMS) producing 2 NADH, 1 NADPH, and 12 NADPH, respectively. Reducing equivalents formed here can have at least three possible fates to dispose of its electrons.

- They can be oxidized via respiration where molecular oxygen (O_2) is the final electron acceptor of the electron transport chain. This case can be ignored in hydrogen production because of the lack of a terminal electron acceptor (O_2) under the anaerobic condition.
- They can be oxidized by pumping electrons into the plastoquinone pool and by using the photosynthetic route to finally evolve molecular hydrogen (H_2) with the help of hydrogenase enzyme (light fermentation).
- They can be oxidized by dissociating into the NAD(P)$^+$ form and producing molecular hydrogen with the help of hydrogenase (dark fermentation).

6.6.1 Light Fermentation

In light fermentation, reducing equivalents get oxidized by feeding electrons into NAD(P)H reductase complex, which further transfers its electrons into the thylakoid membrane with the simultaneous release of CO_2 into the medium [95, 96]. The PQ pool in the photosynthetic electron transport chain accepts the electrons provided by the NAD(P)H reductase complex and releases them into the lumen of thylakoid [97–99], generating an additional proton gradient apart from what was generated during the water splitting.

Protons stored in the lumen exit from thylakoid during ATP synthesis by ATP synthase. This fact supports the argument that formation of molecular hydrogen does not inhibit ATP synthesis, rather ATP production increases along with the increase in molecular hydrogen production [100]. Light fermentative metabolism of *C. reinhardtii* was studied in detail by Gfeller and Gibbs [101]. Two important conclusions need to be highlighted here from this work. First, reducing equivalents produced by conversion of glyceraldehydes 3-P to glycerate 3-P reaction during the glycolysis pathway of starch degradation can feed their electrons at a site in the photosynthetic electron transport chain beyond the 3-(3,4-dichlorophenyl)-1,1-dimethylurea (DCMU) block (which blocks between PSII and the plastoquinone pool); and second, ethanol production is inhibited. Comparing the data obtained during light and dark fermentations, it was found that less ethanol was produced in the light than in the dark [101]. This shows that a major portion of NAD(P)H formed in the metabolic pathways is diverted to produce molecular hydrogen with the help of hydrogenase enzyme since the ethanol formation requires a significant amount of reducing equivalents.

6.6.2 Dark Fermentation

Gfeller and Gibbs [101] found that formate, acetate, and ethanol are produced in the ratio of 2:1:1 during the dark, and they account for more than 90% of the carbon in the starch consumed. The stoichiometry is only possible when metabolic breakdown of glucose follows the Embden–Meyerhof pathway (glycolysis) resulting in the formation of 2 pyruvates per glucose [101]. Pyruvate further follows two routes forming either formate or acetyl-CoA. Acetyl-CoA converts into ethanol by deacylase enzyme with acetaldehyde as the intermediate compound. Acetaldehyde can also be converted into acetate. In the dark, the electron transport chain in green algae is inactive. In this case, to maintain basal level of metabolic activity, the production of hydrogen can take place when NAD(P)H transfers its electron to reduce Fd using PFR1 [29]. The ratio of ethanol to acetate was found to be nearly 1, which shows that ethanol is formed mainly during the dark fermentation and its formation consumes most of the NAD(P)H produced along the glycolysis pathway. The remaining maximum 20% of total NAD(P)H goes to Fd and is used for H_2 production [101]. Therefore during indirect biophotolysis, oxygen and hydrogen evolutions are temporarily and/or spatially separated [82].

6.6.3 Use of Chemicals

PSII activity of the photosynthetic system is responsible for making the culture aerobic, which is undesirable for hydrogen production. However, the product (H^+) of water splitting is also required for the evolution of hydrogen. Using DCMU in the medium followed by sufficient dark incubation of the culture resulted in hydrogen production despite the fact that protons and electrons were not coming from the PSII. DCMU inhibits the activity of PSII by binding at the Q_B site, which blocks transport of electrons and production of protons at the water-splitting enzyme [102]. Mus et al. [103] found that sustained and enhanced hydrogen production takes place by adding an external electron donor such as NAD(P)H into the cell culture medium which maintains the cell growth as the electron flow around PSI does not get hampered.

6.6.4 Sulfur Deprivation

The pioneer work of creating an anaerobic environment by sulfur deprivation was done by Melis et al. [104] and Ghirardi et al. [20], who observed that hydrogen production takes place when culture is deprived of sulfur. Later, it was found that sulfur deprivation had no significant effect on the activity of PSI, rather it was the activity of PSII that was affected. Further investigation of PSII structure showed that sulfur was very much required for the proper functioning of PSII. Turnover of D1 protein present in the PSII decreases under sulfur-deprived conditions. This 32 kDa protein is composed mainly of sulfur-containing amino acids of methionine and cysteine that makes sulfur a crucial element for the synthesis/repair of D1 protein. Again, this implies that PSII activity is inhibited while the PSI activity remains intact when cells of green algae experience sulfur-deprived conditions. Photoinhibition is another phenomenon that is closely linked with the success of the sulfur deprivation process needed for hydrogen production. It can be defined as the selective loss of DI protein of PSII under continuous strong light intensity. In addition, the light energy absorbed by the antenna pigments exceeds the capacity of the photosynthetic system during photoinhibition, leading to the generation of the excited triplet state of chlorophyll. The excited triplet state of chlorophyll interacts with the oxygen-forming reactive oxygen species (ROS). These long-lived ROS can have several damaging effects on the PSII. This happens since the rate of damage to D1 protein exceeds the time needed to repair it [105]. In PSII, it disrupts the structure of manganese cluster, therefore damaging the oxygen-evolving complex [106]. It is to be noted that strong continuous light does not affect the activity of PSI. This is because the presence of sulfur can act as a control factor for the activity of PSII under continuous strong light intensity. But a little activity of PSII is very important for the generation of electrons and protons for the evolution of hydrogen. Cellular respiration is another phenomenon that plays an important role in hydrogen production as it consumes residual oxygen evolved at PSII [107]. Therefore the activity of PSII should be controlled in such a way that oxygen produced at PSII gets consumed in the respiration of cells, hence maintaining the anaerobic environment necessary for the activity of [FeFe] hydrogenase enzyme [26, 108]. In summary, sulfur deprivation has the following effects on green algae [18, 104, 109].

- Light-mediated hydrogen production is a result of the interplay between oxygenic photosynthesis, mitochondrial respiration, catabolism of endogenous substrate, and electron transport via the hydrogenase pathway [18].

- Baseline photosynthetic activity of partially and reversibly inhibited PSII complexes maintains respiratory electron transport activity for the generation of ATP. In the case of *C. reinhardtii*, oxygenic photosynthesis declines quasi exponentially to a value less than 10% of the original value [104].

- It has little effect on cellular mitochondrial respiration. Depending on the species, the cellular respiration rate increases above the photosynthesis rate within a few hours of sulfur deprivation even during continuous illumination.

- Catabolism of endogenous substrates such as starch, protein, and lipid continues to take place, generating much-needed ATP during starvation by oxidative phosphorylation in mitochondria, and photophosphorylation in chloroplast both releasing CO_2 in the medium [18].

- It inhibits the activity of RuBisCO, thus blocking the Calvin cycle.

- Reduced Fd uses the Fe-hydrogenase pathway instead of the Calvin cycle responsible for biomass formation and utilizes protons (H^+) as the terminal electron acceptor, resulting in the production of substantial amounts of hydrogen and ATP [110].

- The morphology of *C. reinhardtii* cells were changed during sulfur deprivation and H$_2$ production. Normal ellipsoid-shaped cells changed to larger and spherical cell shapes in the initial (0–24 h) stage, followed by cell mass reduction at longer (24–120 h) times during sulfur deprivation and H$_2$ production [48].

- The algal cultures progress through five phases: aerobic, O$_2$ consumption, anaerobic, H$_2$ production, and termination [109].

After the termination of hydrogen production in the sulfur deprivation process, the residual activity of PSII can be restored by the addition of a small quantity of sulfur. Oncel and Vardar-Sukan [111] found the enhanced effect on residual PSII activity by the addition of a small amount of sulfate either at the start of the sulfur deprivation period or during the H_2-production phase to be due to a temporary recovery of PSII-catalyzed water-oxidation capacity.

6.6.5   Control of Sulfur Quantity

Another approach for continuous and sustainable hydrogen production is to use the right amount of sulfur in the medium. As was seen earlier, sulfur can act as a control valve for hydrogen production in green algae. A minimum amount of sulfur in the culture is required so that the O_2 production rate by PSII should be just below the rate at which it is being consumed by cellular respiration.

6.6.6   Immobilization

Immobilization of cells in alginate film was found to be another method for creating an anaerobic condition. An immobilized algal hydrogen-producing system was found to be less susceptible to oxygen and the hydrogen production took place even in the presence of oxygen [112]. It was found that the alginate beads were capable of separating entrapped cells from O_2 in the liquid and head space, and inhibited O_2 diffusion in the matrix [113]. Alginate immobilization for hydrogen production requires both phosphorus and sulfur deprivations, as the former caused stabilization of the Ca^{2+} alginate films. It helped in increasing the cell density above values obtained for suspension cultures. Dense cell suspensions have the advantage of becoming anaerobic more quickly than dilute cultures due to active cellular respiration inside the shaded area of the photobioreactor [112].

6.6.7   Molecular Approach

6.6.7.1   Algal Mutants with Attenuated Photosynthesis/Respiration (P/R) Ratio   Under normal growth conditions, the photosynthetic rate is about fourfold to sevenfold higher than the respiration rate. The photosynthesis/respiration (P/R) ratio of wild type *C. reinhardtii* drops below 1, in sulfur-deprived medium, leading to the establishment of an anaerobic condition. In attenuated mutants of *C. reinhardtii,* the P/R ratio drops below 1 mimicking the physiological status of sulfur-deprived cells and their established anaerobic conditions [69, 114]. In the mutant of this strain, the respiration rate increased up to 23%, leading to a sevenfold to eightfold increase in the hydrogen yield [115].

6.6.7.2   Engineering Proton Channels Across the Thylakoid Membranes of Algae   As described earlier, when the CO_2 production rate exceeds the O_2 evolution rate, an anaerobic condition prevails causing hydrogen production as the

hydrogenase enzyme responsible for hydrogen production gets activated. *Chlamydomonas reinhardtii* wild strain 137C showed a distinctive hydrogen evolution profile in the range of 10–5000 ppm of O_2. Maximum hydrogen production was obtained below 100 ppm of O_2 while a drastic decrease was observed thereafter. However, the hydrogenase enzyme responsible for hydrogen production was still active up to 5000 ppm of O_2 [90]. Competitive consumption of reducing equivalents for the reduction of residual O_2 by oxygenase activity of RuBisCO and/or reduced Fd in the Mehler reaction was interpreted as the possible reason. Since reduction in the number of ATP in the system may block these competitive pathways of hydrogen production, it was envisaged to engineer algae with proton channels across the thylakoid membranes to decrease the proton gradient needed for ATP production [90].

6.6.7.3 Engineering Mutant Lacking Sulfate Permease Activity Dependability of the sulfur deprivation process for hydrogen production was ended by the discovery of a mutant lacking sulfate permease activity. The lack of this activity disrupts the transport of sulfate into the chloroplast. For this mutant sulfur deprivation was not required for hydrogen production [116].

6.6.8 Recent Trends in the Field of Hydrogen Production by Green Algae

- Effect of sulfur deprivation was studied on green algae [109].
- Cells grown in synchronized culture (14:10 light/dark cycle) attained an anaerobic condition and produced hydrogen earlier compared to unsynchronized culture [109]. Furthermore, synchronization of algal cell division [117] increased the duration of H$_2$ production [118].
- Small readdition of sulfur (12–50 μmol MgSO$_4$) in sulfur-deprived medium caused an increase in hydrogen production [109, 119].
- Light conditions and pH of medium were optimized for hydrogen production in *C. reinhardtii* [109, 120, 121]. The optimum light condition for hydrogen production by *C. reinhardtii* was found to be 200 μmol m^{-2} s^{-1} [122].
- Hydrogen-producing medium for green algae was optimized [123]. The presence of acetate in the medium was found to be necessary for creating an anaerobic condition. However, it had no role during hydrogen production [124]. Sulfur-deprived culture of *C. reinhardtii* was capable of hydrogen production in all autotrophic, mixotrophic, and heterotrophic conditions. Among these, the mixotrophic condition caused the highest hydrogen production, which shows that the presence of acetate and CO$_2$ can enhance hydrogen production. During the aerobic phase of sulfur-deprived culture, CO$_2$ is needed as a starch accumulator while acetate is required as the direct substrate for cellular respiration for the establishment of anaerobiosis in the culture [125].
- Hydrogen production was coupled to a fuel cell for direct electricity generation [126].

- Sulfur-deprived mutants that are starch overaccumulators and blocked in state transition were discovered to have significant increases in rate and duration of hydrogen production [24]. However, the RuBisCO less mutant of *C. reinhardtii* failed to accumulate starch upon sulfur deprivation and thus failed to photoproduce hydrogen, highlighting the importance of starch accumulation [127].

6.7   PHOTOBIOREACTORS

A suitable photobioreactor is required for CO_2 sequestration and hydrogen production by green algae and cyanobacteria. CO_2 sequestration by microalgae can be done in open or closed system, but for the hydrogen production process, it essentially requires a closed photobioreactor as hydrogen can be generated only in anaerobic conditions. Moreover, it is also mandatory to have a closed system for the collection of produced hydrogen gas. Even in CO_2 sequestration, an open system has several disadvantages, such as poor control over light penetration, mixing, pH, and temperature, leading to less biomass productivity. Also, only a few strains can be grown in the open system in semisterilized conditions because of the contamination susceptibility. A good example of these few strains is *Spirulina,* which is cultivated in several open systems in the world, because it grows well in a culture medium with high pH values, greatly reducing the contamination chances. Although CO_2 in the form of carbonates can be used as a carbon source in open ponds, direct use of flue gas containing CO_2 is of lesser advantage for CO_2 sequestration as the open ponds provide little residence time to the injected CO_2-rich gas and the conversion efficiency of inlet CO_2 concentration into biomass is small.

Photobioreator design takes consideration of all the features of a bioreactor such as high heat and mass transfer, proper agitation, and control over pH and temperature. In addition, it requires a high surface-to-volume ratio (S/V ratio) in order to enhance light penetration into the photobioreactor. In photobioreactors designed for CO_2 sequestration, CO_2-rich gas is used for mixing as well as providing nutrients for the growth of algae. In some of these photobioreactors such as airlift, bubble column, horizontal tubular, and flat panel, a nonmechanical method of agitation is used. However, in a few bioreactors agitation can be done by mechanical means as well as by bubbling of CO_2-rich inlet gas, such as a stirred tank reactor. Furthermore, high mass transfer is the requisite criterion for the bioreactors designed especially for CO_2 sequestration as CO_2 must be siphoned into the algal cells through a liquid phase, thus mass transfer takes place among all three phases—namely, gaseous, liquid, and solid. Some types of lab scale photobioreactors for CO_2 sequestration and H_2 production in operation at the Indian Institute of Technology Kharagpur, India, are shown in the Fig. 6.5. Based on the geometric features of different photobioreactors, their performances for CO_2 sequestration and hydrogen production vary.

6.7.1   Vertical Tubular Photobioreactor

The airlift and bubble column come under the category of vertical tubular photobioreactor (Fig. 6.5A, B). Generally, these are cylindrical vessels having a height greater

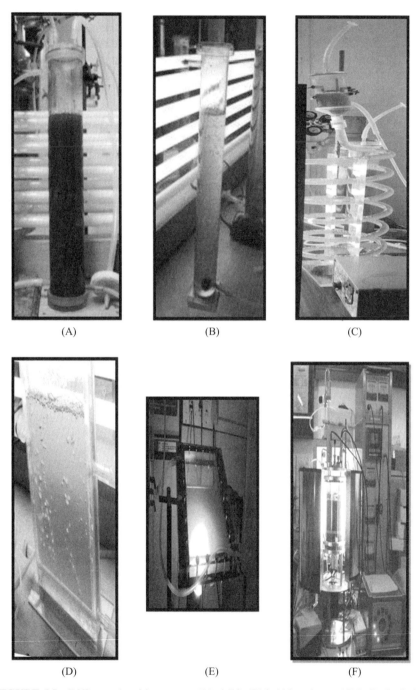

FIGURE 6.5 Different photobioreactors: (A) airlift, (B) bubble column, (C) helical tubular, (D) flat panel, (E) flat panel in rocking mode, and (F) stirred tank under operation at the Indian Institute of Technology Kharagpur, India.

than twice the vessel diameter [5]. Reactors are made up of transparent material like plexiglass to allow the penetration of light. CO_2-rich gas is passed through the bottom of the reactor through a sparger. The sparger can be designed in many ways, giving different bubble sizes, which is one of the parameters controlling mass transfer. Sparging with a gas mixture causes overall mixing and effective mass transfer of CO_2. In addition, it also helps in removal of O_2 produced during photosynthesis. Vertical tubular photobioreactors have advantages such as low capital cost, high S/V ratio, lack of moving parts, satisfactory heat and mass transfer, and efficient disengagement of residual gas mixture including photosynthetically produced O_2 [5]. Light is provided externally. It is the mode of liquid flow which differentiates a bubble column from an airlift reactor (Fig. 6.5A, B). On the basis of light penetration, the bubble column gets divided into two physical regions: outer light zone and central dark zone. Gas flow rate influences the photosynthetic efficiency, which in turn depends on the light and dark cycle as the liquid circulates regularly from the central dark zone to the external photic zone. Airlift reactors are similar to the bubble column ones except for the addition of one external central tube that divides the vessel into two interconnecting zones. The zone containing sparger is called the riser, which lifts the gassed liquid. Whereas another zone is the downcomer, where CO_2-rich liquid moves downward because of gravity. The riser behaves similar to the bubble column where sparged gas moves upward randomly and haphazardly in contrary to the downcomer where liquid moves in a laminar fashion with defined and oriented motion [5, 46]. Upward movement of liquid in the riser is assisted by the gas hold up of the riser, which in turn decreases the density of the liquid. The characteristic advantage of the airlift reactor is its circular mixing pattern, which makes the liquid culture pass continuously through dark and light phases, creating a flashing light effect on algal cells [5, 46, 128].

6.7.2 Horizontal Tubular Photobioreactor

Similar to the vertical tubular design, horizontal tubular reactors are also used for the CO_2 sequestration process. Here, tubes are placed horizontally but may be in different designs such as a loop shape, α shape, parallel set of tubes or tubes, placed in some inclined position for gathering the light intensity under outdoor conditions. A CO_2 gas mixture is introduced into the tube connection from one end of the tube via a dedicated gas exchange system. Poor gas disengagement leading to oxygen buildup and lack of proper temperature control are the main disadvantages of the horizontal tubular photobioreactor [51]. In this type of reactor, temperature is generally controlled by sparging cool water on the surface of the tubes, overlapping of tubes of cooled water and culture, and placing the light-harvesting unit inside a pool of temperature-controlled water [5].

6.7.3 Helical Tubular Photobioreactor

A transparent and flexible tube of small diameter is coiled vertically, giving a helical-like shape (Fig. 6.5C). It can have a separate or attached degassing unit to disengage residual and photosynthetically produced oxygen. Helical reactors can be used for

CO_2 sequestration as well as hydrogen production. For CO_2 sequestration, a CO_2-rich gas stream is introduced from the bottom inlet, which comes out from the top or from the degassing unit after traveling through the helical path. A long helical unit gives not only a high S/V ratio but also a long residence time for the CO_2-rich inlet stream. Helical reactors can be used for hydrogen production by circulating the culture from top and bottom sections using a pump. In addition to better utilization of CO_2 gas, a helical photobioreactor requires a small ground footprint [5]. However, it has several disadvantages such as shear stress during gas introduction or recirculating of culture, fouling inside the reactor, and high energy utilization for the circulation of culture. Helical reactors were further modified by giving a cone shape to the reactor in order to balance the energy input and photosynthetic efficiency [5, 129]. It is recommended that the light source be placed along the axis of the helical reactor for minimum loss of light energy.

6.7.4 Flat Panel Photobioreactor

The flat panel reactor has a cuboidal-shaped vessel made up of transparent materials such as glass, plexiglass, or polycarbonate [5] (Fig. 6.5D). The cuboidal shape increases the S/V ratio and allows better light collection in the reactor. Similar to the airlift and bubble column reactors, mixing is done by sparging with CO_2-rich gas from the bottom of the reactor. This reactor meant for CO_2 sequestration can be converted into a photobioreactor for hydrogen production by stopping gas sparging, mounting it on the stand, and giving a rocking motion to it using a motor (Fig. 6.5E).

6.7.5 Stirred Tank Photobioreactor

A stirred tank reactor is a conventional reactor, which can be used for both CO_2 sequestration and hydrogen production. Mixing is done mechanically using an impeller of different sizes and shapes (Fig. 6.5F). Baffles are used in order to reduce vortex. CO_2-enriched air can be introduced along the coiled sparger at the bottom. Its main disadvantages are poor light penetration because of a low S/V ratio, shear stress due to the impeller, and scale-up problems [5]. Furthermore, light can be provided externally by placing fluorescent lamps outside or using optical fibers or a dedicated light unit surrounding the vessel.

6.7.6 Hybrid Photobioreactor

By combining two reactors, it is possible to take advantage of both of the reactors in the CO_2 sequestration process. Kumar et al. [5] coined the term "hybrid photobioreactor" for the combination of two or more photobioreactors, where one can overcome the disadvantages of the other. For example, the airlift reactor was integrated with a horizontally placed tubular loop photobioreactor by Fernandez et al. [130]. The horizontal loop was for harvesting more sunlight because of its high S/V ratio, whereas the airlift reactor was acting as a degassing unit. The challenge of oxygen buildup in the reactor was reduced by the introduction of the airlift reactor. The α-shaped reactor can also be considered as a hybrid reactor, which acquires the α shape with a combination of horizontal and vertical tubular reactors. The advantage of this reactor was in achieving a high liquid flow rate at relatively low air flow rates [131].

6.8 CONCLUSION

Both green algae and cyanobacteria can be utilized successfully for CO$_2$ sequestration. This process leads to production of algal biomass, which can subsequently be utilized for hydrogen production under suitable conditions. The hydrogen production pathway is different in green algae and both nitrogen-fixing and non-nitrogen-fixing cyanobacteria. Non-nitrogen-fixing cyanobacteria have bidirectional hydrogenase, which can produce and consume hydrogen. However, equilibrium of the reaction can be shifted toward hydrogen production under favorable conditions. Hydrogen production in nitrogen-fixing cyanobacteria takes place in both vegetative cells and heterocysts. The hydrogen production mechanism of vegetative cells of cyanobacteria can be considered similar to green algae. But, unlike the hydrogenase of green algae, the enzymes of cyanobacteria are constitutively expressed in cells. This results in immediate production of hydrogen in the dark just at the onset of glycogen catabolism [29]. However, green algae have the advantage of a unidirectional hydrogenase that is different from the bidirectional hydrogenase of cyanobacteria. The nitrogen-fixing machinery of the heterocyst is utilized for hydrogen production using nitrogenase enzyme, where hydrogen is only the by-product of the nitrogen fixation reaction. However, all the reductants and electrons can be utilized for hydrogen production in the absence of N$_2$. Promising hydrogen producers may be on the order of non-nitrogen-fixing cyanobacteria, green algae and nitrogen-fixing cyanobacteria.

ACKNOWLEDGMENTS

The financial supports received from the Council of Scientific & Industrial Research (CSIR), the Bhabha Atomic Research Centre (BARC), the Department of Biotechnology (DBT), and the Ministry of New and Renewable Energy Sources (MNRE) of the government of India are duly acknowledged.

REFERENCES

1. B. Metz, O. Davidso, H. Coninck, M. Loos, L. Meyer, *IPCC Special Report on Carbon Dioxide Capture and Storage*, International Panel on Climate Change, Cambridge University Press, 2007.

2. Inventory of U.S greenhouse gas emissions and sinks: 1990–2008 (March 2010) (http://www.epa.gov/climatechange/emissions/usinventoryreport.html).

3. P. Tans, NOAA/ESRL (2010) (www.esrl.noaa.gov/gmd/ccgg/trends/).

4. D. E. Allen, B. R. Strazisar, Y. Soong, S. W. Hedges, Modeling carbon dioxide seques-tration in saline aquifers: significance of elevated pressures and salinities, Fuel Process. Technol. 86(14) (2005) 1569–1580.

5. K. Kumar, C. N. Dasgupta, B. Nayak, P. Lindblad, D. Das, Development of suitable photobioreactors for CO_2 sequestration addressing global warming using green algae and cyanobacteria, Bioresour. Technol. 102 (2011) 4945–4953.

6. K. Skjånes, P. Lindblad, J. Muller, $BioCO_2$—a multidisciplinary, biological approach using solar energy to capture CO_2 while producing H_2 and high value products, Biomol. Eng. 24 (2007) 405–413.

7. K. Loubiere, E. Olivo, G. Bougaran, J. Pruvost, R. Robert, J. Legrand, A new photobiore-actor for continuous microalgal production in hatcheries based on external-loop airlift and swirling flow, Biotechnol. Bioeng. 102(1) (2009), 132–147.

8. Canadian energy supply and demand, National Energy Board, 1993–2010.

9. R. E. Billings, *The Hydrogen World View*, 1st ed. American Academy of Science, 1991.

10. D. Das, Advances in biohydrogen production processes: an approach towards commer-cialization, Int. J. Hydrogen Energy. 34 (2009) 7349–7357.

11. J. R. Benemann, N. M Weare, Hydrogen evolution by nitrogen-fixing *Anabaena cylin-drica* cultures, Science. 184 (1974) 174–175.

12. A. Tiwari, A. Pandey, Cyanobacterial hydrogen production—a step towards clean envi-ronment, Int. J. Hydrogen Energy 37 (2012) 139–150.

13. H. Masukawa, M, Mochimaru, H. Sakurai, Disruption of the uptake hydrogenase gene, but not of the bidirectional hydrogenase gene, leads to enhanced photobiological hydro-gen production by the nitrogen-fixing cyanobacterium *Anabaena* sp. PCC 7120, Appl. Microbiol. Biotechnol. 58(5) (2002) 618–624.

14. H. Gaffron, J. Rubin, Fermentative and photochemical production of hydrogen by algae, J. Gen. Physiol. 26 (1942) 219–240.

15. P. G. Rosseler, S. Lien, Activation and de novo synthesis of hydrogenase in *Chlamy-domonas*, Plant Physiol. 76 (1984) 1086–1089.

16. T. Happe, B. Mosler, J. D. Naber, Induction, localization and metal content of hydro-genase in the green algae *Chlamydomonas reinhardtii*, Eur. J. Biochem. 222 (1994) 769–774.

17. M. L. Ghirardi, R. K. Togasaki, M. Seibert, Oxygen sensitivity of algal H_2 production, Appl. Biochem. Biotechnol. 63 (1997) 141–151.

18. A. Melis, Green alga hydrogen production: progress, challenges and prospects, Int. J. Hydrogen Energy 27 (2002) 1217–1228.

19. A. Melis, T. Happe, Hydrogen production. Green algae as a source of energy, Plant Physiol. 127(3) (2001) 740–748.

20. M. L. Ghirardi, S. Kosourov, M. Seibert, Cyclic photobiological algal H_2-production. Proceedings of the DOE Hydrogen Program Review NREL/ CP-570-30535 (2001).

21. R. A. Andersen, Diversity of eukaryotic algae, Biodiverse Conserv. 1 (1992) 267–292.

22. J. Rupprecht, B. Hankamer, J. H. Mussgnug, G. Ananyev, C. Dismukes, O. Kruse, Per-spectives and advances of biological H_2 production in microorganisms, Appl. Microbiol. Biotechnol.72 (2006) 442–449.

23. V. A. Boichenko, P. Hoffmann, Photosynthetic hydrogen-production in prokaryotes and eukaryotes—occurrence, mechanism, and functions, Photosynthetica. 30 (1994) 527–552.

24. O. Kruse, J. Rupprecht, J. H. Mussgnug, G. C. Dismukes, B. Hankamer, Photosynthesis: a blue print for energy capture and conversion technologies, Photochem. Photobiol. 4 (2005) 957–970.

25. J. M. Olson, J. D. Bernstein, Industrial and Chemistry Product Research and Development. 21 (1982) 640.

26. T. Happe, A. Kaminski, Differential regulation of the Fe-hydrogenase during anaerobic adaptation in the green alga *Chlamydomonas reinhardtii*, Eur. J. Biochem. 269 (2002) 1022–1032.

27. B. Tamburic, F. W. Zemichael, G. C. Maitland, K. Hellgardt, Parameters affecting the growth and hydrogen production of the green alga *Chlamydomonas reinhardtii*, Int. J. Hydrogen Energy. 36 (2011) 7872–7876.

28. D. Dutta, D. De, S. Chaudhuri, S. K. Bhattacharya, Hydrogen production by cyanobacteria, Microb. Cell Fact. 4 (2005) 36.

29. K. Srirangan, M. E. Pyne, C. P. Chou, Biochemical and genetic engineering strategies to enhance hydrogen production in photosynthetic algae and cyanobacteria, Bioresour. Technol. 102 (2011) 8589–8604.

30. P. Fay, *The Blue Greens (Cyanophyta – Cyanobacteria)*, 5th ed., Edward Arnold, London, 1983, p. 88.

31. D. Campbell, V. Hurry, A. K. Clarke, P. Gustafsson, G. Öquist, Chlorophyll fluorescence analysis of cyanobacterial photosynthesis and acclimation, Microbiol. Mol. Biol. Rev. 30 (1998) 667–680.

32. L. Cournac, F. Mus, L. Bernard, G. Guedeney, P. Vignais, G. Peltier, Limiting steps of hydrogen production in *Chlamydomonas reinhardtii* and *Synechocystis* PCC 6803 as analysed by light induced gas exchange transients, Int J. Hydrogen Energy 27 (2002) 1229–1237.

33. M. Calvin, A. A. Benson, The path of carbon in photosynthesis, Science. 107 (1948) 476–480.

34. D. L. Nelson, M. M. Cox, *Lehninger Principles of Biochemistry*, 4th ed., W.H. Freeman, New York 2005.

35. M. Giordano, J. Beardall, J. A. Raven, Mechanisms in algae: mechanisms, environmental modulation, and evolution, Annu. Rev. Plant Biol. 56 (2005) 99–131.

36. J. V. Moroney, A. Somanchi, How do algae concentrate CO_2 to increase the efficiency of photosynthetic carbon fixation? Plant Physiol. 119 (1999) 9–16.

37. J. R. Coleman, The molecular and biochemical analysis of CO_2 concentrating mechanisms in cyanobacteria and microalgae, Plant Cell Environ. 14 (1991) 861–867.

38. J. P. Fett, J. R. Coleman, Regulation of periplasmic carbonic anhydrase expression in *Chlamydomonas reinhardtii* by acetate and pH, Plant Physiol. 106 (1994) 103–108.

39. M. H. Spalding, Microalgal carbon-dioxide-concentrating mechanisms: *Chlamydomonas* inorganic carbon transporters, J. Exp. Biol. 59(7) (2008) 1463–1473.

40. C. V. G. López, F. G. A. Fernández, J. M. F. Sevilla, J. F. S. Fernández, M. C. C. García, E. M., Grima, Utilization of the cyanobacteria *Anabaena* sp. ATCC 33047 in CO_2 removal processes, Bioresour. Technol.100 (2009) 5904–5910.

41. C. Yoo, S. Y. Jun, J. Y. Lee, C. Y. Ahn, H. M. Oh, Selection of microalgae for lipid production under high level of carbon dioxide, Bioresour. Technol. 101 (2010) 71–74.

42. M. Packer, Algal capture of carbon dioxide; biomass generation as a tool for greenhouse gas mitigation with reference to New Zealand energy strategy and policy, Energy Policy. 37 (2009) 3428–3437.

43. T. M. Sobczuk, F. G. Camacho, F. C. Rubio, F. G. A. Fernandez, E. M. Grima, Carbon dioxide uptake efficiency by outdoor microalgal cultures in tubular airlift photobioreactors, Biotechnol. Bioeng. 67 (2000) 465–475.

44. S. Y. Chiu, C. Y. Kao, C. H. Chen, T. C. Kuan, S. C. Ong, C. S. Lin, Reduction of CO_2 by a high-density culture of *Chlorella* sp. in a semicontinuous photobioreactor, Bioresour. Technol. 99 (2008) 3389–3396.

45. K. Maeda, M. Owada, N. Kimura, K. Omata, I. Karube, CO_2 fixation from flue gas on coal fired thermal power plant by microalgae, Energy Convers. Mgmt. 36(6-9) (1995) 717–720.

46. K. Kumar, D. Das, Growth characteristics of *Chlorella sorokiniana* in airlift and bubble column photobioreactors, Bioresour. Technol. 116 (2012) 307–313.

47. G. Torzillo, B. Pushparaj, J. Masojidek, A. Vonshak, Biological constraints in algal biotechnology, Biotechnol. Bioprocess Eng. 8 (2003) 338–348.

48. K. Zhang, S. Miyachi, N. Kurano, Evaluation of a vertical flat-plate photobioreactor for outdoor biomass production and carbon dioxide bio-fixation: effects of reactor dimensions, irradiation and cell concentration on the biomass productivity and irradiation utilization efficiency, Appl. Microbiol. Biotechnol. 55(4) (2001) 428–433.

49. Q. Hu, H. Guterman, A. Richmond, A flat inclined modular photobioreactor for outdoor mass cultivation of photoautotrophs, Biotechnol. Bioeng. 51 (1996) 51–60.

50. J. A. Costa, L.M. Colla, P. F. Duarte, Improving *Spirulina platensis* biomass yield—a fed-batch process, Bioresour. Technol. 92 (2004) 237–241.

51. A. S. Miron, A. C. Gomez, F. G. Camacho, E. M. Grima, Y. Chisti, Comparative evaluation of compact photobioreactors for large-scale monoculture of microalgae, J. Biotechnol. 70 (1999) 249–270.

52. D. Carrieri, K. Wawrousek, C. Eckert, J. Yu, P. C. Maness, The role of the bidirectional hydrogenase in cyanobacteria, Bioresour. Technol. 102 (2011) 8368–8377.

53. M. Holmqvist, S. Karin, O. Paulo, P. Lindberg, P. Lindblad, Characterization of the hupSL promoter activity in *Nostoc punctiforme* ATCC 29133, BMC Microbiology, 9 (2009) 54.

54. P. Tamagnini, R. Axelsson, P. Lindberg, F. Oxelfelt, R. Wünschiers, P. Lindblad, Hydrogenases and hydrogen metabolism of *Cyanobacteria*, Microbiol. Mol. Biol. Rev. (2002) 1–20.

55. R. Abdel-Basset, T. Friedl, K. I. Mohr, N. Rybalka, W. Martin, High growth rate, photosynthesis rate and increased Hydrogen(ases) in manganese deprived cells of a newly isolated Nostoc-like cyanobacterium (SAG 2306), Int. J. Hydrogen Energy 36(19) (2011) 12200–12210.

56. L. Cournac, G. Guedeney, G. Peltier, P. M. Vignais, Sustained photoevolution of molecular hydrogen in a mutant of *Synechocystis* sp. strain PCC 6803 deficient in the type I NADPH dehydrogenase complex, J. Bacteriol. 186 (2004) 1737–1746.

57. J. Yu, P. Takahashi, Biophotolysis-based hydrogen production by cyanobacteria and green microalgae, Commun. Curr. Res. Educ. Topics Trends appl. microbiol. (2007) 79–89.

58. L. T. Serebryakova, M. Sheremetieva, A. Tsygankov, Reversible hydrogenase activity of Gloeocapsa alpicola in continuous culture FEMS Microbiol. Lett. 166(1) (1998) 89–94.

59. H. Böhme, Regulation of nitrogen fixation in heterocyst-forming cyanobacteria, Trends Plant Sci. 3(9) (1998) 346–351.

60. P. C. Hallenbeck (Ed.), *Microbial Technologies in Advanced Biofuels Production*, Springer, 2012. doi: 10.1007/978-1-4614-1208-3_2.

61. H. Masukawa, K. Inoue, H. Sakurai, Effects of disruption of homocitrate synthase genes on *Nostoc* sp. strain PCC 7120 photobiological hydrogen production, Appl. Environ. Microbiol. 73(23) (2007) 7562–7570.

62. C. M. Yeager, C. E. Milliken, C. E. Bagwell, L. Staples, P. A. Berseth, H. T. Sessions, Evaluation of experimental conditions that influence hydrogen production among heterocystous cyanobacteria, Int. J. Hydrogen Energy. 36 (2011) 7487–7499.

63. W. D. P. Stewart, *Algal Physiology and Biochemistry*. vol. 10, Blackwell, 1974.

64. Y. Asada, J. Miyake, Photobiological hydrogen production, J. Biosci. Bioeng. 88(1) (1999) 1–6.

65. K. Vijayaraghavan, R. Karthik, S. P. K. Nalini, Hydrogen generation from algae: a review, J. Plant Sci. 5(1) (2010) 1–19.

66. F. Yoshino, H. Ikeda, H. Masukawa, H. Sakurai, High photobiological hydrogen production activity of a *Nostoc* sp. PCC 7422 uptake hydrogenase-deficient mutant with high nitrogenase activity, Marine Biotechnol. 9 (2007) 101–112.

67. J. Mathews, G. Wang, Metabolic pathway engineering for enhanced biohydrogen production, Int. J. Hydrogen Energy 34 (2009) 7404–7416.

68. L. J. Stal, W. E. Krumbein, Nitrogenase activity in the non-heterocystous cyanobacterium *Oscillatoria* sp. grown under alternating light-dark cycles, Arch Microbiol. 143 (1985) 67–71.

69. N. Z. Muradov, T. N. Veziroglu, *Carbon-Neutral Fuels and Energy Carriers*, CRC Press, 2011, Chap. 10, p. 491.

70. D. Madamwar, N. Garg, V. Shah, Cyanobacterial hydrogen production, World J. Microbiol. Biotechnol. 16(8–9) (2000) 757–767.

71. S. A. Kranz, M. Eichner, B. Rost, Interactions between CCM and N₂ fixation in *Trichodesmium*, Photosynth. Res. 109(1-3) (2011) 73–84.

72. A. Makino, C. Miyake, A. Yokota, Physiological functions of the water–water cycle (Mehler reaction) and the cyclic electron flow around PSI in rice leaves, Plant Cell Physiol. 43(9) (2002) 1017–1026.

73. G. D. Smith, G. D. Ewart, W. Tucker, Hydrogen production by cyanobacteria, Int. J. Hydrogen Energy 17 (1992) 695–698.

74. R. Y. Igarashi, L. C. Seefeldt, Nitrogen fixation: the mechanism of the Mo-dependent nitrogenase, Crit. Rev. Biochem. Mol. Biol. 38 (2003) 351–384.

75. D. C. Rees, J. B. Howard, Nitrogenase: standing at the crossroads, Curr. Opin. Chem. Biol. 4 (2000) 559–566.

76. A. Tsygankov, V. B. Borodin, K. K. Rao, D. O. Hall, H₂ photoproduction by batch culture of *Anabaena variabilis* ATCC 29413 and its mutant PK84 in a photobioreactor, Biotechnol. Bioeng. 64(6) (1999) 709–715.

77. A. Demirbas, *Biohydrogen: For Future Engine Fuel Demands. Trabzon*, Springer, 2009.

78. J. C. Weissman, J. R. Benemann, Hydrogen production by nitrogen-starved cultures of *Anabaena cylindrica*, Appl. Environ. Microbiol. 33 (1977) 123.

79. P. C. Hallenbeck, J. R. Benemann, Biological hydrogen production; fundamentals and limiting processes, Int. J. Hydrogen Energy. 27 (2002) 1185–1193.

80. Y. Miura, T. Akano, K. Fukatsu, H. Miyasaka, T. Mizoguchi, K. Yagi, I. Maeda, Y. Ikuta, H. Matsumoto, Stably sustained hydrogen production by biophotolysis in natural day/night cycle, Energy Convers. Mgmt. 38 (1997) S533–S537.

81. M. Seibert, P. King, M. C. Posewitz, A. Melis, M. L. Ghirardi, Photosynthetic water-splitting for hydrogen production. In: J. Wall, C. Harwood, A. Demain (Eds.), *Bioenergy.* ASM Press, Washington DC, 2008, pp. 273–291.

82. J. R. Benemann, The technology of biohydrogen, In: Zaborsky et al. (Eds.), *Biohydrogen,* Plenum Press, New York, 1998, pp. 19–30.

83. J. Prabina, K. Kumar, Studies on the optimization of cultural conditions for maximum hydrogen production by selected cyanobacteria, ARPN J. Agic. Biol. Sci. 5(5) (2010) 23–31.

84. D. M. Rawson, The effects of exogenous amino acids on growth and nitrogenase activity in the cyanobacterium *Anabena cylindrica* PCC 7122, J. Gen. Microbiol. 134 (1985) 2544–2549.

85. M. Datta, G. Nikki, V. Shah, Cyanobacterial hydrogen production, World J. Microbiol. Biotechnol. 16 (2000) 8–9.

86. L. T. Serebryakova, A. A. Tsygankov, Two-stage system for hydrogen production by immobilized cyanobacterium *Gloeocapsa alpicola* CALU 743, Biotechnol. Prog. 23 (2007) 1106–1110.

87. P. C. Hallenbeck, D. Ghosh, Advances in fermentative biohydrogen production: the way forward? Trends Biotechnol. 27(5) (2009) 287–297.

88. S. E. Baker, R. C. Hopkins, C. Blanchette, V. Walsworth, R. Sumbad, N. Fischer, E. Kuhn, M. Coleman, B. Chromy, S. Letant, P. Hoeprich, M. W. W. Adams, P. T. Henderson, Hydrogen production by a hyperthermophilic membrane-bound hydrogenase in water-soluble nanolipoprotein particles, J. Am. Chem. Soc. 131(22) (2009) 7508–7509.

89. J. Schnackenberg, H. Ikemoto, S. Miyachi, Photosynthesis and hydrogen evolution under stress conditions in a CO_2-tolerant marine green alga, *Chlorococcum littorale*, J. Photochem. Photobiol. B. Biol. 34, (1996) 59.

90. J. W. Lee, E. Greenbaum, A new oxygen sensitivity and its potential application in photosyntheic H_2 production, Appl. Biochem. Biotechnol. 105-108, (2003) 303–313.

91. C. P. Spruit, Simultaneous photoproduction of hydrogen and oxygen by *Chlorella.* Meded Landbouwhogeschool Wageningen. 58 (1958) 1–17.

92. E. Greenbaum, R. R. L. Guillard, W. G. Sunda, Hydrogen and oxygen photoproduction by marine algae, Photochem. Photobiol. 37 (1983) 649–655.

93. J. R. Benemann, Hydrogen biotechnology: progress and prospects, Nature Biotechnol. 14 (1996) 1101–1103.

94. K. Belafi-Bako, D. Bucsu, Z. Pientka, B. Balint, Z. Herbel, K. L. Kovacs, M. Wessling, Integration of biohydrogen fermentation and gas separation processes to recover and enrich hydrogen, Int. J. Hydrogen Energy. 31(11) (2006) 1490–1495.

95. E. Kessler, Hydrogenase, photo reduction and anaerobic growth. In: W. D. P Stewart (Ed.), *Algal Physiology and Biochemistry.* Blackwell, 1974, pp. 456–473.

96. E. S. Bamberger, D. King, D. L. Erbes, M. Gibbs, H_2 and CO_2 evolution by anaerobically adapted *Chlamydomonas reinhardtii* F60, Plant Physiol. 69 (1982) 1268–1273.

97. K. Shinozaki, M. Ohme, M. Tanake, T Waksugi, N. Hayashida, T. Matsubayashi, N. Zaita, J. Chunwongse, J. Obokata, K. Yameguchi-Shinozaki, The complete nucleotide sequence of tobacco chloroplast genome: its gene organization and expression, EMBO J. 5(9) (1986) 2043–2049.

98. L. A. Sazanov, P. A. Burrows, P. J. Nixon, The plastid ndh genes code for an NADH specific dehydrogenase isolation of a complex I analogous from pea thylakoid membranes, Proc. Natl. Acad. Sci. U.S.A. 95 (1998) 1319–1324.

99. A. Kubicki, E. Funk, P. Westhoff, K. Steinmuller, Differential expression of plastome encoded ndh genes in mesophyll and bundle–sheath chloroplasts of the C-4 plant sorghum bicolor indicates that the complex I–homologus NAD(P)H–plastoquinone oxidoreductase is involved in cyclic electron transport, Planta. 199(2) (1996) 276–281.

100. B. Mahro, A. C. Kusel, L. H. Grimme, The significance of hydrogenase activity for the energy metabolism of green algae: anaerobiosis favors ATP synthesis in cells of *Chlorella* with active hydrogenase, Arch. Microbiol. 144 (1986) 91–95.

101. R. P. Gfeller, M. Gibbs, Fermentative metabolism of *Chlamydomonas reinhardtii*, Plant Physiol. 75 (1986) 212–218.

102. T. Happe, J. D. Naber, Isolation, characterization and N-terminal amino acid sequence of hydrogenase from the green alga *Chlamydomonas reinhardtii*, Eur. J. Biochem. 214 (1993) 475–481.

103. F. Mus, L. Cournac, V. Cardettini, A. Caruana, G. Peltier Inhibitor studies on non-photochemical plastoquinone reduction and H₂ photoproduction in *Chlamydomonas reinhardtii*, Biochim. Biophys. Acta 1708 (2005) 322–332.

104. A. Melis, L. P. Zhang, M. Forestier, M. L. Ghirardi, M. Seibert, Sustained photobiological hydrogen gas production upon reversible inactivation of oxygen evolution in the green alga *Chlamydomonas reinhardtii*, Plant Physiol. 122(1), (2000) 127–136.

105. D. J. Kyle, I. Ohad, C. J. Arntzen, Membrane protein damage and repair: selective loss of a quinone-protein function in chloroplast membranes, Natl. Acad. Sci. 81 (1984) 4070–4074.

106. E. Tyystjärvi, Photoinhibition of photosystem II and photodamage of the oxygen evolving manganese cluster, Coord. Chem. Rev. 252 (2008) 361–376.

107. D. D. Wykoff, J. P. Davies, A. Melis, A. R. Grossman, The regulation of photosynthetic electron transport during nutrient deprivation in *Chlamydomonas reinhardtii*, Plant Physiol. 117 (1998) 129–139.

108. M. Forestier, P. King, L. Zhang, Expression of two [Fe]-hydrogenases in *Chlamydomonas reinhardtii* under anaerobic conditions, Eur. J. Biochem. 270 (2003) 2750–2758.

109. S. Kosourov, A. Tsygankov, M. Seibert, M. L. Ghirardi, Sustained hydrogen photoproduction by *Chlamydomonas reinhardtii*: effects of culture parameters, Biotechnol. Bioeng. 78(7) (2002) 731–740.

110. L. Zhang, A. Melis, Probing green algal hydrogen production, Philos. Trans. Rev. Soc. Lond. B 357 (2002) 1499–1509.

111. S. Oncel, F. Vardar-Sukan, Photo-bioproduction of hydrogen by *Chlamydomonas reinhardtii* using a semi-continuous process regime, Int. J. Hydrogen Energy 34 (2009) 7592–7602.

112. S. N. Kosourov, M. L. Ghirardi, M. Seibert, A truncated antenna mutant of *Chlamydomonas reinhardtii* can produce more hydrogen than the parental strain, Int. J. Hydrogen Energy 36 (2011) 2044 –2048.

113. S. N. Kosourov, M. Seibert, Hydrogen photoproduction by nutrient-deprived *Chlamydomonas reinhardtii* cells immobilized within thin alginate films under aerobic and anaerobic conditions, Biotechnol. Bioeng. 102(1) (2009) 50–58.

114. A. Melis, Photosynthetic H_2 metabolism in *Chlamydomonas reinhardtii* (unicellular green algae), Planta 226 (2007) 1075–1086.

115. S. Wu, L. Xu, R. Wang, X. Liu, Q. Wang, A high yield mutant of *Chlamydomonas reinhardtii* for photoproduction of hydrogen, Int. J. Hydrogen Energy 36 (2011) 14134–14140.

116. H. C. Chen, A. J. Newton, A. Melis, Role of SulP, a nuclear-encoded chloroplast sulfate permease, in sulfate transport and H_2 evolution in *Chlamydomonas reinhardtii*, Photosynth. Res. 84 (2005) 289–296.

117. A. Tsygankov, S. Kosourov, M. Seibert, M. L. Ghirardi, Hydrogen photoproduction under continuous illumination by sulfur-deprived, synchronous *Chlamydomonas reinhardtii* cultures, Int. J. Hydrogen Energy 27 (2002) 1239–1244.

118. A. S. Fedorov, S. Kosourov, M. L. Ghirardi, M. Seibert, Continuous hydrogen photoproduction by *Chlamydomonas reinhardtii*: using a novel two-stage, sulfate-limited chemostat system, Appl. Biochem. Biotechnol. 121–124 (2005) 403–412.

119. L. Zhang, T. Happe, A. Melis, Biochemical and morphological characterization of sulfur-deprived and H_2-producing *Chlamydomonas reinhardtii* (green alga), Planta 214 (2002) 552–561.

120. J. J. Hahn, M. L. Ghirardi, W. A. Jacoby, Effect of process variables on photosynthetic algal hydrogen production, Biotechnol. Prog. 20 (2004) 989–991.

121. T. Laurinavichene, I. Tolstygina, A. Tsygankov, The effect of light intensity on hydrogen production by sulfur-deprived *Chlamydomonas reinhardtii*, J. Biotechnol. 114 (2004) 143–151.

122. J. P. Kim, C. D. Kang, T. H. Park, M. S. Kim, S. J. Sim, Enhanced hydrogen production by controlling light intensity in sulfur-deprived *Chlamydomonas reinhardtii* culture, Int. J. Hydrogen Energy 31 (2006) 1585–1590.

123. J. H. Jo, D. S. Lee, J. M. Park, Modeling and optimization of photosynthetic hydrogen gas production by green alga *Chlamydomonas reinhardtii* in sulfur-deprived circumstance, Biotechnol. Prog. 22 (2006) 431–437.

124. S. Fouchard, A. Hemschemeier, A. Caruana, J. Pruvost, J. Legrand, T. Happe, G. Peltier, L. Cournac, Autotrophic and mixotrophic hydrogen photoproduction in sulfur-deprived *Chlamydomonas* cells, Appl. Environ. Microbiol. 71(10) (2005) 6199–6205.

125. S. Kosourov, E. Patrusheva, M. L. Ghirardi, M. Seibert, A. Tsygankov, A comparison of hydrogen photoproduction by sulfur-deprived *Chlamydomonas reinhardtii* under different growth conditions, J. Biotechnol. 128 (2007) 776–787.

126. M. Rosenbaum, U. Schroder, F. Scholz, Utilizing the green alga *Chlamydomonas reinhardtii* for microbial electricity generation: a living solar cell, Appl. Microbiol. Biotechnol. 68 (2005) 753–756.

127. A. L. White, A. Melis, Biochemistry of hydrogen metabolism in *Chlamydomonas reinhardtii* wild type and a Rubisco-less mutant, Int. J. Hydrogen Energy. 31 (2006) 455–464.

128. M. J. Barbosa, M. Janssen, N. Ham, J. Tramper, R. H. Wijffels, Microalgae cultivation in air-lift reactors: modeling biomass yield and growth rate as a function of mixing frequency, Biotechnol. Bioeng. 82(2) (2003) 170–179.

129. Y. Watanabe, D. O. Hall, Photosynthetic production of the filamentous *Cyanobacterium Spirulina platensis* in a cone-shaped helical tubular photobioreactor, Appl. Microbiol. Biotechnol. 44 (1996) 693–698.

130. F. G. A. Fernandez, J. M. F. Sevilla, J. A. S. Perez, E. M. Grima, Y. Chisti, Airlift-driven external loop tubular photobioreactors for outdoor production of microalgae: assessment of design and performance, Chem. Eng. Sci. 56 (2001) 2721–2732.

131. Y. K. Lee, S. Y. Ding, C. S. Low, Y. C. Chang, W. L. Forday, P. C. Chew, Design and performance of a α-type tubular photobioreactor for mass cultivation of microalgae, J. Appl. Phycol. 7 (1995) 47–51.

132. M. G. de Morais, J. A. V. Costa, Biofixation of carbon dioxide by *Spirulina* sp. and *Scenedesmus obliquus* cultivated in a three-stage serial tubular photobioreactor, J. Biotechnol. 129 (2007) 439–445.

133. D. A. Sveshnikov, N. V. Sveshnikova, K. K. Rao, D. O Hall, Hydrogen metabolism of mutant forms of *Anabaena variabilis* in continuous cultures and under nutritional stress, FEBS Microbiol. Lett. 147 (1997) 297–301.

134. A. A. Tsygankov, L. T. Serebryakova, K. K. Rao, D. O. Hall, Acetylene reduction and hydrogen photoproduction by wild type and mutant strains of *Anabaena* at different CO_2 and O_2 concentrations, FEMS Microbiol. Lett. 167 (1998) 13–17.

135. S. A. Markov, M. J. Bazin, D. O. Hall, Hydrogen photoproduction and carbon dioxide uptake by immobilized *Anabaena variabilis* in a hollow-fiber photobioreactor, Enzyme Microbial. Technol. 17 (1995) 306–310.

136. D. Howarth, G. Codd, The uptake and production of molecular hydrogen by unicellular cyanobacteria, J. Gen. Microbiol. 131 (1985) 1561–1569.

137. R. Moezelaar, S. M. Bijvank, L. J. Stal, Fermentation and sulfur reduction in the mat-building cyanobacterium *Microcoleus chtonoplastes*, Appl. Environ. Microbiol. 62 (1996) 1752–1758.

138. S. Kumazawa, A. Mitsui, Comparative amperometric study of uptake hydrogenase and hydrogen photoproduction activities between heterocystous cyanobacterium *Anabaena cylindrica* B629 and nonheterocystous cyanobacterium *Oscillatoria* sp. strain Miami BG7, Appl. Environ. Microbiol. 50(2) (1985) 287–291.

139. T. V. Laurinavichene, A. S. Fedorov, M. L. Ghirardi, M. Seibert, A. A. Tsygankov, Demonstration of hydrogen photoproduction by immobilized, sulfur-deprived *Chlamydomonas reinhardtii* cells, Int. J. Hydrogen Energy 31 (2006) 659–667.

140. S. Ohta, K. Miyamoto, Y. Miura, Hydrogen evolution as a consumption mode of reducing equivalents in green algal fermentation, Plant Physiol. 83 (1987) 1022–1026.

141. S. Chader, H. Haceneb, S. N. Agathos, Study of hydrogen production by three strains of *Chlorella* isolated from the soil in the Algerian Sahara, Int. J. Hydrogen Energy 34 (2009) 4941–4946.

142. D. Gielen, CO_2 removal in the iron and steel industry, Energy Conversion and Management, 44(7) (2003) 1027–1037.

143. D. V. Essendelft, J. Taron, M. Fitzgerald, A. Abdel-Hafez, A. Sakti, O. Achimugu, I. Faoro, CO_2 sequestration through deep saline injection and photosynthetic fixation: system design for two plausible CO_2 sequestration strategies, Department of Energy and Geo-environmental Engineering, The Pennsylvania

State University (http://www.ems.psu.edu/~elsworth/courses/egee580/Seqeustration_final_report.pdf).

144. C. D. Keeling, S. C. Piper, R. B. Bacastow, M. Wahlen, T. P. Whorf, M. Heimann, H. A. Meijer, Atmospheric CO_2 and CO_2 exchange with the terrestrial biosphere and oceans from 1978 to 2000: observations and carbon cycle implications. In: J. R. Ehleringer, T. E. Cerling, M. D. Dearing (Eds.), *A History of Atmospheric CO_2 and Its Effects on Plants, Animals, and Ecosystems*, Springer-Verlag, New York, 2005, pp. 83–113.

145. P. M. Vignais, J. P. Magnin, J. C. Willison, Increasing biohydrogen production by metabolic engineering, Int. J. Hydrogen Energy 31 (2006) 1478–1483.

146. U.S.DOE/NETL-2002.

147. http://www.habmigern2003.info/19_flue-gas-contents.html.

148. http://www.eia.gov/pub/oil_gas/natural_gas/analysis_publications/natural_gas_1998_issues_trends/pdf/it98.pdf.

CHAPTER 7

Cyanobacterial Biofuel and Chemical Production for CO$_2$ Sequestration

JOHN W. K. OLIVER and SHOTA ATSUMI

7.1 CARBON SEQUESTRATION BY BIOMASS

Sequestration by photosynthetic bioreactors [1, 2] is an important addition to the toolbox for overall capture of CO$_2$. Biological sequestration has the inherent benefit that photosynthesis chemically cleaves the CO$_2$ molecule, redistributing waste carbon into organic materials. Bioreactors specifically can be built at the location of industrial emissions, do not compete for agricultural land, and can use brine or wastewater unsuitable for other industrial applications. In addition, the metabolic simplicity of single-cell photosynthetic organisms makes them more productive than soil-based crops [3]. Recently, successes in synthetic biology—the application of a genetic toolbox for the engineering of new organisms [4, 5]—have stimulated interest in engineered cyanobacteria as a photosynthetic resource. Synthetic metabolic pathways have been developed for production of valuable captured-carbon products, which aid cost mitigation in bioreactors (Fig. 7.1). This technology redirects metabolites generated by photosynthetic cleavage of CO$_2$, allowing for increased control over the ultimate fate of sequestered carbon.

The capture and storage of CO$_2$ can be divided into three areas: (1) removing CO$_2$ from air by increasing CO$_2$ sinks with reforestation, changing land use practices, direct filtering, or less controlled means such as ocean algal bloom cultivation [6]; (2) removing CO$_2$ from point sources by directly capturing CO$_2$ produced by large installations and either converting it in bioreactors or sequestering it in subterranean or deep-ocean sinks; and (3) removing CO$_2$ from diffuse sources. Diffuse CO$_2$ from engines and small installations cannot be captured directly, however, switching from fossil fuel derived carbon to biologically derived "net-zero emission" fuels can preemptively capture this source.

Natural and Artificial Photosynthesis: Solar Power as an Energy Source, First Edition. Edited by Reza Razeghifard.
© 2013 John Wiley & Sons, Inc. Published 2013 by John Wiley & Sons, Inc.

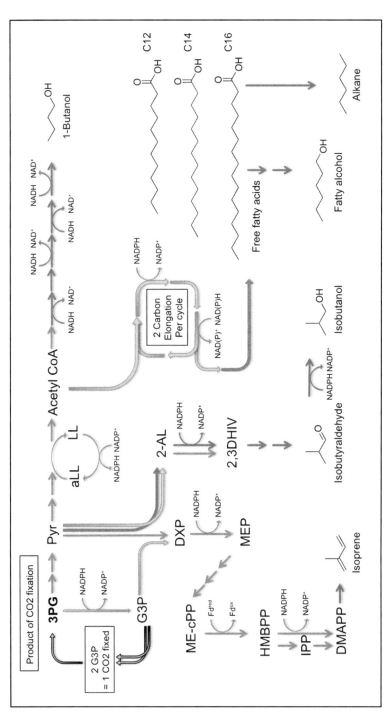

FIGURE 7.1 Metabolic pathways to captured-carbon products. Arrows represent enzymatic steps that are either native to the host or introduced with exogenous genes. Not all steps are shown. Reducing equivalents are included for reference. 3PG, 3-phospho-D-glycerate; G3P, D-glyceraldehyde-3-phosphate; Pyr, pyruvate; DXP, 1-deoxy-D-xylulose-5-phosphate; MEP, 2-C-methyl-D-erythritol-4-phosphate; ME-cPP, 2-C-methyl-D-erythritol-2,4-cyclodiphosphate; Fd, ferredoxin; HMBPP, 1-hydroxy-2-methyl-2-(*E*)-butenyl-4-diphosphate; IPP, isopentenyl diphosphate; DMAPP, dimethylallyl diphosphate; 2-AL, 2-acetolactate; 2,3DHIV, 2,3-dihydroxyisovalerate; aLL, [lipoate acetyltransferase] N^6-(*S*-acetyldihydrolipoyl)lysine; LL, [lipoate acetyltransferase] N^6-(lipoyl)lysine.

Continuous monitoring of CO$_2$ levels in the last fifty years has indicated that accelerating accumulation of atmospheric CO$_2$ is not only due to increased emissions from world growth and intensifying carbon use, but also from a possible attenuation in the efficiency of the world's natural carbon sinks [7]. Many creative solutions have been proposed and argued for carbon capture, each with varied environmental side effects and costs [6]. Through this debate two central concerns for the development of new technology have been identified: (1) minimizing change in natural ecosystems at the site of sequestration and (2) minimizing associated costs to encourage use of technology. While different technologies meet either of these requirements, few fit both as well as bioreactors. Good analysis of conventional technologies is already available [6]. The use of microalgae for CO$_2$ capture at point sources and the production of biodiesel from innately oil-producing algae are also already well characterized and reviewed [3, 8]. This chapter focuses on new applications of metabolic engineering for cyanobacterial production of captured-carbon products.

7.2   INTRODUCTION TO CYANOBACTERIA

Cyanobacteria are single-celled, gram-negative, photosynthetic bacteria found abundantly in the earth's oceans. They are the most plentiful constituents of phytoplankton, which are collectively responsible for almost 50% of global photosynthesis [9] and are found in a wide range of environments, from high-temperature hot springs [10] to permafrost zones [11]. Many cyanobacteria are halophilic and can be grown in water sources from the ocean or brine aquifers not suitable for freshwater use. Notably, natural thermophiles can be grown in high temperatures, making them suitable for use in flue gas exhaust systems [10]. Underlying this diversity is the relatively simple genetic structure of a prokaryotic cell, which makes cyanobacteria accessible for metabolic engineering [12]. Over the last 10 years increasing interest in cyanobacteria for the elucidation of circadian rhythm and the photosynthetic apparatus has resulted in a large body of genetic knowledge [12], and the genomes of more than 50 strains have been sequenced [13, 14, 15].

7.3   CO$_2$ UPTAKE EFFICIENCY OF CYANOBACTERIA

CO$_2$ uptake efficiency has historically been difficult to quantify. It has previously been common to assume that measuring the elemental carbon in cell dry weight can approximate total CO$_2$ uptake [10, 16, 17]. This method gives values ranging from 0.02 to 1.44 gCO$_2$/L/day (Table 7.1). Recently, it has been noted that these levels may represent less than 5% of total carbon fixation [18]. Low partitioning of fixed carbon to biomass is counterintuitive as growth is typically seen as the largest consumer of metabolic energy. Confusion is perpetuated by the fact that studies on cyanobacteria to date have lacked broad quantification of carbon-containing products secreted into the media during growth. When CO$_2$ fixation is measured by comparing the concentration of CO$_2$ in gas inlet and outlet streams, or in solution if dissolved

TABLE 7.1 Representative Measurements of CO_2 Uptake Efficiency by Cyanobacteria, Showing the Comparison Between Total CO_2 Uptake and CO_2 Fixed into Biomass

Strain	CO_2 Uptake (gCO_2/L/day)	Method of Measurement	Culture Setup	Reference
Chlorogloeopsis sp.	0.02	%w/w Carbon in biomass: elemental analysis	CO_2 bubbling, flask	[10]
Synechococcus PCC7942	0.6	%w/w Carbon in biomass: elemental analysis	CO_2 bubbling, flask	[16]
Aphanothece microscopica Nägeli	1.44	%w/w Carbon in biomass: elemental analysis	CO_2 bubbling, bioreactor	[17]
Aphanothece microscopica Nägeli	26.9	Inlet outlet streams: gas chromatography	CO_2 bubbling, bioreactor	[18]
Aphanothece microscopica Nägeli	24.6	Concentration of CO_2 in media: polarographic probe	Dissolved CO_2, bioreactor	[18]

carbon is used, values of 24.6–26.9 gCO_2/L/day have been measured [18]. Additional data on carbon uptake efficiencies in varied units and a detailed analysis of bioreactor design are available [1], however, in-depth analysis of cyanobacteria carbon fixation efficiencies across species is yet to be completed.

The amount of carbon dioxide captured by biological systems consuming light is influenced directly by the amount of light [19, 20]. The values in Table 7.1 are measured using electricity to provide constant direct light in order to maximize capture of CO_2. This is applicable in industrial settings, potentially with costs offset by carbon taxes or credits, where maximum CO_2 uptake is needed to clean waste streams. One overall goal of photosynthetic carbon capture, however, is to utilize the abundant energy provided by sunlight. This raises the challenge of light–dark cycle effects in biological systems. Switching from 24-hour illumination to 12-hour cycles, it would be expected that fixation per 24 hours would fall by at least 50%. One study has suggested that the cyanobacterial strain *Aphanothece microscopica Nägeli* can continue normal photosynthetic rates up to 2 hours into dark cycles by using stored energy [17]. This, however, is offset by natural respiration, which consumes oxygen and rereleases fixed carbon as CO_2 during dark cycles. In nonengineered cyanobacteria this results in a total drop in fixation of about 78% [17].

In continuous light carbon is the limiting factor until the limits of carbon fixation in the photosynthetic machinery are reached. During 12-hour cycles, photons are the limiting factor assuming a constant supply of CO_2 (such as that from waste CO_2 streams). This allows us to assess productivity based on the available photons for photosynthesis and the ability of chlorophyll to capture those photons. Because of the inherent bias in photosynthesis for photons in the range 400–740 nm the term *photosynthetically active radiation* (PAR) has been developed to describe only the incident photons that contribute to water splitting and CO_2 fixation [19]. When considering photosynthesis using sunlight, it is important to be aware that calculations for incident solar radiation on the earth vary and are often noncomparable [21]. Methodic measurements at the earth's surface are available from the National Renewable Energy Laboratory, and allow us to bypass approximations of atmospheric effects [22]. During irradiation by direct sunlight, photosynthesis is also limited by the turnover rate in chlorophyll. Rapid mixing of hyperdense cell solutions to match the turnover rate has been proposed as a method for maximizing utilization of photons [23].

The general efficiency of CO_2 conversion to carbon-captured products using sunlight has been estimated conservatively at about 7% of total photon irradiance [19]. This may seem low, however, in the case of carbon-captured fuel generation this would equate theoretically to 238 MJ of energy being fixed in fuel per square meter per year. Annual energy consumption in the United States from petroleum fuels is about 2.8×10^{13} MJ, which means that an insolation area of 0.12×10^{12} m^2 (roughly half the size of California) would be needed to generate enough fuel for the entire country.

7.4   MITIGATION OF COSTS THROUGH CAPTURED-CARBON PRODUCTS

Captured CO_2 doesn't have to be sequestered as CO_2. Cyanobacteria offer a unique opportunity to redirect fixed CO_2 into useful products. This allows installations that may be built for the purpose of removing CO_2 or treating wastewater to also produce molecules of value, which may significantly offset the costs of CO_2 capture. This can be done by the introduction of novel pathways using synthetic biology, to create photosynthetic strains capable of the overproduction of new molecules.

The list of "captured-carbon products" (CCP) directly converted from CO_2 by engineered cyanobacteria is growing and at the time of writing includes isoprene [24], isobutyraldehyde [25], isobutanol [25], free fatty acids [26], *n*-butanol [27], and lactic acid [28]. As described in more detail in the following sections, these synthetic pathways provide the ability to produce valuable chemical resources from CO_2 by construction of new metabolic pathways. The ability to create a range of novel strains capable of production of chemicals otherwise not produced biologically by photosynthetic organisms, or to increase the production of natural pathways, helps to diversify the options for cost mitigation in CO_2 sequestration using bioreactors. A comprehensive description of the major contributions to CCP engineering in cyanobacteria is outlined below. The pathways highlighted target the major branches of carbon flux from CO_2 fixation each in different ways. Additionally, technology for blind scanning and selection in libraries of cells promises to allow for optimization beyond the rational design ability of synthetic biology.

7.5 CAPTURED-CARBON PRODUCTS FROM ENGINEERED CYANOBACTERIA

7.5.1 Isobutyraldehyde

Isobutyraldehyde is a building block chemical, which can readily be converted into a range of carbon products traditionally synthesized from petroleum, including four-carbon imines and oximes. Notably, it can be converted into isobutanol, a drop-in fuel, in a simple synthetic step or by biological catalysis as described below. Isobutyraldehyde has a low boiling point and high vapor pressure, making it ideal for separation by gas stripping, which can increase titer by decreasing toxicity [29,30]. Biological production of isobutyraldehyde was proposed in previous work in *Escherichia coli*, which elucidated the redirection of amino acid synthesis pathways for the generation of alcohols [31]. In 2009 Atsumi et al. [25] produced isobutyraldehyde in the cyanobacterium *Synechococcus elongatus* PCC 7942 through the addition of a ketoacid decarboxylase gene, *kivd,* from *Lactococcus lactis*. This gene encodes an enzyme that converts 2-ketoisovalerate, which is naturally produced in small amounts from pyruvate as part of the valine biosynthesis pathway, into isobutyraldehyde, creating a branching point in the metabolic pathway. To increase the flux from pyruvate to 2-ketoisovalerate the *alsS* gene from *Bacillus subtilis* and *ilvC* and *ilvD* genes from *E. coli* were also expressed. The pathway could be switched on and off by using the IPTG inducible promoter P_{trc} to control expression of the pathway genes [32]. From this strain a maximum yield of 2500 μg/L/h was achieved [25]. To further improve carbon flux to pyruvate, a second version of the carbon-fixing RuBisCO enzyme from *S. elongatus* PCC6301 was integrated downstream of the endogenous *rbcLS* genes in the production strain, in an attempt to increase the efficiency of carbon fixation. Production in the new strain reached 6230 μg/L/h, surpassing the established production of fuel oils in eukaryotic algae systems [8]. A study of O$_2$ evolution in the final strain did not show increased oxidative photosynthesis with the addition of *rbcLS*. Instead, the authors propose that carbon fixation was improved by more efficient use of reducing power generated by photosynthesis. Increases in yield from engineering such as that from *rbcLS* indicate that there is room for improvement in production systems, and that titers may continue to increase as knowledge of cyanobacterial systems increases.

7.5.2 Isobutanol

Isobutanol has been well characterized as a "drop-in" fuel substitute in gasoline cars. It has a higher energy density, lower hygroscopicity, and lower vapor pressure than ethanol, making it a desirable competitor in alternative fuels. Atsumi et al. [33] were able to biologically produce isobutanol from CO$_2$ by integrating an alcohol dehydrogenase downstream from *kivd* on the same P_{trc} controlled operon. A total of four genes were tested, taken from *S. cerevisiae*, *L. lactis*, and *E. coli,* with the gene *yqhD* (*E. coli*) [33] showing the highest activity in *S. elongatus*. The final strain was able to produce 450 mg/L over 6 days, which approached the toxicity limit for isobutanol in cyanobacteria [25].

7.5.3 Fatty Acids

Fatty acids are abundant in oleaginous (high oil-producing) microalgae in the esterified form of triacylglycerols, but are not stored in significant amounts in cyanobacteria. The methyl esters of these lipids have been investigated in depth as fuel substitutes and are generally considered to be a leading alternative to petroleum fuels. Control of fatty acid composition has been an important focus in oleaginous algae studies [34], and this is an area where pathway control through synthetic biology in cyanobacteria could excel. Triacylglycerol production from oleaginous algae currently comes with high recovery cost of harvesting algae and separating oil from the cells after growth. Production considerations for free fatty acids (FFAs), fatty acid chains that are not connected to glycerol as an ester, differ from that of acylglycerol in the key aspect that high temperatures must be used to convert the less reactive acid moiety. In this process water must also be continually removed during the reaction to prevent inhibition of catalysis. The added cost of lower reactivity is perhaps balanced, however, if free fatty acids are produced continually and can be separated from the culture without energy-intensive cell-harvesting steps.

In cyanobacteria acylglycerols are naturally synthesized in small amounts as part of the cell membrane through the successive addition of acyl units from activated acyl-CoA, to a chain bearing acyl carrier protein. To take advantage of this pathway to produce fuel, a thioesterase gene was expressed in the cyanobacterium *Synechocystis* sp. PCC6803, which allowed for inducible redirection of carbon from membrane esters to FFAs [26]. The thioesterase (TE) concept built on similar work in *E. coli*, which takes advantage of interrupted membrane lipid production [35]. TE enzymes cleave the thioester bond of acyl-acyl carrier proteins (acyl-ACP) producing FFAs and preventing buildup of regulatory acyl-ACP in one step. Intracellular FFAs of medium length are naturally secreted, driving the equilibrium forward. Secretion allows for continuous production and collection from culture media without harvesting or processing cells. The fatty acid secretion system was further diversified by utilizing a markerless transformation method that allows for successive generations of strains to be constructed. To increase C14 chain lengths desirable for enhanced biodiesel properties [36], TEs from different plant sources were inserted. FatB1s from *Cinnamomum camphorum* and *Umbellularia californica,* and FatB2 from *Cuphea hookeriana* produced myristate (14:0) and laurate (12:0) preferentially when expressed in the production strain. FatB2 created the biggest shift with an increase of myristate of 13.3% compared to a strain also containing both FatB1 genes. This system has shown that modification of fatty acid synthesis pathways in cyanobacteria can redirect carbon to secreted biodiesel precursors in measurable amounts, and that these carbon products can be tailored by changing the genes.

7.5.4 Hydrocarbons

Alkanes are constituents of gasoline and diesel, and therefore represent direct production of replacement fuel from renewable sources. Fatty alcohols and aldehydes are similarly valuable as fuel additives, and also serve as chemical building blocks with reactive moieties. The natural pathway for production of fatty aldehydes and alkanes was identified in cyanobacteria in 2010 and cloned into *E. coli* for the production of fuel products [15]. As with FFAs, the hydrophobic product is secreted and separates from the media, allowing it to be continuously collected from cultures. Nonnative pathways were introduced into the cyanobacterium *Synechocystis* sp. PCC6803 to increase production of alkanes directly from CO_2 [37]. The introduced genes catalytically reduce the thioester moiety of a fatty acyl-ACP to a fatty acid, which is then converted to acyl-CoA to allow for further reduction to an alcohol or an alkane. Although production was increased by 50%, yields are still low, reflecting the complexity of this pathway.

7.5.5 1-Butanol

1-Butanol is a drop-in fuel with properties similar to that of gasoline, conceptually making it a competitor to isobutanol as an alternative fuel. 1-Butanol was produced in the cyanobacterium *S. elongatus* PCC 7942 by partial reconstruction of the acetyl-CoA pathway with insertion of five exogenous genes [27]: *hbd*, *crt*, and *adhE2* genes from a natural 1-butanol producing pathway in *Clostridium acetobutylicum*, and *ter* and *atoB* from *Treponema denticola* and *E. coli*, respectively, which were identified for activity in previous work [38]. This pathway combines two acetyl-CoA to acetoacetyl-CoA, which is then dehydrated and hydrogenated to form the alkane thioester butyryl-CoA. The thioester is then cleaved to a terminal aldehyde, which is subsequently reduced to form 1-butanol. 1-Butanol production was challenged by the need for selection of genes from anaerobic organisms, which raised the challenge of oxygen sensitivity. Cyanobacteria are photosynthetic and produce oxygen during the water-splitting process, which prohibits the use of oxygen-sensitive enzymes from anaerobic hosts. Switching cyanobacteria cultures into anaerobic conditions and continuing growth increased production to a final titer of 14.5 mg/L.

7.5.6 Isoprene

Isoprene is a five-carbon, energy-dense biofuel molecule, useful as a chemical feedstock. Like isobutyraldehyde, isoprene has a high vapor pressure and separates into the gas phase during bubbling, allowing for simple capture and separation. Isoprene is naturally synthesized and secreted by many plants and is also the iterative unit used in biological terpenoid synthesis [39]. Cyanobacteria do not possess a gene for isoprene production, however, they do possess pathways for essential isoprenoid synthesis. To engineer bacterial production of isoprene from CO_2, an isoprene synthase gene *ispS* from *Pueraria montana* was inserted by homologous recombination into the cyanobacterium *Synechocystis* sp. PCC6803 [24]. This strain was able to produce 50 μg/gDCW/day, confirming engineered isoprene biosynthesis. The low production titer was theorized to be due to significant carbon flux to sugar biosynthesis (85%) and fatty acid synthesis (10%) rather than isoprenoid synthesis (3–5%) based on unpublished data. It is acknowledged that future work to improve the titer of isoprene production in cyanobacteria will have to target this carbon-partitioning problem.

7.5.7 Hydrogen

Efforts are also being made to increase cyanobacterial production of hydrogen, which is sought after as an alternative fuel [40]. Although hydrogen is not a CCP and therefore does not contribute to CO_2 sequestration directly, it does provide another alternate source of income for CO_2 capturing bioreactor systems where CO_2 would be sequestered in biomass entirely. In 2010, Ducat et al. [41] were able to express an exogenous hydrogenase from *Clostridium acetobutylicum* in *S. elongatus* PCC7942 and demonstrate that the gene was active. This work installed hydrogen production in a well-characterized host strain, increasing production to a rate of 2.8 μmol H_2/h/mg chlorophyll *a* through a synthetic pathway. In a separate approach Cournac et al. [42] measured increased production of 6 μmol H_2/h/mg chlorophyll *a* in a mutant strain defective in Type I NADPH-dehydrogenase complex.

7.5.8 Poly-3-hydroxybutyrate

In addition to rationally engineered pathways, synthetic biology is allowing researchers to create tools for the high-throughput scanning of random pathways in cyanobacteria, both from natural sources and from mutagenesis. In *Synechocystis* sp. PCC6803, a high-throughput scan protocol has been identified which can be used for improved production of poly-3-hydroxybutyrate (PHB) [43]. PHB is a leading candidate for production of renewable and biodegradable plastics. Scanning was achieved by addition of the fluorescent dye nile-red, which changes its emission wavelength when in contact with hydrophobic environments. The specificity of the dye for the desired product was further improved by adjusting ionic buffers in the testing medium. A linear correlation between fluorescence and PHB concentration was observed, with a top production of more than 45% PHB content (% of dry cell weight). By coupling the method with flow cytometry for high-throughput fluorescence measurement, and adjusting conditions to retain viability of the cells, large libraries of mutants could be tested further for increased production, and subsequently cultured as new strains.

7.5.9 Indirect Production Technology

Liu et al. [44] were able to demonstrate further the facility of synthetic biology in cyanobacteria by adapting a CO_2 limitation inducible operon for the expression of lipase genes. Increased expression of inorganic carbon uptake proteins *ndhF3*, *sbtA*, and *cmpA*, after rapid CO_2 depletion of culture media, had been reported previously by McGinn et al. [45]. By inserting genes between the promoter sequence and ATG start codon of these operons, McGinn et al. [45] identified the P_{cmp} promoter sequence as active for the expression of exogenous genes on a CO_2 concentration-dependent basis in cyanobacteria. Three lipase genes were tested: *fol*, *shl*, and *gpl*, from a fungus, bacteria, and mammal, respectively. A combination of *fol* and *shl* under P_{cmp} promotion showed the highest activity allowing for 36.1×10^{-12} mg/cell of FFAs to be recovered. Analysis of the lipids showed high unsaturation providing evidence that the free fatty acids result from increased degradation of cell membranes.

TABLE 7.2 Production of Captured-Carbon Products from Synthetic Pathways in Cyanobacteria

Product	Photons Required	Theoretical Yield (kg/m²/yr^a)	Experimental Titer	Production Rate	Host Organism	Reference
Isobutyraldehyde	48	16.0	1100 mg L^{-1}	6.23 mg L^{-1} h^{-1}	*S. elongatus* PCC7942	[25]
Isobutanol	48	16.5	450 mg L^{-1}	3.12 mg L^{-1} h^{-1}	*S. elongatus* PCC7942	[25]
1-Butanol	48	16.5	14.5 mg L^{-1}	86.31 µg L^{-1} h^{-1b}	*S. elongatus* PCC7942	[27]
Fatty acid (total)	NA	NA	197 mg L^{-1}	NA	*Synechocystis* sp. PCC6803	[26]
C12	144	14.8	39.23 mg L^{-1b}	NA	*Synechocystis* sp. PCC6803	[26]
C14	168	14.5	41.27 mg L^{-1b}	NA	*Synechocystis* sp. PCC6803	[26]
C16	192	14.2	85.3 mg L^{-1b}	NA	*Synechocystis* sp. PCC6803	[26]
Hydrocarbon	72^c	12.8	161.562 ± 10.315 µg OD^{-1}L^{-1}	NA	*Synechocystis* sp. PCC6803	[37]
Fatty alcohol	72^c	15.1	200.44 ± 8.07 µg L^{-1}	11.14 µg L^{-1} day^{-1b}	*Synechocystis* sp. PCC6803	[37]
Isoprene	48	15.1	NA	50 µg g^{-1} (DCW)day^{-1}	*Synechocystis* sp. PCC6803	[24]

^{a}Values calculated as per method of Atsumi et al. [25]. NA = Data is not available.
^{b}Values calculated from data published in the reference.
^{c}Represents an arbitrary average of different chain lengths.

This promotion strategy provides an ecological alternative to previously developed methods using heavy metals such as nickel. Additionally, this method lowers the energy needed for industrial separation and recovery of materials such as lipids from postculture cells.

7.6 CONCLUSION

The ability of synthetic biology to aid in rational design of pathways from CO_2 to chemical products has been demonstrated, opening the door to a wide range of captured-carbon products. Titers high enough to compete with existing standards in algae biofuel production have already been achieved [8, 25], however, in-depth studies on regulation and carbon partitioning are needed to increase overall titers of most products. The general metabolic map in cyanobacteria is well characterized with remaining details continually being elucidated [46]. At the same time, significant progress is being made in understanding the flux of carbon between pathways [47]. Cyanobacteria are capable of sequestering large amounts of CO_2 into captured carbon (Table 7.2).

It remains to be seen which products will provide the best cost-to-sequestration ratio. Products such as isobutyraldehyde and hydrogen solve the significant challenge of product recovery from media by having a high vapor pressure compared to water. This allows for techniques such as gas stripping to be highly efficient. Scaling of laboratory-tested production to industrial scale will be a challenge, however, a number of companies have already raised capital with business plans based on cyanobacterial systems. As bioreactor technology moves forward, the promise of cyanobacterial engineering should not be overlooked, for its potential role both in carbon sequestration and in the renewable production of chemical feedstocks.

REFERENCES

1. K. Kumar, C. N. Dasgupta, B. Nayak, P. Lindblad, D. Das, Development of suitable photobioreactors for CO_2 sequestration addressing global warming using green algae and cyanobacteria, Bioresour. Technol. 102 (2011) 4945–4953.

2. E. Jacob-Lopes, S. Revah, S. Hernández, K. Shirai, T. T. Franco, Development of operational strategies to remove carbon dioxide in photobioreactors, Chem. Eng. J. 153 (2009) 120–126.

3. Y. Chisti, Biodiesel from microalgae, Biotechnol. Adv. 25 (2007) 294–306.

4. S. K. Lee, H. Chou, T. S. Ham, T. S. Lee, J. D. Keasling, Metabolic engineering of microorganisms for biofuels production: from bugs to synthetic biology to fuels, Curr. Opin. Biotechnol. 19 (2008) 556–563.

5. M. R. Connor, S. Atsumi, Synthetic biology guides biofuel production, J. Biomed. Biotechnol. 2010 (2010) 541–698.

6. H. Herzog, D. Golomb, Carbon capture and storage from fossil fuel use. In: C. J. Cleveland (Ed.), *Encyclopedia of Energy,* Elsevier Science, New York, 2004, pp. 277–287.

7. M. R. Raupach, C. B. Field, E. T. Buitenhuis, P. Ciais, J. G. Canadell, C. L. Que, T. J. Conway, N. P. Gillett, R. A. Houghton, G. Marland, Contributions to accelerating atmospheric CO_2 growth from economic activity, carbon intensity, and efficiency of natural sinks, Proc. Natl. Acad. Sci. U.S.A. 20 (2007) 18866–18870.

8. J. Sheehan, Engineering direct conversion of CO_2 to biofuel, Nat. Biotechnol. 27 (2009) 1128–1129.

9. C. B. Field, Primary production of the biosphere: integrating terrestrial and oceanic components, Science 281 (1998) 237–240.

10. E. Ono, J. Cuello, Carbon dioxide mitigation using thermophilic cyanobacteria, Biosyst. Eng. 96 (2007) 129–134.

11. T. A. Vishnivetskaya, Viable cyanobacteria and green algae from the permafrost darkness, Soil. Biol. 16 (2009) 73–84.

12. A. M. Ruffing, Engineered cyanobacteria: teaching an old bug new tricks, Bioeng. Bugs 2 (2011) 136–149.

13. T. Kaneko, S. Sato, H. Kotani, A. Tanaka, E. Asamizu, Y. Nakamura, et al., Sequence analysis of the genome of the unicellular cyanobacterium *Synechocystis* sp strain PCC6803. II. Sequence determination of the entire genome and assignment of potential protein-coding regions (supplement), DNA Res. 3 (1996) 185–209.

14. C. Sugita, K. Ogata, M. Shikata, H. Jikuya, J. Takano, M. Furumichi, M. Kanehisa, T. Omata, M. Sugiura, M. Sugita, Complete nucleotide sequence of the freshwater unicellular cyanobacterium *Synechococcus elongatus* PCC 6301 chromosome: gene content and organization, Photosynth. Res. 93 (2007) 55–67.

15. A. Schirmer, M. A. Rude, X. Li, E. Popova, S. B. del Cardayre, Microbial biosynthesis of alkanes, Science 329 (2010) 559–562.

16. S. Kajiwara, H. Yamada, N. Ohkuni, K. Ohtaguchi, Design of the bioreactor for carbon dioxide fixation by *Synechococcus* PCC7942, Energy Convers. Mgmt. 38 (1997) S529–S532.

17. E. Jacoblopes, C. Scoparo, L. Lacerda, T. Franco, Effect of light cycles (night/day) on CO_2 fixation and biomass production by microalgae in photobioreactors, Chem. Eng. Process 48 (2009) 306–310.

18. E. Jacob-Lopes, C. H. Gimenes Scoparo, M. I. Queiroz, T. T. Franco, Biotransformations of carbon dioxide in photobioreactors, Energy Convers. Mgmt. 51 (2010) 894–900.

19. X.-G. Zhu, S. P. Long, D. R. Ort, What is the maximum efficiency with which photosynthesis can convert solar energy into biomass? Curr. Opin. Biotechnol. 19 (2008) 153–159.

20. X.-G. Zhu, S. P. Long, D. R. Ort, Improving photosynthetic efficiency for greater yield, Annu. Rev. Plant. Biol. 61 (2010) 235–261.

21. D. E. Robertson, S. A. Jacobson, F. Morgan, D. Berry, G. M. Church, N. B. Afeyan, A new dawn for industrial photosynthesis, Photosynth. Res. 107 (2011) 269–277.

22. S. Wilcox, M. Anderberg, W. Beckman, A. DeGaetano, R. George, C. Gueymard, N. Lott, W. Marion, D. Myers, R. Perez, D. Renné, P. Stackhouse, F. Vignola, T. Whitehurst, *National Solar Radiation Database 1991–2005 Update: User's Manual*, 2007.

23. H. Qiang, Y. Zarmi, A. Richmond, Combined effects of light intensity, light-path and culture density on output rate of *Spirulina platensis* (Cyanobacteria), Eur. J. Phycol. 33 (1998) 37–41.

24. P. Lindberg, S. Park, A. Melis, Engineering a platform for photosynthetic isoprene production in cyanobacteria, using *Synechocystis* as the model organism, Metab. Eng. 12 (2010) 70–79.

25. S. Atsumi, W. Higashide, J. C. Liao, Direct photosynthetic recycling of carbon dioxide to isobutyraldehyde, Nat. Biotechnol. 27 (2009) 1177–1180.

26. X. Liu, J. Sheng, R. Curtiss, Fatty acid production in genetically modified cyanobacteria, Proc. Natl. Acad. Sci. U. S. A. 108 (2011) 6899–6904.

27. E. I. Lan, J. C. Liao, Metabolic engineering of cyanobacteria for 1-butanol production from carbon dioxide, Metab. Eng. 13 (2011) 353–363.

28. H. Niederholtmeyer, B. T. Wolfstädter, D. F. Savage, P. A. Silver, J. C. Way, Engineering cyanobacteria to synthesize and export hydrophilic products, Appl. Environ. Microbiol. 76 (2010) 3462–3466.

29. K. Inokuma, J. C. Liao, M. Okamoto, T. Hanai, Improvement of isopropanol production by metabolically engineered *Escherichia coli* using gas stripping, J. Biosci. Bioeng. 110 (2010) 696–701.

30. A. Baez, K.-M. Cho, J. C. Liao, High-flux isobutanol production using engineered *Escherichia coli*: a bioreactor study with in situ product removal, Appl. Microbiol. Biotechnol. 90 (2011) 1681–1690.

31. S. Atsumi, T. Hanai, J. C. Liao, Non-fermentative pathways for synthesis of branched-chain higher alcohols as biofuels, Nature 451 (2008) 86–89.

32. J. Brosius, M. Erfle, J. Storella, Spacing of the −10 and −35 regions in the tac promoter: effect on its *in vivo* activity, J. Biol. Chem. 260 (1985) 3539–3541.

33. S. Atsumi, T.-Y. Wu, E.-M. Eckl, S. D. Hawkins, T. Buelter, J. C. Liao, Engineering the isobutanol biosynthetic pathway in *Escherichia coli* by comparison of three aldehyde reductase/alcohol dehydrogenase genes, Appl. Microbiol. Biotechnol. 85 (2010) 651–657.

34. M. A. Danielewicz, L. A. Anderson, A. K. Franz, Triacylglycerol profiling of marine microalgae by mass spectrometry, J. Lipid. Res. 52 (2011) 2101–2108.

35. H. Cho, Defective export of a periplasmic enzyme disrupts regulation of fatty acid synthesis, J. Biol. Chem. 270 (1995) 4216–4219.

36. G. R. Stansell, V. M. Gray, S. D. Sym, Microalgal fatty acid composition: implications for biodiesel quality, J. Appl. Phycol. 24 (2011) 791–801.

37. X. Tan, L. Yao, Q. Gao, W. Wang, F. Qi, X. Lu, Photosynthesis driven conversion of carbon dioxide to fatty alcohols and hydrocarbons in cyanobacteria, Metab. Eng. 13 (2011) 169–176.

38. S. Atsumi, A. F. Cann, M. R. Connor, C. R. Shen, K. M. Smith, M. P. Brynildsen, K. J. Y. Chou, T. Hanai, J. C. Liao, Metabolic engineering of *Escherichia coli* for 1-butanol production, Metab. Eng. 10 (2008) 305–311.

39. L. Ruzicka, The isoprene rule and the biogenesis of terpenic compounds 1953, Experientia 50 (1994) 395–405.

40. M. Z. Jacobson, W. G. Colella, D. M. Golden, Cleaning the air and improving health with hydrogen fuel-cell vehicles, Science 308 (2005) 1901–1905.

41. D. C. Ducat, G. Sachdeva, P. A. Silver, Rewiring hydrogenase-dependent redox circuits in cyanobacteria, Proc. Natl. Acad. Sci. U. S. A. 108 (2011) 3941–3946.

42. L. Cournac, G. Guedeney, G. Peltier, P. M. Vignais, Sustained photoevolution of molecular hydrogen in a mutant of *Synechocystis* sp strain PCC 6803 deficient in the Type I NADPH-dehydrogenase complex, J. Bacteriol. 186 (2004) 1737–1746.

43. K. E. Tyo, H. Zhou, G. N. Stephanopoulos, High-throughput screen for poly-3-hydroxybutyrate in *Escherichia coli* and *Synechocystis* sp strain PCC6803, Appl. Environ. Microbiol. 72 (2006) 3412–3417.

44. X. Liu, S. Fallon, J. Sheng, R. Curtiss, CO_2-limitation-inducible green recovery of fatty acids from cyanobacterial biomass, Proc. Natl. Acad. Sci. U. S. A. 108 (2011) 6905–6908.

45. P. J. McGinn, G. D. Price, R. Maleszka, M. R. Badger, Inorganic carbon limitation and light control the expression of transcripts related to the CO_2-concentrating mechanism in the cyanobacterium *Synechocystis* sp strain PCC6803, Plant. Physiol. 132 (2003) 218–229.

46. S. Zhang, D. A. Bryant, The tricarboxylic acid cycle in cyanobacteria, Science 334 (2011) 1551–1553.

47. J. D. Young, A. A. Shastri, G. Stephanopoulos, J. A. Morgan, Mapping photoautotrophic metabolism with isotopically nonstationary (13)C flux analysis, Metab. Eng. 13 (2011) 656–665.

CHAPTER 8

Hydrogen Production by Microalgae

HELENA M. AMARO, M. GLÓRIA ESQUÍVEL, TERESA S. PINTO, and
F. XAVIER MALCATA

8.1 INTRODUCTION

The world has been confronted with an energy crisis in the last decades arising from irreversible depletion of fossil fuels, and concomitant atmospheric accumulation of greenhouse effect gases. The urgent need to replace traditional fuels with others from novel and sustainable sources has led to the emergence of biofuels—encompassing biohydrogen in particular, produced by microalgae [1].

Microalgae offer indeed a promising alternative, and likely a highly efficient route to produce biohydrogen that does not compete with land for food crops; and the corresponding cultivation is easy to implement and control [2]; they can capture solar energy and store it as chemical energy, via conversion of water to molecular hydrogen. Release of hydrogen by unicellular green algae was originally reported by Hans Gaffron and J. Rubin [3] in the 1940s as a by-product of carbohydrate synthesis, although microalgae (and cyanobacteria) can also perform both oxygenic photosynthesis and hydrogen production [4]. If seawater can be utilized, then the water-splitting bioprocess carried out by marine microalgae will presumably lead to one of the most abundant and accessible energies on earth [5].

As aquatic organisms, microalgae lay at the bottom of the food chain; they take up H_2O and atmospheric CO_2 that, at the expense of absorbed sunlight, are then converted to complex organic compounds or simple electron acceptors (as is the case for molecular hydrogen). Furthermore, microalgae are well-adapted microorganisms to survive under a large range of environmental stresses, for example, heat, cold, drought, salinity, photooxidation, anaerobiosis, osmotic pressure, and UV radiation [6], owing to their huge metabolic plasticity. Microalgae indeed combine a few features regularly found in higher plants, such as efficient oxygenic photosynthesis and simple nutritional requirements, with biotechnological attributes typical of microorganisms with fast growth rates and ability to accumulate or secrete

Natural and Artificial Photosynthesis: Solar Power as an Energy Source, First Edition. Edited by Reza Razeghifard.
© 2013 John Wiley & Sons, Inc. Published 2013 by John Wiley & Sons, Inc.

metabolites. This useful combination backs up prospects for microalgal biotechnology in the near future. The metabolic versatility is complemented with the large number of species of microalgae (estimated as $>100,000$) that constitutes a unique reservoir of biodiversity [7].

Currently, the microalga *Chlamydomonas reinhardtii* is considered one of the best eukaryotic producers of H_2; this realization derives from knowledge of its genome, coupled with a remarkable metabolism encompassing aerobic respiration, fermentation, and the possibility of changing the accumulation rates of organic acids, ethanol, CO_2, and H_2 [1]. Moreover, its remarkable metabolic flexibility allows acclimation to hypoxic and anoxic conditions, even when exposed to uncontrolled environments [8]. Therefore it appears as a particularly useful model for studies focusing on enhancement and optimization of H_2 release.

The ability of microalgae to produce H_2 under anaerobic conditions is based on expression of a hydrogenase enzyme that recombines H^+ and e^- resulting from the photosynthetic electron transfer chain, or fermentative degradation of endogenous substrates instead [9]. *Chlamydomonas reinhardtii* exhibits an intrinsically high specific activity of that enzyme 10–100-fold that of most other wild type species [10].

Despite the aforementioned useful features, *C. reinhardtii* has not yet attained commercial success, because of an unfavorable combination of several biochemical and engineering issues. For instance, oxygen inhibits transcription and action of hydrogenase(s), so anoxia is required for H_2 production; however, anoxia is, in turn, delayed due to breakdown of H_2O into H^+ and e^- via photosynthesis. This is because the photosystem II (PSII) electron transport chain accounts for oxygen release, thus constraining the overall hydrogen productivity [1].

Melis et al. [11] reported a two-phase photobiological process of H_2 production, with deliberate depletion of sulfur in the culture medium that led to a loss in the oxygen evolution activity mediated by PSII, thus overcoming hydrogenase inhibition in *C. reinhardtii*. In fact, sulfur depletion apparently interferes with repair of the D1 subunit of PSII. Respiration consumes the remaining O_2 until anaerobic conditions are eventually attained, thus improving hydrogenase activity and extending H_2 production for 4–5 d [11, 12]. However, the H_2 productivity in practice is consistently lower than its theoretical maximum: this is due to depletion of electron sources in the absence of photosynthetic activity, because PSII activity is lost upon sulfur deprivation [9]; and to accumulation of toxic fermentation compounds during anaerobic fermentation [13]. The hypothesized mechanisms underlying hydrogen photoproduction in sulfur-deprived *C. reinhardtii* were critically revised elsewhere [14].

Several attempts to affect DNA insertional mutagenesis in *C. reinhardtii* were carried out by finding high H_2-yield mutants, while trying to identify the relevant genes involved therein. Several proteins were accordingly found to be critical for maturation of the active Fe hydrogenase, for example, HydE, HydF, and HydG [15]; hence both genetic and metabolic engineering will likely be called upon in attempts to further improve biohydrogen production. A number of mutants have indeed been obtained targeting hydrogenase, sulfate permease, ribulose-1,5-bisphosphate carboxylase/oxygenase (RuBisCO), and photosystems I and II [1], as well as starch reserves and respiration rates; these topics will be addressed next in further detail.

8.2   HYDROGENASE ENGINEERING

Increasing the tolerance of hydrogenase to O_2 is a major challenge toward commercial feasibility of microalga-mediated production of H_2; this would permit full exploitation of photobiological release of H_2 using water as the sole electron donor. This deed was recently addressed in silico and in vitro via experimental engineering of hydrogenases for strict accessibility of oxygen to the catalytic site, thus leading to development of O_2-tolerant phenotypes in H_2-producing strains [16, 17, 18].

Recall that oxygen acts as a transcriptional repressor by inhibiting the hydrogenase assembly, and as an irreversible inhibitor of catalysis thereby [17]. Elucidation of the highly complex biosynthesis/maturation process associated with hydrogenases started with [FeFe] and [NiFe] hydrogenases, since these enzymes are phylogenetically distinct—as the former belongs to green algae and the other to cyanobacteria [17]. [NiFe] hydrogenases possess a lower catalytic turnover number than [FeFe] hydrogenases, and may be involved in either H_2 uptake or production (bidirectional). In green algae, hydrogenases are encoded by *HydA1* and *HydA2* nuclear genes, but the mature enzyme is localized in the chloroplast; after translocation, the transit sequence is cleaned out, which is a common step of protein compartmentation. Conserved maturation proteins (HYDE, HYDF, and HYDG) are essential for making functional hydrogenases [17, 19]. In cyanobacteria, the biosynthesis and maturation of hydrogenases are quite complex phenomena that require several proteins to carry out incorporation of ligands in the active center (large subunit—LSU), insertion of FeS clusters (small subunit—SSU), cleavage of C-terminal polypeptide of LSU, and assembly of mature subunits to form the final enzyme [17, 19]. Prokaryotic microorganisms harbor bidirectional hydrogenases, so further research is needed at this stage to develop enzymes that catalyze only hydrogen release.

It is well known that the primary electron source for proton reduction carried out by hydrogenases is reduced ferredoxin (Fd); however, the specific interface between the two proteins was not analyzed in detail until recently. Mutant variants of both hydrogenase and ferredoxin were accordingly used, and the amino acid residues essential for their mutual interaction were duly identified [18, 20]; HydA1 has eight highly conserved basic residues (lysine and arginine) on the surface, with Lys396 being especially important for interaction. On the other hand, the residues Glu122, Asp56, and Phe93 in ferredoxin (Fd) are the most critical for hydrogenase/ferredoxin interaction [21]. It is likely that rationalization of how ferredoxin interacts with partner proteins will enable scientists to introduce modifications with the goal of enhancing electron flow to the hydrogenase.

8.3   METABOLIC REPROGRAMING

Strategies to induce H_2 production without the stressful impact of nutrient deprivation have been selected as platforms for research in this field [4, 22]. One approach has focused on attenuation of the photosynthesis/respiration (P/R) ratio able to induce anoxia; the strain with the lowest P/R ratio found ("apr1") was characterized by a two- to three-fold increase in H_2 production rate relative to the wild type [22].

Hydrogen release is often viewed as a safety valve to protect the photosynthetic electron transport chain from overreduction; disruption of the regulatory process might lead to redox imbalance, thus stimulating H_2 photoproduction. All cellular processes dependent on electron transfer from Fd may be targets for maximization of intracellular reductant flow toward H_2-producing enzymes; the main competitor is ferredoxin $NADP^+$ oxidoreductase (FNR) that is responsible for regeneration of NADPH. To elucidate such a competition, a ferredoxin hydrogenase was engineered by fusion, and the kinetics of hydrogen release and electron transfer to both FNR and $NADP^+$ were monitored. Replacing the hydrogenase with a Fd hydrogenase fusion switched the bias of electron transfer from FNR to hydrogenase, and resulted in an increased rate of hydrogen photoproduction [23]. Similar approaches to increase the specificity of electron flow toward H_2 production can be applied; for instance, complexes in which a Fd and a heterologous hydrogenase were either immobilized in a modular protein scaffold or directly attached to each other via an amino acid linker were recently expressed in *Escherichia coli* [24].

Another approach takes advantage of the Calvin cycle (a carbon reduction cycle) in which glyceraldehyde 3-phosphate dehydrogenase (GAPDH) uses the regenerated NADPH for carbon reduction. The Calvin cycle (specifically via its key enzymes RuBisCO, which is responsible for CO_2 fixation [25], and GAPDH) may provide important targets to improve hydrogen production [26]. Hence H_2 photoproduction was reported concomitantly with impairment of CO_2 fixation [27, 28], and it was significantly reduced when the level of exogenous CO_2 was increased [28].

Other competitors for reduced power are nitrogen and sulfur metabolisms: nitrite reductase (NiR) and sulfite reductase (SiR) depend on electron transfer from reduced Fd, so nitrogen deprivation [29] or sulfur deprivation [11] produces an improvement of hydrogen production [30]. In fact, sustainability of hydrogen production by green algae was originally achieved in sulfur-deprived media [11], not only due to competition for reducing power by SiR but also due to PSII inactivation, and consequently to inhibition of oxygen evolution. Mimicking sulfur deprivation conditions by reducing sulfur uptake capacity was also achieved through attenuation of expression of a sulfate permease (SulP) with the purpose of enhancing hydrogen production [31].

Other enzymes that depend on electron transfer from reduced Fd are glutamate synthase, fatty acid desaturase, thioredoxin reductase, and xanthine reductase [13, 21, 32], thus making them potential targets in attempts to reprogram the metabolism toward enhancing hydrogen production. Cyclic electron flow around PSI (CEF) also competes with hydrogenases for photosynthetic reductant from Fd; this negative relationship between hydrogen photoproduction and CEF has frequently been reported [14, 29, 32, 33]. A *C. reinhardtii* mutant (pgrl1), impaired in CEF due to a defect in the proton gradient regulation like1 (PGRL1) protein, was accordingly isolated, and H_2 photoproduction was strongly enhanced during both short- and long-term operations [34]. The significant metabolic pathways competing for reduced power from Fd in the chloroplast are summarized in Fig. 8.1.

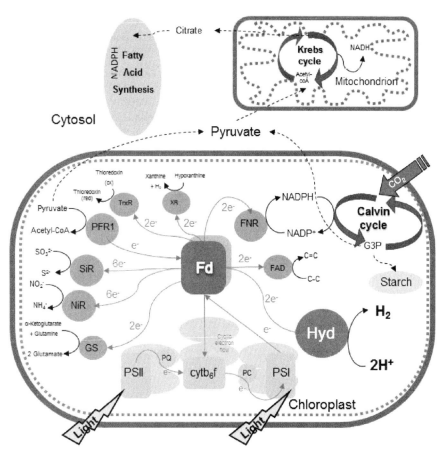

FIGURE 8.1   Ferredoxin-dependent pathways in *Chlamydomonas reinhardtii* chloroplast of photosynthetic electron transport chain. Fd (ferredoxin) accepts electrons from PSI (photosystem I) and is a redox partner of various enzymes, such as NiR (nitrite reductase), SiR (sulfite reductase), FNR (ferredoxin-NADP$^+$ oxidoreductase), GS (glutamate synthase), FAD (fatty acid desaturase), TrxR (ferredoxin-thioredoxin-reductase), and XR (xanthine reductase). Hyd (FeFe hydrogenase) and PFR1 (pyruvate-ferredoxin-oxidoreductase) are also ferredoxin-dependent enzymes in the chloroplast. PSII, photosystems II; cyt b_6f, cytochrome b_6f; PQ, plastoquinone; PC, plastocyanin; G3P, Glyceraldehyde 3-phosphate. (Scheme based on Hemschemeier and Happe [29] and Seibert et al. [32]).

Preservation of anaerobic conditions via constraining O_2 evolution by PSII is another approach to improve H_2 production; however, inhibiting the performance of PSII via mutation(s) in D1 protein simultaneously caused a decrease in the rate of H_2 release by *C. reinhardtii* [35]. These results suggest that PSII activity is indispensable for accumulation of starch reserves required for cell growth, and its reduction also limits use of water as the electron source for H_2 production. An additional metabolic problem to be solved stems from the decrease of photosynthetic electron flow under conditions where the proton gradient is not properly dissipated—a situation that occurs under anaerobic culturing when ATP synthase is not operating [33, 36].

Conditions under which the genes of anaerobic metabolism related with hydrogenase are upregulated (such as copper deficiency [37]) may also serve to induce H_2 production. Studying multiple alternative fermentative pathways of *C. reinhardtii* [38] showed strong induction of hydrogen metabolism genes (e.g. *HYD1, HYDEF, HYDG,* and *PFR1*). The fermentative pathways in *C. reinhardtii* have been elucidated comprehensively at the level of transcriptome, proteome, and metabolome [38, 39, 40]; this microalga exhibits formate, ethanol, acetate, CO_2, molecular hydrogen, and traces of glycerol, lactate, malate, and succinate as fermentation products—but their relative stoichiometry varies with growth conditions. The fermentation compounds are important targets for metabolic engineering, in attempts to have energy stored and substrate(s) supplied for hydrogenases.

A metabolomic study of anaerobic H_2 photoproduction in *C. reinhardtii* (strain stm6glc4) unfolded several pathways, including the synthesis of fatty acids, neutral lipids, and fermentative products [41]; new perspectives were consequently opened toward improvement of hydrogen production. Fatty acids and neutral lipids are possible competitors for H_2 production, so a goal of future research efforts is to diminish their extent of synthesis in favor of metabolic pathways leading to accumulation of energy-rich storage compounds (e.g. starch [41, 42]). Fatty acids synthesis appears to be regulated by the key enzyme acetyl-CoA carboxylase [42] that catalyzes conversion of acetyl-CoA to malonyl-CoA.

8.4 LIGHT CAPTURE IMPROVEMENT

In commercial production of H_2 via microalga-mediated processes, operating costs have to be affordable so as to successfully compete with other forms of renewable energies, as well as other sources of hydrogen. One of the major challenges for effective scale-up in production of biohydrogen is photobioreactor performance in terms of light penetration; for dense cultures, it may indeed compromise light utilization efficiency, and thus photosynthetic productivity [43].

Recall that H_2 production by green algae can be divided into two stages: an aerobic stage, when cells are cultivated photoautotrophically to produce biomass (chiefly via synthesis of carbohydrates from CO_2); and an anaerobic phase, induced by sulfur deprivation that promotes H_2 production due to starch catabolism [44]. It is estimated that production of 1 mol of H_2 via the hydrogenase pathway requires utilization of a minimum of 5 mol of photons by the photosynthetic apparatus [45]. Therefore, beyond the sulfur deprivation needed for inactivation of PSII, another important factor that precludes a high yield in H_2 production is the low-light utilization efficiency by the mass culture [46]. The rate of photon absorption by the chlorophyll antennas of chloroplasts in the cells occupying the top layers of ponds or photobioreactors usually exceeds the rate of photosynthesis, so at least 80% of light will eventually be wasted [47]. Conversely, intense incident light may lead to photoinhibition at low culture densities. Recent evidence has revealed that the minimum number of required chlorophyll molecules is 37 for PSII and 95 for PSI [48]; hence a good solution for the light-limited regime should entail reduction of antenna sizes, thus leading to a better efficiency in light capture provided that the efficiency of photosynthesis can concomitantly be increased.

In view of the above, genetic manipulation aimed at truncating the antenna size was attempted in *C. reinhardtii*, leading (among many others) to a mutant *T7* bearing a 10–17% reduction in antenna size that resulted in ca. 50% increase in photosynthetic efficiency at saturating light. Reduction of the antenna size by less than 20% may be enough to bring about a sufficient increase in photosynthetic efficiency, along with an almost insensitivity against photoinhibition [49]; however, saturating light intensity was assumed as a basis for that claim, which does not correspond to the conditions normally prevailing in photobioreactors. Another improvement was attained by Kousorov et al. [46] with *C. reinhardtii* mutant *tla1*; it was characterized by enhanced production of H_2 under several light conditions, with higher maximum specific rates of H_2 production recorded at low and medium light intensities (i.e., 19 and 184 μE m^{-2} s^{-1}) than the parental strain. However, upon immobilization in ca. 300 mm thick alginate films and deprived of sulfur and phosphorus, it exhibited 4- and 8.5-fold higher maximum specific rates, at 285 and 350 μE m^{-2} s^{-1}, respectively, under similar cell density—besides a higher conversion of light to H_2 (0.08 ± 0.04%).

Other efforts toward enhancement of microalgal H_2 production have also met with success. For example, based on realization that the maximum solar energy conversion efficiency by *C. reinhardtii* attained in a 20 cm thick flat panel bioreactor was a mere 0.061% [50], and taking advantage of exposure to increasing light levels instead of constant light, both culture density and H_2 release could be increased [48, 50]. Furthermore, hydrogen fractions of ca. 90% could be attained when a mutant of *C. reinhardtii*, known for its efficiency of photon-to-H_2 conversion (i.e., 2% at 20 W m^{-2} irradiance, in the presence of acetate and absence of sulfur), was grown under homogenous LED light distribution inside the reactor at moderate cell densities using a novel self-desulfurization method at the end of the anaerobic stage [44].

Very recently, Hoshino et al. [51] established a new strategy for photosynthetic H_2 production associated with application of spectral-selective PSI activating/PSII deactivating radiation (or PSI light). This procedure drives a steady flow of electrons in the electron transport chain for delivery to hydrogenase toward H_2 production; oxygen production is then reduced via water photolysis below the respiratory oxygen consumption, so anoxic conditions are maintained as required by hydrogenase. This strategy was implemented with LED hight at 692 nm (which is optimal for PSI), and led to a sustained H_2 production of 0.108 mL H_2 mg^{-1} Chl by *C. reinhardtii*—thus exceeding by 0.066 mL H_2 mg^{-1} Chl the amount obtained under white light. This also permitted the microalga cells to alternately switch between the H_2 production and recovery periods by simply turning the PSI light on or off.

ACKNOWLEDGMENTS

This work received partial funding from project *MICROPHYTE* (ref. PTDC/EBB-EBI/102728/2008), under POCI 2010 with support of FSE (III Quadro Comunitário de Apoio), coordinated by author F. X. Malcata. A PhD fellowship (ref. SFRH/BD/62121/2009, also supervised by author F. X. Malcata) was granted to H. M. Amaro, under the auspices of ESF and the Portuguese State.

REFERENCES

1. M. G. Esquível, H. M. Amaro, T. S. Pinto, P. S. Fevereiro, F. X. Malcata, Efficient H_2 production via *Chlamydomonas reinhardtii*, Trends Biotechnol. 29 (2011) 595–600.

2. Y. Chisti, Biodiesel from microalgae, Biotechnol. Adv. 25 (2007) 294–306.

3. H. Gaffron, J. Rubin, Fermentative and photochemical production of hydrogen in algae, J. Gen. Physiol. 26 (1942) 219–240.

4. A. Hemschemeier, A. Melis, T. Happe, Analytical approaches to photobiological hydrogen production in unicellular green algae, Photosynth. Res. 102 (2009) 523–540.

5. M. He, L. Li, J. Liu, Isolation of wild microalgae from natural water bodies for high hydrogen producing strains, Int. J. Hydrogen Energy 37 (2012) 4046–4056.

6. A. C. Guedes, H. M. Amaro, F. X. Malcata, Microalgae as a source of carotenoids, Marine Drugs 9 (2011) 625–644.

7. R. León, M. Martín, J. Vigara, C. Vilchez, J. Veja, Microalgae-mediated photoproduction of β-carotene in aqueous organic two phase systems, Biomol. Eng. 20 (2003) 177–182.

8. S. S. Merchant, S. E. Prochnik, O. Vallon, E. H. Harris, S. J. Karpowicz, G. B. Witman, A. Terry, A. Salamov, L. K. Fritz-Laylin, L. Maréchal-Drouard, W. F. Marshall, L. H. Qu, D. R. Nelson, A. A. Sanderfoot, M. H. Spalding, V. V. Kapitonov, Q. Ren, P. Ferris, E. Lindquist, H. Shapiro, S. M. Lucas, J. Grimwood, J. Schmutz, P. Cardol, H. Cerutti, G. Chanfreau, C. L. Chen, V. Cognat, M. T. Croft, R. Dent, S. Dutcher, E. Fernández, H. Fukuzawa, D. González-Ballester, D. González-Halphen, A. Hallmann, M. Hanikenne, M. Hippler, W. Inwood, K. Jabbari, M. Kalanon, R. Kuras, P. A. Lefebvre, S. D. Lemaire, A. V. Lobanov, M. Lohr, A. Manuell, I. Meier, L. Mets, M. Mittag, T. Mittelmeier, J. V. Moroney, J. Moseley, C. Napoli, A. M. Nedelcu, K. Niyogi, S. V. Novoselov, I. T. Paulsen, G. Pazour, S. Purton, J. P. Ral, D. M. Riaño-Pachón, W. Riekhof, L. Rymarquis, M. Schroda, D. Stern, J. Umen, R. Willows, N. Wilson, S. L. Zimmer, J. Allmer, J. Balk, K. Bisova, C. J. Chen, M. Elias, K. Gendler, C. Hauser, M. R. Lamb, H. Ledford, J. C. Long, J. Minagawa, M. D. Page, J. Pan, W. Pootakham, S. Roje, A. Rose, E. Stahlberg, A. M. Terauchi, P. Yang, S. Ball, C. Bowler, C. L. Dieckmann, V. N. Gladyshev, P. Green, R. Jorgensen, S. Mayfield, B. Mueller-Roeber, S. Rajamani, R. T. Sayre, P. Brokstein, I. Dubchak, D. Goodstein, L. Hornick, Y. W. Huang, J. Jhaveri, Y. Luo, D. Martínez, W. C. Ngau, B. Otillar, A. Poliakov, A. Porter, L. Szajkowski, G. Werner, K. Zhou, I. V. Grigoriev, D. S. Rokhsar, A. R Grossman, The *Chlamydomonas* genome reveals the evolution of key animal and plant functions, Science 318 (2007) 245–250.

9. A. Melis, Photosynthetic H_2 metabolism in *Chlamydomonas reinhardtii*, Planta 226 (2007) 1075–1086.

10. L. Florin, A. Tsokoglou, T. Happe, A novel type of iron hydrogenase in the green alga *Scenedesmus obliquus* is linked to the photosynthetic electron transport chain, J. Biol. Chem. 276 (2001) 6125–6132.

11. A. Melis, L. Zhang, M. Forestier, M. L. Ghirardi, M. Seibert, Sustained photobiological hydrogen gas production upon reversible inactivation of oxygen evolution in the green alga *Chlamydomonas reinhardtii*, Plant Physiol. 122 (2000) 127–136.

12. M. L. Ghirardi, M. C. Posewitz, P.-C. Maness, A. Dubini, J. Yu, M. Seibert, Hydrogenases and hydrogen photoproduction in oxygenic photosynthetic organisms, Annu. Rev. Plant. Biol. 58 (2007) 71–91.

13. A. Hemschemeier, T. Happe, The exceptional photofermentative hydrogen metabolism of the green alga *Chlamydomonas reinhardtii*, Biochem. Soc. Trans. 33 (2005) 39–41.

14. T. K. Antal, T. E. Krendeleva, A. B. Rubin, Acclimation of green algae to sulfur deficiency: underlying mechanisms and application for hydrogen production, Appl. Microbiol. Biotechnol. 89 (2010) 3–15.

15. F. Vallese, P. Berto, M. Ruzzene, L. Cendron, S. Sarno, E. De Rosa, G. M. Giacometti, P. Costantini, Biochemical Analysis of the Interactions between the Proteins Involved in the [FeFe]-Hydrogenase Maturation Process, J. Biol. Chem. 287 (2012) 36544–36555.

16. T. Flynn, M. L. Ghirardi, M. Seibert, Accumulation of O_2-tolerant phenotypes in H_2-producing strains of *Chlamydomonas reinhardtii* by sequential applications of chemical mutagenesis and selection, Int. J. Hydrogen Energy 27 (2002) 1421–1430.

17. M. L. Ghirardi, P. W. King, M. C. Posewitz, P. C. Maness, A. Fedorov, K. Kim, J. Cohen, K. Schulten, M. Seibert, Approaches to developing biological H_2-photoproducing organisms and processes, Biochem. Soc. Trans. 33 (2005) 70–72.

18. H. Long, P. W. King, M. L. Ghirardi, K. Kim, Hydrogenase/ferredoxin charge-transfer complexes: effect of hydrogenase mutations on the complex association, J. Phys. Chem. A 113 (2009) 4060–4067.

19. M. L. Ghirardi, L. Zhang, J. W. Lee, T. Flynn, M. Seibert, E. Greenbaum, A. Melis, Microalgae: a green source of renewable H_2, Trends Biotechnol. 18 (2000) 506–511.

20. R. Wünschiers, M. Batur, P. Lindblad, Presence and expression of hydrogenase specific C-terminal endopeptidases in cyanobacteria, BMC Microbiol. 3 (2003) 8.

21. M. Winkler, A. Hemschemeier, J. Jacobs, S. Stripp, T. Happe, Multiple ferredoxin isoforms in *Chlamydomonas reinhardtii*—their role under stress conditions and biotechnological implications, Eur. J. Cell. Biol. 89 (2010) 998–1004.

22. T. Ruhle, A. Hemschemeier, A. Melis, T. Happe, A novel screening protocol for the isolation of hydrogen producing *Chlamydomonas reinhardtii* strains, BMC Plant Biol. 8 (2008) 107–112.

23. I. Yacoby, S. Pochekailov, H. Toporik, M. L. Ghirardi, P. W. King, S. Zhang, Photosynthetic electron partitioning between [FeFe]-hydrogenase and ferredoxin:NADP β-oxidoreductase (FNR) enzymes *in vitro*, PNAS 108 (2011) 9396–9401.

24. C. M. Agapakis, D. C. Ducat, P. M. Boyle, E. H. Wintermute, J. C. Way, P. A. Silver, Insulation of a synthetic hydrogen metabolism circuit in bacteria, J. Biol. Eng. 4 (2010) 3.

25. R. J. Spreitzer, M. E. Salvucci, Rubisco: structure, regulatory interactions, and possibilities for a better enzyme, Annu. Rev. Plant Biol. 53 (2002) 449–475.

26. J. Marín-Navarro, M. G. Esquível, J. Moreno, Hydrogen production by *Chlamydomonas reinhardtii* revisited: rubisco as a biotechnological target, World J. Microbiol. Biotechnol. 26 (2010) 1785–1793.

27. A. Hemschemeier, S. Fouchard, L. Cournac, G. Peltier, T. Happe, Hydrogen production by *Chlamydomonas reinhardtii*: an elaborate interplay of electron sources and sinks, Planta 227 (2008) 397–407.

28. M. C. Posewitz, A. Dubini, J. E. Meuser, M. Seibert, M. L. Ghiraldi, Hydrogenases, hydrogen production and anoxia. In: D. B. Stern (Ed.), *The Chlamydomonas Sourcebook.* Vol. 2, Academic Press, Oxford, 2009, pp. 217–255.

29. A. Hemschemeier, T. Happe, Alternative photosynthetic electron transport pathways during anaerobiosis in the green alga *Chlamydomonas reinhardtii*, Biochim. Biophys. Acta. 1807 (2011) 919–926.

30. G. Philipps, T. Happe, A. Hemschemeier, Nitrogen deprivation results in photosynthetic hydrogen production in *Chlamydomonas reinhardtii*, Planta 235 (2012) 729–745.

31. H. C. Chen, A. J. Newton, A. Melis, Role of SulP, a nuclear-encoded chloroplast sulfate permease, in sulfate transport and H_2 evolution in *Chlamydomonas reinhardtii*, Photosynth. Res. 84 (2005) 289–296.

32. M. Seibert, P. W. King, M. C. Posewitz, A. Melis, M. L. Ghirardi, Photosynthetic water-splitting for hydrogen production. In: J. Wall (Ed.), *Bioenergy*, ASM Press, Washington DC 2008.

33. O. Kruse, J. Rupprecht, K. P. Bader, S. Thomas-Hall, P. M. Schenk, G. Finazzi, B. Hankamer, Improved photobiological H_2 production in engineered green algal cells, J. Biol. Chem. 280 (2005) 34170–34177.

34. D. Tolleter, B. Ghysels, J. Alric, D. Petroutsos, I. Tolstygina, D. Krawietz, T. Happe, P. Auroy, J. M. Adriano, A. Beyly, S. Cuiné, J. Plet, I. M. Reiter, B. Genty, L. Cournac, M. Hippler, G. Peltier, Control of hydrogen photoproduction by the proton gradient generated by cyclic electron flow in *Chlamydomonas reinhardtii*, Plant Cell 23 (2011) 2619–2630.

35. V. V. Makarova, S. Kosourov, T. E. Krendeleva, B. K. Semin, G. P. Kukarskikh, A. B. Rubin, R. T. Sayre, M. L. Ghirardi, M. Seibert, Photoproduction of hydrogen by sulfur-deprived *C. reinhardtii* mutants with impaired photosystem II photochemical activity, Photosynth. Res. 94 (2007) 79–89.

36. T. K. Antal, T. E. Krendeleva, T. V. Laurinavichene, V. V. Makarova, M. L. Ghirardi, A. B. Rubin, A. A. Tsygankov, M. Seibert, The dependence of algal H_2 production on photosystem II and O_2 consumption activities in sulfur-deprived *Chlamydomonas reinhardtii* cells, Biochim. Biophys. Acta. 1607 (2003) 153–160.

37. M. Castruita, D. Casero, S. J. Karpowicz, J. Kropat, A. Vieler, S. I. Hsieh, W. Yan, S. Cokus, J. A. Loo, C. Benning, M. Pellegrini, S. S. Merchant, Systems biology approach in *Chlamydomonas* reveals connections between copper nutrition and multiple metabolic steps, Plant Cell 23 (2011) 1273–1292.

38. F. Mus, A. Dubini, M. Seibert, M. C. Posewitz, A. R. Grossman, Anaerobic acclimation in *Chlamydomonas reinhardtii*, anoxic gene expression, hydrogenase induction, and metabolic pathways, J. Biol. Chem. 282 (2007) 25475–25486.

39. A. Dubini, F. Mus, M. Seibert, A. R. Grossman, M. C. Posewitz, Flexibility in anaerobic metabolism as revealed in a mutant of *Chlamydomonas reinhardtii* lacking hydrogenase activity, J. Biol. Chem. 284 (2009) 7201–7213.

40. M. Terashima, M. Specht, B. Naumann, M. Hippler, Characterizing the anaerobic response of *Chlamydomonas reinhardtii* by quantitative proteomics, Mol. Cell Proteom. 9 (2010) 1514–1532.

41. A. Doebbe, M. Keck, M. la Russa, J. H. Mussgnug, B. Hankamer, E. Tekçe, K. Niehaus, O. Kruse, The interplay of proton, electron, and metabolite supply for photosynthetic H_2 production in *Chlamydomonas reinhardtii*, J. Biol. Chem. 285 (2010) 30247–30260.

42. R. Radakovits, R. E. Jinkerson, A. Darzins, M. C. Posewitz, Genetic engineering of algae for enhanced biofuel production, Eukaryot. Cell 9 (2010) 486–501.

43. C. N. Dasgupta, J. J. Gilbert, P. Lindblad, T. Heidorn, S. A. Borgvang, K. Skjanes, D. Das, Recent trends on the development of photobiological processes and photobioreactors for the improvement of hydrogen production, Int. J. Hydrogen Energy 35 (2010) 10218–10238.

44. F. Lehr, C. Posten, G. Schaub, O. Kruse. Photobiotechnological hydrogen production with microalgae. In: D. Stolten, T. Grube (Eds.), *18th World Hydrogen Energy Conference 2010—WHEC 2010 Parallel Sessions Book 2: Hydrogen Production Technologies, Part 1.* May 16–21, EssenSchriften des Forschungszentrums, 2010.

45. A. Melis, T. Happe, Hydrogen production. Green algae as a source of energy, Plant Physiol. 127 (2001) 740–748.

46. S. N. Kosourov, M. L. Ghirardi, M. Seibert, A truncated antenna mutant of *Chlamydomonas reinhardtii* can produce more hydrogen than the parental strain, Int. J. Hydrogen Energy 36 (2011) 2044–2048.

47. A. Melis, J. Neidhardt, J. R. Benemann, *Dunaliella salina* (Chlorophyta) with small chlorophyll antenna sizes exhibit higher photosynthetic productivities and photon use efficiencies than normally pigmented cells, J. Appl. Phycol. 10 (1999) 515–525.

48. S. Wahal, S. Viamajala, Maximizing algal growth in batch reactors using sequential change in light intensity, Appl. Biochem. Biotechnol. 161 (2010) 511–522.

49. J. Beckmann, F. Lehr, G. Finazzi, B. Hankamer, C. Posten, L. Wobbe, O. Kruse, Improvement of light to biomass conversion by de-regulation of light-harvesting protein translation in *Chlamydomonas reinhardtii*, J. Biotechnol. 142 (2009) 70–77.

50. H. Berberoğlu, L. Pilon, Maximizing the solar to H_2 energy conversion efficiency of outdoor photobioreactors using mixed cultures, Int. J. Hydrogen Energy 35 (2010) 500–510.

51. T. Hoshino, D. J. Johnson, J. L. Cuello, Design of new strategy for green algal photo-hydrogen production: spectral-selective photosystem I activation and photosystem II deactivation, Biores. Technol. 120 (2012) 233–240.

CHAPTER 9

Algal Biofuels

ARCHANA TIWARI and ANJANA PANDEY

9.1 INTRODUCTION

Environmental pollution and exhaustive depletion of nonrenewable energy sources demand the exploration of alternate energy sources. Scientists are exploring the options for biological fuel production. The choice for the most suitable energy carrier to be produced from algae is a promising option (Fig. 9.1). Algae for biodiesel production have been researched the most [1–9]. Both open and closed land-based cultivation systems appear suitable for this option. The conversion of the extracted lipids to biodiesel is relatively easy, and the product price can easily be compared with fossil fuel prices. Most commercially aimed pilot installations also choose this pathway. Since nutrient limitation is often used as a lipid stimulation strategy, this technology requires strict nutrient input control; therefore using manure or wastewater as a nutrient source may be relatively complicated.

Hydrogen production by algae needs commercial implementation. Although yield improvement options are being investigated, a breakthrough is likely to occur in the next decade. From all energy carriers produced from algae, biodiesel has received the most attention and is the only initiative which is on the border of pilot-scale and full-scale deployment. Using algae for ethanol production and future development deserve attention. The main product of *Botryococcus braunii* resembles compounds from fossil fuel, which offers exciting possibilities [10]. Bio-oils from photosynthetic algae could be used to manufacture a full range of fuels including gasoline, diesel fuel, and jet fuel that meet the same specifications as today's products.

9.2 ADVANTAGES OF ALGAE

- Algae can be grown using land and water unsuitable for plant or food production, unlike some other first- and second-generation biofuel feedstocks.
- Algae require only sunlight, water, and carbon dioxide.

Natural and Artificial Photosynthesis: Solar Power as an Energy Source, First Edition. Edited by Reza Razeghifard.
© 2013 John Wiley & Sons, Inc. Published 2013 by John Wiley & Sons, Inc.

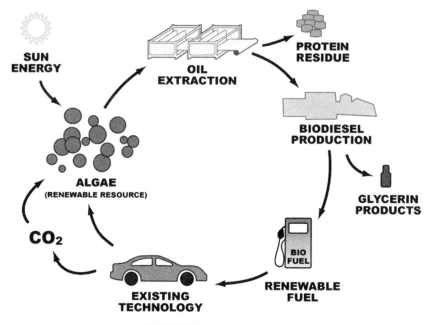

FIGURE 9.1 Algal biofuel.

- Growing algae consume carbon dioxide; this provides greenhouse gas mitigation benefits.
- Bio-oil produced by photosynthetic algae and the resultant biofuel will have molecular structures that are similar to the petroleum and refined products we use today.
- Algae have the potential to yield greater volumes of biofuel per acre of production than other biofuel sources.
- Algae used to produce biofuels are highly productive. As a result, large quantities of algae can be grown quickly, and the process of testing different strains of algae for their fuel-making potential can proceed more rapidly than for other crops with longer life cycles.

Algae are photosynthetic organisms with unique properties. There are many microalgae reported to yield biofuels. Algae are photosynthetic autotrophs and inhabit a variety of habitats. They are tiny biological factories that use photosynthesis to transform carbon dioxide and sunlight into energy so efficiently that they can double their weight several times a day. As part of the photosynthesis process algae produce oil and can generate 15 times more oil per acre than other plants used for biofuels, such as corn and switchgrass. Algae can grow in salt water, fresh water, or even contaminated water, at sea or in ponds, and on land not suitable for food production [11]. The photosynthesis process occurs in almost all algae, and in fact much of what is known about photosynthesis was first discovered by studying the green alga *Chlorella*.

Photosynthesis comprises both light reactions and dark reactions or the Calvin cycle (Fig. 9.2). During the dark reactions, carbon dioxide is bound to ribulose

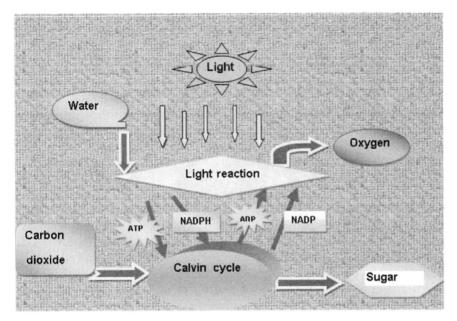

FIGURE 9.2   Photosynthesis in algae.

bisphosphate, a five-carbon sugar with two phosphate groups, by the enzyme ribulose bisphosphate carboxylase/oxygenase. This is the initial step of a complex process leading to the formation of sugars. During the light reactions, light energy is converted into the chemical energy (ATP and NADPH) needed for the dark reactions. Chloroplasts are the sites of photosynthesis where the complex set of biochemical reactions that use the energy of light to convert carbon dioxide and water into sugars take place. Each chloroplast contains flattened, membranous sacs called thylakoids that contain photosystems I and II, cyt b_6f complex, and ATP synthase.

In summary, advantages of algae over plants for biofuel production are outlined below.

Algae	Plants
Algae can grow directly on combustion gas (typically containing 4–15% CO_2).	Plants take up CO_2 from the atmosphere (open air concentration 0.036%).
Algae can contain large concentrations of the desired product.	Desired products are often concentrated in the seeds, as in corn, soy, and rapeseed.
Algae are much more uniform than higher plants. Entire biomass can be processed.	Entire biomass cannot be processed.
Algae growth is dependent on climatic conditions, and thus on the seasons.	Plants have an annual growth cycle of sprouting in the spring and with harvest in autumn.
Algae could yield more than 2000 gallons of fuel per acre per year of production.	Palm: 650 gallons/acre/year
	Sugarcane: 450 gallons/acre/year
	Corn: 250 gallons/acre/year
	Soy: 50 gallons/acre/year

9.3   ALGAL STRAINS AND BIOFUEL PRODUCTION

Algae have been explored for their unique potential to yield a variety of biofuels. Many algal strains have been reported to produce biofuels (Table 9.1). The algal strains to be cultivated would be selected based on many criteria, of which oil content, productivity, and harvestability would be primary, but also resistance to contamination, tolerance of high oxygen levels and temperature extremes, and adaptation to the local water chemistry and other local conditions experienced by the algal cells in the growth ponds.

The potential of algae in the production of biofuel and long-chain polyunsaturated fatty acids (PUFAs) was determined to be promising [13]. Algae are rich in oil and can grow extremely rapidly by doubling their biomass within 24 h; however, in the exponential growth phase, the doubling time is $\sim$3.5 h. Oil content in microalgae can exceed 80% by weight of dry biomass [14,15]. The presence of lipids, hydrocarbons, and other complex oils is dependent on the algal species [10, 16, 17]. The green alga *Chlorella prototothecoides* has shown a high cell density (51.2 g/L) and oil content (>50%, dry basis) in fed-batch heterotrophic culture conditions [18]. The biodiesel made from this heterotrophically grown alga was of good quality [19]. *Botryococcus braunii* produces high amounts of hydrocarbons, especially oils in the form of triterpenes that are typically around 30–40% of their dry weight. The vast majority of these hydrocarbons are botryococcus oils: botryococcenes, alkadienes, and

TABLE 9.1   **Chemical Composition of Selected Microalgae**

Source	Carbohydrates (%)	Proteins (%)	Lipids (%)
Anabaena cylindrica	25–30	43–56	4–7
Chalmydomonas reinhardii	17	48	21
Chlorella vulgaris	12–17	51–58	14–22
Dunaliella salina	32	57	6
Porphyridium cruentum	40–57	28–39	9–14
Spirulina maxima	13–16	60–71	6–7
Chaetoceros muelleri	11–19	44–65	22–44
Chaetoceros calcitrans	10%	58%	30
Isochrysis galbana	7–25	30–45	23–30
Chlorella sp.	38–40	12–18	28–32
Chlorella prototothecoides[a]	10.62–15.43	10.28–52.64	14.57–55.20
Nannochloropsis sp.	n.a.[b]	n.d.[b]	31–68
Neochloris oleoabundans	n.a.	n.d.	35–54
Schizochytrium sp.	n.a.	n.d.	50–77
Scenedesmus obliquus	10–17	50–56	12–14
Scenedesmus quadricauda	—	47	1.9

[a]Values are for two types of reactors used and not the range.
[b]n.a., not available; n.d., not determined.
Source: Satyanarayana et al. [12]. Reproduced with permission of John Wiley & Sons.

alkatrienes [10]. The heterotrophic alga *Schizochytrium limacinum* was investigated for its potential of producing omega-3 polyunsaturated fatty acid (DHA, C22:6, n-3) using biodiesel-derived crude glycerol as a less expensive substrate. In addition to DHA, *S. limacinum* also contains a high level of total fatty acid (nearly 50% of dry biomass), which is ideal for making biodiesel [20].

9.4 ALGAL BIOFUELS

9.4.1 Complete Cell Biomass

Dry algal biomass can be combusted directly to produce heat and electricity, or used in high-temperature high-pressure processes like pyrolysis, gasification, and hydrothermal upgrading (HTU) to produce fuel gas or fuel oil, respectively. These technologies require dry biomass. Drying requires a great deal of energy, which has a strong negative effect on the energy balance and capital costs of required equipment [21] (drying with solar heat would compete for solar light with algae production). Thermochemical liquefaction is a high-temperature, high-pressure treatment in which a wet biomass stream can be applied [10, 22, 23], but this technology is still under development and is likely to require at least five years before it can be applied commercially [24]. A biochemical way to process the whole biomass is anaerobic digestion (Fig. 9.3). This produces biogas from the wet stream and requires much less energy input than the thermochemical options. There is 55–75% methane in biogas [25], which can be combusted to produce heat and/or electricity, or upgraded to replace natural gas.

9.4.2 Lipids

They are one of the main components of microalgae (Fig. 9.4). These lipids can be used as a liquid fuel in adapted engines as *straight vegetable oil* (SVO). Triglycerides and free fatty acids, a fraction of the total lipid content, can be converted into biodiesel. In comparison with SVO, algal oil is unsaturated to a larger degree, making it less appropriate for direct combustion in sensitive engines.

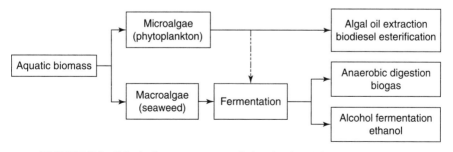

FIGURE 9.3 Principal energy processes being developed for aquatic biomass.

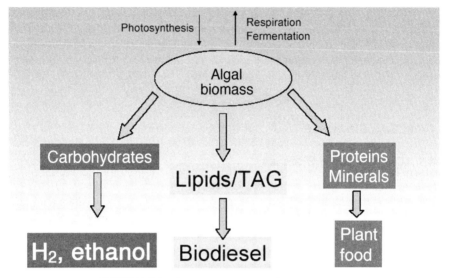

FIGURE 9.4 Complex utilization of algal biomass.

9.4.3 Biodiesel

Biodiesel is a nontoxic and biodegradable alternative fuel derived from renewable sources [26]. Biodiesel provides comparable engine performance to petroleum diesel fuel, while reducing sulfur and particulate matter emissions [27, 28]. During the manufacturing process, triacylglycerols (TAGs) are transesterified with an acid or alkali catalyst to produce biodiesel and glycerol [29, 30, 31]. While biodiesel is an acceptable substitute for petroleum diesel, it is unlikely that its production using terrestrial biomass feedstocks will displace annual demand for petroleum diesel. Chisti [29] noted that 326% of available cropping area in the United States would be required for soy-based biodiesel to meet 50% of the annual demand for transportation fuel. In contrast, it has been estimated that algal biodiesel could satisfy this demand with reduced land use effects [29]. Despite this potential, presently there are no large-scale operations producing algal-derived biodiesel [32].

In the algal biodiesel production processes, fatty acid methyl esters (FAMEs) are commonly prepared by transesterification of algal oil using either acid or alkali as a catalyst [27]. Biodiesel can be produced through direct transesterification of algal biomass [33] or by a two-step process by which lipids are extracted, collected, and transesterified [30]. Either process requires lipid extraction using combinations of solvents and alcohols, such as chloroform/methanol, hexane/ isopropanol, or petroleum ethers and methanol [30, 34]. The direct method is advantageous, because it combines lipid extraction and transesterification into one process, making it less time consuming than extraction–transesterification processes [30]. In order to efficiently

TABLE 9.2   Oil Content of Some Microalgae

Species	Oil Content (% dry weight)
Botryococcus braunii	25–75
Chlorella sp.	28–32
Chlorella emersonii	63
Chlorella minutissima	57
Chlorella prototchecoides	23
Chlorella sorokiniana	22
Chlorella vulgaris	40, 56.6
Cylindrotheca	16–37
Crypthecodinium cohnii	20
Dunaliella primolecta	23
Isochrysis sp.	25–33
Monodus subterraneus	39.3
Monallanthus salina	>20
Nitzschia laevis	69.1
Nannochloris sp.	20–35
Nitzchia sp.	45–47
Parietochloris incisa	62
Phaeodactylum tricornutum	20–30
Schizochytrium sp.	50–77
Tetraselmis sueica	15–23

Source: Satyanarayana et al. [12]. Reproduced with permission of John Wiley & Sons.

produce biodiesel from algae, strains have to be selected with a high growth rate and oil content (Table 9.2).

9.4.4   Advantages of Biodiesel from Algae Oil

Producing biodiesel from algae has been touted as the most efficient way to make biodiesel fuel [35]. The main advantages of deriving biodiesel from algae oil include:

- Algae have rapid growth rates and a high per-acre yield (7 to 31 times greater than the next best crop—palm oil).
- The majority of microalgal strains can be harvested daily and biofuel contains no sulfur.
- Algae biofuel is nontoxic and highly biodegradable.
- Algae consume carbon dioxide as they grow, so they could be used to capture CO_2 from power stations and other industrial plants that would otherwise go into the atmosphere.

9.4.5 Hydrocarbons

Algal species also produce hydrocarbons. Like petroleum, these hydrocarbons can be turned into gasoline, kerosene, and diesel. One species of algae, *Botryococcus braunii*, is well known for its ability to produce hydrocarbons, which have loosely been described as equivalent to the "gas–oil fraction of crude oil" [36]. *Botryococcus braunii* lives in fresh water but can also adapt to a large range of (sea) salt concentrations [37]. At present, the highest known salt concentration that a *Botryococcus* species can survive is 3 M NaCl, the optimum salinity being around 0.2 M NaCl [37], seawater contains about 0.6 M NaCl [38, 39].

While other algal species usually contain less than 1% hydrocarbons, in *B. braunii* they typically make 20–60% of its dry matter, with a reported maximum of >80% [21]. Depending on the strain, these hydrocarbons are either C30 to C37 alkenes or C23 to C33 odd numbered alkenes [38]. These hydrocarbons mainly accumulate on the outside of the cell, making extraction easier than when the cell wall has to be passed to reach the organics inside the cell [21].

9.4.6 Hydrogen

Algae have a unique potential to make hydrogen directly from sunlight and water, although only in the complete absence of oxygen. As an energy carrier, hydrogen offers great promise as a fuel of the future, since it can be applied in mobile applications with only water as exhaust product and no NO_x emissions when used in a fuel cell. One major bottleneck for the full-scale implementation of hydrogen-based technology is the absence of a large-scale sustainable production method for hydrogen. Currently, hydrogen gas is produced by the process of steam reformation of fossil fuels. Large-scale electrolysis of water is also possible, but this production method costs more electricity than can be generated from the hydrogen it yields. Hydrogen can be produced biologically by a variety of means, including the steam reformation of bio-oils [40], dark and photofermentation of organic materials [41], and photolysis of water catalyzed by special microalgal species [41, 42]. Several bacteria can extract hydrogen from carbohydrates in the dark, a group called purple nonsulfur bacteria can use energy from light to extract more hydrogen gas (H_2) from a wider range of substrates, while green sulfur bacteria can make H_2 from H_2S. These options are only interesting if a wastewater with these compounds is available [43]. Other algae can make hydrogen directly from sunlight and water, although only in the complete absence of oxygen. In practice, this means that hydrogen formation is only possible under conditions that either cost a great deal of energy, or prevent storing solar energy [41] and a closed culture system is required. At the moment, it is only possible to produce a fraction of the theoretical maximum of 20 g $H_2/m^2/d$, making bulk-scale hydrogen production by algae not yet viable.

Cyanobacteria also known as blue green algae, are potential microbial species for hydrogen production via direct and indirect photolysis. They are ideal microbes for photobiological H_2 production, since they require the simplest nutritional conditions. Hydrogen production has been studied in a very wide variety of cyanobacterial species. Cyanobacteria use two distinct enzymes to generate hydrogen gas—*nitrogenase* and *hydrogenase*. The nitrogenase enzyme is found in the heterocysts of filamentous cyanobacteria when they grow under nitrogen-limiting conditions, catalyzing the production of hydrogen concomitantly with the reduction of N_2 to NH_3. Hydrogenase has two types, an uptake hydrogenase, catalyzing the consumption of hydrogen produced by the nitrogenase and mutating this uptake hydrogenase is reported to increase yield of hydrogen production [44], and a bidirectional hydrogenase, which has the capacity to both take up and produce hydrogen. There has been a lot of research on cyanobacterial hydrogen production and many statergies have been employed to enhance productivity and early commercialization of this future fuel [45].

More knowledge of the organisms that can produce hydrogen (only a few have been investigated) and the required conditions is necessary, as well as optimization of the biological route of solar energy to hydrogen, through genetic modification. If these improvements prove to be possible, this would constitute a profitable and renewable hydrogen production [46]. While many algal species show different interesting characteristics for biofuel production, there is still plenty of potential for modern biotechnology tools to increase their effectiveness.

9.4.7   Ethanol

Bioethanol can be used as a biofuel, which can replace part of the fossil-derived petrol. Currently, bioethanol is produced by fermenting sugars, which in the case of corn are derived from hydrolyzing starch. Algae species with starch contents over 50% have been reported. With new technologies, cellulose and hemicellulose can be hydrolyzed to sugars [47], creating the possibility of converting an even larger part of algal dry matter to ethanol. Algae have some beneficial characteristics compared to woody biomass, the traditional target for this technology. Most notable is the absence of lignin in algae.

Furthermore, algae composition is generally much more uniform and consistent than biomass from terrestrial plants, because algae lack specific functional parts such as roots and leaves. Algal cell walls are made up largely of polysaccharides, which can be hydrolyzed to sugar. Another algae-specific technology for ethanol production is being developed, in which green algae are genetically modified to produce ethanol from sunlight and CO_2 [48]. Ethanol production from or by algae has very interesting prospects, but is currently only in the preliminary phase of research. More development is needed to analyze a full-scale production system.

9.4.8   Unique Products

There are many initiatives on energy generation from algae; some ideas have been around for many decades, some are currently at the pilot stage, but so far there is no commercial implementation. Alga culture, however, is performed worldwide to produce products with a higher economical value than energy. Sometimes the entire alga is the product, but often compounds are extracted which are very difficult or impossible to produce in other ways. Some examples of these "unique products" include food, food additives, and health food, feed for fish, shrimp, and shellfish, colorants, and omega-3 fatty acids [49, 50].

9.5 ALGAL CULTIVATION FOR BIOFUEL PRODUCTION

Optimum yield of algal biofuels requires a number of physical and chemical factors. Growing algae for the purpose of energy production requires analysis of these streams to be able to choose the most economical and environmentally friendly options.

9.5.1 Carbon Dioxide Capture

Algae use CO_2 as a carbon source like all photosynthetic organisms. No growth can occur in the absence of CO_2, and an insufficient supply of CO_2 is often the limiting factor in productivity. Based on the average chemical composition of algal biomass, approximately 1.8 tonnes of CO_2 is needed to grow 1 tonne of biomass. Natural dissolution of CO_2 from the air into the water is not enough. This could be improved by bubbling air through the water, but, since air contains about 0.0383% CO_2, all of the CO_2 in about 37,000 m^3 air is needed for 1 tonne of dry algae. Other options are using pure CO_2, which is rather expensive, or a waste source of CO_2 like flue gas from a boiler [51]. Flue gas typically contains about 4–15% of CO_2. After studying the culture of algae with flue gas, it was reported that the SO_2 and NO_x in the flue gas inhibited algal growth, but a few years later it was reported that there was barely any difference with growth on pure CO_2 [52, 53]. More recently, others have also confirmed that flue gas can be used to grow algae without harmful effects [54, 55, 56] and at least one commercial algae cultivator on Hawaii is using CO_2 from a small power plant [57]. In addition, dissolved NO_x can be used by algae as a nitrogen source. The amount of flue gas needed per hectare will differ per species of algae and will also vary throughout the day with light intensity and temperature, and thus needs to be optimized for each specific application. High dissolved concentrations of CO_2 (and also SO_2) will affect the pH, which thus needs to be controlled or buffered. Solubility of CO_2 in a salt water system (higher pH) is higher than when sweet water is used. Also, extra CO_2 can be sequestered by growing algae that produce hard scales around their cells, made of calcium carbonate ($CaCO_3$) [58].

9.5.2 Light

Algae rely on (sun) light for photosynthesis and thus growth. Light gets absorbed by the algae, so the higher the algae concentration, the less deeply light enters into the algae broth. Therefore all alga culture systems are shallow and optimized to catch as much light as possible. Light is available in different quantities in differ- ent geographical locations. Only a part (about 45%) of the total light spectrum is photosynthetically active radiation (PAR, ~400–700 nm), and thus can be used by algae to capture CO_2 during photosynthesis, a process with a maximum efficiency of 27%. Multiplying these two factors gives the maximum theoretical conversion of light energy to chemical energy by photosynthesis: about 11% [59]. At night (or other dark conditions) photosynthesis cannot occur, so algae consume stored energy for respiration. Depending on the temperature and other conditions, up to 25% of the biomass produced during the day may be lost again at night [29]. Algae have evolved under conditions where light is often limiting, therefore they harvest as much of the available light as they can, but under good light conditions, this characteristic makes algae waste up to 60% of the absorbed irradiance as heat [46].

9.5.3 Nutrient Removal

Besides CO_2 and light, algae require nutrients to grow, nitrogen (N) and phosphorus (P) being the most important ones. These can be supplied in the form of agricultural fertilizer, which is simple and easily available but can be a significant cost factor [60, 61]. There are several options for cheaper sources of these nutrients. Aresta et al. [62] mention wastewater effluent, for example, from a fishery, and Olguín *et al.* [63] describe a system where 84–96% of nitrogen and 72–87% of phosphorus was removed from the anaerobic effluent of piggery wastewater by growing algae, thereby reducing eutrophication in the environment. Another option is nutrient recycling within the process, depending on the treatment technology chosen like nutrient recycling after anaerobic digestion [60] or gasification [64]. The combination of nutrients and light will cause organisms capable of photosynthesis to produce chlorophyll in any part of the world. Because of its reflection of green light, chlorophyll concentrations can be measured by satellite, which is a good indication of the best geographic locations for (unfertilized) algae growth.

9.5.4 Temperature

Temperature has profound effects on algal growth. In temperate and subtropical regions, algae have a growth season (e.g., in The Netherlands from April to November); in the winter season algae only grow at a fraction of the summertime growth rate. In many industrial processes, heat is produced. Sometimes this heat is used somewhere else in the process, sold to neighboring industries, or used to heat neighboring houses or other buildings. But often this surplus heat has no further use. This heat could be employed to optimize the temperature of the medium in which the algae grow. Especially interesting is the combination with CO_2 of power plants, since power plants are important sources of this surplus heat. On the other hand, closed systems can get too hot and require cooling, which can be done with heat exchangers [29] or spraying water on the outside [65].

9.5.5 Biomass Harvesting

Harvesting of the biomass is required by every system. The most commonly used method is a specially developed harvesting vessel, which cuts the seaweed and hauls it inside [66]. Microalgae concentrations always remain very low while growing, typically 0.02–0.05% dry matter in raceways and between 0.1% and 0.5% dry matter in tubular reactors [67]. The size of algae is only a few micrometers. These two aspects make the harvesting and further concentration of algae difficult and therefore expensive. Harvesting has been claimed to contribute 20–30% to the total cost of producing the biomass [49]. In order to produce energy from algae as economically as possible, the cheapest way of concentrating the algal biomass to a water content that is low enough for oil pressing is essential. The technically simplest option is the use of settling ponds. Another way of separating algae from the water is filtration. Many options have been described, including different materials, vacuum, and pressured and rotating filtering. Some acceptable results have been obtained for colonial microalgae, but not for unicellular species [49, 68]. Furthermore, filtration is a slow process [67], so a very large total capacity system would be required to keep up with the production of a large algae farm. Centrifugation is the best known method of concentrating small unicellular algae [49] and is cost effective [66]. To acquire a dry matter content of above 30%, some form of thermal technology is needed [60]. Algae produce oils as a storage for energy. Thus when removed from their reactor, the algae may consume part of the stored energy for maintenance.

9.6   PHOTOBIOREACTORS EMPLOYED FOR ALGAL BIOFUELS

The PBR systems are used to cultivate algae in bags, tubing, or other transparent materials that are not exposed directly to the atmosphere [69, 70]. Because algae are grown in a closed system, there are fewer concerns regarding species contamination and evaporative losses when compared with open pond systems [71, 72]. The concentration of algal biomass grown in PBRs typically is greater than open pond systems, because algae are provided maximum exposure to sunlight [1]. Similar to raceway ponds, unless wastewater is used, PBR systems will require the addition of fertilizers to prepare the nutrient broth.

9.6.1   Tubular Photobioreactors

These are commonly used to cultivate algae, because they have large surface-to-volume ratios and can be arranged in a number of different configurations [70]. These options include helical tubular PBRs, fence-like PBRs, and horizontal PBRs. Each of these systems must maintain a minimum flow rate to prevent sedimentation of algal biomass within the tubing [1]. However, it was noted that dense algal cultures and associated photosynthetic activity generate dissolved oxygen levels within the system that are well beyond saturation levels [29]. Excessive concentrations of dissolved oxygen will inhibit photosynthesis, leading to the photooxidative destruction of algal cells [1]. Because the dissolved oxygen cannot be removed inside the tubes, it is recommended that degassing vessels be used to prevent excessive accumulation of photosynthetically generated oxygen [29, 73]. Furthermore, external cooling or heat exchangers are needed to prevent excessive temperatures within the system.

9.6.2 Flat Panel Photobioreactors

These are thin rectangular containers composed of transparent materials [70]. These units often are configured with tilt angles necessary for maximum solar exposure [74]. This allows flat panel PBRs to have their entire surface illuminated by direct beam radiation, whereas various portions of horizontal reactors are shaded partially throughout the day [74]. Despite the advantage of increased sunlight exposure, Shen et al. [70] noted that flat panel and tubular PBRs generally yield similar algal productivity.

9.6.3 Offshore Membrane Enclosure for Growing Algae (OMEGA)

This system was developed by the National Aeronautics and Space Administration (NASA) (Washington DC). It is a novel PBR system consisting of semipermeable bags designed to float in the ocean. The offshore placement of the OMEGA system eliminates competition for terrestrial resources and may enhance marine ecosystems by simulating a reef-like environment. The OMEGA process cultivates freshwater algae using wastewater effluent as the growth medium [75]. This system is advantageous because mixing, cooling, and structural support are provided by the ocean, rather than with expensive mechanical components that are typical of conventional PBRs [75]. The osmotic gradient between the wastewater and marine environment enables the growth media to diffuse through the semipermeable OMEGA material, leaving concentrated algal biomass inside the bag [75]. The PBR material used for this system is also gas permeable, eliminating the need for degassing vessels. After an incubation period of 10–20 days, the algal biomass is harvested and processed into fuel [75].

9.7 RECENT ACHIEVEMENTS IN ALGAL BIOFUELS

There has been exploration regarding alternate fuels for civilization around the globe. Scientific interest in producing fuel from algae has been aroused since the 1950s. The U.S. Department of Energy did pioneering research on it from 1978 to 1996. A number of companies are attempting to take advantage of the fact that algae naturally produce oil. But growing algae and extracting their oil efficiently is difficult, time consuming, and expensive. While some companies are focusing on better growing and harvesting methods, others, such as Origin Oil, are focused on finding new ways to access the oil.

Origin Oil, an algae biofuel company based in Los Angeles, has developed a simpler and more efficient way to extract oil from algae. The process combines ultrasound and an electromagnetic pulse to break the algal cell walls. Then the algae solution is force-fed carbon dioxide, which lowers its pH, separating the biomass from the oil. Each algal cell has a sturdy cell wall protecting it, making the oil hard to get at. The algae also have to be separated from the water that they are grown in and dried out before the oil can be removed. Typically, the oil is expelled from algae by using a press to physically squeeze it out. The leftover mashed up pulp is then treated with a solvent to remove any remaining oil. While the combination removes about 95% of the oil, it is energy intensive. Another method does away with the press and treats the algae pulp with supercritical fluids that can remove nearly all the oil, but the process requires special machinery, adding to the expense. Other researchers are genetically engineering algae that secrete oil. In Origin Oil's process, the algae solution is channeled through a pipe to which an electromagnetic field and ultrasound are applied, rupturing the cell walls and releasing the oil. Carbon dioxide is bubbled through, which lowers the pH. The resulting solution is then piped into another container. The lowered pH separates the biomass from the oil, and the oil floats to the top, while the biomass sinks to the bottom. The oil can be skimmed off, the biomass can be further processed, and the water is recycled. The whole process takes a matter of minutes as reported [76].

Solazyme, a startup based in South San Francisco, California, has developed a new way to convert biomass into fuel using algae, and the method could lead to less expensive biofuels. The company recently demonstrated its algae-based fuel in a diesel car, and in January 2013, it announced a development and testing agreement with Chevron. Late last year, the company received a $2 million grant from the National Institute of Standards and Technology to develop a substitute for crude oil based on algae. The new process combines genetically modified strains of algae with an uncommon approach to growing algae to reduce the cost of making fuel. Rather than growing algae in ponds or enclosed in plastic tubes that are exposed to the sun, as other companies are trying to do, Solazyme grows the organisms in the dark, inside huge stainless-steel containers. The company's researchers feed algae sugar, which the organisms then convert into various types of oil. The oil can be extracted and further processed to make a range of fuels, including diesel and jet fuel, as well as other products. The company uses different strains of algae to produce different types of oil. Some algae produce triglycerides such as those produced by soybeans and other oil-rich crops. Others produce a mix of hydrocarbons similar to light crude petroleum. Solazyme's method has advantages over other approaches that use microorganisms to convert sugars into fuel. The most common approaches use microorganisms such as yeast to ferment sugars, forming ethanol. The oils made by Solazyme's algae can then be used for a wider range of products than ethanol.

The algae have a particular advantage over many other microorganisms when it comes to processing sugars from cellulosic sources, such as grass and wood chips. Such cellulosic sources require less energy, land, and water to grow than corn grain, the primary source of biofuel in the United States. But when biomass is broken down into sugars, it still contains substances such as lignin that can poison other microorganisms. In most other processes, lignin has to be separated from the sugars to keep the microorganisms healthy. But the tolerance of the algae to lignin makes it possible to skip this step, which can reduce costs. The process also has significant advantages over a quite different way of using algae to create biofuels–one that makes use of algae's ability to employ sunlight to produce their own supply of sugar, using photosynthesis. In these approaches, the algae are grown in ponds or bioreactors where they are exposed to sunlight and make their own sugar. In Solazyme's approach, the researchers deliberately turn off photosynthetic processes by keeping the algae in the dark. Instead of getting energy from sunlight, the algae get energy from the sugars that the researchers feed them.

Enhanced Biofuels & Technologies [77] developed multiple vegetable oil biofuel technologies. The EBT algae process combines a bioreactor with an open pond, both using waste CO_2 from coal-fired power plant flue gases as a fertilizer for the algae. The biodiesel and ethanol produced can be sold, or used as an alternative fuel. Emissions are reduced up to 82%. EBTs headquarters are in London, U.K. and the company has a biofuel R&D center in India.

Green Fuel Technologies' [78] *Emissions-to-Biofuels*^{TM} process harnesses photosynthesis to grow algae, capture CO_2, and produce high-energy biomass. Retrofitting fossil-fired power plants and other anthropogenic sources of carbon dioxide, the algae can be converted economically to solid fuel, methane, or liquid transportation fuels such as biodiesel and ethanol. In December 2005, Green Fuel Technologies secured $11 million in the second round of financing from Draper Fisher Jurvetson and Access Private Equity LLC. In April 2006 the company raised a further $7 million from Polaris Venture Partners.

GreenShift [79] has a license agreement with Ohio University for its patented bioreactor process based on a newly discovered iron-loving cyanobacterium (blue green algae), through their subsidiary Veridium, for the purpose of air pollution control of exhaust gas streams from electrical utility fossil-fueled power generation facilities. Once the algae grow to maturity, they fall to the bottom of the bioreactor and are harvested for fuel or fertilizer. GreenShift's shares are traded on the OTC-BB in New York. The company owns majority stakes in the following subsidiary companies—GS CleanTech Corporation, GS AgriFuels Corporation, GS EnviroServices, GS Carbon Corporation, and GS Energy Corporation.

LiveFuels [80] is a national alliance of labs and scientists dedicated to transforming algae into biocrude. It is working on breeding various strains of algae, driving down the costs of harvesting algae, and extracting fats and oils from the algae. In May 2007 LiveFuels announced a first round of $10 million investment, led by David Gelbaum of the Quercus Trust—a major donor to conservation advocacy and environmental organizations.

Valcent Products [81] has developed a high density vertical bioreactor for the mass production of oil-bearing algae while removing large quantities of CO_2 from the atmosphere. This new bioreactor is tailored to grow a species of algae that yields a large volume of high-grade vegetable oil, which is very suitable for blending with diesel to create a biodiesel fuel.

Aquaflow Bionomics Corporation [82] is New Zealand based and has set itself the objective to be the first company in the world to economically produce biofuel from wild algae harvested from open-air environments and to market it.

Infinifuel Biodiesel [83], located in Wabuska, Nevada, is home to a unique biodiesel project under development and is being touted as the world's first geothermal-powered and heated biodiesel plant. The existing geothermal power plant features two production wells and seven power production units creating more than 5 MW of electricity, according to Infinifuel. The power plant will provide 2 MW of electricity and 104 °C (220 °F) steam to the biodiesel facility, which is nearing completion. The company has over 300 acres to grow oil-seed and develop algae ponds on site.

Solix Biofuels [84] is a developer of massively scalable photobioreactors for the production of biodiesel and other valuable biocommodities from algae oil. Solix's closed photobioreactors allow fossil-fuel power plant exhaust to be captured through the growing system. The algae growth rates increase in the presence of the carbon dioxide that would otherwise be emitted into the atmosphere. Solix Biofuels has solid backing from a local private investor and says it plans to develop its technology as far as it can on its own before seeking venture capital. Solix believes it can build a system that's competitive on a small commercial scale with between $5 million and $15 million.

Algoil [85] is a pioneer project focusing on the production of biodiesel/biomass from microalgae. The target is to use the rest of the extracted biomass to make food, biofuel, hydrogen, and paper, or simply burning it like charcoal.

PetroAlgae [86] is commercializing environmentally friendly algae developed by a research team at Arizona State University that generate over 200 times more oil per acre than crops like soybeans. Using a cost-effective, modular cultivation process that can be massively scaled, PetroAlgae will produce renewable feedstock oils for use in applications such as transportation fuels, heating oil, and plastics. PetroAlgae is a 95%-owned subsidiary of XL TechGroup, a conglomerate diversified in biotechnologies, life sciences, and environmental technologies. XL TechGroup has been listed since October 2006 on London's AIM market and secured a new mezzanine borrowing facility of $20 million in January 2007.

Aurora BioFuels [87] is a California-based renewable energy company exploring new sources of feedstock for the production of biofuels. In particular, Aurora focuses on utilizing microalgae to generate bio-oil, which can be converted into biodiesel.

Many universities in different countries are working on different algae-based biofuels. A University of Virginia research team has recently been funded by a new University of Virginia. Collaborative Sustainable Energy Seed Grant worth about $30,000 to determine exactly how promising algae biofuel production can be by tweaking the inputs of carbon dioxide and organic matter to increase algae oil yields.

The U.S. Department of Energy did pioneering research on algal biofuels from 1978 to 1996. Most previous and current research on algae biofuel has used the algae in a manner similar to its natural state—essentially letting it grow in water with just the naturally occurring inputs of atmospheric carbon dioxide and sunlight. This approach results in a rather low yield of oil—about 1% by weight of the algae.

9.8   STRATEGIES FOR ENHANCEMENT OF ALGAL BIOFUEL PRODUCTION

9.8.1   Biorefinery: The High-Value Coproduct Strategy

Biorefinery is described as the production of a wide range of chemicals and biofuels from biomasses by the integration of bioprocessing and appropriate low environmental impact chemical technologies in a cost-effective and environmentally sustainable manner [3, 29]. Examples include the two-phase conversion reaction of fructose to 5-hydroxymethylfurfural [88], fermentative production of ethanol from sugars derived from cellulose and semicellulose [89] and bio oils and/or biosyngas by the pyrolysis/gasification of woods or other types of biomasses [90].

The economical feasibility of microalgal biofuel production should be significantly enhanced by a high-value coproduct strategy, which would, conceptually, involve sequentially the cultivation of microalgae in a microalgal farming facility (CO_2 mitigation), extracting bioreactive products from harvested algal biomass, thermal processing (pyrolysis, liquefaction, or gasification), extracting high-value chemicals from the resulting liquid, vapor, and/or solid phases, and reforming/upgrading biofuels for different applications. The employment of a high-value coproduct strategy through the integrated biorefinery approach is expected to significantly enhance the overall cost effectiveness of microalgal biofuel production [3].

9.8.2   Exploration of Growth Conditions and Nutrients

Feeding the algae more carbon dioxide and organic material could boost the oil yield to as much as 40% by weight as hypothesized by the University of Virginia team. Proving that the algae can thrive with increased inputs of either carbon dioxide or untreated sewage solids will confirm its industrial ecology possibilities—to help with wastewater treatment, where dealing with solids is one of the most expensive challenges, or to reduce emissions of carbon dioxide, such as coal power-plant flue gas, which contains about 10–30 times as much carbon dioxide as normal air [11].

The effect of iron on the growth and lipid accumulation in the marine microalga *Chlorella vulgaris* resulted in an increase in cell density but did not induce lipid accumulation in the late growth phase. The total lipid content in cultures of late-exponential growth phase when supplemented with 1.2×10^{-5} mol/L $FeCl_3$ resulted in an increase in biomass by dry weight (56.6%), which was three-fold to seven-fold higher than that in F/2-Si medium supplemented with a lower iron concentration [5].

Schizochytrium limacinum, when grown on glycerol as a substrate, resulted in a docosahexaenoic acid (DHA, 22:6 n-3) with a yield of 75–100 g/L [91]. The *S. limacinum* algal biomass was comprised of 45–50% lipid, 14–20% protein, 25% carbohydrate, and 8–13% ash and was rich in lysine and cysteine, while palmitic acid (C16:0) and DHA were the major components in the biodiesel [20].

9.8.3   Design of Advanced Photobioreactors

The choice of cultivation system is another key aspect that significantly affects the efficiency and cost effectiveness of a microalgal biofuel production process. There have been reports on both open and closed systems [29, 92, 93, 94]. Lee [92] has discussed a few open systems and systematically compared them with closed systems over different geographical regions. Pulz [93] focused more on process parameters and suggested a number of open systems. Useful conceptual diagrams for some of the discussed closed systems and described new systems to be examined, including the use of optical fiber to enhance lighting, were offered [95]. Even though the open pond systems seem to be favored for commercial cultivation of microalgae at present due to their low capital costs, closed systems offer better control over contamination, mass transfer, and other cultivation conditions. The combination of the closed photobioreactor and open pond combines the benefits of the two and has been demonstrated to be effective at a 2-ha scale [96].

9.8.4 Biotechnological Tools

While many algal species show different interesting characteristics for biofuel production, there is still plenty of potential for modern biotechnology tools to increase their effectiveness. Two possible improvements have already received significant attention in research.

- *Reduction of Antenna Size*. Under natural conditions, algae have to compete for light with surrounding photosynthetic organisms. Therefore the so-called antenna part of chlorophyll, which receives light, has evolved to catch more light than can be photosynthetically processed under optimal lighting conditions, wasting up to 60% of the received light energy [46]. Scientists have been able to reduce the chlorophyll antenna size, resulting in more efficient use of high intensity light [97].
- *Triggering of Lipid Production*. Certain algae species accumulate large amounts of intracellular lipids, usually after being exposed to some form of stress. The exact biochemical pathway has received significant attention, in order to find out what triggers this accumulation. Genetic tools could exploit this trigger to modify algal species to produce more lipids and throughout the whole growth cycle [9]. No significant progress has been reported in this area.

9.8.5 Cost-Effective Technologies for Biomass Harvesting and Drying

Selection of cost-effective techniques are essential for commercialization of biofuels. The selection of appropriate harvesting technology depends on the value of the target products, the biomass concentration, and the size of microalgal cells of interest. Biomass drying before further lipid/bioproduct extraction and/or thermochemical processing is another step that needs to be taken into consideration. Sun drying is probably the cheapest drying method that has been employed for the processing of microalgal biomass [98, 99]. But this method requires a long drying time and a large drying surface, and risks the loss of some bioreactive products. Low-pressure shelf drying is another low-cost drying technology that has been investigated [99]. It is nevertheless also of low efficiency. More efficient but more costly technologies for drying microalgae include drum drying [97], spray drying [100, 101], fluidized bed drying [101], freeze drying [98], and refractance window dehydration technology [102]. The balance between the drying efficiency and cost effectiveness to maximize the net energy output of the fuels from microalgae strategy is very significant.

9.9 CONCLUSION

Algae technology has enormous potential, not only for algae- based biofuels, but also for food, feed, renewable chemicals, and many other products that are critical for a more sustainable society. Microalgal farming can be coupled with flue gas CO_2 mitigation and wastewater treatment. It can also be carried out with seawater as the medium in which marine microalgal species are adapted, providing a feasible alternative for biofuel production to populous and dry coastal regions. Microalgae can produce a large variety of novel bioproducts with wide applications in medicine, food, and cosmetic industries. Combining microalgal farming and the production of biofuels using biorefinery strategy is expected to significantly enhance the overall cost effectiveness of biofuel from the microalgae approach.

To prove the viability of algae concepts, more information is needed on the economics of the process: optimized costs of the different inputs, and also the market value and market size of the outputs, not only for fuels but also higher-value compounds. Especially in developing countries, if economical viability and robustness can be proved, many projects can be deployed rapidly, through microcredits or similar measures. Economic sustainability as well as environmental safety are vital. Adequate attention and support should be given to both basic and applied research on algae for biofuel applications and the engineering of sustainable microalgal systems. Technoeconomic analysis indicates that algal productivity is the primary production cost determinant and so efforts should be focused on various aspects of algal biology that can have the greatest impact on growth rate and lipid biosynthesis. This work cannot be done in isolation, and it would be a mistake to equate progress in productivity made at the bench scale with success in large-scale cultivation. Hence attention must be paid to growth under conditions that model commercial production (including climate and input sources), with data exchanged between biologists and process engineers.

Technological developments, including advances in photobioreactor design, microalgal biomass harvesting, drying, and other downstream processing technologies, are important areas that may lead to enhanced cost effectiveness and therefore effective commercial implementation of the biofuel from microalgae strategy. Improvisation in biofuels mediated by algae will pave the path for commercialization of these fuels as environment friendly energy sources.

REFERENCES

1. Y. Chisti, Biodiesel from microalgae beats bioethanol, Trends Biotechnol. 26 (2008) 126–131.
2. S. Y. Chiu, C. Y. Kao, M. T. Tsai, S. C. Ong, C. H. Chen, C. S. Lin, Lipid accumulation and CO_2 utilization of *Nannochloropsis oculata* in response to CO_2 aeration, Bioresour. Technol. 100 (2009) 833–838.
3. Y. Li, M. Horsman, N. Wu, C. Q. Lan, N. Dubois-Calero, Biofuels from microalgae, Biotechnol. Prog. 24 (2008) 815–820.

4. Y. Q. Li, M. Horsman, B. Wang, N. Wu, C. Q. Lan, Effects of nitrogen sources on cell growth and lipid accumulation of green alga *Neochloris oleoabundans*, Appl. Microbiol. Biotechnol. 81 (2008) 629–636.

5. Z. Y. Liu, G. C. Wang, B. C. Zhou, Effect of iron on growth and lipid accumulation in *Chlorella vulgaris*, Bioresour. Technol. 99 (2008) 4717–4722.

6. X. Meng, J. M. Yang, X. Xu, L. Zhang, Q. J. Nie, M. Xian, Biodiesel production from oleaginous microorganisms, Renew. Energy 34 (2009) 1–5.

7. W. Mulbry, S. Kondrad, J. Buyer, Treatment of dairy and swine manure effluents using freshwater algae: fatty acid content and composition of algal biomass at different manure loading rates, J. Appl. Phycology 20 (2008) 1079–1085.

8. N. Sazdanoff, *Modeling and Simulation of the Algae to Biodiesel Fuel Cycle*, College of Engineering, Department of Mechanical Engineering, The Ohio State University, 2006.

9. J. Sheehan, T. Dunahay, J. Benemann, P. Roessler, *Look Back at the U.S. Department of Energy's Aquatic Species Program: Biodiesel from Algae*, 1998, Close-Out Report.

10. A. Banerjee, R. Sharma, Y. Chisti, U. C. Banerjee, *Botryococcus braunii*: a renewable source of hydrocarbons and other chemicals, Crit. Rev. Biotechnol. 22 (2002) 245–279.

11. www.sciencedaily.com.

12. K. G. Satyanarayana, A. B. Mariano, J. V. C. Vargas, A review on microalgae, a versatile source for sustainable energy and materials, Int. J. Energy Res. 35 (2011) 291–311.

13. J. L. Harwood, I. A. Guschina, The versatility of algae and their lipid metabolism, Biochimie 91 (2009) 679–684.

14. F. B. Metting, Biodiversity and application of microalgae, J. Ind. Microbiol.17, (1996) 477–489.

15. P. Spolaore, C. Joannis-Cassan, E. Duran, A. Isambert, Commercial applications of microalgae, J. Biosci. Bioeng. 101 (2006) 87–96.

16. P. Metzger, C. Largeau, *Botryococcus braunii*: a rich source for hydrocarbons and related ether lipids, Appl. Microbiol. Biotechnol. 66 (2005) 486–496.

17. I. A. Guschina, J. L. Harwood, Lipids and lipid metabolism in eukaryotic algae, Prog. Lipid Res. 45 (2006) 160–186.

18. W. Xiong, X. F. Li, J. Y. Xiang, Q. Y. Wu, High-density fermentation of microalga *Chlorella protothecoides* in bioreactor for microbiodiesel production, Appl. Microbiol. Biotechnol. 78 (2008) 29–36.

19. H. Xu, X. L. Miao, Q. Y. Wu, High quality biodiesel production from a microalga *Chlorella protothecoides* by heterotrophic growth in fermenters, J. Biotechnol. 126 (2006) 499–507.

20. D. J. Pyle, R. A. Garcia, Z. Wen, Producing docosahexaenoic acid (DHA)-rich algae from biodiesel-derived crude glycerol: effects of impurities on DHA production and algal biomass composition, J. Agric. Food Chem. 56 (2008) 3933–3939.

21. R. Wijffels, Presentation, Microalgae for production of energy, 2007.

22. Y. Dote, S. Sawayama, S. Inoue, T. Minowa, S. Yokoyama, Recovery of liquid fuel from hydrocarbon-rich microalgae by thermochemical liquefaction, Fuel 73 (1994) 1855–1857.

23. K. Tsukahara, S. Sawayama, Liquid fuel production using microalgae, J. Jpn. Petrol. Inst. 48 (2005) 251–259.

24. B. Meuleman, Personal communication on biomass conversion technology, 2007.

25. T. Z. D. Mes, A. J. M. Stams, J. H. Reith, G. Zeeman, Methane production by anaerobic digestion of wastewater and solid wastes. In: J. H. Reith, R. H. Wijffels, H. Barter (Eds.), *Biomethane & Bio-hydrogen: Status and Perspectives of Biological Methane and Hydrogen Production*, Dutch Biological Hydrogen Foundation, Petten, The Netherlands, 2003, pp. 58–102.

26. A. Hossain, A. Salleh, A. Boyce, P. Chowdhury, M. Naqiuddin, Biodiesel fuel production from algae as renewable energy, Am. J. Biochem. Biotechnol. 4 (2008) 250–254.

27. X. L. Miao, Q. Y. Wu, Biodiesel production from heterotrophic microalgal oil, Bioresour. Technol. 97 (2006) 841–846.

28. A. Scragg, A. Illman, A. Carden, S. Shales, Growth of microalgae with increased calorific values in a tubular bioreactor, Biomass Bioenergy, 23 (2002) 67–73.

29. Y. Chisti, Biodiesel from microalgae, Biotechnol. Adv. 25 (2007) 294–306.

30. M. Johnson, Z. Wen, Production of biodiesel fuel from the microalga *Schizochytrium limacinum* by direct transesterification of algal biomass, Energy Fuels 23 (2009) 294–306.

31. J. Van Gerpen, Biodiesel processing and production, Fuel Process. Technol. 86 (2005) 1097–1107.

32. L. Lardon, A. Hélias, B. Sialve, J. Steyer, O. Bernard, Life-cycle assessment of biodiesel production from microalgae, Environ. Sci. Technol. 43 (2009) 6475–6481.

33. T. Lewis, P. D. Nichols, T. A. McMeekin, Evaluation of extraction methods for recovery of fatty acids from lipid-producing micro heterotrophs, J. Microbiol. Methods 43 (2000) 107–116.

34. W. Mulbry, S. Kondrad, J. Buyer, D. Luthria, Optimization of an oil extraction process for algae from the treatment of manure effluent, J. Am. Oil Chem. Soc. 86 (2009) 909–915.

35. L. Wagner, *Biodiesel from Algal Oil*, Research report, 2007.

36. L. W. Hillen, G. Pollard, L. V. Wake, N. White, Hydrocracking of the oils of *Botryococcus braunii* to transport fuels, Biotechnol. Bioeng. 24 (1982) 193–205.

37. J. Qin, Bio-hydrocarbons from algae: impacts of temperature, light and salinity on algae growth, Barton, Rirdc, 2005.

38. A. G. Dickson, C. Goyet, *Handbook of Methods for the Analysis of the Various Parameters of the Carbon Dioxide System in Sea Water*. Version 2; 1994.

39. A. Ranga Rao, G. A. Ravishankar, Influence of CO_2 on growth and hydrocarbon production in *Botryococcus braunii*, J. Microbiol. Biotechnol. 17 (2007) 414–419.

40. Z. Wang, Y. Pan, T. Dong, X. Zhu, T. Kan, L. Yuan, Y. Torimoto, M. Sadakata, Q. Li, Production of hydrogen from catalytic steam reforming of bio- oil using C12A7-O-based catalysts, Appl. Catal. A 320 (2007) 24–34.

41. I. K. Kapdan, F. Kargi, Bio-hydrogen production from waste materials, Enzyme Microb Technol. 38 (2006) 569–582.

42. C. Q. Ran, Z. A. Chen, W. Zhang, X. J. Yu, M. F. Jin, Characterization of photo- biological hydrogen production by several marine green algae, *Wuhan Ligong Daxue Xuebao* 28 (2006) 258–263.

43. J. Rupprecht, B. Hankamer, J. H. Mussgnug, G. Ananyev, C. Dismukes, O. Kruse, Perspectives and advances of biological H_2 production in microorganisms, Appl. Microbiol. Biotechnol. 72 (2006) 442–449.

44. A. Pandey, A. Pandey, P. Srivastava, A. Pandey, Using reverse micelle as microreactor for hydrogen production by coupled system of *Nostoc/P4* and *Anabaena/P4*, World J. Microbiol. Biotech. 23 (2007) 269–274.

45. A. Tiwari, A. Pandey, Cyanobacterial hydrogen production—a step towards clean environment, Int. J. Hydrogen Energy 37 (2012) 139–150.

46. A. Melis, T. Happe, Hydrogen production. Green algae as a source of energy, Plant Physiol. 127 (2001) 740–748.

47. C. N. Hamelinck, G. van Hooijdonk, A. P. C. Faaij, Ethanol from lignocellulosic biomass: techno-economic performance in short-, middle- and long-term, Biomass Bioenerg. 28 (2005) 384–410.

48. M. D. Deng, J. R. Coleman, Ethanol synthesis by genetic engineering in cyanobacteria, Appl. Environ. Microbiol. 65 (1999) 523–528.

49. E. M. Grima, E. H. Belarbi, F. G. A. Fernandez, A. R. Medina, Y. Chisti, Recovery of microalgal biomass and metabolites: process options and economics, Biotechnol. Adv. 20 (2003) 491–515.

50. J. H. Reith, Duurzame co-productie van fijnchemicaliën en energie uit micro-algen: openbaar eindrapport E.E.T. project K99005/398510–1010. Energieonderzoek Centrum Nederland, Petten, The Netherlands, 2004.

51. FAO, Algae-based biofuels: a Review of challenges and opportunities for developing countries (www.fao.org/bioenergy/aquaticbiofuels), 2009.

52. M. Negoro, N. Shioji, K. Miyamoto, Y. Miura, Growth of microalgae in high CO_2 gas and effects of SO_x and NO_x, Appl. Biochem. Biotechnol. 28 (1991) 877–886.

53. M. Negoro, A. Hamasaki, Y. Ikuta, T. Makita, K. Hirayama, S. Suzuki, Carbon-dioxide fixation by microalgae photosynthesis using actual flue-gas discharged from a boiler, Appl. Biochem. Biotechnol. 39 (1993) 643–653.

54. K. Brown, The Utility of Remote Sensing Technology for Carbon Sequestration, Winrock International, 1611 N. Kent St., Suite 600, Arlington, VA 22209, USA, 1996.

55. J. T. Hauck, S. J. Scierka, M. B. Perry, Effects of simulated flue gas on growth of microalgae. In: *Proceedings of 212th ACS National Meeting*, 25–30 August, Orlando, Florida 41(4) (1996), 1391–1396.

56. J. Doucha, F. Straka, K. Livansky, Utilization of flue gas for cultivation of microalgae (*Chlorella sp.*) in an outdoor thin-layer photobioreactor, J. Appl. Phycology 17 (2005) 403–412.

57. P. Pedroni, J. Davidson, H. Beckert, P. Bergman, J. Benemann, A Proposal to Establish an International Network on Biofixation of CO_2 and Greenhouse Gas Abatement with Microalgae, 2001.

58. N. R. Moheimani, M. A. Borowitzka, Limits to productivity of the alga *Pleurochrysis carterae* (Haptophyta) grown in outdoor raceway ponds, Biotechnol. Bioeng. 96 (2007) 27–36.

59. K. S. Gao, Y. P. Wu, G. Li, H. Y. Wu, V. E. Villafane, E. W. Helbling, Solar UV radiation drives CO_2 fixation in marine phytoplankton: a double-edged sword, Plant Physiol. 144 (2007) 54–59.

60. A. Braun, J. Reith, Algen in de Nederlandse energiehuishouding, in opdracht van het programma, Energiewinning uit Afval en Biomassa (EWAB) van Novem, Braun Consultants, Utrecht, 1993.

61. Y. Chisti, Response to Reijnders: do biofuels from microalgae beat biofuels from terrestrial plants? Trends Biotechnol. 26 (2008) 351–352.

62. M. Aresta, A. Dibenedetto, G. Barberio, Utilization of macro-algae for enhanced CO_2 fixation and biofuels production: development of a computing software for an LCA study, Fuel Process. Technol. 86 (2005) 1679–1693.

63. E. J. Olguín, S. Galicia, G. Mercado, T. Pérez, Annual productivity of *Spirulina* (*Arthrospira*) and nutrient removal in a pig wastewater recycling process under tropical conditions, J. Appl. Phycology 15 (2003) 249.

64. T. Minowa, S. Sawayama, A novel microalgal system for energy production with nitrogen cycling, Fuel 78 (1999) 1213.

65. G. Chini Zittelli, F. Lavista, A. L. Bastianini, L. Rodolfi, M. Vincenzini, M. R. Tredici, Production of eicosapentaenoic acid by *Nannochloropsis* sp cultures in outdoor tubular photobioreactors, J. Biotechnol. 70 (1999) 299–312.

66. J. H. Reith, E. P. Deurwaarder, K. Hemmes, A. P. W. M. Curvers, P. Kamermans, W. Brandenburg, G. Zeeman, Bio-offshore: grootschalige teelt van zeewieren in combinatie met offshore windparken in de Noordzee, Energieonderzoek Centrum Nederland, Petten, The Netherlands, 2005.

67. M. R. Tredici, Microalgae for oil: strain selection, induction of lipid synthesis and outdoor mass cultivation in a low-cost photobioreactor, Biotechnol. Bioeng. 102 (2009) 100–112.

68. J. R. Benemann, W. J. Oswald, Systems and economic analysis of microalgae ponds for conversion of CO_2 to biomass, Final report, 1996.

69. F. Lehr, C. Posten, Closed photo-bioreactors as tools for biofuel production, Curr. Opin. Biotechnol. 20 (2009) 280–285.

70. Y. Shen, W. Yuan, Z. Pei, Q. Wu, E. Mao, Microalgae mass production methods, Trans. ASABE 52 (2009) 1275–1287.

71. C. Posten, G. Schaub, Microalgae and terrestrial biomass as source for fuels—a process view, J. Biotechnol. 142 (2009) 64–69.

72. K. Yeang, Biofuel from algae, Architectural Des. 78 (2008) 118–119.

73. M. Tredici, P. Carlozzi, G. Chini Zittelli, R. A. Materassi, Vertical alveolar panel (VAP) for outdoor mass cultivation of microalgae and cyanobacteria, Bioresour. Technol. 38 (1991) 153–159.

74. Q. Hu, H. Guterman, A. A. Richmond, Flat inclined modular photobioreactor for outdoor mass cultivation of photoautotrophs, Biotechnol. Bioeng. 51(1996) 51–60.

75. J. Trent, Wind sea algae. In: J. Trent (Ed.), *International Workshop on Offshore Algae Cultivation*, Maribo, Denmark, April 20–22; Baltic Sea Solutions and STP Productions, Lolland, Denmark, 2009, p. 274.

76. www.technologyreview.com/energy.

77. www.ebtplc.com.

78. www.greenfuelonline.com.

79. www.greenshift.com/news.php?id=97.

80. www.livefuels.com.

81. www.valcent.net/news_detail.sstg?id=36.

82. aquaflowgroupcom.axiion.com.

83. www.infinifuel.com.

84. www.solixbiofuels.com.

85. www.algoil.com.

86. www.petroalgae.com.

87. www.aurorabiofuels.com.

88. B. Kamm, Production of platform chemicals and synthesis gas from biomass, Angew. Chem. Int. Ed. 46 (2007), 5056–5058.

89. A.B.C. Ltd., Company news: biodiesel from algae makes debut. Manure effluents using freshwater algae: fatty acid content and composition, Fuels Lubes Int. 13 (2007): 28.

90. D. Mohan, C. U. Pittman, Jr., P. H. Steele, Pyrolysis of wood/ biomass for bio-oil: a critical review, Energy Fuels 20 (2006) 848–889.

91. Z. Chi, D. Pyle, Z. Wen, C. Frear, S. Chen, A laboratory study of producing docosahexaenoic acid from biodiesel-waste glycerol by microalgal fermentation, Process Biochem. 42 (2007) 1537–1545.

92. Y. K. Lee, Microalgal mass culture systems and methods: their limitation and potential, J. Appl. Phycol. 13 (2001) 307–315.

93. O. Pulz, Photobioreactors: production systems for phototrophic microorganisms, Appl. Microbiol. Biotechnol. 57 (2001) 287–293.

94. A. P. Carvalho, L. A. Meireles, F. X. Malcata, Microalgal reactors: a review of enclosed system designs and performances, Biotechnol. Prog. 22 (2006) 1490–1506.

95. M. Janssen, J. Tramper, L. R. Mur, R. H. Wijffels, Enclosed outdoor photo- bioreactors: light regime, photosynthetic efficiency, scale-up, and future prospects, Biotechnol. Bioeng. 81 (2003) 193–210.

96. M. E. Huntley, D. G. Redalje, CO_2 mitigation and renewable oil from photosynthetic microbes: a new appraisal, Mitigation Adapt Strat Global Change 12 (2007) 573–608.

97. M. Mitra, A. Melis, Optical properties of microalgae for enhanced biofuels production, Opt. Express. 16 (2008) 21807–21820.

98. O. M. Millamena, E. J. Aujero, I. G. Borlongan, Techniques on algae harvesting and preservation for use in culture and as larval food, Aquacult Eng. 9 (1990) 295–304.

99. J. Prakash, B. Pushparaj, P. Carlozzi, G. Torzillo, E. Montaini, R. Materassi, Microalgal biomass drying by a simple solar device, Int. J. Solar Energy 18 (1997) 303–311.

100. H. Desmorieux, N. Decaen, Convective drying of *Spirulina* in thin layer, J. Food Eng. 77 (2006) 64–70.

101. G. Leach, G. Oliveira, R. Morais, Spray-drying of *Dunaliella salina* to produce a β-carotene rich powder, J. Ind. Microbiol. Biotechnol. 20 (1998) 82–85.

102. C. I. Nindo, J. Tang, Refractance window dehydration technology: a novel contact drying method, Drying Technol. 25 (2007) 37–48.

CHAPTER 10

Green Hydrogen: Algal Biohydrogen Production

ELA EROGLU, MATTHEW TIMMINS, and STEVEN M. SMITH

10.1 INTRODUCTION

Photosynthetic molecular hydrogen (H_2) production was first reported in green microalgae by Gaffron [1] and Gaffron and Rubin [2]. Biological H_2 production can be achieved through various processes including (1) direct photolysis of water (in eukaryotic algae), (2) during nitrogen fixation (in photosynthetic bacteria including cyanobacteria), (3) nonphotosynthetic hydrogen production from organic compounds (in obligate anaerobic bacteria), and (4) fermentative nitrogen fixing (in fermentative bacteria) [3, 4, 5]. This chapter focuses on eukaryotic algal biohydrogen production.

Green unicellular algae have a rare capability among organisms in that some have an ability to switch to an H_2-producing metabolism when environmental conditions become anaerobic. The process of H_2 production is greatly accentuated in the light due to the role of the photosynthetic apparatus directing electron flow to hydrogenase enzymes located in the chloroplast. Initial chapters of this volume cover the basic mechanisms behind H_2 production from green algae and diversity of algal species suitable for H_2 production. As hydrogen production through biological systems needs improvements to reach commercial targets, recent advancing technologies are included that cover bioreactor design, immobilized cultures, integrated systems, and metabolic pathways of H_2 generation.

10.2 HYDROGEN PRODUCTION BY ALGAE

H_2 production from biological means relies mainly on the photosynthetic apparatus. Some green microalgae are able to produce H_2 by utilizing the primary products of photosynthesis (i.e., NADPH, reduced ferredoxin, and ATP) under anaerobic conditions. During normal growth, these systems act together to provide NADPH (reduced

Natural and Artificial Photosynthesis: Solar Power as an Energy Source, First Edition. Edited by Reza Razeghifard.
© 2013 John Wiley & Sons, Inc. Published 2013 by John Wiley & Sons, Inc.

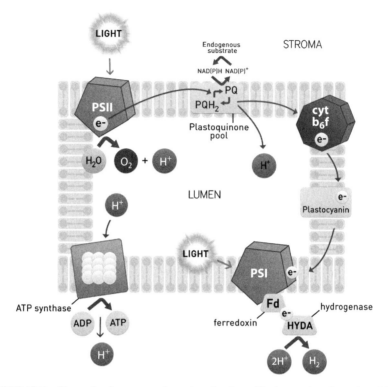

FIGURE 10.1 H_2 production occurs through reduction of hydrogen ions by reduced ferredoxin. This reaction is catalyzed by hydrogenase. The electron source may come from water oxidation or from endogenous substrate following reduction of the plastoquinone pool via NAD(P)H plastoquinone oxidoreductase complex [12, 16, 17].

nicotinamide adenine dinucleotide) and ATP (adenosine triphosphate), which are used in the Calvin cycle to power the formation of sugars. When photosystem II (PSII) absorbs light at 680 nm, an electron from a reaction center chlorophyll molecule is excited sufficiently to be transferred to a quinone molecule (Fig. 10.1). The oxidized reaction center (P680) is then capable of oxidizing water, resulting in the formation of hydrogen ions and O_2. Electrons from the quinone molecule are transferred through the cytochrome $b_6 f$ complex to photosystem I (PSI), where they are excited by light and transferred through to ferredoxin. Under typical oxygenic photosynthetic conditions, electrons may cycle around PSI to generate additional ATP or they may be used to reduce $NADP^+$. In some algae there is an additional electron route from ferredoxin. Under anaerobic conditions, [Fe] hydrogenase enzymes are synthesized that catalyze the reduction of hydrogen ions to produce H_2, according to the following equation [6]:

$$2H^+ + 2 \text{ ferredoxin}_{(red)} \rightarrow H_2 + 2 \text{ ferredoxin}_{(ox)}$$

The process occurs only in a hypoxic environment as O_2 inhibits hydrogenase activity [6, 7] and transcription of the hydrogenase gene [8].

It was more than 70 years ago that it was discovered that green algae possessed the ability to uptake and produce H_2 [1, 2]. The process was found to be possible in the dark but largely enhanced upon light exposure. Light is required for both water oxidation at PSII and ferredoxin reduction at PSI. Water oxidation can provide a source of electrons and hydrogen ions that can be used for H_2 production but the inherent problem with the system is the simultaneous O_2 generation, which prohibits continual H_2 production. In order to continually produce photosynthetically derived H_2, oxygen must be removed, either by purging the reaction mixture with inert gases [9, 10] or by consumption of oxygen through mitochondrial respiration [6, 7].

The source of electrons for H_2 production has been proposed to come from both water and endogenous substrate oxidation [2, 11]. Electrons from water may pass through both photosystems prior to reaching hydrogenase. If an endogenous substrate is involved then it is believed that electrons enter the photosynthetic chain at the level of the plastoquinone pool [12]. Endogenous substrate has been proposed to involve the actions of an NAD(P)H/plastoquinone oxidoreductase complex [11, 13]. In this scenario, oxidation of endogenous substrate results in reduction of $NAD(P)^+$; these molecules are then reoxidized by reducing the plastoquinone pool in the thylakoid membrane (Fig. 10.1). PSI is still required for H_2 production using endogenous substrate as an electron source but does not require PSII. The contribution of electron source from water and endogenous substrate has remained a difficult problem to dissect. Evidence that both sources contribute comes from studies showing higher starch levels lead to higher H_2 production while lower starch levels lead to lower H_2 production [14, 15] and blocking PSII with DCMU (3-(3,4-dichlorophenyl)-1,1-dimethylurea) leads to lower levels of H_2 production [8, 16]. It is probable that the two systems of H_2 production act together in a concerted fashion to allow for enhanced survival under anaerobic conditions.

10.3 HYDROGENASE ENZYME

Discovery of a hydrogenase enzyme responsible for the reduction of protons into molecular hydrogen dates backs to 1931 [18]. H_2 production efficiency is closely linked to hydrogenase enzyme expression and activity. There are several phylogenetically distinct forms of hydrogenases: (1) [NiFe] uptake (hup) hydrogenases, (2) [NiFe] bidirectional (hox) hydrogenases, (3) [FeFe] hydrogenases, (4) [NiFeSe] hydrogenase, and (5) [Fe]-only hydrogenases [5].

[NiFe] hydrogenases of bacteria and cyanobacteria, and [FeFe] hydrogenases of green microalgae and obligate anaerobic fermentative bacteria are the best characterized hydrogenase enzymes [19, 20, 21]. [Fe]-only hydrogenases of hydrogenotrophic methanogenic archaea are different from algal [FeFe] hydrogenases, as they lack the [Fe-S] cluster required for electron transport to soluble mediators; these have previously been cited as "iron sulfur cluster-free" hydrogenase enzymes [20]. Commonly,

[NiFe] hydrogenases consist of two peptides and catalyze the uptake of H_2 while [FeFe] hydrogenases are better known to function in the production of H_2 [22].

[FeFe] hydrogenases are localized in the chloroplast of green microalgae and work as an electron valve under anaerobic oxygenic photosynthesis [23]. The active site of [FeFe] hydrogenases is known as the H cluster; it is located in the interior of the protein and consists of a [4Fe4S] center that is easily oxidized. Green algal [FeFe] hydrogenases have catalytic activities between 10 and 100 times higher than [NiFe] hydrogenases [8, 24]. [FeFe] hydrogenases reversibly catalyze the reduction of protons into molecular hydrogen, according to the following equation [6]:

$$2H^+ + 2FD_{(red)} \leftrightarrow H_2 + 2FD_{(oxd)}$$

Green microalgae such as *Chlamydomonas reinhardtii, Chlorella fusca, Chlorococcum littorale, Scenedesmus obliquus*, and *Platymonas subcordiformis* are genetically endowed with the [FeFe] hydrogenase genes [25, 26]. Two accessory genes, *HydEF* and *HydG*, are encoded in *C. reinhardtii* that have been found to be required for hydrogenase assembly and activation [27, 28].

The first cell-free preparation of [FeFe] hydrogenase from *Chlamydomonas* was performed by Abeles [29] though partial sequencing of the enzyme was performed by Happe and Naber [30]. This enzyme has since been named *HydA1*. A second [FeFe] hydrogenase in *C. reinhardtii* was identified by Forestier et al. [27]; this is named *HydA2* and shares 68% amino acid identity with *HydA1*. Both enzymes are nuclear encoded, expressed following the imposition of anaerobic conditions, transported to the chloroplast stroma, and highly O_2 sensitive [7, 27]. The oxygen sensitivity of hydrogenases makes continual systems of H_2 production a challenge.

10.4 DIVERSITY OF HYDROGEN-PRODUCING ALGAE

Algae are a diverse group of uni- or multicellular organisms that are generally phototrophic, many can uptake reduced carbon sources and some live exclusively on external energy sources. Apart from blue green algae (cyanobacteria), all algae are eukaryotic. They differ from land plants most notably in the lack of specialized structures such as roots, leaves, and vascular systems. Algal habitats range from oceans to fresh water, soil, and snow; they can live on bare rock and survive airborne [31]. Though vastly different in cellular machinery, both prokaryotic and eukaryotic algae often occupy similar niches.

Due to enormous diversity, algae are difficult to classify taxonomically. Traditional classification was based on the type of chlorophyll and accessory pigments present in the chloroplast. Other criteria used can include number and structure of flagella, cell wall chemistry, habitat, motility, life-cycle, or any specialized structures. The first algal classification system based on pigmentation was performed by W. H. Harvey (1840) who divided algae into four groups: red algae (Rhodophyta), brown algae (Heterokontophyta), green algae (Chlorophyta), and Diatomaceae [32]. Since that time more elaborate taxonomical systems have been adopted to encompass the vast

**TABLE 10.1   Groups of Organisms in Which Most Algal Lineages Reside;
Major Pigments, and Some Commonly Known Model Specimens are Shown**

Group	Example Genera	Defining Pigments
Alveolates	*Alexandrium, Gymnodinium, Symbiodinium*	Chlorophyll *a* and *c*
Chlorarachniophytes	*Lotharella, Bigelowiella, Chlorarachnion*	Chlorophyll *b*
Cryptomonads	*Cryptomonas, Rhodomonas, Chroomonas*	Chlorophyll *a* and *c* and phycobilins
Cyanobacteria	*Anabaena, Nostoc, Spirulina*	Phycobilin
Euglenids	*Astasia, Hyalophacus, Euglena*	Chlorophyll *b*
Glaucophytes	*Cyanophora, Glaucocystis, Gloeochaete*	Phycobilin
Haptophytes	*Isochrysis, Pavlova, Emiliania*	Chlorophyll *a* and *c*
Rhodophytes	*Rhodella, Porphyra, Rhodosorus*	Phycobilin
Stramenopiles	*Cymbella, Eustigmatos, Chaetocerus*	Chlorophyll *a* and *c*
Viridaeplantae	*Chlamydomonas, Chlorella, Dunaliella*	Chlorophyll *b*

Source: Compiled from Patterson et al. [34].

differences among these organisms. Algal systems of nomenclature are constantly changing and so far no definitive system has been decided upon. Recently, molecular classification has been performed based on RuBisCO or ribosomal RNA sequence information [33]. In Table 10.1 ten groups of organisms are listed in which most algal lineages reside, compiled from Algaebase, The Tree of Life Web Project, and Patterson et al. [34]. Their major pigments and some commonly known model specimens are given in Table 10.1. These ten groupings encompass the majority of known algae and are based largely on pigmentation, motility, and cell structure.

Analyzing pigments from water sources has also been performed for purposes of wastewater monitoring and profiling of algal communities [35]. While it is possible to derive taxonomical information from pigment profiles for various water sources, it has been shown that for detailed information further microscopic or genetic analysis is also required.

The diversity of algae and difficulties in culturing some species in the laboratory have made it difficult to determine which algae are best suited for H_2 production. A number of studies have been performed on assaying strains kept in culture collections and from the wild to determine which lineages are best suited to this task [33, 36, 37, 38, 39]. The ability to produce H_2 was found to lie predominantly with unicellular green algae, though it has been found in some yellow green algae, diatoms, and a few multicellular or colonial strains [39]. Of all tested species, the highest H_2-producing genera (in order) were found to be *Chlamydomonas, Chlorococcum, Scenedesmus, Dictyosphaerium, Coelastrum,* and *Chlorella.* These genera all belong to phylum Chlorophyta, the division of green algae that gave rise to land plants. Of all algae tested to date, *C. reinhardtii* has proved to be capable of generating the highest levels of continual H_2 production under a variety of conditions [6, 17, 33, 39].

10.5 MODEL MICROALGAE FOR H₂ PRODUCTION STUDIES: *CHLAMYDOMONAS REINHARDTII*

Chlamydomonas reinhardtii has proved to be capable of sustaining the highest rates of H_2 production among all tested algae. This unicellular alga has been used extensively for studying a vast range of biological functions and consequently much knowledge has been gained in regard to processes that can be related to H_2 production; such as photosynthesis, nutrient metabolism, molecular biology, and biochemistry. In addition, the genome has been sequenced, methods of mating have been documented, and genetic manipulations are routinely performed [40, 41, 42].

Chlamydomonas reinhardtii is typically propagated in its haploid state. In this state cells exist as either mating-type (+) or mating-type (−). Haploid cells are ellipsoidal in shape and measure approximately 10 μm in length (Fig. 10.2A). There is a single cup-shaped chloroplast that occupies approximately 40% of the cell volume. Under certain stress conditions, such as sulfur deprivation, cells swell in correlation with the starch accumulation [43], change their appearance from ellipsoidal to spherical, and can even lose their flagella (Fig. 10.2B). A large amount of starch accumulation has been reported by several researchers for sulfur-deprived *C. reinhardtii* within the first couple of hours after sulfur limitation [6, 43, 44]; catabolism of starch provides a source of reductant for respiration, oxidative phosphorylation, and H_2 production [16].

Chlamydomonas reinhardtii has nuclear, chloroplast, and mitochondrial genomes. During mating, nuclear genes segregate according to Mendelian rules. The chloroplast genome is inherited uniparentally from the (+) mating-type and the mitochondrial from the (−) mating-type [41]. The nuclear genome contains approximately 120 megabases and has been predicted to give rise to ∼15,000 protein-encoding genes.

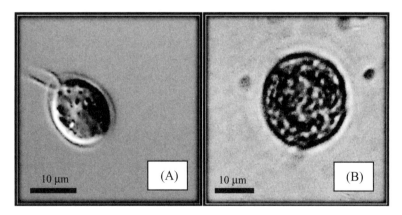

FIGURE 10.2 (A) Ellipsoidal shaped haploid cells of C. *reinhardtii* (CC125) under aerobic conditions (diameter of an individual cell is approximately 10 μm). (B) After 24 hours of sulfur deprivation, a significant increase in cell size and starch accumulation is seen [43] and changes in cell morphology from ellipsoidal to spherical and loss of flagella is observed. (Images taken in Melis–Lab at the University of California Berkeley, E. Eroglu.)

Chloroplast DNA consists of approximately 80 copies of circular molecules of 196 kilobases. The mitochondrial genome is smaller in comparison and consists of copies of 15.8 kilobase linear fragments [45].

Chlamydomonas reinhardtii cells are able to survive without mitochondrial function provided that photosynthetic function is intact [46]. They are also able to survive without photosynthetic function, provided a reduced carbon source is supplied. These features, in addition to the ability to transform all three genetic systems, make *C. reinhardtii* ideally suited to a broad range of studies, in particular, on the photosynthetic apparatus, primary metabolism, and H_2 production [47, 48, 49].

10.6   APPROACHES FOR ENHANCING HYDROGEN PRODUCTION

Systematic approaches undertaken to increase the efficiency of the hydrogen production process are outlined below. Photosynthetic H_2 production rates are usually given in units of volume of H_2, either (1) per culture volume per time ($mL_{H_2} \cdot L^{-1} \cdot h^{-1}$), or (2) per culture mass per time ($mL_{H_2} \cdot g^{-1} \cdot h^{-1}$) based usually on dry cell weight. Direct comparisons between laboratories can be difficult due to multiple variables, such as substrate source, reactor geometry, and illumination conditions.

10.6.1   Immobilization Processes

Immobilization of cells on solid surfaces has several advantages over free cells since the immobilized material occupies smaller space, less amount of growth medium can be used, they are easier to process, and repetitive use of microorganisms during product generation is possible. Immobilization of cells can be effective for protecting the cell cultures from harsh environments such as salinity, metal toxicity, and sudden pH variations [50, 51]. Entrapment is one of the most common immobilization methods; it involves capturing the cells in a three-dimensional gel lattice, mostly made of polymeric materials or inorganic spheres [51]. Both synthetic (such as acrylamide, polyurethane, polyvinyl) and natural polymers (such as collagen, agar, cellulose, alginate, carrageenan) have been used for cell immobilization [52]. The main objectives are to retain algal cells within the matrix, while the pores inside the gel lattice can allow diffusion of substrates and products to and from the cells [53]. Several solid matrices have been utilized successfully for photosynthetic cell immobilization, such as porous glass [54], carrageenan [55], agar gel [56], and even clay surfaces [57].

H_2 production was enhanced by immobilized *Chlamydomonas reinhardtii* cultures upon sulfur deprivation; rates almost doubled from 2.5 to 4.3 $mL \cdot L^{-1} \cdot h^{-1}$ after being immobilized in/on porous glass sheets of Al-borosilicate [58]. Hahn et al. [59] investigated the immobilization of *C. reinhardtii* algal cells on silica particles for the production of H_2 in a two-step cycle; attached cells shifted easily between growth and H_2 production cycles.

Immobilization processes within three-dimensional gel lattices need improvements due to the issues caused by unbalanced light and poor mass diffusion rates resulting in nutrient gradients within the system [60, 61]. Attempts to increase mass diffusion include the entrapment of microalgal cells on thin film surfaces such electrospun nanofibers [62] and C_a^{+2} alginate matrix films [63].

10.6.2 Increasing the Resistance of Algal Cells to Stress Conditions

[FeFe] hydrogenases are highly sensitive to O_2. Attempts have been made to enhance photosynthetic H_2 production by lowering the concentration of O_2 in the culture. Purging a culture with inert gases in a continuous manner aids the removal of O_2 from the culture medium, while allowing the synthesis and function of the hydrogenase enzyme and simultaneous H_2 production by the photosynthetic apparatus [9, 10]. While this method is efficient for H_2 production, it is difficult to scale-up to an economical commercial scale. Alternate approaches to reduce oxygen concentration to a point where it no longer inhibits the hydrogenase enzymes include (1) reversible inactivation of O_2 production from PSII, (2) enrichment of endogenous cellular respiration, and (3) decreasing the O_2 partial pressure by photochemical methods [64].

A significant advance in H_2 production studies was achieved by Melis and his colleagues [6] by using sulfur depletion as a tool to inactivate PSII. The sulfur-depletion process downregulates PSII activity and so decreases O_2 generation [6, 65]. All photosynthetically generated O_2 is consumed by cellular respiration and the culture becomes anaerobic [6]. The sulfur deprivation method directs photosynthetic electron transport toward the [FeFe] hydrogenase. This method of inducing algae into an anaerobic H_2 production status has become very popular due to H_2 production being possible for several days [6, 7, 44, 66]. By depriving a culture of sulfur, protein biosynthesis is impaired. In particular, de novo biosynthesis of the D1 protein in PSII does not occur to a sufficient degree following photooxidative damage and this leads to a decline in water oxidation and O_2 generation. Sulfate transporter mutants have been created that limit sulfate supply and allow only low-level turnover of PSII [67]. These have led to a continuous hypoxic H_2 production environment.

According to a recent study by Faraloni et al. [68], sulfur-deprived cultures of *Chlamydomonas reinhardtii* produced increased amounts of H_2 when grown in media supplemented with olive mill wastewater that was rich in organic acids and sugars. It was reported that sulfur-deprived cultures grown in wastewater and TAP media produced 150 mL·H_2·L^{-1}; this was 50% higher than control cells grown in TAP media only (100 mL·H_2·L^{-1}), with average H_2 production rates of 1.29 mL·L^{-1}·h^{-1} and 1.03 mL·L^{-1}·h^{-1}, respectively.

According to a current manuscript by Batyrova et al. [69], culturing *Chlamydomonas reinhardtii* cultures in phosphorus-deprived media also generated sustained H_2 production. In contrast to sulfur deprivation studies, resuspension of algae in phosphorus-deprived media is not sufficient to cause phosphate deficiency within the algal cells due to high internal phosphate reserves. For this reason, inoculating high concentrations of washed algae into the P-deprived medium was avoided, as only low initial cell densities (less than 2 mg Chl·L^{-1} culture solution) generated H_2. This process has been proposed as an alternative to sulfur deprivation for species where natural media with a high sulfate concentration would be necessary to use (i.e., seawater).

10.6.3   Optimization of Bioreactor Conditions

Design parameters for a H_2-producing photobioreactor are pH, temperature, composition of the atmosphere above the cultures, intensity of light irradiance, dissolved O_2 level, mixing method, geometry, and material selection. It has been difficult to design systems to induce H_2 production from algae due to the fundamental problem of both light and anaerobicity being required for continual productions: these conditions are essentially mutually exclusive in eukaryotic photosynthetic organisms. System development for overcoming this problem is challenging.

The first experiments that could quantitatively measure H_2 production from algae were performed by either dark anaerobic incubation or purging a culture with N_2 [2]. In these experiments, vessels were used that could measure gas evolution or consumption by pressure changes in the vessel side arm. Identification of H_2 was by absorption with palladium black and discoloration of methylene blue in the presence of platinum. As technology progressed, improved systems were designed including a flow-through system that could continuously purge a culture with an inert gas in the light and measure evolved gases simultaneously [37]. H_2 could be quantitatively measured using a zirconium oxide fuel cell and electric current was converted to gas produced using Faraday's law of electrochemical equivalence.

Light absorbance is an essential parameter of photosynthesis that affects the efficiency and yield, especially for the concentrated cell culture due to the self-shading effect. It is essential to avoid overabsorption of sunlight and wasting absorbed irradiance [70, 71, 72]. Several researchers have reported an inverse relationship between the light conversion efficiency (η) to H_2 and the incident light intensities [73, 74]. Light conversion efficiency depends on the incident light intensity, irradiated area, reactor design, duration of H_2 production, and amount of H_2 accumulated; it is calculated as the ratio of the energy stored in H_2 over the total energy input of the photobioreactor.

Mixing of culture media is another important design parameter as it provides homogeneity of nutrients and cells, minimizes light and temperature fluctuations, and aids removal of gas [5]. Homogeneous media improve culture conditions by increasing substrate uptake and light exposure [61]. Mechanical mixing is possible for smaller-scale systems but it is not economical for scale-up processes. The operation of continuously stirred reactors causes back-mixing with high residence-time distribution of the cells inside the reactors, yielding lower product efficiencies [75]. Plug-flow type reactors or purging of gases through the reactor is usually preferred for larger-scale systems; this is understood to be less harmful to cells by avoiding destructive shearing forces caused by mechanical mixing.

Geometry and building material are considered to be important factors to achieve high rates of H_2 production, especially in relation to light availability. A high surface to volume ratio is a requirement for an efficient photobioreactor [71] but this has ramifications on size requirements and cost. Various geometrical shapes have been

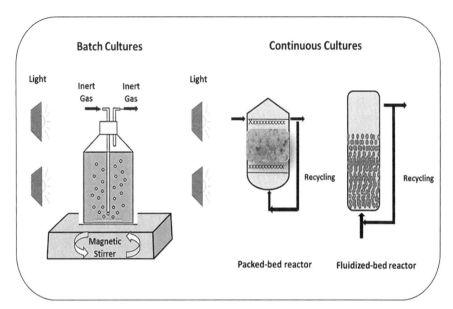

FIGURE 10.3 Examples of various photobioreactor designs with microalgal cultures, under batch and continuous culture growth conditions.

used for photosynthetic H_2 production studies, including flat-plate, tubular, or helical photobioreactors [71, 76].

Basic types of photobioreactor designs for batch and continuous process conditions are illustrated in Fig. 10.3. Illumination usually takes place under an artificial light source and stirring is by a laboratory shaker or magnetic stirrer. Purging of a culture with an inert gas is sufficient to induce H_2 production from some microalgal species. Generated gases can be collected in a gas collection system, such as a reversed burette standing inside a beaker filled with water. Under continuous systems, a column or flat type reactor can be used while agitation of the process is provided by the direction of the fluid. A packed, fixed-bed type photobioreactor with down-flow direction and a fluidized-bed type photobioreactor with up-flow direction are shown in Fig. 10.3, as examples for continuous process. During system scale-up, outdoor and continuous-flow conditions are the main targets; secondary are temperature and light control facilities. Maintaining an anaerobic atmosphere still remains the largest challenge for large-scale systems.

10.6.4 Integrated Photosynthetic Systems

Even though the range of solar radiation is very broad, not every wavelength can be used by every organism. Particularly algae and plants have limited absorption capacity for visible light [77]. Green algae can use the visible region of the light spectrum from 400 to 700 nm, which equates to about 45% of total solar radiation. The near-infrared region of light between 700 and 900 nm can also be used by

FIGURE 10.4 (A) Green microalgae cultures (*Chlamydomonas reinhardtii*) and (B) anoxygenic photosynthetic bacteria (*Rhodospirillum rubrum*) growth in photobioreactors under artificial light illumination. *Chlamydomonas reinhardtii* cells were precultured in TAP medium, *R. rubrum* was grown in Ormerod medium. Generated gases were collected in a burette standing upside-down inside a beaker filled with water. Water displacement gives a measure of gas volume. (Photographs taken in Melis–Lab at the University of California Berkeley, E. Eroglu.)

anoxygenic photosynthetic bacteria [4]. Only a percentage of the visible spectrum of light can be turned into photochemical energy.

In order to improve solar energy utilization, Melis and Melnicki [4] integrated green *Chlamydomonas reinhardtii* microalgae cultures (Fig. 10.4A) with *Rhodospirillum rubrum* as anoxygenic photosynthetic bacteria (Fig. 10.4B). This integration increases the absorption range of the solar spectrum by including wavelengths from 400 to 900 nm. Both organisms can coexist as green microalgae use CO_2 and H_2O to generate organic carbon, which can be utilized by photosynthetic bacteria. Melis and Melnicki [4] also investigated the potential to use a dual photosynthetic system (algae and photosynthetic bacteria) with dark fermentative bacteria, which can generate small organic acids and would be beneficial during the utilization of waste-based nutrient sources.

10.6.5 Genetic Engineering Approaches to Improve Photosynthetic Efficiency

Several microalgal mutants have been isolated in different laboratories with altered photosynthetic antenna size. Increasing antenna size is an evolutionary survival mechanism for photosynthetic organisms under limited light conditions. However, a large antenna complex can also cause overabsorption of sunlight, which can decrease photosynthetic productivity due to losses from heat or fluorescence [78]. It has been reported that there is an inverse relationship between the size of the light-harvesting antenna and the light energy conversion efficiency [70, 72, 79, 80]. It was reported that H_2 productivity of green algal cultures was increased by a truncated antenna size due to minimized fluorescence and heat dissipation of absorbed sunlight [73]. Cloning of the genes for truncating Chl antenna size in green microalgae has been achieved and their functions analyzed [81, 82]. Polle et al. [73] reported that the *tla1* mutant of *Chlamydomonas reinhardtii* (50% truncated antenna size) achieved two times higher photosynthetic yield compared to its wild type. In a recent study, Kirst and colleagues [83] used a new *Chlamydomonas reinhardtii* mutant with a truncated light-harvesting antenna (*tla2*), and reported a lower amount of chlorophyll (Chl) per cell, higher Chl *a/b* ratio, and a lighter-green phenotype than its wild type strain. They also showed that deleting the *cpftsy* gene in green algae can be used for the formation of *tla* mutants.

Analyzing mutant strains for altered levels of H_2 production requires a high-throughput method. This has successfully been established by Posewitz et al. [15] by using chemochromic sensors [84] to monitor insertional mutants grown on plates following a dark anaerobic induction period. This work led to the identification of a mutant defective in starch accumulation and consequently H_2 production. A method to screen new algal cultures, existing as axenic or mixed cultures, was employed that involved purging a culture with N_2 and then sealing in a culture vessel fitted with a gas collection tube in the light; new algal species were identified from the wild that were able to produce H_2 [33]. Detailed approaches for monitoring and regulating photosynthesis, respiration, cellular metabolism, and H_2 production of *Chlamydomonas* are further discussed by Hemschemeier and her colleagues [16].

10.6.6 Metabolic Pathways of H_2 Production

Metabolites can be the end points of metabolism, intermediates in central metabolic pathways, signaling molecules, defense molecules, and regulatory molecules. Metabolomic studies aim at the quantification of all metabolites from an organism and a number of these studies have been performed on *C. reihnardtii* to better understand its metabolism. Bolling and Fiehn [85] performed a large-scale metabolomic study of *C. reihnardtii* under a variety of nutrient-limiting stresses, including sulfur, to reveal a wide variety of changes occurring at the metabolic level. Timmins et al.

[86] performed a metabolomic study on *C. reihnardtii* under sulfur-depleted anaerobic H_2-producing conditions over 120 h. This study revealed that the cells entered a largely fermentative metabolic state with cells breaking down starch, but surprisingly also increasing in their triacylglycerol (TAG) content. Their findings point to the further breakdown of substantial energy reserves within the cells being a possible approach to increasing H_2 output. Additional targeted analysis of certain fermentative compounds has been performed during H_2 production in *C. reinhardtii* [37, 87] and metabolic pathways in operation during H_2 production are now becoming clearer. The knowledge of how algal cells restructure metabolic pathways involved in energy metabolism during H_2 production will provide a framework for which genetic engineering projects can be modeled and investigated.

10.7 CONCLUSION

There is potential to develop a carbon neutral energy source from green algae in the form of H_2. Currently, the process of H_2 production involves growth of algae in an aerobic environment followed by the imposition of anaerobic conditions in the light. Under these current conditions levels of hydrogen production are not yet economically feasible. The switch from an aerobic growth regime into an anaerobic H_2-producing regime involves distinct changes in energy metabolism; studying these changes at genomic and metabolomic levels offers the opportunity to understand, model, and manipulate the cellular energy machinery. Together with redesigned antenna architecture and reactor design to optimize light use, it may be possible to ultimately increase H_2 production to levels viable to supply areas of the energy sector.

REFERENCES

1. H. Gaffron, Reduction of CO_2 with H_2 in green plants, Nature 143 (1939) 204–205.
2. H. Gaffron, J. Rubin, Fermentative and photochemical production of hydrogen in algae, J. Gen. Physiol. 26 (1942) 219–240.
3. J. Benemann, Hydrogen biotechnology: progress and prospects, Nat. Biotechnol. 14 (1996) 1101–1103.
4. A. Melis, M. Melnicki, Integrated biological hydrogen production, Int. J. Hydrogen Energ. 31 (2006) 1563–1573.
5. E. Eroglu, A. Melis, Photobiological hydrogen production: recent advances and state of the art, Bioresour. Technol. 102 (2011) 8403–8413.
6. A. Melis, L. Zhang, M. Forestier, M. L. Ghirardi, M. Seibert, Sustained photobiological hydrogen gas production upon reversible inactivation of oxygen evolution in the green alga *Chlamydomonas reinhardtii*, Plant Physiol. 122 (2000) 127–133.
7. M. L. Ghirardi, L. Zhang, J. W. Lee, T. Flynn, M. Seibert, E. Greenbaum, A. Melis, Microalgae: a green source of renewable H_2, Trends Biotechnol. 18 (2000) 506–511.

8. L. Florin, A. Tsokoglou, T. Happe, A novel type of iron hydrogenase in the green alga *Scenedesmus obliquus* is linked to the photosynthetic electron transport chain, J. Biol. Chem. 276 (2001) 6125–6132.

9. E. Greenbaum, Photosynthetic hydrogen and oxygen production: kinetic studies, Science 196 (1982) 879–880.

10. E. Greenbaum, Energetic efficiency of hydrogen photoevolution by algal water-splitting, Biophys. J. 54 (1988) 365–368.

11. R. P. Gfeller, M. Gibbs, Fermentative metabolism of *Chlamydomonas reinhardtii*: I. Analysis of fermentative products from starch in dark and light, Plant Physiol. 75 (1984) 212–218.

12. T. S. Stuart, H. Gaffron, The mechanism of H_2 photoproduction by several algae. II The contribution of photosystem II, Planta 106 (1972) 101–112.

13. P. Bennoun, Chlororespiration and the process of carotenoid biosynthesis, Biochim. Biophys. Acta Bioenergetics 1506 (2001) 133–142.

14. C. Schonfeld, L. Wobbe, R. Borgstadt, A. Kienast, P. J. Nixon, O. Kruse, The nucleus-encoded protein MOC1 is essential for mitochondrial light acclimation in *Chlamydomonas reinhardtii*, J. Biol. Chem. 279 (2004) 50366–50374.

15. M. C. Posewitz, S. L. Smolinski, S. Kanakagiri, A. Melis, M. Seibert, M. L. Ghirardi, Hydrogen photoproduction is attenuated by disruption of an isoamylase gene in *C. reinhardtii*, Plant Cell 16 (2004) 2151–2163.

16. A. Hemschemeier, A. Melis, T. Happe, Analytical approaches to photobiological hydrogen production in unicellular green algae, Photosynth. Res. 102 (2009) 523–540.

17. O. Kruse, B. Hankamer, Microalgal hydrogen production, Curr. Opin. Biotechnol. 21 (2010) 238–243.

18. M. Stephenson, L. H. Stickland, Hydrogenase: a bacterial enzyme activating molecular hydrogen, Biochem. J. 25 (1931) 205–214.

19. E. C. Hatchikian, N. Forget, V. M. Fernandez, R. Williams, R. Cammack, Further characterization of the Fe]-hydrogenase from *Desulfovibrio desulfuricans* ATCC 7757, Eur. J. Biochem. 203 (1992) 357–366.

20. S. Shima, O. Pilak, S. Vogt, M. Schick, M. S. Stagni, W. Meyer-Klaucke, E. Warkentin, R. K. Thauer, U. Ermler, The crystal structure of Fe]-hydrogenase reveals the geometry of the active site, Science 321 (2008) 572–575.

21. P. M. Vignais, B. Billoud, J. Meyer, Classification and phylogeny of hydrogenases, FEMS Microbiol. Rev. 25 (2001) 455–501.

22. P. Tamagnini, R. Axelsson, P. Lindberg, F. Oxelfelt, R. Wunschiers, P. Lindblad, Hydrogenases and hydrogen metabolism of cyanobacteria, Microbiol. Mol. Biol. Rev. 66 (2002) 1–20.

23. M. Winkler, A. Hemsehemeier, C. Gotor, A. Melis, T. Happe, Fe]-hydrogenase in green algae: photo-fermentation and hydrogen evolution under sulfur deprivation. Int. J. Hydrogen Energy 27 (2002) 1431–1439.

24. M. Frey, Hydrogenases: hydrogen-activating enzymes, ChemBioChem 3 (2002) 153–160.

25. A. Melis, Photosynthetic H_2 metabolism in *Chlamydomonas reinhardtii* (unicellular green algae), Planta 226 (2007) 1075–1086.

26. D. Das, T. N. Veziroglu, Advances in biological hydrogen production processes, Int. J. Hydrogen Energy 33 (2008) 6046–6057.

27. M. Forestier, P. King, L. Zhang, M. Posewitz, S. Schwarzer, T. Happe, M. L. Ghirardi, M. Seibert, Expression of two Fe]-hydrogenases in *Chlamydomonas reinhardtii* under anaerobic conditions, Eur. J. Biochem. 270 (2003) 2750–2758.

28. M. C. Posewitz, P. W. King, S. L. Smolinski, L. Zhang, M. Seibert, M. L. Ghirardi, Discovery of two novel radical S-adenosylmethionine proteins required for the assembly of an active Fe] hydrogenase, J. Biol. Chem. 279 (2004) 25711–25720.

29. F. B. Abeles, Cell-free hydrogenase from *Chlamydomonas*, Plant Physiol. 39 (1964) 169–176.

30. T. Happe, J. D. Naber, Isolation, characterization and N-terminal amino acid sequence of hydrogenase from the green alga *C. reinhardtii*, Eur. J. Biochem. 214 (1993) 475–481.

31. M. T. Madigan, J. M. Martinko, J. Parker, *Brock Biology of Microorganisms*, 9th ed., Prentice Hall, Englewood Cliffs, NJ, 2000.

32. P. S. Dixon, *Biology of the Rhodophyta*, Oliver & Boyd, Edinburgh, 1973.

33. M. Timmins, S. R. Thomas-Hall, A. Darling, E. Zhang, B. Hankamer, U. C. Marx, P. M. Schenk, Phylogenetic and molecular analysis of hydrogen-producing green algae, J. Exp. Bot. 60 (2009) 1691–1702.

34. D. J. Patterson, A. G. Simpson, N. Weerakoon, Free-living flagellates from anoxic habitats and the assembly of the eukaryotic cell, Biol. Bull. 196 (1999) 381–384.

35. J. Stoń, A. Kosakowska, M. Łotocka, Pigment composition in relation to phytoplankton community structure and nutrient content in the Baltic Sea, Oceanologia, 44 (2002) 419–437.

36. N. I. Bishop, M. Frick, L. W. Jones, Photohydrogen production in green algae: water serves as the primary substrate for hydrogen and oxygen production. In: A. Mitsui, S. Miyachi, A. San Pietro, S. Tamura (Eds.), *Biological Solar Energy Conversion*, Academic Press, New York, 1977, pp. 3–22.

37. E. Greenbaum, R. R. L. Guillard, W. G. Sunda, Hydrogen and oxygen photoproduction by marine algae, Photochem. Photobiol. 37 (1983) 649–655.

38. J. J. Brand, J. N. Wright, S. Lien, Hydrogen production by eukaryotic algae, Biotechnol. Bioeng. 33 (1989) 1482–1488.

39. V. A. Boichenko, P. Hoffmann, Photosynthetic hydrogen production in prokaryotes and eukaryotes: occurrence, mechanism, and function, Photosynthetica 30 (1994) 527–552.

40. S. S. Merchant, S. E. Prochnik, O. Vallon, E. H. Harris, S. J. Karpowicz, G. B. Witman, et al., The *Chlamydomonas* genome reveals the evolution of key animal and plant functions, Science 318 (2007) 245–250.

41. J. D. Rochaix, *C. reinhardtii* as the photosynthetic yeast, Annu. Rev. Genet. 29 (1995) 209–230.

42. K. L. Kindle, O. A. Sodeinde, Nuclear and chloroplast transformation in *C. reinhardtii*: strategies for genetic manipulation and gene expression, J. Appl. Phycol. 6 (1994) 231–238.

43. E. Eroglu, A. Melis, Density equilibrium method for the quantitative and rapid in situ determination of lipid, hydrocarbon, or biopolymer content in microorganisms, Biotechnol. Bioeng. 102 (2009) 1406–1415.

44. L. Zhang, T. Happe, A. Melis, Biochemical and morphological characterization of sulfur-deprived and H_2-producing *Chlamydomonas reinhardtii* (green alga), Planta 214 (2002) 552–561.

45. E. H. Harris, *The C. reinhardtii Sourcebook: A Comprehensive Guide to Biology and Laboratory Use*, Academic Press, San Diego, 1989.

46. N. W. Gillham, J. E. Boynton, E. H. Harris, Specific elimination of mitochondrial DNA from *Chlamydomonas* by intercalating dyes, Curr. Genet. 12 (1987) 41–47.

47. J. E. Boynton, N. W. Gillham, E. H. Harris, J. P. Hosler, A. M. Johnson, A. R. Jones, et al., Chloroplast transformation in *Chlamydomonas* with high velocity microprojectiles, Science 240 (1988) 1534–1538.

48. B. L. Randolph-Anderson, J. E. Boynton, N. W. Gillham, E. H. Harris, A. M. Johnson, M. P. Dorthu, R. F. Matagne, Further characterization of the respiratory deficient dum-1 mutation of *C. reinhardtii* and its use as a recipient for mitochondrial transformation, Mol. Gen. Genet. 236 (1993) 235–244.

49. J. D. Rochaix, J. van Dillewijn, Transformation of the green alga *Chlamydomonas reinhardii* with yeast DNA, Nature 296 (1982) 70–72.

50. L. Hall-Stoodley, J. W. Costerton, P. Stoodley, Bacterial biofilms: from the natural environment to infectious diseases, Nature Rev. Microbiol. 2 (2004) 95–108.

51. Y. Liu, M. H. Rafailovich, R. Malal, D. Cohn, D. Chidambaram, Engineering of bio-hybrid materials by electrospinning polymer-microbe fibers, PNAS 34 (2009) 14201–14206.

52. M. S. A. Hameed, O. H. Ebrahim, Biotechnological potential uses of immobilized algae, Int. J. Agric. Biol. 9 (2007) 183–192.

53. N. Mallick, Biotechnological potential of immobilized algae for wastewater N, P and metal removal: a review, BioMetals 15 (2002) 377–390.

54. A. A. Tsygankov, Y. Hirata, M. Miyake, Y. Asada, J. Miyake, Photobioreactor with photosynthetic bacteria immobilized on porous glass for hydrogen photoproduction, J. Ferment. Bioeng. 77 (1994) 575–578.

55. N. Francou, P. M. Vignais, Hydrogen production by *Rhodopseudomonas capsulata* cells entrapped in carrageenan beads, Biotechnol. Lett. 6 (1984) 639–644.

56. H. Zhu, T. Suzuki, A. A. Tsygankov, Y. Asada, J. Miyake, Hydrogen production from tofu wastewater by *Rhodobacter sphaeroides* immobilized in agar gels, Int. J. Hydrogen Energy 24 (1999) 305–310.

57. C. Y. Chen, J. S. Chang, Enhancing phototropic hydrogen production by solid-carrier assisted fermentation and internal optical-fiber illumination, Process Biochem. 41 (2006) 2041–2049.

58. V. Laurinavichene, A. S. Fedorov, M. L. Ghirardi, M. Seibert, A. A. Tsygankov, Demonstration of sustained hydrogen photoproduction by immobilized, sulfur-deprived *Chlamydomonas reinhardtii* cells, Int. J. Hydrogen Energy 31 (2006) 659–667.

59. J. J. Hahn, M. L. Ghirardi, W. A. Jacoby, Immobilized algal cells used for hydrogen production, Biochem. Eng. J. 37 (2007) 75–79.

60. A. Tsygankov, Hydrogen production by purple bacteria: immobilized versus suspension cultures. In: J. Miyake, T. Matsunaga, A. S. Pietro (Eds.), *Bio-hydrogen II*, Elsevier Science Ltd., Amsterdam, 2001, pp. 229–243.

61. H. Koku, I. Eroglu, U. Gunduz, M. Yucel, L. Turker, Aspects of metabolism of hydrogen production by *Rhodobacter sphaeroides*, Int. J. Hydrogen Energy 27 (2002) 1315–1329.

62. E. Eroglu, V. Agarwal, M. Bradshaw, X. Chen, S. M. Smith, C. L. Raston, K. S. Iyer, Nitrate removal from liquid effluents using microalgae immobilized on chitosan nanofiber mats, Green Chem. 14 (2012) 2682–2685.

63. S. N. Kosourov, M. Seibert, Hydrogen photoproduction by nutrient-deprived *Chlamydomonas reinhardtii* cells immobilized within thin alginate films under aerobic and anaerobic conditions, Biotechnol. Bioeng. 102 (2009) 50–58.

64. O. Kruse, J. Rupprecht, J. H. Mussgnug, G. C. Dismukes, B. Hankamer, Photosynthesis: a blue print for energy capture and conversion technologies, Photochem. Photobiol. Sci. 4 (2005) 957–970.

65. T. Happe, A. Hemschemeier, M. Winkler, A. Kaminski, Hydrogenases in green algae: do they save the algae's life and solve our energy problems? Trends Plant Sci. 7 (2002) 246–250.

66. A. S. Fedorov, S. Kosourov, M. L. Ghirardi, M. Seibert, Continuous H_2 photoproduction by *Chlamydomonas reinhardtii* using a novel two-stage, sulfate-limited chemostat system, Appl. Biochem. Biotechnol. 121–124 (2005) 403–412.

67. A. Melis, H. C. Chen, Chloroplast sulfate transport in green algae—genes, proteins and effects, Photosynth. Res. 86 (2005) 299–307.

68. C. Faraloni, A. Ena, C. Pintucci, G. Torzillo, Enhanced hydrogen production by means of sulfur-deprived *Chlamydomonas reinhardtii* cultures grown in pretreated olive mill wastewater, Int. J. Hydrogen Energy 36 (2011) 5920–5931.

69. K. A. Batyrova, A. A. Tsygankov, S. N., Kosourov, Sustained hydrogen photoproduction by phosphorus-deprived *Chlamydomonas reinhardtii* cultures, Int. J. Hydrogen Energy 37 (2012) 8834–8839.

70. A. Melis, J. Neidhardt, J. R. Benemann, *Dunaliella salina* (Chlorophyta) with small chlorophyll antenna sizes exhibit higher photosynthetic productivities and photon use efficiencies than normally pigmented cells, J. Appl. Phycol. 10 (1999) 515–525.

71. I. Akkerman, M. Janssen, J. Rocha, R. H. Wijfels, Photobiological hydrogen production: photochemical efficiency and bioreactor design, Int. J. Hydrogen Energ. 27 (2002) 1195–1208.

72. A. Melis, Solar energy conversion efficiencies in photosynthesis: minimizing the chlorophyll antennae to maximize efficiency, Plant Sci. 177 (2009) 272–280.

73. J. E. W. Polle, S. Kanakagiri, A. Melis, Tla1, a DNA insertional transformant of the green alga *Chlamydomonas reinhardtii* with a truncated light-harvesting chlorophyll antenna size, Planta 217 (2003) 49–59.

74. K. Nath, A. Kumar, D. Das, Hydrogen production by *Rhodobacter sphaeroides* strain O.U.001 using spent media of *Enterobacter cloacae* strain DM11, Appl. Microbiol. Biotechnol. 68 (2006) 533–541.

75. B. Hankamer, F. Lehr, J. Rupprecht, J. H. Mussgnug, C. Posten, O. Kruse, Photosynthetic biomass and H_2 production by green algae: from bioengineering to bioreactor scale-up, Physiol. Plant. 131 (2007) 10–21.

76. A. A. Tsygankov, D. O. Hall, J. Liu, K. K. Rao, An automated helical photobioreactor incorporating cyanobacteria for continuous hydrogen production. In: O. R. Zaborsky (Ed.), *Bio-hydrogen*, Plenum Press, London, 1998, pp. 431–440.

77. R. E. Blankenship, D. M. Tiede, J. Barber, G. W. Brudvig, G. Fleming, M. Ghirardi, M. R. Gunner, W. Junge, D. M. Kramer, A. Melis, T. A. Moore, C. C. Moser, D. G. Nocera, A. J. Nozik, D. R. Ort, W. W. Parson, R. G. Prince, R. T. Sayre, Comparing photosynthetic and photovoltaic efficiencies and recognizing the potential for improvement, Science 332 (2011) 805–809.

78. C. N. Dasgupta, J. J. Gilbert, P. Lindblad, T. Heidorn, S. A. Borgvang, K. Skjanes, D. Das, Recent trends on the development of photobiological processes and photobioreactors for the improvement of hydrogen production, Int. J. Hydrogen Energy 35 (2010) 10218–10238.

79. A. Tanaka, A. Melis, Irradiance-dependent changes in the size and composition of the chlorophyll *a–b* light-harvesting complex in the green alga *Dunaliella salina*, Plant Cell Physiol. 38 (1997) 17–24.

80. M. Mitra, A. Melis, Optical properties of microalgae for enhanced biofuels production, Opt. Express. 16 (2008) 21807–21820.

81. S. D. Tetali, M. Mitra, A. Melis, Development of the light-harvesting chlorophyll antenna in the green alga *Chlamydomonas reinhardtii* is regulated by the novel Tla1 gene, Planta 225 (2007) 813–829.

82. M. Mitra, A. Melis, Genetic and biochemical analysis of the TLA1 gene in *Chlamydomonas reinhardtii*, Planta 231 (2010) 729–740.

83. H. Kirst, J. G. García-Cerdán, A. Zurbriggen, A. Melis. Assembly of the light-harvesting chlorophyll antenna in the green alga *Chlamydomonas reinhardtii* requires expression of the TLA2-CpFTSY gene, Plant Physiol. 158 (2012) 930–945.

84. M. Seibert, D. K. Benson, T. Flynn, Method and Apparatus for Rapid Bio-hydrogen Phenotypic Screening of Microorganisms Using a Chemochromic Sensor, U.S. Patent 6, 277 (2001) 589 B1.

85. C. Bolling, O. Fiehn, Metabolite profiling of *Chlamydomonas reinhardtii* under nutrient depletion, Plant Physiol. 139 (2005) 1995–2005.

86. M. Timmins, W. Zhou, L. Lim, S. R. Thomas-Hall, B. Hankamer, U. C. Marx, S. M. Smith, P. M. Schenk, The metabolome of *Chlamydomonas reinhardtii* following induction of anaerobic H_2 production by sulphur deprivation, J. Biol. Chem. 284 (2009) 23415–23425.

87. F. Mus, A. Dubini, M. Seibert, M. C. Posewitz, A. R. Grossman, Anaerobic acclimation in *C. reinhardtii*: anoxic gene expression, hydrogenase induction, and metabolic pathways, J. Biol. Chem. 282 (2007) 25475–25486.

CHAPTER 11

Growth in Photobioreactors

NIELS THOMAS ERIKSEN

11.1 INTRODUCTION

Photobioreactors are transparent bioreactors where light can penetrate into the reactors and be used for photosynthesis and production of phototrophic organisms. These bioreactors are used for cultivation of phototrophic bacteria, mosses, plant cells, and particularly microalgae. Several eukaryote phyla as well as cyanobacteria are recognized as microalgae, and their phylogenetic variability is therefore enormously large. Most microalgae depend on light for growth although some are also able to take up organic carbon and combine photosynthesis with heterotrophic nutrition. Only a minority of the microalgae grow well heterotrophically with no need for light [1, 2].

Phototrophic microalgae are produced in open ponds or raceways or in closed photobioreactors for use in foods and health foods, for feeds used in the aquaculture industry, and for production of pigments, polyunsaturated fatty acids, or other compounds [3, 4]. In ponds and raceways the culture is in direct contact with air and contaminating microorganisms and compounds have direct access to the cultures. Although ponds and raceways are the dominating systems for production of microalgae, only a few, competitive species can successfully be maintained in open systems and often under harsh, selective conditions. Photobioreactors are usually closed, meaning that liquids and aeration gases are sterilized before being added to the cultures. This allows cultivation of a greater variety of species in either axenic or monoalgal cultures.

Photobioreactors have been constructed in numerous different ways but they are generally designed to maximize the light supply and the utilization of light by the culture inside. Mixing and gas exchange are also important for the performance of microalgal cultures. Light and CO_2 supplies are essential to photosynthetic processes as demonstrated by the overall reaction stoichiometry of photosynthetic synthesis of carbohydrates.

$$CO_2 + H_2O + \text{light} \rightarrow CH_2O + O_2 \tag{11.1}$$

Natural and Artificial Photosynthesis: Solar Power as an Energy Source, First Edition. Edited by Reza Razeghifard.
© 2013 John Wiley & Sons, Inc. Published 2013 by John Wiley & Sons, Inc.

Light cannot be mixed homogeneously inside cultures of microalgae, and the supply of light in combination with inhomogeneous distribution of light inside photobioreactors is the most prominent rate-limiting factor in microalgal cultures [5]. The productivity of microalgal cultures can also be CO_2 limited. Light and CO_2 limitations are kinetic limitations. Both are supplied during cultivation but not necessarily at rates that saturate the needs of the culture. If internal light intensities or CO_2 concentrations are insufficient to saturate the microalgal cells, specific growth rates will be lower than maximal. In batch cultures, light limitation results in linear growth curves with slopes depending on the incident light intensity [6].

11.2   DESIGN OF PHOTOBIOREACTORS

Many photobioreactors have been designed to maximize light collection, minimize dark zones, and enable sufficient mixing and gas exchange. Small photobioreactors are often illuminated by artificial light sources while natural sunlight is always the source of light in large photobioreactors. In order to collect as much light per volume of microalgal culture as possible, the surface area of photobioreactors in relation to volume must be high while still allowing mixing and aeration of the cultures inside. Most photobioreactors can be categorized as tubular, panel, or column photobioreactors. Detailed descriptions on how these photobioreactors are constructed and operated have been reviewed in Pulz [7], Janssen et al. [8], Tredici [9], Carvalho et al. [10],Chisti [11], Eriksen [12], and Posten [13].

In tubular photobioreactors, the microalgal cultures are pumped through long, narrow transparent tubes. The tubes can be organized horizontally [14, 15], vertically [16, 17, 18], inclined [19], or as a helix [20, 21, 22, 23, 24]. Mechanical pumps or airlifts create the pumping force. Airlifts also allow CO_2 and O_2 to be exchanged between the liquid medium and the aeration gas [14, 17, 21, 23, 24], while little gas exchange takes place in the tubes. Although tubular photobioreactors are the most commonly used closed photobioreactor type for large-scale cultivation of microalgae [11], O_2 accumulation, CO_2 depletion, and pH variations put a limit to the length of the tubes. Scale-up of tubular photobioreactors must therefore rely partly on multiplication of reactor units [8].

Panel photobioreactors are made of two transparent plates with a short distance in between [25, 26, 27]. These reactors support the highest densities of photoautotrophic cells. Biomass concentrations have exceeded 80 g L^{-1} [25] and highest productivities of microalgal cultures of up to 4–8 g L^{-1} day^{-1} [25, 28] have been obtained in panel photobioreactors. In alveolar panel photobioreactors the microalgal cultures are enclosed in thin, transparent tubes running in parallel [29, 30]. This construction allows better control of the fluid motion inside the panel.

Column photobioreactors can be stirred tank reactors [31, 32, 33, 34] but have more often been constructed as bubble columns [35, 36, 37] or airlifts [38, 39, 40]. The columns are placed vertically, aerated from below, and illuminated through transparent walls. Light sources have also been installed internally [39, 41]. In vertical column photobioreactors, it is particularly difficult to keep the inner dark zone of the reactor small and at the same time have a large reactor surface for light collection. In annular column photobioreactors the inner, unproductive dark zone is avoided or kept at a minimal volume by installation of an air-filled inner tube [42]. Still, scale-up of column photobioreactors remains a challenge. Undesirable long internal light paths are created if the diameter of the column is increased, while high hydrostatic pressures in the bottom of the reactors are developed if column height is increased. Scale-up of column photobioreactor plants must therefore also rely on multiplication of reactor units [42, 43].

11.3   LIMITATIONS TO PRODUCTIVITY OF MICROALGAL CULTURES

In most microalgal cultures, the biomass or biomass components are the product of interest. The productivity of microalgal cultures is therefore best described in terms of biomass productivity. The overall productivity of a light-limited microalgal culture depends on the rate by which photons are supplied to the photobioreactor multiplied by the net photosynthetic yield, that is, the efficiency by which photons are utilized to fix inorganic carbon into biomass.

It costs a minimum of 8 moles of photons for the light reactions of photosynthesis to generate the 2 moles of reduced NADPH and also the ATP that are subsequently utilized by Calvin's cycle to reduce and incorporate 1 mole of CO_2 into glyceraldehyde-3-phosphate or carbohydrate, the primary products of photosynthesis (Fig. 11.1). This corresponds to a maximal photosynthetic yield of 0.125 mol carbon (mol photon)$^{-1}$ based on the reaction in Eq. 11.1. However, actual photosynthetic yields are lower. NADPH and ATP are also consumed by other biochemical reactions and some of the carbon once fixed may also be lost again as CO_2 [44]. For example, fatty acids and some amino acids are synthesized via pathways that involve decarboxylation reactions. Lipids are also more reduced than carbohydrates and their synthesis depends on additional consumption of NADPH. Reduced ferredoxin or NADPH is also used to reduce, for example, nitrate and sulfate when these are used as nitrogen and sulfur sources. The different biosynthesis as well as maintenance processes that also take place in living cells create an additional demand for ATP that is met by cyclic electron flow around photosystem I, a process that depends on photons but does not generate NADPH. Respiratory processes also take place during darkness as well as in the light [45] and these may reoxidize a considerable fraction of the produced biomass back into CO_2 [46].

The photosynthetic yields of microalgal cultures are not determined by the biochemical reactions alone since photons absorbed by photosynthetic pigments may not all be used for photosynthesis. A variable fraction of the absorbed photons will be lost as fluorescence or heat. The efficiency by which microalgae utilize absorbed photons for photosynthesis and growth is strongly dependent on the light intensity experienced by the cells. Figure 11.2A shows how the specific growth rate μ is expected to depend on light intensity. At low light intensities, μ is negative due to respiratory processes but increases with increasing light intensity as long as the light intensity remains below saturation intensity. Above this intensity, photosynthesis is saturated

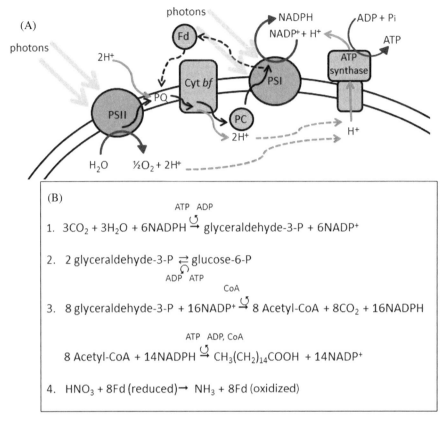

FIGURE 11.1 (A) Schematic representation of the light reactions of photosynthesis taking place at thylakoid membranes. A total of 2×4 photons (light gray arrows) are needed to excite chlorophyll a molecules in photosynthetic reaction centers II and I (PSII and PSI) and transfer 4 electrons (black arrows) to $NADP^+$ from the water molecules oxidized to O_2. The proton gradient that is created across the thylakoid membrane (dark gray arrows) is used for ATP synthesis. Other major components of the electron transport chain are plastoquinone (PQ), cytochrome b/f (cyt b/f), and plastocyanine (PC). PSI can also catalyze cyclic electron flow via ferredoxin (Fd) and build up the proton gradient without forming NADPH. (B) Examples of reactions of photosynthesis. Reaction 1: In Calvin's cycle, NADPH and ATP formed in the light reactions are used to reduce and incorporate CO_2 into glyceraldehyde-3-phosphate. Reaction 2: Two molecules of glyceraldehyde-3-phosphate can be condensed into glucose-6-phosphate. Reaction 3: Transformation of glyceraldehyde-3-phosphate into fatty acids (e.g., C16 palmitic acid) involves decarboxylations and additional NADPH dependent reductions. Reaction 4: Reduction of one nitrate molecule is at the expense of 8 electrons delivered by ferredoxin.

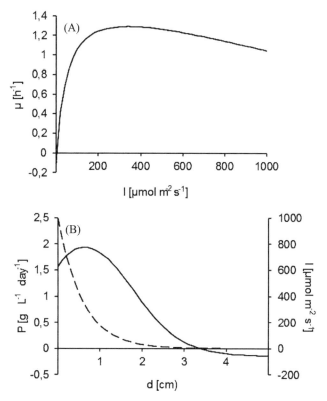

FIGURE 11.2 (A) Schematic representation of specific growth rate (μ) as function of light intensity. (B) Light intensity (I, - - - -) and predicted local productivity (P, ———) as function of distance (d) from surface of microalgal culture.

and μ no longer stimulated by increased light intensity. The rate by which photons are absorbed by the photosynthetic pigments is proportional to the light intensity but the photosynthetic reaction centers can process only one photon at a given time and only reaction centers that are not already excited are able to process newly absorbed photons into electron transport. Since each reaction center is connected to an antenna complex comprised of hundreds of pigment molecules, photons will at high light intensities be absorbed at faster rates than they can be processed by the reaction centers. Excess photons are dissipated as heat or fluorescence, and may even work in formation of singlet oxygen and thereby cause photoinhibition and inactivation of the photosynthetic apparatus [47]. In many microalgae, photosynthesis saturates at 100–200 μmol photons m^{-2} s^{-1}, while full sunlight causes photoinhibition and productivities below maximal levels.

The performance of photobioreactors and microalgal cultures is also commonly reported as photosynthetic efficiency, PE, which is the ratio between chemical energy stored in the produced biomass compared to the light energy supplied to the photobioreactor. In Fig. 11.2A, the highest PE would be found at low, limiting light intensities and the lowest PE at high, saturating light intensities. When it is taken into consideration that less than half of the energy in sunlight is outside the spectrum of the photosynthetic active radiation ($\sim$400–665 nm), that all photons generate the same amount of NADPH and ATP although photons in the blue part of the photosynthetic active spectrum carry more energy than photons in the red part, that there is a loss of energy associated with each step in the electron transport chain in the thylakoid membrane of the chloroplasts (Fig. 11.1A) as well as each biochemical step in the carbon reactions (Fig. 11.1B), the maximal obtainable PE for microalgal cultures grown in sunlight is less than 10% [48].

11.4 ACTUAL PRODUCTIVITIES OF MICROALGAL CULTURES

The effect of light limitation on productivity can be illustrated using the 3 L bubble column photobioreactor in Fig. 11.3 as an example. The average incident light intensity on the surface of this photobioreactor is 200 μmol photons m^{-2} s^{-1}. If all photons were utilized for fixation of CO_2 into carbohydrates, the maximal rate of carbon fixation and biomass production in this photobioreactor would be 0.8 or 1.5 g L^{-1} day^{-1}, respectively (calculations on productivity are described in Table 11.1). In a bubble column photobioreactor designed similarly to the photobioreactor in Fig. 11.3, and illuminated by 205 μmol photons m^{-2} s^{-1}, Eriksen et al. [49] measured the steady-state biomass productivity in a light-limited phototrophic continuous flow culture of *Rhodomonas* sp., a marine microalga used to feed filter feeding invertebrates, to be 0.4 g organic carbon L^{-1} day^{-1} or roughly half the theoretical maximal productivity at this incident light intensity.

Outdoor photobioreactors illuminated by the sun can, during daytime, be exposed to incident light intensities above 2000 μmol m^{-2} s^{-1}. Still, productivities have in most studies remained below 1–2 g L^{-1} day^{-1} and have only in a few examples been higher. Table 11.2 summarizes biomass productivities measured in various types of photobioreactors and compares productivities to cultures grown in open ponds and raceways. For comparison, biomass productivities in cultures of heterotrophic bacteria and yeast can be as high as 100–230 g L^{-1} day^{-1} [50]. The highest productivity seen in a microalgal culture is probably 50 g biomass L^{-1} day^{-1}, but this was in a heterotrophic culture of *Galdieria sulphuraria* [51].

Especially in large-scale photobioreactors, productivity is conveniently described as biomass productivity per unit of reactor surface area, per unit of land area covered by the reactor, or per unit of land area covered by the reactor and auxiliary installations [13]. The land area covered by the photobioreactor is related to the amount of light supplied to the photobioreactor. Productivity per area of land covered by the reactor therefore describes how efficiently light energy is utilized for biomass synthesis at a given location, and it enables comparisons to agricultural crops. The different photobioreactor types—tubular, panel, and column photobioreactors—have all provided biomass productivities of 20–40 g m^{-2} day^{-1} (Table 11.2) and PE of 5–9%. These areal productivities are not very much higher than those found in open ponds and raceways, while the volumetric productivities are highest in closed photobioreactors

FIGURE 11.3   Bubble column photobioreactor (3 L) containing continuous flow culture of *Rhodomonas salina*. Estimation of maximal biomass productivity is presented in Table 11.1.

TABLE 11.1   Estimation of Maximal Biomass Productivity Potential in the Light-Limited 3 L Bubble Column Photobioreactor in Fig. 11.1

Biomass composition[a]	$CH_{1.8}O_{0.5}N_{0.2}$
Biomass molecular weight[b]	$MW_x = 26$ g C-mol^{-1}
Maximal photosynthetic efficiency	$PE_{max} = 0.125$ C-mol photon^{-1}
Reactor volume	$V = 3$ L
Reactor surface area	$A_s = 0.08$ m^2
Surface photon flux density	$PFFD_s = 200$ µmol photons m^{-2} s^{-1}
Volumetric photon flux	$PFFD_V = (PFFD_s \cdot A_s)/V = 0.0192$ mol photons L^{-1} h^{-1}
Maximal cell carbon productivity	$r_{c,max} = PE_{max} \cdot PFFD_V = 0.0024$ C-mol L^{-1} h^{-1}
Maximal biomass productivity	$r_{x,max} = r_{c,max} \cdot MW_x = 1.5$ g L^{-1} day^{-1}

[a]Average biomass composition of microorganisms taken from Roels [97].
[b]Molecular weight includes 5% ash in biomass.

TABLE 11.2   Biomass Concentrations and Volumetric and Areal Productivities[a] in Photoautotrophic Cultures Grown in Different Types of Open Systems or Enclosed Photobioreactors

Reactor Type	Species	Light Source	Biomass $(g\,L^{-1})$	Productivity $(g\,L^{-1}\,day^{-1})$	Productivity $(g\,m^{-2}\,day^{-1})$	Reference
Open systems						
Raceway	*Spirulina*	Sun	0.47	0.05	14.0	[98]
Raceway	*Arthrospira*	Sun	0.9	0.15	12.2	[99]
Raceway	*Anabaena*	Sun	0.23	0.24	23.5	[100]
Open panel	*Chlorella*	Sun	—	*3.8*	22.8	[101]
Open panel	*Chlorella*	Sun	—	*3.2*	19.4	[101]
Closed photobioreactors						
Tubular	*Spirulina*	Artificial	—	0.42	—	[17]
Tubular	*Arthrospira*	Sun	2.37	1.15	25.4	[16]
Tubular	*Phaeodactylum*	Sun	2.29	1.15	19.1	[14]
Tubular	*Phaeodactylum*	Sun	4.1	1.52	25.3	[14]
Panel	*Nannochloropsis*	Sun	—	0.24	12.1	[102]
Panel	*Chlorococcum*	Artificial	20	8.0	16.7	[25]
Alveolar panel	*Spirulina*	Sun	6.9	1.26	15.8	[29]
Column	*Rhodomonas*	Artificial	*1.3*	*0.8*	*8.5*	[49]
Column	*Arthrospira*	Sun	3.0	0.91	—	[103]
Column	*Tetraselmis*	Sun	1.7	0.42	38.2	[42]

[a]Values in italics are calculated based on information in the references.

since cells in these systems are suspended in smaller volumes. Highest productivities are obtained at high light intensities while the PE is highest at low light intensities. In outdoor cultures, highest PE is seen in the morning and in the afternoon [15].

11.5   DISTRIBUTION OF LIGHT IN PHOTOBIOREACTORS

The productivity of a phototrophic culture is influenced by the incident light intensity on the surface of the reactor as well as the way the light is distributed to the cells inside the reactor. At the culture surface, light intensities are high but absorption and scattering result in rapid attenuation of light intensities as function of the distance from the surface. A few millimeters to centimeters below the culture surface it will be virtually dark [52].

If light travels radially in one direction from the surface toward the center of the reactor and is attenuated by absorption by the cells, the light intensity will decrease exponentially with distance from the surface. Scattering by cells and other particles has a randomizing effect on the direction of the light [53] and cells in microalgal cultures are exposed to dispersed light coming from all directions. Also air bubbles affect light intensity gradients [43] and can increase the light penetration depth [54]. In column photobioreactors illuminated from all sides and plate bioreactors, light distribution profiles may still be well described when only radial light paths are taken into consideration [18, 32]. Light distribution models have also been expanded to take into account the dispersed light derived from scattering by cells and other particles, reactor geometry and orientation, and the position of the sun have also been developed and used to describe light intensities in outdoor photobioreactors of different geometries [36, 55, 56].

Figure 11.2B illustrates how light intensity and local productivity are expected to change with distance from the reactor surface in a microalgal culture [57]. At the culture surface, photosynthesis is saturated but productivity may be negatively affected by light inhibition. In central parts, light intensities may fall below the compensation point where photosynthesis balances respiration and production is negative. Highest productivity is found in a narrow zone at intermediate light intensities where photosynthesis is saturated but the cells are not inhibited. Since the PE is low when photosynthesis is saturated, the PE will be low in this zone. If dark zones develop, the culture productivity is reduced as respiration exceeds production in these zones [58]. Takache et al. [59] demonstrated that the volumetric productivity of *Chlamydomonas reinhardtii* cultures grown in different photobioreactor types is mainly a function of incident light intensity on the reactor surface, the ratio between the illuminated surface area and the culture volume, and the fraction of the photobioreactor where light intensity is greater than the compensation point. The productivity was highest in cultures where the light intensity in the darkest parts of the photobioreactors reached the compensation point but was not decreased below this value. In cyanobacterial cultures, the productivity may be less sensitive to respiration in dark zones and maximal productivities are reached when light intensities are decreased below the compensation point in the darkest zones of the photobioreactors [60].

Figure 11.2B does not take into account that currents in liquid media move cells through the differently illuminated zones. The individual cells will therefore, because of mixing, experience fluctuations in light intensity. Many studies have shown that these fluctuations affect PE and thereby productivities of microalgal cultures, see for example, Grobbelaar [5] and Richmond [61, 62], although there may be exceptions [58]. Mixing is also important for homogeneous distribution of cells, nutrients, heat, and transfer of gasses across gas–liquid interfaces in microalgal cultures.

Fluctuating light intensities affect productivity when incident light intensities are high enough to saturate or inhibit photosynthesis. If the fluctuations between high light intensities and darkness occur at a frequency faster than $1 \ s^{-1}$, the specific growth rates and productivities of microalgal cultures are enhanced [57, 63, 64]. Excitation energy can then be carried into the dark zones and there do photosynthetic work at rates similar to what would have been found in continuous light while saturation of the photosynthetic apparatus is minimized [65]. Also the inactivation of the photosynthetic apparatus may proceed at a lower rate when intense light is supplied intermittently and not continuously [63, 65]. In outdoor cultures where photosynthetic photon flux densities rise above $1000 \ \mu mol \ m^{-2} \ s^{-1}$, light exposure times should preferentially be as short as 10 ms to maintain maximal PE [66]. Still, light/dark cycles of 94/94 ms were sufficiently short to increase PE in cultures of *Dunaliella tertiolecta*, while light/dark cycles of 3/3 s were too long and resulted in a lower PE than in continuous light in *D. tertiolecta* [66] as well as in *Chlamydomonas reinhardtii* [67].

Panel photobioreactors are well designed to create rapid fluctuations in light intensities experienced by the cells and to keep dark zones at a minimum. These reactors have short light paths and steep light gradients if operated at high cell densities. This enables rapid circulation of cells between illuminated and dark zones, and the cells are only exposed to high surface light intensities for fractions of a second. Richmond et al. [27] found that the PE of *Nanochloropsis* sp. cultures was almost doubled when the light path length was shortened from 9 to 1 cm and the cell density increased from 3.9 to 43.5 g L^{-1}. The biomass productivity in a panel photobioreactor was also improved by installation of stationary baffles that increase the rate of medium circulation through the light gradient [26].

In tubular photobioreactors, cells are also exposed to fluctuating light intensities when the liquid medium is pumped through the tubes at sufficiently high velocities to create turbulent flow, where eddies in the liquid will cause the cells to travel between illuminated zones near the wall of the pipes and darker zones in the center of the pipes. Turbulent flow is therefore a necessity for high biomass productivity in tubular photobioreactors [61].

In airlift photobioreactors, the frequency of light exposure is largely determined by the circulation time through riser and down-comer, while it is the turbulent eddies alone that circulate cells between illuminated and dark zones in bubble columns. Airlifts have often been regarded superior to bubble columns because of their well-defined flow patterns and circulation times, and some studies also report higher productivity in airlift compared to bubble column photobioreactors [38, 40, 68]. However, the circulation times in airlifts are in the order of several seconds, which would be too slow to diminish light saturation and photoinhibitory effects [66, 67]. Different studies also do not unambiguously support that airlifts are superior to bubble columns. Mirón et al. [69] and Barbosa et al. [70] found that bubble column photobioreactors were comparable or even superior to airlifts with regards to productivity.

11.6 GAS EXCHANGE IN PHOTOBIOREACTORS

The exchange of gasses between gas and liquid phases can be as important to microalgal cultivation as the supply of light. In particular, the magnitude of the CO_2 transfer rate is of concern. The solubility of CO_2 in aqueous media is low compared to the need for CO_2 in a typical microalgal culture but many species can use also bicarbonate ions as the carbon source. Membrane bound bicarbonate transporters allow HCO_3^- uptake into cells and chloroplasts [71] or bicarbonate ions can be converted into CO_2 by extracellular carbonic anhydrase enzymes [72]. The primary role of the CO_2 concentrating mechanisms may be to minimize photorespiration but they also increase the amount of available carbon in growth media. The relationship between the different inorganic carbon species in water is described by the carbonate system ($pK_{a,1} \approx 6.3$ and $pK_{a,2} \approx 10.3$)

$$CO_2 + H_2O \rightleftarrows H_2CO_3 \rightleftarrows HCO_3^- + H^+ \rightleftarrows CO_3^{2-} + 2H^+ \qquad (11.2)$$

In neutral and alkaline growth media, HCO_3^- is the dominating inorganic carbon species, and the available reservoir of dissolved inorganic carbon is much larger than the dissolved CO_2 alone. However, consumption and utilization of HCO_3^- in microalgal cultures are at the expense of a concurrent proton consumption and may cause considerable increases of pH [73]. CO_2 is therefore supplied to microalgal cultures to provide carbon as well as to prevent changes of pH.

The CO_2 transfer rate (CTR) from the aeration gas to the liquid medium is described by

$$CTR = k_L a_{CO_2}(c_{CO_2}^* - c_{CO_2}) \tag{11.3}$$

where $k_L a_{CO_2}$ is the mass transfer coefficient of CO_2 and $c_{CO_2}^*$ and c_{CO_2} are saturation dissolved CO_2 and dissolved CO_2 concentrations, respectively. In a steady-state situation, CTR will balance the CO_2 uptake rate $(-r_{CO_2})$ in the photobioreactor

$$-r_{CO_2} = CTR \tag{11.4}$$

In a microalgal culture with a maximal productivity of $1\ g\ L^{-1}\ day^{-1}$, the need for inorganic carbon is roughly $0.5\ g\ L^{-1}\ day^{-1}$ or $1.8\ g\ CO_2\ L^{-1}\ day^{-1}$. The solubility of CO_2 in equilibrium with air containing $0.04\%\ CO_2$ is just $0.7\ mg\ L^{-1}$ at $20\ °C$ (e.g., calculated from Weiss [74]) and therefore 2570 times more CO_2 must be supplied each day. Figure 11.4 is an illustration of the relationships between $k_L a_{CO_2}$, CTR, and c_{CO_2} in such a culture (calculations are described in Table 11.3). If the culture is aerated by air, only photobioreactors with $k_L a_{CO_2}$ values above $110\ h^{-1}$ are able to supply 1.8 g

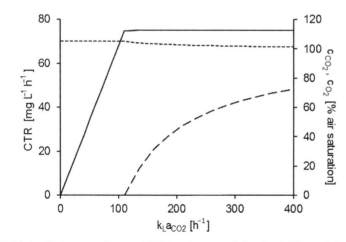

FIGURE 11.4 Carbon transfer rate (CTR) and expected dissolved CO_2 and O_2 concentrations (c_{CO_2} and c_{O_2}, respectively) as function of the mass transfer coefficient of CO_2 ($k_L a_{CO_2}$) in photobioreactor at steady state where maximal biomass production rate is $1\ g\ L^{-1}$, corresponding to a maximal CO_2 uptake rate of $75\ mg\ L^{-1}\ h^{-1}$. Data estimated as described in Table 11.3.

TABLE 11.3   Estimation of CO_2 Transfer Rate and Dissolved CO_2 and O_2 Concentrations as Function of Mass Transfer Coefficient for CO_2, $k_La_{CO_2}$ During Steady State[a]

Saturation dissolved CO_2 concentration (20 °C)[b]	$c^*_{CO_2} = 0.68$ mg L^{-1}
CO_2 consumption rate	$-r_{CO_2} = CTR$
CO_2 transfer rate	$CTR = k_La_{CO_2} \cdot c^*_{CO_2}$ ($0 \leq CTR \leq 75$ mg L^{-1} h^{-1})
Dissolved CO_2 concentration	$c_{CO_2} = c^*_{CO_2} - r_{CO_2}/k_La_{CO_2}$ ($c_{CO_2} \geq 0$)
Saturation dissolved O_2 concentration (20 °C)	$c^*_{O_2} = 9.1$ mg L^{-1}
Photosynthetic quotient	$PQ = -r_{CO_2}/r_{O_2} = 1$
O_2 evolution rate[c]	$r_{O_2} = -r_{CO_2} \cdot PQ \cdot MW_{O_2}/MW_{CO_2}$
O_2 mass transfer coefficient[d]	$k_La_{O_2} = 1.1 \cdot k_La_{CO_2}$
Dissolved O_2 concentration	$c_{O_2} = c^*_{O_2} + r_{O_2}/k_La_{O_2}$

[a]Maximal biomass productivity is set at 1 g L^{-1} day^{-1} corresponding to a maximal CO_2 consumption, $-r_{CO_2} = 75$ mg L^{-1} h^{-1}, assuming 50% carbon content in the biomass. Results are shown in Fig. 11.5.
[b]Calculated according to Weiss [74].
[c]Molecular weights; $MW_{O_2} = 32$ g mol^{-1}, $MW_{CO_2} = 44$ g mol^{-1}.
[d]Calculated according to von Schalien et al. [82].

CO_2 L^{-1} day^{-1}. At lower $k_La_{CO_2}$ values, cultures aerated by air will be carbon limited with $c_{CO_2} \approx 0$, and c_{CO_2} will remain at values considerably lower than $c^*_{CO_2}$ even at 3–4 times higher $k_La_{CO_2}$ values. This is also what was observed in cultures of *Dunaliella tertiolecta* and *Chlorella vulgaris* grown in bubble column photobioreactors [75]. In reality microalgal cultures can remain carbon limited also at higher c_{CO_2} values despite their ability to utilise HCO_3^-. For example, *Chlamydomonas reinhardtii*, *Chlorella pyrenoidosa*, and *Scenedesmus obliquus* did not achieve maximal specific growth rates unless CO_2 concentrations were above 1–4 mg L^{-1} [76], which is even above $c^*_{CO_2}$.

Zhang et al. [77] and Ugwu et al. [78] compared k_La values in different photobioreactors and found that they ranged between 0.36 and 72 h^{-1}. Some of the lowest k_La values were found in tubular and panel photobioreactors, while the highest k_La values were in airlift photobioreactors. In all these photobioreactors, the k_La values were too low to support the CO_2 requirements of microalgal cultures if aerated by air, unless productivities for other reasons were much below 1 g L^{-1} day^{-1}. Carbon limitation is therefore a potential growth-limiting factor in all types of photobioreactors aerated by air, although a couple of studies have indicated that actual CTRs in microalgal cultures can be higher than predicted by Eq. 11.3 [79, 80]. Fully satisfactory explanations for these unexpectedly high CTRs were not provided but it was suggested that the microalgal cultures could have affected the properties of the liquid–gas interphase.

Oxygen produced by photosynthesis has to be released from the medium and into the gas phase. If not, high dissolved O_2 concentrations will build up. As a result, photorespiration will increase causing a decrease in PE [81]. The relationship between rates of O_2 production (r_{O_2}) and O_2 transfer between liquid and gas phases (OTR) in a steady-state situation can be described by

$$r_{O_2} = \text{OTR} = k_1\, a_{O_2}(c^*_{O_2} - c_{O_2}) \tag{11.5}$$

where the mass transfer coefficient of oxygen $k_L a_{O_2}$ is approximately 10% larger than $k_L a_{CO_2}$ since O_2 is a smaller molecule [82], and $c^*_{O_2}$ and c_{O_2} are saturation dissolved O_2 and dissolved O_2 concentrations, respectively. If the relationship between $-r_{CO_2}$ and r_{O_2} follows the stoichiometry of Eq. 11.1, c_{O_2} will obtain a steady-state value at 105% air saturation at $k_L a$ values below 110 h^{-1} (Fig. 11.4). This dissolved O_2 concentration is probably not high enough to impose major negative effects on the productivity of microalgal cultures [81]. However, in tubular photobioreactors with no or little gas exchange in the tubes, oxygen can build up to concentrations of 200–300% that of air saturation [14, 55]. In this situation photorespiration can certainly affect PE and productivity, and it has also been observed that such high dissolved oxygen concentrations are associated with relative low productivities [14].

Because of the low $k_L a$ values in photobioreactors, 1–10% CO_2 is often added to the aeration gas to increase $c^*_{CO_2}$ and thereby CTR. However, CTR will not increase proportionally to $c^*_{CO_2}$ since also c_{CO_2} will increase unless the cultures remain carbon limited (Fig. 11.4). Elevated CO_2 concentrations in the aeration gas will therefore reduce the efficiency by which CO_2 is transferred into the medium and utilized by the microalgae. If a microalgal culture with a CO_2 consumption rate of 1.8 g L^{-1} day^{-1} is aerated at a rate of 0.1 L of gas per L of medium per min (0.1 VVM), and the aeration gas contains 1% CO_2, a total of 26 g of CO_2 will daily pass through each liter of culture, but only 7% of the CO_2 will be incorporated into new biomass. CO_2 can also be added to the aeration gas to keep CO_2 at predetermined levels in the photobioreactor or used as acidic titrant to maintain constant pH [6, 34]. In such cases, CO_2 is added to the aeration gas only when CO_2 is below a set point or pH is increased above a desired value. CO_2 has then been utilized at more than 70% efficiency in systems open to air [83] and more than 90% efficiency in closed photobioreactors [84]. A particularly efficient way to increase CTR has been to separate the CO_2 supply from the remainder aeration gas. In so-called dual sparging photobioreactors, large air bubbles ($\sim$4 mm in diameter) are supplied continuously through one sparger in order to mix the culture, while pure CO_2 is added through a different, perforated rubber membrane sparger that creates small bubbles ($\sim$1 mm in diameter) with only little mixing power but high surface-to-volume ratios and approximately 2500 times higher CO_2 partial pressure than in air [85]. Separation of CO_2 from the remainder aeration gas resulted in 5 times higher CTR in a 1.7 L bubble column photobioreactor [49].

11.7 SHEAR STRESS IN PHOTOBIOREACTORS

Rapid mixing, meaning high liquid velocities inside bioreactors, are needed to expose cells to rapid fluctuations in light intensity and to maximize gas transfer rates. Mixing depends on the power input from aeration, impellers, or mechanical pumps [80]. Sierra et al. [86] and Hulatt and Thomas [75, 87] used power inputs of 0.1–0.53 W L^{-1} for aeration and mixing in bubble column and panel photobioreactors. Considerably higher power inputs of 25–34 W L^{-1} have been reported for tubular photobioreactors [24, 88].

It is of general concern that high liquid velocities and high degrees of turbulence can damage shear-sensitive microalgal cells. However, shear damage does not seem to occur in tubular photobioreactors despite the fact that these are the photobioreactors exposed to highest power inputs [86]. More striking is the observation that growth and glucose uptake are only moderately inhibited in the heterotrophic dinoflagellate *Crypthecodinium cohnii* when exposed to power inputs as high as 5.9 kW L^{-1} [89]. Bursting air bubbles are probably more damaging to the microalgal cells than the turbulent eddies in the liquid medium [90, 91]. Shear-sensitive animal cell cultures are routinely supplemented with the nonionic surfactant, Pluronic F-68, which prevents cell adhesion to gas bubbles and reduces their shear damage; see, for example, Ma et al. [92]. Pluronic F-68 as well as carboxymethyl cellulose also reduce adherence of microalgal cells to gas bubbles and it also makes microalgal cells less vulnerable to shear damage [33, 93]. This strongly suggests that gas bubbles are the major cause for shear damage also to microalgal cells, and shear stress can be as problematic in pneumatically mixed as in mechanically mixed microalgal photobioreactors.

11.8 CURRENT TRENDS IN PHOTOBIOREACTOR DEVELOPMENT

The low productivities and the challenges regarding the supplies and utilization of light and CO_2 in photobioreactors have nourished long-term research in microalgal cultivation. The rate- and yield-limiting processes in microalgal cultures are now well understood, and efficient photobioreactors have been developed for many years. Different constructions have been employed in order to maximize light harvesting and utilization, and photobioreactors have probably become more variable in their designs than the bioreactors used for cultivation of heterotrophic microorganisms.

In recent years there has been a strong focus on the use of microalgal cultures for production of renewable energy, particularly biodiesel [94, 95]. This interest has launched the need for novel low-cost photobioreactors that can be installed and operated at very large scales [96]. Novel types of bioreactors, such as the Accordion photobioreactor (Fig. 11.5) made in thin, single use plastic film, are therefore being developed in order to minimize investment and maintenance costs. Productivity will remain an important issue with regard to design and operation of novel photobioreactors but economy and net energy return (i.e., the difference between the energy content of the harvested biomass and the energy costs of running the process) may be even more important when renewable energy is the product of interest. If the energy content of algal biomass is 20 kJ g^{-1} [77] and biomass productivity is 1 g L^{-1} day^{-1}, the accumulation of energy in the biomass will proceed at a rate corresponding to 0.2 W L^{-1}, while power inputs for mixing and aeration in photobioreactors are of similar or higher values [24, 75, 86, 87, 88]. It makes little sense to invest more energy in mixing, aeration, and processing than is returned from the produced biodiesel, and Hulatt and Thomas [75] actually found higher net energy return at a power input of 0.1

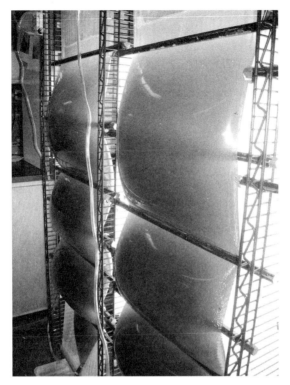

FIGURE 11.5   Accordion photobioreactor made in single use plastic film. (Courtesy of Roald Flo, Biopharmia AS.)

W L^{-1} compared to 0.5 W L^{-1} although highest PE and productivities were obtained at the highest power input. Power input and costs seem therefore to become variables of similar or even higher importance than photosynthetic efficiency, productivity, or gas transfer rate when novel photobioreactors are being designed and constructed.

ACKNOWLEDGMENT

This chapter was prepared as part of the MarBioShell project supported by the Danish Agency for Science, Research and Innovation.

REFERENCES

1. G.-Q. Chen, F. Chen, Growing phototrophic cells without light, Biotechnol. Lett. 28 (2006) 607–616.
2. N. T. Eriksen, Heterotrophic microalgae in biotechnology. In: M. N. Johansen (Ed.), *Microalgae: Biotechnology, Microbiology and Energy*, Nova Science Publishers, New York, 2012, pp. 387–412.

3. P. Spolaore, C. Joannis-Cassan, E. Duran, A. Isambert, Commercial applications of microalgae, J. Biosci. Bioeng. 101 (2006) 87–96.

4. L. Brennan, P. Owende, Biofuels from microalgae—a review of technologies for production, processing, and extractions of biofuels and co-products, Renew. Sustain. Energy Rev. 14 (2010) 557–577.

5. J. U. Grobbelaar, Physiological and technological considerations for optimising mass algal cultures, J. Appl. Phycol. 12 (2000) 201–206.

6. N. T. Eriksen, T. Geest, J. J. L. Iversen, Phototrophic growth in the Lumostat: a photobioreactor with on-line optimization of light intensity, J. Appl. Phycol. 8 (1996) 345–352.

7. O. Pulz, Photobioreactors: production systems for phototrophic microorganisms, Appl. Microbiol. Biotechnol. 57 (2001) 287–293.

8. M. Janssen, J. Tramper, L. R. Mur, R. H. Wijfells, Enclosed outdoor photobioreactors: light regime, photosynthetic efficiency, scale-up, and future prospects, Biotechnol. Bioeng. 81 (2002) 193–210.

9. M. R. Tredici, Mass production of microalgae: photobioreactors. In: A. Richmond (Ed.), *Handbook of Microalgal Culture. Biotechnology and Applied Phycology*, Blackwell Science Ltd., Oxford, 2004, pp. 178–214.

10. A. P. Carvalho, L. A. Meireles, F. X. Malcata, Microalgal reactors: a review of enclosed system designs and performances, Biotechnol. Prog. 22 (2006) 1490–1506.

11. Y. Chisti, Microalgae as sustainable cell factories, Environ. Eng. Man. J. 5 (2006) 261–274.

12. N. T. Eriksen, The technology of microalgal culturing, Biotechnol. Lett. 30 (2008) 1525–1536.

13. C. Posten, Design principles of photo-bioreactors for cultivation of microalgae, Eng. Life Sci. 9 (2009) 165–177.

14. E. Molina, J. Fernández, F. G. Acién, Y. Chisti, Tubular photobioreactor design for algal cultures, J. Biotechnol. 92 (2001) 113–131.

15. P. Carlozzi, B. Pushparaj, A. Degl'Innocenti, A. Capperucci, Growth characteristics of *Rhodopseudomonas palustris* cultured outdoors, in an underwater tubular photobioreactor, and investigation of photosynthetic efficiency, Appl. Microbiol. Biotechnol. 73 (2006) 789–795.

16. P. Carlozzi, Hydrodynamic aspects and *Arthrospira* growth in two outdoor tubular undulating row photobioreactors, Appl. Microbiol. Biotechnol. 54 (2000) 14–22.

17. A. Converti, A. Lodi, A. Del Borghi, C. Solisio, Cultivation of *Spirulina platensis* in a combined airlift-tubular system, Biochem. Eng. J. 32 (2006) 13–18.

18. I. Perner-Nochta, C. Posten, Simulations of light intensity variation in photobioreactors, J. Biotechnol. 131 (2007) 276–285.

19. G. Vunjak-Novakovic, Y. Kim, X. Wu, I. Berzin, J. C. Merchhuk, Air-lift bioreactors for algal growth on flue gas: mathematical modelling and pilot-plant studies, Ind. Eng. Chem. Res. 44 (2005) 6154–6163.

20. T. Hai, H. Ahlers, V. Gorenflo, A. Steinbüchel, Axenic cultivation of anoxygenic phototrophic bacteria, cyanobacteria, and microalgae in a new closed tubular glass photobioreactor, Appl. Microbiol. Biotechnol. 53 (2000) 383–389.

21. L. Travieso, D. O. Hall, K. K. Rao, F. Benítez, E. Sánchez, R. Borja, A helical tubular photobioreactor producing *Spirulina* in a semicontinuous mode, Int. Biodeterioration Biodegradation 47 (2001) 151–155.

22. A. H. Scragg, A. M. Illman, A. Carden, S. W. Shales, Growth of microalgae with increased calorific values in a tubular bioreactor, Biomass Bioeng. 23 (2002) 67–73.

23. F. G. A Fernandéz, D. O. Hall, E. C. Guerrero, K. K. Rao, E. Molina Grima, Outdoor production of *Phaeodactylum tricornutum* biomass in a helical reactor, J. Biotechnol. 103 (2003) 137–152.

24. D. O. Hall, F. G. A. Fernandéz, E. C. Guerrero, K. K. Rao, E. Molina Grima, Outdoor helical tubular photobioreactors for microalgal production: modelling of fluid-dynamics and mass transfer and assessment of biomass productivity, Biotechnol. Bioeng. 82 (2003) 62–73.

25. Q. Hu, N. Kurano, M. Kawachi, I. Iwasaki, A. Miyachi, Ultrahigh-cell-density culture of a marine alga *Chlorococcum littorale* in a plate photobioreactor, Appl. Microbiol. Biotechnol. 49 (1998) 655–662.

26. J. Degen, A. Uebele, A. Retze, U. Scmid-Staiger, W. Trösch, A novel photobioreactor with baffles for improved light utilization through the flashing light effect, J. Biotechnol. 92 (2001) 89–94.

27. A. Richmond, Z. Cheng-Wu, Y. Zarmi, Efficient use of strong light for high photosynthetic productivity: interrelationships between the optical path, the optimal population density and cell-growth inhibition, Biomol. Eng. 20 (2003) 229–239.

28. N. Zou, C. Zhang, Z. Cohen, A. Richmond, Production of cell mass and eicosapentaenoic acid (EPA) in ultrahigh cell density cultures of *Nannochloropsis* sp. (Eustigmatophyceae), Eur. J. Phycol. 35 (2000) 127–133.

29. M. R. Tredici, P. Carlozzi, G. C. Zittelli, R. Materassi, A vertical alveolar panel (VAP) for outdoor mass cultivation of microalgae and cyanobacteria, Biores. Technol. 38 (1991) 153–159.

30. M. R. Tredici, R. Materassi, From open ponds to vertical alveolar panels: the Italian experience in the development of reactors for the mass cultivation of phototrophic microorganisms, J. Appl. Phycol. 4 (1992) 221–231.

31. J. Li, N. Shou, W. W. Su, Online estimation of stirred-tank microalgal photobioreactor cultures based on dissolved oxygen measurements, Biochem. Eng. J. 14 (2003) 51–65.

32. J. K. Sloth, M. G. Wiebe, N. T. Eriksen, Accumulation of phycocyanin in heterotrophic and mixotrophic cultures of the acidophilic red alga *Galdieria sulphuraria*, Enzyme Microb. Technol. 38 (2006) 168–175.

33. T. M. Sobczuk, F. G. Camacho, E. Molina Grima, Y. Chisti, Effects of agitation on the microalgae *Phaeodactylum triconutum* and *Porphyridium cruentum*, Bioproc. Biosyst. Eng. 28 (2006) 243–250.

34. N. T. Eriksen, F. K. Riisgård, W. Gunther, J. J. L. Iversen, On-line estimation of O_2 production, CO_2 uptake, and growth kinetics of microalgal cultures in a gas tight photobioreactor, J. Appl. Phycol. 19 (2007) 161–174.

35. H.-S. Lee, M.-W. Seo, Z.-H. Kim, C.-G. Lee, Determining the best specific light uptake rates for the lumostatic cultures of bubble column photobioreactors, Enzyme Microb. Technol. 39 (2006) 447–452.

36. R. Bosma, E. van Zessen, J. H. Reith, J. Tramper, R. H. Wijffels, Prediction of volumetric productivity of an outdoor photobioreactor, Biotechnol. Bioeng. 97 (2007) 1108–1120.

37. M. G. de Morais, J. A. V. Costa, Biofixation if carbon dioxide by *Spirulina* sp. and *Scenedesmus obliquus* cultivated in a three-stage serial tubular photobioreactor, J. Biotechnol. 129 (2007) 439–445.

38. J. C. Merchuk, M. Gluz, I. Mukmenev, Comparison of photobioreactors for cultivation of the red microalga *Porphyridium* sp., J. Chem. Technol. Biotechnol. 75 (2000) 1119–1126.

39. I. S. Suh, S. B. Lee, Cultivation of a cyanobacterium in an internally radiating air-lift photobioreactor, J. Appl. Phycol. 13 (2001) 381–388.

40. S. Krichnavaruk, S. Powtongsook, P. Pavasant, Enhancd productivity of *Chaetoceros calcitrans* in airlift photobioreactors, Biores. Technol. 98 (2007) 2123–2130.

41. Z. Csögör, M. Herrenbauer, K. Schmidt, C. Posten, Light distribution in a novel photobioreactor—modelling for optimization, J. Appl. Phycol. 13 (2001) 325–333.

42. G. C. Zitelli, L. Rodolfi, N. Biondi, M. R. Tredici, Productivity and photosynthetic efficiency of outdoor cultures of *Tetraselmis suecica* in annular columns, Aquaculture 261 (2006) 932–943.

43. A. S. Mirón, A. C. Gómez, F. G. Camacho, E. Molina Grima, Y. Chisti, Comparative evaluation of compact photobioreactors for large-scale monoculture of microalgae, J. Biotechnol. 70 (1999) 249–270.

44. C. Wilhelm, T. Jakob, From photons to biomass and biofuels: evaluation of different strategies for the improvement of algal biotechnology based on comparative energy balances, Appl. Microbiol. Biotechnol. 92 (2011) 909–919.

45. A. Lewitus, T. Kana, Light respiration in six eustarine phytoplankton species: contrast under photoautotrophic and mixotrophic growth conditions, J. Phycol. 31 (1995) 754–761.

46. A. M. J. Kliphuis, M. Janssen, E. J. van den End, D. E. Martens, R. H. Wijffels, Light respiration in *Chlorella sorokiniana*, J. Appl. Phycol. 23 (2011) 935–947.

47. A. Melis, Photosystem-II damage and repair cycle in chloroplasts: what modulates the rate of photodamage *in vivo*? Trends Plant Sci. 4 (1999) 130–135.

48. X.-G. Zhu, S. P. Long, D. R. Ort, What is the maximum efficiency with which photosynthesis can convert solar energy into biomass? Curr. Opini. Biotechnol. 19 (2008) 153–159.

49. N. T. Eriksen, B. R. Poulsen, J. J. L. Iversen, Dual sparging photobioreactor for continuous production of microalgae, J. Appl. Phycol. 10 (1998) 377–382.

50. D. Riesenberg, R. Guthke, High-cell-density cultivation of microorganisms, Appl. Microbiol. Biotechnol. 51 (1999) 422–430.

51. O. S. Graverholt, N. T. Eriksen, Heterotrophic high cell-density fed-batch and continuous flow cultures of *Galdieria sulphuraria* and production of phycocyanin, Appl. Microbiol. Biotechnol. 77 (2007) 69–75.

52. A. Gitelson, H. Qiuang, A. Richmond, Photic volume in photobioreactors supporting ultrahigh population densities of the photoautotroph *Spirulina platensis*, Appl. Environ. Microbiol. 62 (1996) 1570–1573.

53. T. Katsuda, T. Arimoto, K. Igarashi, M. Azuma, J. Kato, S. Takakuwa, H. Ooshima, Light intensity distribution in the illuminated cylindrical photo-bioreactor and its application to hydrogen production by *Rhodobacter capsulatus*, Biochem. Eng. J. 5 (2000) 157–164.

54. H. Berberoglu, J. Yin, L. Pilon, Light transfer in bubble sparged photobioreactors for H_2 production and CO_2 mitigation, Int. J. Hydrogen Energy 32 (2007) 2273–2285.

55. E. Molina Grima, F. G. A. Fernández, F. G. Camacho, Y. Chisti, Photobioreactors: light regime, mass transfer, and scaleup, J. Biotechnol. 70 (1999) 231–247.

56. F. García Camacho, A. C. Gómez, F. G. A. Fernández, J. F. Sevilla, E. Molina Grima, Use of concentric-tube airlift photobioreactors for microalgal outdoor mass culture, Enzyme Microb. Technol. 24 (1999) 164–172.

57. J. C. Ogbonna, H. Tanaka, Light requirement and photosynthetic cell cultivation—developments of processes for efficient light utilization in photobioreactors, J. Appl. Phycol. 12 (2000) 207–218.

58. A. M. J. Kliphuis, L. de Wintre, C. Vejrazka, D. E. Martens, M. Janssen, R. Wijffels, Photosynthetic efficiency of *Chlorella sorokiniana* in a turbulent mixes short light-path photobioreactor, Biotechnol. Prog. 26 (2010) 687–696.

59. H. Takache, G. Christophe, J.-F. Cornet, J. Pruvost, Experimental and theoretical assessment of maximum productivities for the microalgae *Chlamydomonas reinhardtii* in two different geometries of photobioreactors, Biotechnol. Prog. 26 (2010) 431–440.

60. J.-F. Cornet, C.-G. Dussap, A simple and reliable formula for assessment of maximum volumetric productivities in photobioreactors, Biotechnol. Prog. 25 (2009) 424–435.

61. A. Richmond, Microalgal biotechnology at the turn of the millennium: a personal view, J. Appl. Phycol. 12 (2000) 441–451.

62. A. Richmond, Biological principles of mass cultivation. In: A. Richmond (Ed.) *Handbook of Microalgal Culture. Biotechnology and Applied Phycology*, Blackwell Science Ltd., Oxford, 2004, pp. 178–214.

63. L. Nedbal, V. Tichý, F. Xiong, J. U. Grobbelaar, Microscopic green algae and cyanobacteria in high-frequency intermittent light, J. Appl. Phycol. 8 (1996) 325–333.

64. N. Yoshimoto, T. Sato, Y. Kondo, Dynamic discrete model of flashing light in photosynthesis of microalgae, J. Appl. Phycol. 17 (2005) 207–214.

65. F. Camacho Rubio, F. G. Camacho, J. M. F. Sevilla, Y. Chisti, E. Molina Grima, A mechanistic model of photosynthesis in microalgae, Biotechnol. Bioeng. 81 (2003) 559–473.

66. M. Janssen, P. Slenders, J. Tramper, L. R. Mur, R. H. Wijffels, Photosynthetic efficiency of *Dunaliella tertiolecta* under short light/dark cycles, Enzyme Microb. Technol. 29 (2001) 298–305.

67. M. Janssen, M. Janssen, M. de Winther, J. Tramper, L. R. Mur, J. Snel, R. H. Wijffels, Efficiency of light utilization of *Chlamydomonas reinhardtii* under medium light/dark cycles, J. Biotechnol. 78 (2000) 123–137.

68. K. Kaewpintong, A. Shotipruk, S. Powtongsook, P. Pavasant, Photoautotrophic high-density cultivation of vegetative cells of *Haematococcus pluvialis* in airlift bioreactor, Biores. Technol. 98 (2007) 288–295.

69. A. S. Mirón, M.-C. C. García, F. G. Camcho, E. Molina Grima, Y. Chisti, Growth and biochemical characterization of microalgal biomass produced in bubble column and airlift photobioreactors: studies in fed-batch culture, Enzyme Microb. Technol. 31 (2002) 1015–1023.

70. M. J. Barbosa, M. Janssen, N. Ham, J. Tramper, R. W. Wijffels, Microalgae cultivation in air-lift reactors: modelling biomass yield and growth rate as a function of mixing frequency, Biotechnol. Bioeng. 82 (2003) 170–179.

71. M. Giordano, J. Beardall, J. A. Raven, CO_2 concentrating mechanisms in algae: mechanisms, environmental modulation, and evolution, Annu. Rev. Plant Biol. 56 (2005) 99–131.

72. D. Sültemeyer, Carbonic anhydrase in eukaryotic algae: characterization, regulation, and possible function during photosynthesis, Can. J. Bot. 76 (1998) 962–972.

73. E. Granum, S. M. Myklestad, A photobioreactor with pH control: demonstration by growth of the marine diatom *Skeletonema costatum*, J. Plankton Res. 24 (2002) 557–563.

74. R. F. Weiss, Carbon dioxide in water and seawater: the solubility of a non-ideal gas, Mar. Chem. 2 (1974) 203–215.

75. C. J. Hulatt, D. N. Thomas, Productivity, carbon dioxide uptake and net energy return of microalgal bubble column photobioreactors, Biores. Technol. 102 (2011) 5775–5787.

76. Y. Yang, K. Gao, Effects of CO_2 concentrations on the freshwater microalgae, *Chlamydomonas reinhardtii*, *Chlorella pyrenoidosa* and *Scenedesmus obliquus* (Chlorophyta), J. Appl. Phycol. 15 (2003) 379–389.

77. K. Zhang, N. Kurano, S. Miyachi, Optimized aeration by carbon dioxide gas for microalgal production and mass transfer characterization in a vertical flat-plate photobioreactor, Bioprocess Biosyst. Eng. 25 (2002) 97–101.

78. C. U. Ugwu, H. Aoyagi, H. Uchiyama, Photobioreactors for mass cultivation of algae, Biores. Technol. 99 (2008) 4021–4028.

79. Y.-K. Lee, S. J. Pirt, CO_2 absorption rate in algal cultures: effect of pH, J. Chem. Tech. Biotechnol. 34B (1984) 28–32.

80. E. Molina Grima, J. A. S. Pérez, F. G. Camacho, A. R. Medina, Gas–liquid transfer of atmospheric CO_2 in microalgal cultures, J. Chem. Tech. Biotechnol. 56 (1993) 329–337.

81. S. Raso, B. van Genugten, M. Vermuë, R. H. Wijffels, Effect of oxygen concentration on the growth of *Nannochloropsis* sp. at low light intensity, J. Appl. Phycol. (2011) doi: 10.1007/s10812-012-9706-z.

82. R. von Schalien, K. Gafervik, B. Saxén, K. Ringbom, M. Rudström, Adaptive on-line model for aerobic *Saccharomyces cerevisiae* fermentation, Biotechnol. Bioeng. 48 (1995) 631–638.

83. J. Doucha, K. Lívanský, Productivity, CO_2/O_2 exchange and hydraulics in outdoor open high density microalgal (*Chlorella* sp.) photobioreactors operated in a Middle and Southern European climate, J. Appl. Phycol. 18 (2006) 812–826.

84. J. L. G. Sánchez, M. Berenguel, F. Rodríguez, J. M. F. Sevilla, C. B. Alias, F. G. A. Fernández, Mimimization of carbon losses in pilot-scale outdoor photobioreactors by model-based predictive control, Biotechnol. Bioeng. 84 (2003) 533–543.

85. B. R. Poulsen, J. J. L. Iversen, Membrane sparger in bubble column, airlift, and combined membrane-ring sparger bioreactors, Biotechnol. Bioeng. 64 (1999) 452–458.

86. E. Sierra, F. G. Acién, J. M. Fernandez, J. L. García, C. González, E. Molina, Characterization of a flat plate photobioreactor for the production of microalgae, Chem. Eng. J. 138 (2008) 136–147.

87. C. J. Hulatt, D. N. Thomas, Energy efficiency of an outdoor microalgal photobioreactor sited at mid-temperate latitude, Biores. Technol. 102 (2011) 6687–6695.

88. F. Camacho, F. F. G. Acién, J. A. Sánchez, F. García, E. Molina, Prediction of dissolved oxygen and carbon dioxide concentration profiles in tubular photobioreactors for microalgal culture, Biotechnol. Bioeng. 62 (1999) 71–86.

89. W. Hu, W. G. Lowrie, R. Gladue, J. Hansen, C. Wojnar, J. J. Chalmers, Growth inhibition of dinoflagellate algae in shake flasks: not due to shear this time! Biotechnol. Prog. 26 (2010) 79–87.

90. M. J. Barbosa, M. Albrecht, R. H. Wijffels, Hydrodynamic stress and lethal events in sparged microalgae cultures, Biotechnol. Bioeng. 83 (2003) 112–120.

91. J. Vega-Estrada, M. C. Montes-Horcasitas, A. R. Domínguez-Bocanegra, R. O. Cañizares-Villanueva, *Haematococcus pluvialis* cultivation in split-cylinder internal-loop airlift photobioreactor under aeration conditions avoiding cell damage, Appl. Microbiol. Biotechnol. 68 (2005) 31–35.

92. N. Ma, J. J. Chalmers, J. G. Auniņš, W. Zhou, L. Xie, Quantitative studies of cell-bubble interactions and cell damage at different Pluronic F-68 and cell concentrations, Biotechnol. Prog. 20 (2004) 1183–1191.

93. F. García Camacho, E. Molina Grima, A. S. Mirón, V. G. Pascual, Y. Chisti, Carboxymethyl cellulose protects algal cells against hydrodynamic stress, Enzyme Microb. Technol. 29 (2001) 602–610.

94. Y. Chisti, Biodiesel from microalgae, Biotechnol. Adv. 25 (2007) 294–306.

95. Y. Chisti, Biodiesel from microalgae beats bioethanol, Trends Biotechnol. 26 (2008) 126–131.

96. F. Lehr, C. Posten, Closed photo-bioreactors as tools for biofuel production, Cur. Opini. Biotechnol. 20 (2009) 280–285.

97. J. A. Roels, Applications of macroscopic principles to microbial metabolism, Biotechnol. Bioeng. 22 (1980) 2457–2514.

98. C. Jiménez, B. R. Cossío, D. Labella, F. X. Niell, The feasibility of industrial production of *Spirulina* (*Arthrospira*) in Southern Spain, Aquaculture 217 (2003) 179–190.

99. B. Pushparaj, E. Pelosi, M. R. Tredici, E. Pinzani, R. Materassi, An integrated culture system for outdoor production of microalgae and cyanobacteria, J. Appl. Phycol. 9 (1997) 113–119.

100. J. Moreno, M. A. Vargas, H. Rodríguez, J. Rivas, M. G. Guerrero, Outdoor cultivation of a nitrogen-fixing marine cyanobacterium, *Anabaena* sp. ATCC 33047, Biomol. Eng. 20 (2003) 191–197.

101. J. Doucha, F. Straka, K. Lívanský, Utilization of flue gas for cultivation of microalgae (*Chlorella* sp.) in an outdoor open thin-layer photobioreactor, J. Appl. Phycol. 17 (2005) 403–412.

102. A. Richmond, Z. Cheng-Wu, Optimization of a flat plate glass reactor for mass production of *Nannochloropsis* sp. outdoors, J. Biotechnol. 85 (2001) 259–269.

103. G. C. Zitelli, V. Tomasello, E. Pinzani, M. R. Tredici, Outdoor culture of *Arthrospira platensis* during autumn and winter in temperate climate, J. Appl. Phycol. 8 (1996) 293–301.

CHAPTER 12

Industrial Cultivation Systems for Intensive Production of Microalgae

GIUSEPPE OLIVIERI, PIERO SALATINO, and ANTONIO MARZOCCHELLA

12.1 INTRODUCTION

The culture of microalgae has been investigated since 1953 [1]. Except for some cultures devoted to niche applications, commercial processes based on microalgae suffer from the typical low productivity when compared with those based on bacteria, yeasts, and animal cells. The scenario may change definitively when autotrophic processes such as bioremediation of wastewaters or renewable energy-vector production are considered. However, autotrophic cultures have to be optimized in terms of biomass production, efficient light utilization, and efficient use of carbon dioxide. The key issues relevant for the optimization of microalgal cultures may be grouped into biological and engineering aspects of the culture system.

Biological issues include the morphology and composition of microalgae (including wall composition), photosynthesis rate, and growth metabolism. The reader may refer to chapters of this book and to other comprehensive reviews and books available in the scientific literature [2, 3].

The systems for mass culture may be classified in two categories: open and closed with respect to the environment. The first class includes open ponds and raceways. The second class includes photobioreactors. Typically, the second class is adopted for massive production of monoseptic cultures of microalgae.

The present chapter focuses on engineering issues of microalgal cultivation systems. Section 12.2 focuses on issues relevant to design and operation of culture systems. Sections 12.3 and 12.4 report general features of the two classes of culture systems for mass culture: open and closed systems with respect to the environment. Section 12.5 focuses on innovative configuration of photobioreactors. A technoeconomic analysis related to photobioreactors is reported in Section 12.6.

Natural and Artificial Photosynthesis: Solar Power as an Energy Source, First Edition. Edited by Reza Razeghifard.
© 2013 John Wiley & Sons, Inc. Published 2013 by John Wiley & Sons, Inc.

12.2 RELEVANT ISSUES FOR DESIGN AND OPERATION OF SYSTEMS FOR MICROALGAL CULTURES

12.2.1 Stoichiometry of Microalgal Growth

Readers interested in detailed models of the metabolism of photosynthetic microorganisms may refer to Cornet et al. [4]. A single stoichiometric equation model is adopted for the assessments reported in this chapter.

From the standpoint of mass balance, the growth of microalgae may be described by means of unstructured models according to the stoichiometry

$$CO_2 + Y_{N/CO_2}(\text{N-source}) + Y_{S/CO_2}(\text{S-source}) + Y_{P/CO_2}(\text{P-source}) \rightarrow$$
$$\rightarrow Y_{X/CO_2}\text{microalgal} + Y_{O_2/CO_2}O_2 \qquad (12.1)$$

where Y_{i/CO_2} is the yield coefficient referred to the i-species. Even though models based on a single stoichiometric equation cannot describe changes of microalgal composition and conversion yields under unbalanced feeding and stress conditions, the overall characterization of performances of culture systems may be assessed with satisfactory accuracy. In particular, the single stoichiometric model Eq. (12.1) characterizes the microalgal growth with a univocal conversion yield Y_{x/CO_2}.

12.2.2 Microalgal Kinetics

Readers interested in detailed models of the kinetics related to the growth of photosynthetic microorganisms may refer to Cornet et al. [4], Masojidek et al. [5], and Richmond [2]. As a general rule, the specific microalgal growth rate (μ) may be expressed according to Eq. (12.2):

$$\frac{1}{X}\frac{dX}{dT} = \mu = \mu\left(X, C_{CO_2}, C_{O_2}, T, I, C_P, C_N, C_S, \ldots\right) \qquad (12.2)$$

where X is the local concentration of microalgae, I the irradiance, C_i the concentration of the species i, and T the temperature. The effects of each single variable must be carefully assessed provided that the process is not in a limited or inhibited regime because of other variables.

Microalgal growth kinetics should take into account the photosynthetic activity, function of incident photon flux, as well as the subsequent "dark" reactions. Moreover, both the frequency of the light/dark cycle and the dark-fraction of the cycle are key issues for the effective growth rate of microalgae. Overall models are reported in the literature and they may be adopted for the assessment of the performance of culture systems. According to results reported by Goldman [6] and later by Posten [7], three growth conditions may be identified regarding the specific growth rate versus the incident photon flux (I) (Fig. 12.1):

- Light-limited growth—μ increases linearly with I for $I < I_S$.
- Light-saturated growth—μ is constant with I.
- Light-inhibited growth—μ decreases with I for $I > I_P$.

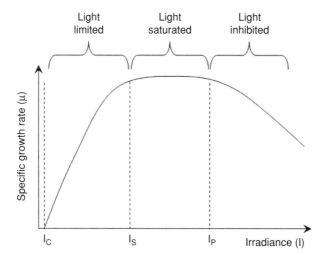

FIGURE 12.1 Specific growth rate as a function of the irradiance.

Two characteristic values of I may be indentified: the saturation threshold I_S and the photoinhibition threshold I_P. The compensation photon flux (I_C) is defined as the minimum flux necessary to balance microalgal maintenance.

The adoption of the growth rate model reported in Fig. 12.1 for microalgal mass cultures requires the distribution function of the irradiance throughout the culture volume. The incident light may reduce along the optical path as a consequence of both microalgal absorption and scattering phenomena. Figure 12.2 shows two typical profiles of intensity light decay: (A) the irradiance is almost constant along the optical path under microalgal dilute conditions resulting in low light conversion efficiency; and (B) the irradiance extinguishes in a fraction of the optical path under microalgal concentrated conditions, which is a prerequisite for a high light conversion efficiency.

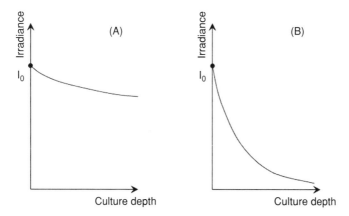

FIGURE 12.2 Two typical profiles of irradiance versus optical depth profiles: (A) dilute culture and (B) concentrated culture.

For concentrated suspensions, microalgae grow under conditions caused by the combination of phenomena described in Figs. 12.1 and 12.2. Moving from the inner to irradiated region of the culture volume, microalgae experience both the light-limited and the light-inhibited growth regimes. Moreover, nonuniformity of irradiance at the culture surface may be present as a consequence of the shape of the culture vessel (parallelepiped vessel, cylindrical vessel, etc.) and biofouling, or light coming from only one side.

The reported issues affect the local rate of microalgal growth and a detailed model/simulation of the microalgae growth should take them into account. Simplified macromodels have also been proposed where the specific growth rate is reported as a function of the average irradiance (I_{av}). Molina Grima et al. [8] proposed one of the first macromodels that was particularly accurate when photoinhibition conditions did not apply:

$$\mu = \frac{\mu_{max} I_{av}^n}{I_K^n + I_{av}^n} \tag{12.3}$$

where μ_{max} is the maximum value of μ, and I_K and n are constants depending on both microalgal species and culture conditions.

The average irradiance may be assessed according to the relationship developed by Alfano et al. [9]:

$$I_{av} = \frac{I_0}{\phi_{eq} K_a X} \left(1 - \ell^{\phi_{eq} K_a X}\right) \tag{12.4}$$

where ϕ_{eq} is the equivalent length of the light path in the culture system, K_a the extension coefficient of the biomass, and X the bulk microalgal concentration.

More complex macromodels for microalgal growth have been proposed to catch the limited-inhibited growth phenomenology [10, 11, 12]. A lumped Andrew-type model was proposed by Eiler and Peeters [10]:

$$\mu = \mu_{max} \frac{I}{K_{sl} + I + I^2/K_{il}} \tag{12.5}$$

where μ_{max}, K_{sl}, and K_{il} are parameters of the model. According to Eq. (12.5) the specific growth rate is maximum at $I_{opt} = \sqrt{K_{sl} K_{il}}$.

It should be noted that the reported models describe the behavior of photosynthetic microalgae operated under continuous and constant irradiance. A quite different scenario may exist in dark/light cycles (Fig. 12.3). Philips and Myers [13] reported on tests carried out under stroboscopic light conditions. They pointed out that a high irradiance level can be sustained if a short light phase (tens of milliseconds)—typical of the fast activation of photosynthetic units (PSUs) by photon flux—alternates with a long dark phase (hundreds of milliseconds) during which the acquired electrical potential of the PSU is spent in the dark reaction of the photosynthetic metabolism.

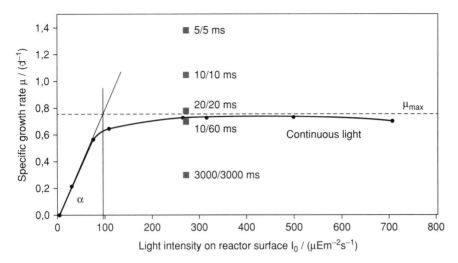

FIGURE 12.3 Specific growth rate of *Porphyridium purpureum* under both continuous irradiance and light/dark cycles. Reproduced from Posten [7] with permission of John Wiley & Sons.

Carbon dioxide concentration in the microalgal cultures may change over a wide interval depending on its concentration in the gas phase, microalgal concentration, and gas–liquid mass transport among other factors. The effect of the carbon concentration on the growth rate depends on the strain investigated. By increasing the CO_2 concentration, the growth rate may change gradually from CO_2 limited conditions to CO_2 inhibited conditions [14]. Typically, studies available in the literature refer to the CO_2 concentration in the gas phase ranging from air to about 15–20%, the maximum concentration that may be found in exhaust gases.

Oxygen produced during the microalgal growth according to Eq. (12.1) may accumulate in the liquid phase. As the oxygen concentration increases over critical values, the microalgal growth is progressively inhibited depending on strains [15,16].

To the best of our knowledge, kinetics models of the microalgal growth rate based on the concentration of both CO_2 and O_2 species still require more investigation. Typically, tests aimed at the assessment of the growth rate at different concentrations of both CO_2 and O_2 were performed using operating conditions and culture modality that span over wide ranges [14, 17, 18]. Therefore the lack of systematic collection of data makes the assessment of kinetic models a difficult task.

With respect to minerals (nitrate, phosphate, sulfate, etc.), the classical Monod-type law is adopted to model the substrate-limited growth [4]:

$$\mu = \mu(I, C_{CO_2}, C_{O_2}) \prod_{i=1}^{n} \frac{C_i}{K_i + C_i} \qquad (12.6)$$

where C_i is the concentration of mineral i, and K_i the related Monod constant.

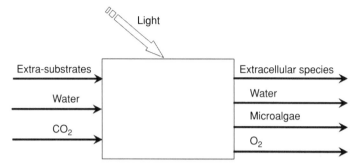

FIGURE 12.4   Sketch of mass and energy fluxes associated with microalgal cultures. Microalgal inoculum is neglected.

12.2.3   Mass Balance

Figure 12.4 shows the main mass and energy fluxes associated with microalgal cultures, whatever the cultivation system adopted. Substrates for microalgae have been lumped in extra substrates except for carbon dioxide. The mass balance referred to the main species and extended to the cultivation system yields the following:

Carbon

$$\text{(mass of carbon in } CO_2)_{IN} - \text{(mass of carbon in } CO_2)_{OUT} = M_{C,m.a.} + M_{C,e.c}$$
$$(12.7)$$

Oxygen

$$\text{(mass of oxygen in } O_2)_{IN} - \text{(mass of oxygen in } O_2)_{OUT} = M_{O,m.a.} + M_{O,e.c}$$
$$(12.8)$$

where $M_{i,j}$ is the content of the element i (read C or O) of the stream j, e.c. is the extracellular fraction of the broth, and m.a. is the microalgae. Equation (12.7) or (12.8) coupled with yield coefficient Y_{i/CO_2} reported in Eq. (12.1) allow assessing material balance for all species involved in the microalgal growth.

12.2.4   Energy Balance

The energy balance extended to the cultivation system (Fig. 12.4) yields

$$(E_{light})_{IN} - (E_{light})_{OUT} = H_{broth} + H_{m.a.} + H_{gas,out} - H_{water} - H_{CO_2} - H_{e.s.} - q_{dis}$$
$$(12.9)$$

where $(E_{light})_{IN}$ is the energy associated with the light (sun or artificial source), $(E_{light})_{OUT}$ the reflected/refracted light, H_{broth} the enthalpy of the broth, $H_{m.a.}$ the enthalpy of the microalgae, $H_{gas,out}$ the enthalpy of the stream flowing out of the culture system (included the steam), H_{water} the enthalpy of the fed water, H_{CO_2} the enthalpy of the CO_2 bearing gas stream, $H_{e.s.}$ the enthalpy of the extra substrates fed to the culture system and q_{dis} the total heat added to the cultivation system (heat exchanged with the environment). Assuming the liquid water at the feeding temperature as reference state and neglecting the contributions of Π_{CO_2} and $H_{e.s.}$, Eq. (12.9) yields

$$(E_{light})_{IN} - (E_{light})_{OUT} = H_{broth} + H_{m.a.} + H_{gas,out} - q_{dis} \qquad (12.10)$$

Equation (12.10) provides the upper limit of the energy content of microalgae cultivated under autothrophic conditions. Assuming the contribution of

$$(E_{light})_{OUT} + H_{broth} + H_{gas,out} - q_{dis} \qquad (12.11)$$

to be negligible with respect to the other terms, Eq. (12.10) becomes

$$(E_{light})_{IN} = H_{m.a.} \qquad (12.12)$$

Equation (12.12) yields the maximum energy content of the microalgal suspension, that is, the maximum energy content of microalgae provided that the microalgal concentration is known.

The energy balance applied to outdoor microalgal culture suffers the unsteady dynamics of both sunlight and environment temperature. This behavior should be taken into account when the explicit form of the each term of the reported equations is developed.

A drawback of autotrophic microalgal cultures is the need for temperature control. Sun irradiance heats up cultures and cooling systems must be adopted. In particular, the cooling expedient/strategy strongly depends on the type of culture systems adopted. They will be discussed in the next sections.

12.2.5 Basic System Design of Microalgal Cultivation

Whatever the cultivation system, productivity maximization of microalgal cultures is typically required for industrial application. Some applications may have different goals such as maximization of carbon dioxide fixation rate or lipid production. This chapter focuses on the maximization of microalgal cultures, which also implies maximization of the carbon dioxide fixation rate for autotrophic cultures.

The design of the cultivation systems depends on the operation mode—batch versus (quasi)continuous—and on the typology of the reactor system. A brief theoretical framework of the reactors will hereby be proposed to support the analysis. The reader is referred to textbooks [19, 20, 21] for a detailed treatment of the subject.

For batch cultures, the time necessary to reach a set concentration X_F of microalgae is

$$X_F = \frac{1}{V_F} \int \int_{V,t} X \cdot \mu \left(X, C_{CO_2}, C_{O_2}, T, I, C_{\text{extra substrates}} \right) dV \, dt \qquad (12.13)$$

FIGURE 12.5   Time schedule program of a typical culture system operated under quasicontinuous conditions.

provided a microalgal inoculum (concentration X_0) at time $t = 0$. In Eq. (12.13) C_i and X are the instantaneous local concentration of the species i and microalgae, T is the local temperature at time t, I the local irradiance at time t, V the volume of the culture system at time t, and V_F the culture volume at the end of the batch process (time t_F). The average microalgal productivity during cultivation is

$$W_x = \frac{V_F X_F}{t_F + t_d} \qquad (12.14)$$

where t_d is the dead time between two successive batches, and the inoculum concentration has been neglected. The average rate of CO_2 fixation is

$$W_{CO_2} = \frac{1}{Y_{X/CO_2}} \frac{V_F X_F}{t_F + t_d} \qquad (12.15)$$

where Y_{X/CO_2} is the yield coefficient biomass to carbon dioxide.

Quasicontinuous cultures may be carried out according to the sequential batch reactor (SBR) conditions: a vessel of given volume V_F is operated under batch conditions according to the time schedule reported in Fig. 12.5. The symbols Fill, Growth, and Draw refer to the typical sequential phases of operation: loading, microalgal growth, and suspension discharging. The microalgal conversion measured in the discharged suspension depends on the time evolution of the growth rate with the progressive change of the conditions in the culture (e.g., X increases continuously). The culture systems are operated at constant volume (V_F) and are characterized in terms of the cycle time (t_c, defined as the time for Fill, Growth, and Draw steps), the dead time (t_d, defined as the time for Fill and Draw steps), and of renewal volumetric fraction f. An average dilution rate $\overline{D}$ may be assessed as

$$\overline{D} = \frac{f}{t_c} \qquad (12.16)$$

Models to describe the behavior of a SBR culture system depend on the hydrodynamics of the adopted system: stirred tank vessel versus plug flow system may be assumed as an extreme case.

The stirred tank reactor (STR) model given in Eq. (12.13) may still be adopted for SBR culture systems behaving like STRs. Under these conditions it is assumed that $V(=V_F)$ is constant, $X_0 = f X_F$, and the culture growth lasts $t = (t_c - t_d)$. According to these assumptions, the microalgal productivity is

$$W_X = \frac{f V_F X_F}{t_c} \tag{12.17}$$

and the average rate of CO_2 fixation is

$$W_{CO_2} = \frac{1}{Y_{X/CO_2}} \frac{f V_F X_F}{t_c} \tag{12.18}$$

Culture systems characterized by constant cross geometrical section and that behave like a plug flow reactor (PFR) (Fig. 12.6A) may be described according to the relationship

$$\frac{iL}{u} = \int_{fX_F}^{X_F} \frac{dX}{X \cdot \mu \left(X, C_{CO_2}, C_{O_2}, T, I, C_{\text{extra substrates}} \right)} \tag{12.19}$$

where i is a parametric value (see later), L the length of the vessel, and u the axial volumetric velocity. For the sake of simplicity, the microalgal concentration and all variables affecting the growth rate have been assumed constant at each section of the vessel (e.g., an appropriate cross-section-averaged value is adopted). Say S is the

(A)

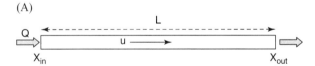

(B)

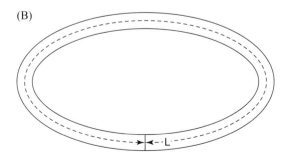

FIGURE 12.6 Sketch of plug flow reactors: (A) once through configuration and (B) closed loop configuration.

cross section of the vessel, $Q = u \cdot S$ is the volumetric flow rate through the vessel, and $L \cdot S = V_F$. Two possible configurations may be adopted.

- **Single Long Reactor.** The concentration of the culture increases from $f\, X_F$ to X_F along the vessel. Under these conditions $i=1$.
- **Closed Loop Reactor Configuration** (Fig. 12.6B). A circular system is run several times to increase the microalgal concentration, $i > 1$. Even though the culture volume is constant and equal to V_F, the culture time for this kind of system is iL/u.

For continuous cultures, two ideal classes of systems must be considered: continuous stirred tank reactor (CSTR) and plug flow reactor (PFR), with or without recycle. For the sake of simplicity, an appropriate averaged value of the irradiance is assumed in the culture system: throughout the volume for the CSTR, and across each section of the PFR.

The maximum concentration of microalgae, the volume (V_F) of the culture system, and the volumetric flow rate (Q) of the broth are related to each other through the material balance referred to the microalgae and extended to the culture system volume. Assuming an unstructured model for the microalgae, the material balance under steady-state conditions gives

$$V = \frac{Q}{\mu|_{out}} \tag{12.20}$$

or

$$D = \mu|_{out} \tag{12.21}$$

where $\mu|_{out}$ is the microalgal specific growth rate at the conditions established in the culture system (X_F, CO_2, O_2, I, pH, ...), and $D(= Q/V)$ is the dilution rate.

A plug flow reactor with recycle is characterized by a nonzero value of R, that is the ratio between the mass flow rate of the recycled stream and the feeding rate Q. The material balance for this case reads

$$L = (R + 1)\, u \int_{\frac{R \cdot X_F + f X_F}{R+1}}^{X_F} \frac{dX}{X \cdot \mu \left(X, C_{CO_2}, C_{O_2}, T, I, C_{\text{extra substrates}} \right)} \tag{12.22}$$

The microalgal productivity of continuous photobioreactors is provided ($X_F \gg X_o$) thus:

$$W_X = Q X_F \tag{12.23}$$

Performances of the above systems depend on the field of concentration of all species (substrates and products) and, in particular, on the intensity of light.

12.2.6 Gas–Liquid Mass Transport

The role of both CO_2 and O_2 on the microalgal growth, reported in Section 12.2.2, asks for the design of solution devices that keep the concentration of CO_2 and O_2 within admissible intervals. In particular, the gas–liquid mass transport rate must be tuned to enhance the CO_2 pumping into the liquid phase and the O_2 stripping from the liquid phase.

The concentrations of CO_2 and O_2 in the liquid phase depend on their fractions in the gas phase and on the hydrodynamics of the culture systems. The mass transport rate of a gas species between the gas and liquid phases with reference to the volume unit of liquid phase is

$$\text{For } CO_2 \quad \frac{dC_{CO_2}}{dt} = K_L a \left(C^*_{CO_2} - C_{CO_2} \right) \tag{12.24}$$

$$\text{For } O_2 \quad \frac{dC_{O_2}}{dt} = K_L a \left(C^*_{O_2} - C_{O_2} \right) \tag{12.25}$$

where C^*_i is the concentration of the i species in the liquid phase under equilibrium conditions with gas phase, K_L is the gas–liquid mass transport coefficient, and a is the specific exchange surface. The C^*_i depends on both the temperature and the composition of the liquid phase. The product $K_L a$ depends on the hydrodynamics of the culture system and is approximately 10% higher for O_2 than CO_2. Except for the driving force—the concentration difference—it is interesting to note that the rate of both CO_2 pumping and O_2 stripping are enhanced by the same factor: $K_L a$. From this point of view, all solution/devices adopted to enhance one of the rates also act on the other rate.

Equations (12.24) and (12.25) must be coupled with models reported in the previous sections to assess the concentration field of both O_2 and CO_2 in the culture systems.

12.2.7 Mixing

Culture mixing is required to:

- Prevent microalgae settlement
- Increase gas–liquid mass transport phenomena
- Renew exposition of microalgae to a light source

The type of expedient/device adopted to improve mixing of the cultures depends on the typology of culture systems and the hydrodynamic flow pattern desiderated (even mixing vs. plug flow). Their applications will be presented and discussed in the next sections.

12.3 OPEN SYSTEMS

The first open systems for microalgal culture were developed by Oswald in the 1950s [22]. Whatever the design of the open system, low biomass density characterizes these systems (<1 g/L). Typically open systems can be classified with respect to the agitation device/strategy.

12.3.1 Typologies

Borowitzka and Borowitzka [22] classified open systems in four categories: large open pond, circular open pond, raceway pond, and large bags. Whatever the typology, open systems are typically operated under either batch or semicontinuous conditions. Raceway ponds are the main systems adopted for commercial application of microalgal productions (Fig. 12.7). Generally, a meandering closed loop configuration is adopted for ponds to optimize land utilization [24]. The size of commercial ponds ranges between 1000 and 5000 m².

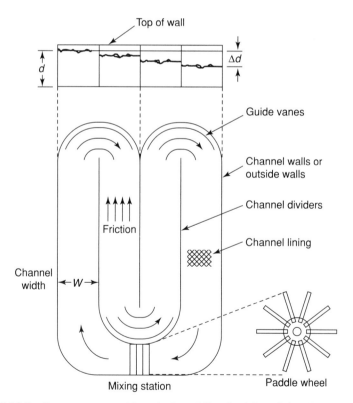

FIGURE 12.7 Open raceways with a single paddle wheel for mixing. Reproduced from Molina Grima [23] with permission of John Wiley & Sons.

Typical design criteria adopted to promote continuous suspension flow and to keep shear mixing as low as possible include suitable bottom lining and a particular geometry of the walls. To induce the suspension mixing and to prevent microalgae sedimentation, it is necessary to assure continuous culture flow throughout the raceway. To this aim, one or more paddle wheels are commonly located along the raceway and the depth of the pond decreases along the liquid path. Oswald [24] reported detailed engineering characterization of pond hydrodynamics. He proposed a model to relate the culture depth (d), channel width (w) and length (L), liquid velocity (v), and power consumption (P):

$$L = \frac{\Delta d \, (1/d + 2/w)^{4/3}}{v^2 \cdot n^2} \tag{12.26}$$

$$P = \frac{v \cdot d \cdot w \cdot \rho \cdot \Delta d}{\eta} \tag{12.27}$$

where Δd is the set depth reduction, n the Manning friction factor, and ρ the liquid culture density. Low culture velocity (<30 cm/s) is adopted to avoid intense shear stress on the microalgal cells.

CO_2 is pumped into the microalgal suspension by gas sparging devices localized in one or more sections of the ponds. Due to the low value of d, the efficiency of the mass transfer process is quite low.

12.3.2 Mass Balances

The mass balances reported in Section 12.2 and extended to raceway systems must take into account the peculiar geometry of these systems: free gas-suspension surface extended throughout the culture system. The balance may be worked out assuming the liquid phase as control volume. Accordingly, liquid streams fed/withdrawn to/from the photobioreactor should be characterized in terms of concentrations and volume/fluxes.

Batch Cultures Assuming the initial contents of carbon and oxygen in the liquid phase are negligible, the mass balances for both species when extended to the batch time t_F are

$$Carbon \quad t_F \frac{MW_C}{MW_{CO_2}} \int_{V_{Pond}} K_L a \left(C^*_{CO_2} - C_{CO_2} \right) dV = V_{Pond} X_F \omega_C + V_{Pond} C_{e.c.} \tag{12.28}$$

$$Oxygen \quad t_F \frac{MW_O}{MW_{O_2}} \int_{V_{Pond}} K_L a \left(C^*_{O_2} - C_{O_2} \right) dV = V_{Pond} X_F \omega_O + V_{Pond} O_{e.c.} \tag{12.29}$$

where V_{Pond} is the volume culture, A_{Pond} the gas–liquid free surface, ω_O and ω_C are the microalgal mass fraction of carbon and oxygen, and ΦCO_2 and ΦO_2 are the local

CO_2 and O_2 flux at the gas–liquid interface, respectively. They may not be constant throughout the free surface. It is expected that the mass transfer rate between the gas and liquid phases may change along the culture flow as a consequence of the local hydrodynamic conditions and species concentrations. Large fluctuations of the rate are typical of quite long ponds.

Continuous Cultures   Assuming the amounts of both carbon and oxygen fed to the culture system in the liquid stream (volumetric flow rate Q) are negligible, the mass balances of both species when extended to the observation time t_{obs} are

$$Carbon \quad t_{obs} \frac{MW_C}{MW_{CO_2}} \int_{V_{Pond}} K_L a \left(C^*_{CO_2} - C_{CO_2} \right) dV = \int_{t_{obs}} (QX_F \omega_C + QC_{e.c.}) \, dt$$

$$(12.30)$$

$$Oxygen \quad t_{obs} \frac{MW_O}{MW_{O_2}} \int_{V_{Pond}} K_L a \left(C^*_{O_2} - C_{O_2} \right) dV = \int_{t_{obs}} (QX_F \omega_O + QO_{e.c.}) \, dt$$

$$(12.31)$$

Equations (12.28) and (12.31) must be coupled with hydrodynamic models to assess the concentration field of both O_2 and CO_2 in the culture systems.

12.3.3 Energy Balance

The energy balance Eq. (12.10) extended to an open system should take into account the energy irradiated from the sun and the loss due to water evaporation. The latter is dependent on weather conditions: temperature, wind, humidity, and so on. In particular, $H_{gas,out}$ is quite high for the water evaporation contribution and it becomes the dominant term of Eq. (12.10):

$$(E_{light})_{IN} - (E_{light})_{OUT} \approx H_{gas,out} \tag{12.32}$$

The temperature of open systems is the result of a delicate equilibrium depending on weather conditions.

12.3.4 Gas–Liquid Mass Transfer

In an open system three components are exchanged with gas phase: carbon dioxide, oxygen, and water. Typically, two dominant gas–liquid mass transfer mechanisms may be found: the natural convection at the free air-suspension surface, and the forced convection within the liquid culture due to gas sparging. Obviously direct CO_2 sparging could enhance the CO_2 mass transfer rate but the low depth of the culture results in a very low contact time between bubbles and liquid.

12.4 CLOSED SYSTEMS: PHOTOBIOREACTORS

Photobioreactors are cultivation systems characterized by stricter control of the growth environment than the open systems. These systems may be classified with respect to the orientation of the illuminated surface (vertical, horizontal, inclined), the shape of the vessel (circular column, flat columns, etc.), the unit adopted for the gas–liquid mass transfer, or the system adopted to circulate the culture. Detailed reviews of photobioreactors have been reported by Jansenn et al. [25] and Chen et al. [26].

Liquid and gas feedings as well as suspension and gas effluents in photobioreactors typically take place via a discrete number of ports. Moreover, the confinement of the culture suspension in a closed system allows easier control of the flow pattern. According to these features, the behavior and performances of cultures in photobioreactors may be predicted theoretically. However, modeling of photobioreactors is still a challenging enterprise [27,28,29,30]. Bitog et al. [31] reported an interesting review of the models of photobioreactors based on computational fluid dynamics.

12.4.1 Photobioreactor Typologies

Figure 12.8 illustrates some sketch of photobioreactors proposed in the literature. The base case photobioreactor is a vertical column (cylindrical or flat) with injection of the CO_2 bearing gas stream at the bottom (Fig. 12.8A, B). The bubble flow mixes the suspension and promotes the gas–liquid mass transfer (CO_2 pumping and O_2 stripping). Internal and external loop airlifts are reported in Fig. 12.8C and Fig. 12.8D, respectively. Tubular photobioreactors (Fig. 12.8E) are the classical systems where the microalgal suspension flows without the use of gas bubbles. An external device provides the energy for the flow, it pumps CO_2 into the suspension, and it strips O_2 from the suspension.

Photobioreactors may be oriented toward the sun so as to maximize irradiance. Typically, inclined photobioreactors compared with the vertical ones can operate without the need of gas bubbles (Fig. 12.8E). In these cases, the microalgae suspension is pumped through the photobioreactor by means of an external system/device. Inclined bubble columns have also been proposed [18, 28, 29].

12.4.2 Mass Balances

The mass balances reported in Section 12.2 applied to photobioreactors must take into account that the closed features of these systems and that all fluxes in/out from vessels are localized at the entrance/exit sections. The gas and liquid streams fed/withdrawn to/from the photobioreactor should be characterized in terms of concentrations and volume/fluxes. In particular, the balances refer to: (1) the suspension volume, when batch systems are considered; and (2) fluid fluxes, when continuous cultures are considered. The terms in Eqs. (12.7) and (12.8) may be represented in explicit forms. Assuming that both CO_2 and O_2 in either the initial culture of batch cultures or fed

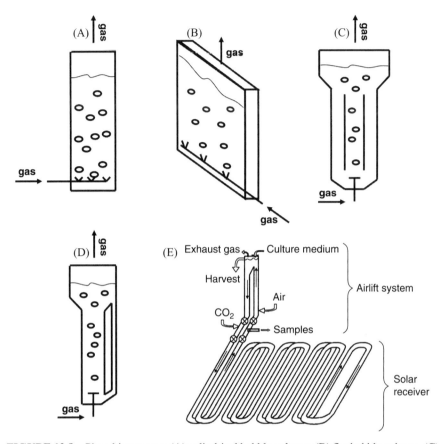

FIGURE 12.8   Photobioreactors: (A) cylindrical bubble column, (B) flat bubble column, (C) internal loop airlift, (D) external loop airlift, and (E) airlift driven tubular photobioreactor. Reproduced from Molina et al. [16] with permission of Elsevier.

liquid stream of continuous cultures are negligible, the mass balances Eqs. (12.7) and (12.8) become

$$
Carbon \quad \frac{C}{CO_2} \left\{ \left[\left(\sum_k (Gy_{CO_2})_k \right)_{IN} - \left(\sum_k (Gy_{CO_2})_k \right)_{OUT} \right] \right.
$$
$$
\left. + \left[(VC_{CO_2})_{IN} - (VC_{CO_2})_{OUT} \right] \right\} = VX_F\omega_C + VC_{e.c.} \quad (12.33)
$$

$$
Oxygen \quad \frac{O}{O_2} \left\{ \left[\left(\sum_k (Gy_{O_2})_k \right)_{IN} - \left(\sum_k (Gy_{O_2})_k \right)_{OUT} \right] \right.
$$
$$
\left. + \left[(VC_{O_2})_{IN} - (VC_{O_2})_{OUT} \right] \right\} = VX_F\omega_O + VO_{e.c.} \quad (12.34)
$$

where the summation is extended to all k gas fed/withdrawn, G is the volume of gas, V the volume of the suspension, y_i the volumetric fraction of the species i in the gas phase, and $C_{e.c.}$ and $O_{e.c.}$ are the volumetric concentration of carbon and oxygen, respectively, in the extracellular species. For continuous systems G and V may be referred to the unit of time; that is, G and V are the volumetric gas and liquid rates, respectively.

12.4.3 Energy Balance

The energy balance extended to a photobioreactor is definitively simplified with respect to the general case Eq. (12.10). The enthalpy of the gas stream flowing out of the culture system may be assumed negligible with respect to the other terms on the right-hand side of the equation. Indeed, the low vapor content of this stream drastically reduces the enthalpy. A drawback of the reduced rate of vapor formation from the culture is the increase of the suspension temperature. Therefore it is necessary to cool the suspension to keep the temperature within the operating range. The q_{dis} term becomes the tuning flux variable to control the temperature of the suspension.

12.4.4 Cultivation System Design

Two geometric features should be set for a photobioreactor: the diameter or depth (D_{ph}) of the vessel and its length (L_{ph}). The former it is also known as the light path (LP).

The diameter and the depth are set for a cylindrical column and for parallelepiped-like column, respectively. For the latter design typology, the geometrical cross size should also be set. For the sake of simplicity, a unit cross size is adopted. Typically, photobioreactor performances may change with D_{ph}/L_{ph}. From the hydraulic point of view, the diameter/depth affects both the volumetric flow rate and the hydrodynamics of the suspension. Keeping the suspension velocity constant as the diameter/depth of the pipes increases, the pressure drop over the photobioreactor decreases while the volumetric flow rate of the suspension increases. Moreover, turbulent flow along the photobioreactor may occur as the diameter/depth increases. From the photosynthetic point of view, the increase of diameter/depth affects the fraction of microalgae getting exposed to sufficient irradiance needed for photosynthesis. The gradual decrease of the irradiance with the suspension depth (Fig. 12.2) does not allow exploiting the inner dark region of the photobioreactor. As a rule of thumb, the light intensity reduces by about one order of magnitude after a few millimeters (photic zone) when biomass concentration is about 1 g/L. Except for a suspension layer close to the photobioreactor surface, the decrease of the light intensity reduces the microalgal growth rate according to Eqs. (12.3) and (12.5). The decrease of light intensity nearby the photobioreactor surface may enhance the microalgal growth rate when the irradiance is higher than the photoinhibition threshold (see Section 12.2.2). Altogether, the diameter/depth of the bioreactor is a design variable to be finely tuned to improve photobioreactor performances. A proper setting of both D_{ph} and volumetric flow rate

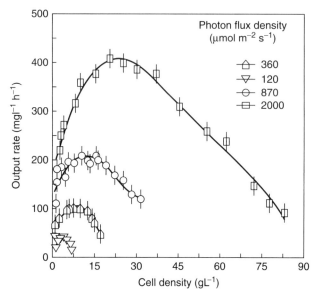

FIGURE 12.9 Biomass productivity as a function of cell density at different irradiance levels in a flat photobioreactor. Reproduced from Hu et al. [33] with permission of Taylor & Francis.

of the suspension may improve the renewal of microalgae population at the irradiated surface and high light conversion efficiency may be approached.

Richmond [2, 32] was able to take advantage of the coupling of the light/dark cycle in a flat column photobioreactor. He produced an interesting map of volumetric productivity of microalgae versus cell density for several levels of irradiance. The volumetric productivity of microalgae is characterized by a maximum, whatever the irradiance level (Fig. 12.9). When cell density is increased, a decrease in the light conversion efficiency should be expected since a significant (inner) region of the photobioreactor receives a negligible level of irradiance (dark zone). On the contrary, the light conversion efficiency increases at low cell density because of the renewal microalgae at the photic zone and their subsequent protection from the excess of photon flux for the time spent in the dark zone.

Effects of the vessel length on the microalgal growth depend on the typology of the photobioreactors. In general, this feature may affect the concentration of both soluble gases (CO_2 and O_2) and microalgae in the liquid phase, the amount of CO_2/O_2 transferred between gas and liquid phases (if two-phase flow is adopted in the irradiated photobioreactor, see next section), and the pressure drop along the photobioreactor.

An upper limit of the length of tubular photobioreactor for processing suspensions without gas supplements in the irradiated area may be assessed as a unit for gas–liquid mass exchange distinct from the photobioreactor adopted. Suspension operated without gas supplements accumulates oxygen along the flow in the photic area and the inhibition level $(C_{O_2})_{crit}$ may be approached. According to Molina et al. [16], the maximum L_{ph} is

$$L_{ph,max} = \frac{v_L}{R_{O_2}} \left[(C_{O_2})_{crit} - (C_{O_2})_{IN} \right] \qquad (12.35)$$

where v_L and R_{O_2} are the axial velocity of the gas-free suspension and the average volumetric rate of oxygen released from the microalgal suspension, respectively. The v_L depends on both the pressure drop along the photobioreactor and on the power of the pumping system adopted for the suspension. R_{O_2} may be assessed by working out Eqs. (12.2) through (12.6) and the yield coefficient Y_{X/O_2}. The average value of R_{O_2} should take into account the spatial distribution of light irradiance, chemical concentrations, microalgal concentration, and the concentrations of CO_2 and O_2.

In conclusion, the setting of the diameter/depth of the vessel and its length should take into account both photosynthetic features of the cultivated strain and the hydrodynamics of photobioreactors. Combined effects of both fluid features (density, viscosity, surface tension) and solid-microalgal features (density, diameter, motility) on hydrodynamics should also be taken into account [25, 34, 35, 36, 37, 38, 39, 40, 41, 42].

12.4.5 Gas–Liquid Mass Transfer

Two strategies are typically adopted for both CO_2/O_2 pumping/stripping to/from the microalgal cultures. The first strategy is based on direct gas–liquid contact in the irradiated vessel (e.g., bubble columns). The second strategy is based on CO_2/O_2 exchange between gas and liquid phases in vessels distinct from irradiated vessels (e.g., tubular photobioreactors equipped with airlifts).

Direct pumping of the CO_2 bearing stream into the microalgal culture is a common strategy adopted in vertical photobioreactors. The gas stream is the source of CO_2, the sink of the O_2, and may function as the mixing/circulation engine of the culture (Fig. 12.8A–C). The concentration of both the CO_2 and O_2 in the microalgal suspension may be controlled by tuning the gas flow rate, the mass transfer coefficient for volume unit of the liquid phase $K_L a$ (see Section 12.2.5), and the volumetric (flux) ratio between the gas and liquid phases.

The irradiated vessels equipped with units dedicated to the CO_2/O_2 exchange between gas and liquid phases are characterized by CO_2 accumulation in the liquid phase in the exchange unit and its conversion in the irradiated area while the exchange unit strips the O_2 accumulated in the irradiated area from the liquid phase . The separation of the unit dedicated to the uptake/accumulation of CO_2/O_2 from that dedicated to their pumping/stripping introduces a delicate system regarding the time spent by the suspension in the irradiated area, the microalgal concentration, the specific microalgal growth rate, and the critical inhibiting concentration of O_2. In particular, the concentration of O_2 accumulated in the liquid phase during the time spent in the irradiated area must be lower than the critical value $(C_{O_2})_{crit}$ (see previous section).

Whatever the strategy adopted for CO_2/O_2 pumping/stripping, the correct design of the gas exchanger—either separate from the irradiated area or coincident with it—requires the assessment of $K_L a$. The roles of both fluid and solid-microalgal features recalled at the end of the previous section still apply.

Devices adopted for CO_2/O_2 pumping/stripping often support circulation of the microalgal suspension through the cultivation systems. The momentum lost by the suspension flowing along the cultivation systems is recovered in the apparatus/devices.

12.5 NOVEL PHOTOBIOREACTOR CONFIGURATIONS

Classical open and closed configurations of photobioreactors are frequently designed according to rules of thumb typically adjusted on the basis of both simple macroscopic approach and on-site experiences. Performances of classical photobioreactor configurations are not always satisfactory. Therefore researchers have developed many novel configurations on a laboratoryscale but fewer on a pilot scale.

Table 12.1 has a list of novel photobioreactor configurations developed on both lab and pilot scales. Performances of photobioreactors are reported in terms of volumetric and areal productivities, and biomass concentration. Light penetration, mixing, gas–liquid mass transfer rate, and microalgae shear damage are the key issues on which researchers focused for developing novel photobioreactor configurations.

The optimization of the light capture by photosynthetic microorganisms is often the strategy adopted to maximize the photosynthesis rate, and to limit the photoinhibition and the light energy loss. To achieve this objective, different strategies were adopted depending on the type of light source.

Photobioreactors were designed to minimize light dispersion out of the microalgal culture. In the case of indoor cultures the goal was achieved by placing the light source inside the culture volume, as fluorescent tubes, optical fibers, metal halide lamps, or LED units [43, 53, 55, 56, 61, 69, 70, 71]. In the case of outdoor cultures, the sun light can be concentrated by light collector units (e.g., Fresnel lens) and distributed inside closed photobioreactors by means of focusing systems such as optical fibers and quartz tubes [25, 43, 57, 62, 72, 73, 74, 75, 76, 77, 78]. Figure 12.10 reports a detail of the SOLARGLASS™ system of Glaverbel Czech Ltd. [57, 79, 80] designed and constructed on a greenhouse. As a consequence of the sunlight concentration, the photobioreactor tubes were able to receive up to 7000 $\mu E/m^2 \cdot s$ of irradiance for most of the daily cycle.

Different typologies of photobioreactors have been developed to optimize the efficiency of light capture, but taking into account the particular microalgal features of the growth rate versus irradiance. The growth of microalgae is maximized at rather modest light intensities. Moreover, further increase of irradiance can cause low photosynthetic efficiencies due to photoinhibition, or even photobleaching (see Section 12.2.2). To increase process efficiencies, the photobioreactors have to be designed to distribute light over a large surface area so that only moderate light intensities are provided for the cells ("light dilution"). Carlozzi [46, 81] arranged tubular reactors in a fence-like construction (Fig. 12.11). The fences were oriented

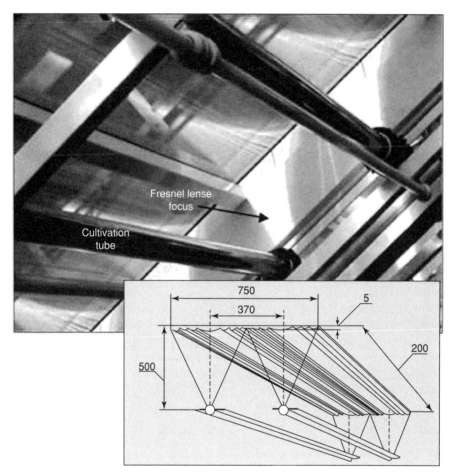

FIGURE 12.10   Details of a Fresnel lens focusing solar irradiance on cultivation tubes of the SOLARGLASS™ system. Reproduced from Masojídek et al. [57] with permission of Springer.

in a north/south direction to prevent direct bright light hitting the surface. Sunlight is then "diluted" along both horizontal and vertical directions. The advantage is that the high photon flux density in outdoor conditions during the summer period can be diluted by increasing the angle between the photon direction and the perpendicular to the irradiated surface. They reported a productivity of 48 g/(m²·day) with a tubular photobioreactor arranged in ten undulating horizontal rows.

Several innovative photobioreactors have been developed to take advantage of operating the microalgal cultures under dark/light cyclic conditions (see Section 12.2.2). Two classes of innovative photobioreactors are presented depending on the flow conditions—single-phase or two-phase.

TABLE 12.1 Design Features and Performances of Novel Laboratory-and Pilot-Scale Photobioreactors

	Configuration	Culture Conditions[a]	V (L)	A (m²)	LP (cm)	Strain	Qg (vvm)	CO_2 (%)	I (µE/m² s)	W_X (g/L·d)	F_X (g/m²·d)	X (g/L)	μ (h⁻¹)
Anand Kim [43]	Integrated solar optical fiber collector	I–O	15	0.084	3.7	Chlorobium thiosul-phatophilum			4–45				0.014
Bayless et al. [44]	Membrane flue gas solar collector optical fiber	O				Chlorogloeopsis		14%	120		55		
Briassoulis et al. [45]	Helical tubular photobioreactor	O	588		10	Nannochloris			700	2.57		0.35×10^9 cell/mL	
Carlozzi [46]	Tubular undulating row photobioreactor	O	11		1	Arthrospira platensis				1.15	25	2.4	0.0132
Chini Zittelli et al. [47]	Modular flat alveolar panel	I	123	20.4	0.6	Nannochloris	0.5	3%	115–230	1.45	8.7	4.4	0.014
Chini Zittelli et al. [48]	Annular photobioreactor	I–O	117–137	5.3–9.3	3–4.5	Nannochloris	0.1	1%	89–259	0.35	5.2	2.3	0.063
Garcia Camacho et al. [49]	Internal loop airlift	O	12	0.6	1	Phaeodactylum tricornutum			20–80	0.95	19	3.3	0.024
Grobbelaar and Kurano [50]	Multicompartment flat photobioreactor	I	0.75–1.5	0.0675	10–20	Synechocystis aquatilis		43%	240	2	67	1.8	1.1
Hall et al. [51]	Helical tubular photobioreactor	O	75	10	3	P. tricornutum A. siamensis				1.4	10	3.03	0.035
Hu et al. [52]	Flat inclined modular photobioreactor	O	6–50	0.63	1.3–10.4	Spirulina platensis M. subterraneus	2.1			4	51.3	14	0.021
Javanmardian and Palsson [53]	Optical fiber photobioreactor with liquid UF	I	0.6	0.192	5.21	C. vulgaris	0.5		76	1.5	472	6	0.000625
Keethesan and Nirmalakhandan [54]	Open pond with airlift pump	I	23		6.9	Scenedesmus	1.04		110	0.19		1.39	0.0056
Lee and Palsson [55]	LED flat photobioreactor with liquid UF	I	0.052–0.08	0.01	1–1.55	C. vulgaris	1.5		277–1000	1.9×10^{11} cell/L·d	1.9×10^{14} cell/m²·d	6.6%v	0.058
Lee and Palsson [56]	LED flat photobioreactor with liquid UF	I	0.052–0.08	0.01	1–1.55	C. vulgaris	1.5		277–1000	1.2×10^{11} cell/L·d	1.2×10^{14} cell/m²·d	6.6%v	0.058

Reference	Reactor type	I/O[a]				Species							
Masojídek et al. [57]	Greenhouse photobioreactor with solar collector	O	65	1.2	4.8	*S. platensis*			700	0.5	32.5	1.2–2.2	0.012
Morita et al. [58]	Conical helicoidal tubular photobioreactor	I	14–36	1–4	1.6	*Chlorella* sp. *HA-1*		10%	484	1.72	31		
Morita et al. [59]	Conical helicoidal tubular photobioreactor	O	14	1	1.6	*Chlorella sorokiniana*	0.13	10%		1.2	33	1–2	0.076
Muller-Fuega et al. [60]	Annular swirling flow photobioreactor	I	100	1.5	3	*Porphyridium cruentum*	0.8		200	0.19	12.5		0.01
Ogbonna et al. [61]	Mechanically stirred internally irradiated	I	2.8	0.072	7	*C. pyrenoidosa*	0.6	5%	205	0.28	14	2.2	
Ogbonna and Tanaka [62]	Solar collector with optical fiber	I–O	3.5	0.086	2.3	*C. pyrenoidosa*	0.3	5%	>60	0.3	12.2	3	
Pushparaj et al. [63]	Open pond with flat alveolar panel	O	282	5	1.2[b] 8[c]	*A. platensis*	0.6–1			0.32	19.4	1.6	
Travieso et al. [64]	Helicoidal tubular photobioreactor	I	21	1.32	1.6	*S. platensis* *Chlorella* *Spirulina*		4%	157	0.4	6.45	5.8	0.078
Tsoglin et al. [65]	Mechanically stirred inside irradiated	I	8–12	0.33	3.0	*Porphyridium* *Dunaliella* *Chlorella* *Spirulina*			1750	4.5	136	3.8	
Tsoglin et al. [65]	Oscillatory pond, up surface irradiated	I	6–14	0.21	4.8	*Porphyridium* *Dunaliella*			125	9.2			
Ugwu et al. [66]	Inclined tubular static mixer photobioreactor	O	6	0.18	3.8	*C. sorokiniana*	0.125–1.25	5%		1.1	37	0.5	0.092
Ugwu et al. [67]	Inclined tubular static mixer photobioreactor	O	58	1	12.5	*S. aquatilis*	0.25	5%		0.95	35	1	0.039
Ugwu et al. [68]	Inclined tubular static mixer photobioreactor	O	6–19–58	0.3–0.6–1	3.8–7.5–12.5	*C. sorokiniana*	0.25	5%		1.22	45.7	1.5	0.034

[a] I, indoor; O, outdoor.
[b] Flat panel.
[c] Open pond.

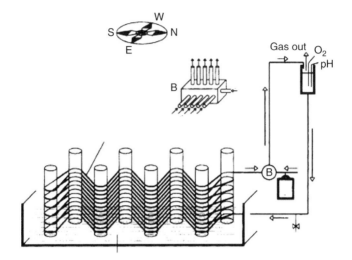

FIGURE 12.11 Tubular undulating row photobioreactor with recirculation loop and degasser section. Reproduced from Carlozzi [46] with permission of Springer.

In single-phase flow in a tubular photobioreactor, microalgae may experience a sequence of light/dark cycles because the local liquid turbulence may induce a microalgal flow from the wall to the center of the tube. However, the time scale of the turbulent flow that may occur in classical photobioeractors is too long with respect to that required to take advantage of the light/dark cycle. The adoption of a static mixer (Fig. 12.12A) or another type of swirling flow inducer (Fig. 12.12B) can enhance the turbulence mixing degree in the photobioreactor [60, 66, 67, 83, 84, 85, 86, 87].

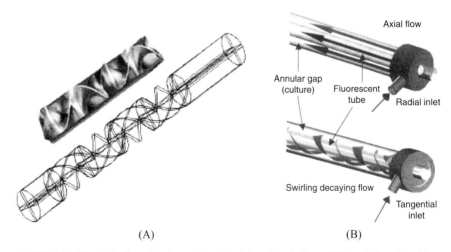

(A)   (B)

FIGURE 12.12 Kuhnii static mixer (A) and swirling flow inducer (B) in tubular photobioreactor. Reproduced from Pruvost et al. [82] with permission of John Wiley & Sons.

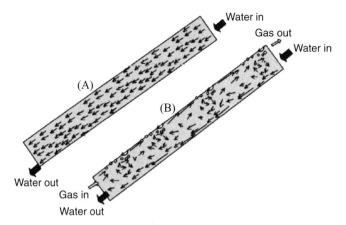

FIGURE 12.13   Flow pattern in inclined photobioreactor without (A) and with gas injection (B). Reproduced with permission from Vunjak-Novakovic et al. [28], copyright © 2005 American Chemical Society.

In photobioreactors operated with gas–liquid two-phase flow, the turbulent mixing is bubble induced [88]. Bubbles induce macroeddies, which are characterized by time- and length-scale of bubble wakes, and a dispersion coefficient of tens of mm^2/s. Accordingly, the frequency of renewal of microalgae at the photic zone increases up to the order of 1 Hz, for a light path of about 1 cm. Nevertheless, the advantage related to the light/dark cycle experienced by microalgae is often made fruitless by the microalgal shear damage due to the presence of bubbles [89, 90, 91]. In order to apply the advantage of the two-phase flow and mitigate the disadvantage of shear damage, several researchers have proposed new designs of two-phase photobioreactors. The segregation of bubbly flow on one side of the photobioreactor induces a large-scale circulation but leaves a large fraction of it free of bubbles. Hu et al. [52], Vunjak-Novakovic et al. [28], and Merchuk et al. [29] observed an increase of the photobioreactor productivities when operated as inclined flat photobioreactors (Fig. 12.13).

The light/dark cycle was also accomplished in flat airlift configurations where the riser was the photic zone [92,93]. However, the mixing times were only a few seconds and advantages in the photosynthesis rate were often not observed. An interesting exception was the triangular reactor from MIT (Fig. 12.14) [94, 95]. It combined the airlift principle with static mixers in an external photic downcomer. The 3DMS reactor exhibited a surprising average productivity of 98 g/(m^2·day).

Degen et al. [96] developed an airlift photobioreactor with horizontal baffles located in the riser zone. The hydrodynamic behavior was characterized by a well-defined macrocirculation cell: the length scale was approximately the light path of the photobioreactor, and the circulation time was less than 1 s. The flat panel airlift was successfully developed on a pilot scale by Subitec (Fig. 12.15).

Despite the numerous novel configurations developed on the laboratoryscale, very few examples of industrial applications can be encountered for photobioreactors

(A)   (B)

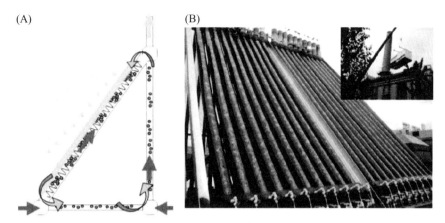

FIGURE 12.14   GreenFuel's 3DMS reactor: (A) U.S. Patent drawing, and (B) industrial plant in the Red Hawk Power Plant, Arizona, U.S.A. Reproduced with permission from Vunjak-Novakovic et al. [28], copyright © 2005 American Chemical Society.

[97]. Some designs are not easy to scale-up and their costs still remain too high. In addition, very few designs are described from the mechanistic point of view. A challenge for researchers is to obtain the design principles from a complete model, coupling hydrodynamic behavior with the photosynthesis kinetic mechanism and mass transfer phenomena.

FIGURE 12.15   Flat panel airlift: outdoor Subitec plant equipped with baffles. Reproduced from Schenk et al. [95] with permission of Springer and from Posten [7] with permission of John Wiley & Sons.

12.6   CASE STUDY: INTENSIVE PRODUCTION OF BIO-OIL

In this section a case study of a photobioreactor adopted for intensive production of bio-oil is reported. Two issues are addressed: (1) the maximum productivity and (2) a rough cost estimation of the plant required for the production. The first issue is hereinafter presented from both energetic and kinetic points of view.

12.6.1   Assessment of Maximum Productivity

Some basic relationships for microalgal production rates are simply given below. Detailed and more complex models may be found in the literature [12].

Energetic Issue   Under autotrophic conditions the microalgal growth is based solely on incident sunlight energy and on the CO_2 captured from the air, with feeding minerals provided. Therefore the energetic content of a microalgal suspension is just a fraction of energy received from the sunlight irradiance (E_{SL}).

The overall efficiency (ε_{ov}) of the conversion of sunlight to chemical energy and microalgal production may be assessed as

$$\varepsilon_{ov} = \text{PAR} \cdot \gamma \tag{12.36}$$

where PAR is the photosynthetically active radiation, and γ the efficiency of conversion of PAR to chemical energy. The latter embodies the photosynthetic efficiency and the optical efficiency of the system.

The biomass energy content of the microalgae (E_{DM}) may be assessed taking into account the energy content of the lipidic fraction (E_{lip}), that of the nonlipidic fraction (E_{nl}), and the lipidic mass fraction of the microalgae (ω_{lip}):

$$E_{DM} = E_{lip}\omega_{lip} + E_{nl}(1 - \omega_{lip}) \tag{12.37}$$

The ratio between the sunlight energy available and the E_{DM} yields the rate of microalgal production per unit of irradiated area ($F_{X,en}$), on a mass basis:

$$F_{X,en} = E_{SL}\varepsilon_{ov}/E_{DM} \tag{12.38}$$

Table 12.2 reports data of microalgae and oil productivity assessed with reference to an irradiance typical of the southern counties of Italy. The data of γ, ω_{lip}, E_{lip}, and E_{nl} reported in the table are average among values reported in the literature. The theoretical productivity (F_X and F_{lip}) should be considered the maximum value achievable with reference to the assumed conditions.

Kinetics Issue   Assuming that the bioreactor conforms to a chemostat and an unstructured model may represent the biomass growth, Eq. (12.21) can still hold. For

TABLE 12.2   Theoretical Productivity of Microalgae and Bio-oil: Energy-Based Assessment

Global irradiation for surface inclined to South at 45°, Napoli (E_{SL})	6500 MJ/yr·m² [98]
Photosynthetically active radiation (PAR)	43%
Efficiency of conversion of PAR to chemical energy (γ)	10%
Microalgal oil fraction (ω_{lip})	25%$_W$
Energy content of lipids (E_{lip})	38 MJ/kg$_{DM}$
Energy content of no-lipid components (E_{nl})	17 MJ/kg$_{DM}$
Energy content of microalgae (E_{DM})	22 MJ/kg$_{DM}$
Theoretical productivity (based on energetic consideration)	
Microalgae ($F_{X,en}$)	13 kg$_{DM}$/yr·m²
Oil fraction (F_{lip})	3.2 kg$_{lip}$/yr·m²

the sake of simplicity, no relationship was assumed between growth conditions and accumulation of lipids. Accordingly, μ represents the growth rate of the microalgae at a constant lipidic mass fraction (ω_{oil}). The mass production rate of microalgae per unit volume of the bioreactor (W'_X) is

$$W'_x = DX_F \tag{12.39}$$

The microalgal production rate per unit irradiated area (F_X) may be estimated taking into account the average depth of photobioreactors (δ_{pr}) [99, 100]:

$$F_X = \delta_{pr}DX \tag{12.40}$$

Under autotrophic conditions and assuming $\delta_{pr} = 0.04$ m, $\mu = 0.25$ L/d, and $X = 3$ kg/m³, Eq. (12.40) yields $F_X = 0.03$ kg/d·m² (=11 kg/yr·m²). Accordingly, the volumetric flow rate of microalgal suspension produced is about $q = 0.010$ m³/ (d·m²$_{pr}$). It should be noted that F_X is about the maximum productivity estimated based on energetic arguments ($F_{X,en}$).

12.6.2   Economic Assessment

The main steps of a process aimed at producing bio-oil by microalgae are: (1) intensive cultures of microalgae, (2) harvesting of microalgae, and (3) separation of the lipidic fraction to be employed as bio-oil. A synoptic chart of the process is reported in Fig. 12.16. But, treatment processes for wastes and wastewaters are not included.

The process to exploit the nonlipidic fraction of microalgae is also not included in Fig. 12.16. The fate of this fraction will critically depend on the microalgal strain. The spectrum of possible paths to exploitation of this fraction is quite broad with examples such as gasification/combustion, food supplements, cosmetics, or substrate for fermentations devoted to produce energetic carriers. In any case, exploitation of this fraction requires for additional processes characterized by capital, costs, and incomes that have not been included in the present analysis.

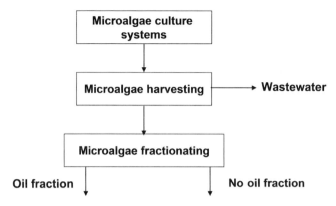

FIGURE 12.16   Synoptic chart of bio-oil production by microalgae.

The apparatus adopted in each operation unit reported in Fig. 12.16 are listed hereinafter.

Microalgal Cultivation System   Photobioreactors are adopted as suggested in the published literature on the subject [99, 101].

Microalgal Harvesting   The unit operations typically adopted for microalgal harvesting are based on mechanical separations where size and/or density difference drive the fractionation process. Filtration, sedimentation, and centrifugation are typically used to harvest microalgae, the latter being the most expensive. The sizes of the harvesting units are controlled by the volumetric flow rate issuing from the photobioreactor. A vacuum continuous filter has been adopted.

Microalgal Processing   The postharvest processing depends on the fate of microalgal components. It should be pointed out that the sizes of the units adopted in the postharvest processes are definitively smaller than the photobioreactor. In fact, the flow rates of the streams fed to the postharvest units are typically at least one to two orders of magnitude smaller than those of streams issuing from the photobioreactors. For the sake of simplicity the contribution of postharvest processing has been ignored in the present analysis.

Cost Estimation   The procedure proposed by Happel and Jordan [102] was adopted for cost estimation of the conceptual design of the flow sheet dedicated to the production of bio-oil from microalgae. The total capital requirement was estimated considering the purchase cost of the main units and adopting the Lang factor (f_L) for ancillary apparatus and installation costs [102]. In particular, the photobioreactor item was assumed to be "special equipment." The purchase cost of the photobioreactor was estimated on the basis of the very limited data available in the literature [101, 103].

The purchase cost for the filter was estimated in agreement with correlations reported in Peters et al. [104].

The peculiarity of the photobioreactor suggests that the purchase cost scales linearly with the exposed area. The present (2012) cost of an up-to-date photobioreactor—based on the plant cost index—is of about 120 €/m^2.

The filtration unit used to harvest the microalgae should be sized to filter the q stream at a microalgal concentration of about 3 kg/m^3. The area of a vacuum continuous filter has been estimated assuming a pressure drop across the filter of 60 kPa and a specific cake resistance of 10^{12} m/kg [105].

The economical sustainability of the process has been assessed in terms of economical potential (EP):

$$EP = \text{Product value} - \text{Raw materials cost} \qquad (12.41)$$

and of the yearly rate (L) of depreciation of the main fixed investment (I_F) required for the plant. L has been estimated as

$$L = eI = e\left(I_{pr} + f_L I_F\right) = e\left(I_{pr} + f_L \sum I_i\right) \qquad (12.42)$$

where e is [1/yr] the yearly fractional depreciation rate, I_F the fixed investment in the complete system, I_i the cost of the major processing equipment [104], and I_{pr} the cost of the photobioreactor. The depreciation rate e has been assessed according to the sinking fund method: $e = i/[\exp(i \cdot n) - 1]$, where n [yr] is the expected project life and i [1/yr] is the rate of return to the firm. Tentative reference values of $f_L = 3$, n = 10, $i = 0.10$ 1/yr, and $e = 0.058$ 1/yr have been used for this preliminary economic assessment [106].

Table 12.3 reports the input reference data used in the present analysis, which has been referred to a cultivation area of 1 km^2. As regards the major processing equipment, only the photobioreactor and the harvesting system have been taken into account. When estimating the EP, the cost of the raw materials (e.g., CO$_2$, and nitrogen and phosphorus supplements) has been neglected and the "Product value" is just the bio-oil cost (s_{oil}) for the yearly production rate (M_{oil}). It should be underlined that the

TABLE 12.3 Selected Input/Output Variables of the Economic Analysis

Photobioreactor area	1 km^2
Bio-oil unit selling price on mass basis (s_{oil})	550 €/t
Volumetric flow rate of the microalgal suspension	0.12 m^3/s
Bio-oil productivity (M_{oil})	2750 t/yr
EP	1500 k€/yr
Photobioreactor cost	120,000 k€
Filter cost	870 k€
L	7000 k€/yr

cost of CO_2 would be a negative cost in Eq. (12.41) in agreement with the current regulations on CO_2 sequestration.

The analysis of data reported in Table 12.3 highlights that under the economical scenario adopted:

- The process is characterized by L larger that the EP.
- The capital investment related to the photobioreactor is the largest contribution to L.

Results suggest that one key to the development of economically sustainable bio-oil production is the development of inexpensive photobioreactors. In particular, the photobioreactor capital cost per unit area should be reduced by more than one order of magnitude. Moreover, studies aimed at increasing the efficiency γ and the microalgal oil fraction will allow improvement of bio-oil productivity, and then the process sustainability [107].

ACKNOWLEDGMENTS

The authors are grateful to Prof. Gabriele Pinto and Prof. Antonino Pollio (Biological Science Department, Università degli Studi di Napoli *Federico II*) for interesting discussions regarding microalgal physiology and metabolisms. Professor Roberto Andreozzi and Prof. Raffaele Marotta (Department of Chemical Engineering, Università degli Studi di Napoli *Federico II*) are also acknowledged for useful discussions regarding biodiesel production from microalgae.

REFERENCES

1. J. S. Burley, *Algal Cultures—From Laboratory to Pilot Plant*, Carniege Institute of Washington Publications, 1953.

2. A. Richmond, *Handbook of Microalgal Cultures*, Blackwell Science, Oxford, UK, 2004.

3. P. G. Falkowski, J. A. Raven, *Aquatic Photosynthesis*, Princeton University Press, Princeton, NJ, 1997.

4. J. F. Cornet, C. G. Dussap, J. B. Gros, Kinetics and energetics of photosynthetic microorganism in photobioreactors—application to *spirulina* growth, Adv. Biochem. Eng. Biotechnol. 59 (1998) 153–224.

5. J. Masojídek, M. Koblížek, G. Torzillo, Photosynthesis in microalgae. In: A. Richmond (Ed.), *Handbook of Microalgal Culture*, Blackwell Science, Oxford, UK, 2004, pp. 20–39.

6. J. C. Goldman, Outdoor algal mass cultures. II. Photosynthetic field limitations, Water Res. 13 (1979) 119–160.

7. C. Posten, Design principles of photobioreactors for cultivation of microalgae, Eng. Life Sci. 9 (2009) 165–177.

8. E. Molina Grima, F. García Camacho, J. A. Sánchez Perez, J. A. Fernández, J. M. Fernández Sevilla, F. G. Acién Fernández, A. Contreras Gómez, A mathematical model of microalgal growth in light limited chemostat culture, J. Chem. Technol. Biotechnol. 61 (1994) 167–173.

9. O. M. Alfano, R. L. Romero, A. E. Cassano, Radiation field modelling in photoreactors. I. Homogeneous media, Chem. Eng. Sci. 41 (1986) 421–444.

10. P. H. C. Eilers, J. C. H. Peeters, A model for the relationship between light intensity and the rate of photosynthesis in phytoplankton, Ecol. Model. 42 (1998) 199–215.

11. P. H. C. Eilers, J. C. H. Peeters, Dynamic behaviour of a model for photosynthesis and photoinhibition, Ecol. Model. 69 (1993) 113–133.

12. F. G. Acien Fernández, F. García Camacho, J. A. Sánchez Perez, J. M. Fernández Sevilla, E. Molina Grima, Modelling of biomass productivity in tubular photobioreactors for microalgal cultures: effects of dilution rate, tube diameter and solar irradiance, Biotechnol. Bioeng. 58 (1998) 605–616.

13. J. N. Phillips, J. N. Myers, Growth rate of *Chlorella* in flashing light, Plant Physiol. 29 (1954) 152–161.

14. M. G. de Morais, J. A. V. Costa, Carbon dioxide fixation by *Chlorella kessleri, C. vulgaris, Scenedesmus obliquus* and *Spirulina* sp. cultivated in flasks and vertical tubular photobioreactors, Biotechnol. Lett. 29 (2007) 1349–1352.

15. H. Märkl, M. Mather, Mixing and aeration of shallow open ponds, Arch. Hydrobiol. Beih. Ergebn. Limnol. 20 (1985) 85–93.

16. E. Molina, J. Fernández, F. G. Acién, Y. Chisti, Tubular photobioreactors design for algal cultures, J. Biotechnol. 92 (2001) 113–131.

17. J. Doucha, F. Straka, K. Lívanský, Utilization of flue gas for cultivation of microalgae (*Chlorella* sp.) in an outdoor open thin-layer photobioreactor, J. Appl. Phycol. 17 (2005) 403–412.

18. G. Olivieri, I. Gargano, R. Andreozzi, R. Marotta, A. Marzocchella, G. Pinto, A. Pollio, Effects of CO_2 and pH on *Stichococcus bacillaris* in laboratory scale photobioreactors, Chem. Eng. Trans. 27 (2012) 127–132.

19. J. E. Bailey, D. F. Ollis, *Biochemical Engineering Fundamentals*, McGraw-Hill, New York, 1986.

20. O. Levenspiel, *Chemical Reaction Engineering*, 3rd ed., Wiley, Hoboken, NJ, 1999.

21. J. Nielsen, J. Villadsen, *Bioreaction Engineering Principles*, 2nd ed., Kluwer Academic/Plenum Publishers, Dordrecht, The Netherlands, 2002.

22. M. A. Borowitzka, L. J. Borowitzka, *Micro-algal Biotechnology*. Cambridge University Press, Cambridge, UK, 1988.

23. E. Molina Grima, Microalgae, mass culture methods. In: M. C. Flickinger, S. W. Drew (Eds.), *The Encyclopedia of Bioprocess Technology: Fermentation, Biocatalysis, and Bioseparation*, Wiley, Hoboken, NJ 1999.

24. W. J. Oswald, Large-scale algal culture systems (engineering aspects). In: M. A. Borowitzka, L. J. Borowitzka (Eds.), *Microalgal Biotechnology*, Cambridge University Press, Cambridge,UK, 1988, pp. 357–394.

25. M. Jansenn, J. Tramper, L. R. Mur, R. H. Wijffels, Enclosed outdoor photobioreactors: light regime, photosynthetic efficiency, scale-up, and future prospects, Biotechnol. Bioeng. 81 (2003) 193–210.

26. C. Y. Chen, K. L. Yeh, R. Aisyah, D. J. Lee, J. S. Chang, Cultivation, photobioreactor design and harvesting of microalgae for biodiesel production: a critical review, Bioresour. Technol. 102 (2011) 71–81.

27. F. G. Acién Fernàndez, J. M. Fernàndez Sevilla, J. A. Sànchez Pérez, E. Molina Grima, Y. Chisti, Airlift-driven external-loop tubular photobioreactors for outdoor production of microalgae: assessment of design and performance, Chem. Eng. Sci. 56 (2001) 2721–2732.

28. G. Vunjak-Novakovic, Y. Kim, X. Wu, I. Berzin, J. C. Merchuk, Air-lift bioreactors for algal growth on flue gas: mathematical modeling and pilot-plant studies, Ind. Eng. Chem. Res. 44 (2005) 6154–6163.

29. J. C. Merchuk, Y. Rosenblat, I. Berzin, Fluid flow and mass transfer in a counter-current gas–liquid inclined tubes photo-bioreactor, Chem. Eng. Sci. 62 (2007) 7414–7425.

30. O. Bernard, Hurdles and challenges for modelling and control of microalgae for CO_2 mitigation and biofuel production, J. Proc. Control 21 (2011) 1378–1389.

31. J. P. Bitog, I. B. Lee, C. G. Leec, K. S. Kimd, H. S. Hwang, S. W. Hong, I. H. Seoa, K. S. Kwon, E. Mostafa, Application of computational fluid dynamics for modeling and designing photobioreactors for microalgae production: a review, Comp. Electron. Agric. 76 (2011) 131–147.

32. A. Richmond, Growth characteristics of ultrahigh-density microalgal cultures, Biotechnol. Bioproc. Eng. 8 (2003) 349–353.

33. Q. Hu, Y. Zarmi, A. Richmond, Combined effects of light intensity, light-path, and culture density on output rate of *Spirulina platensis* (Cyanobacteria), Eur. J. Phycol. 33 (1998) 165–171.

34. L. S. Fan, *Gas–Liquid–Solid Fluidization Engineering*, Butterworths, Stoneham, MA, 1989.

35. J. Zahradník, M. Fialová, M. Ruzicka, J. Drahoš, F. Kaštánek, N. H. Thomas, Duality of the gas–liquid flow regimes in bubble column reactors, Chem. Eng. Sci, 52 (1997) 3811–3826.

36. Y. Chisti, Penumatically agitated bioreactors in industrial and environmental bioprocessing: hydrodynamics, hydraulics and transport phenomena, Appl. Mech. Rev. 51(1998) 33–98.

37. G. Olivieri, A. Marzocchella, P. Salatino, Hydrodynamics and mass transfer in a lab-scale three-phase internal loop airlift. Chem. Eng. J. 96 (2003) 49–58.

38. P. Cozma, M. Gavrilescu, Airlift reactors: hydrodynamics, mass transfer and applications in environmental remediation, J. Environ. Eng. Manage. 9 (2010) 681–702.

39. G. Olivieri, A. Marzocchella, J. R. Van Ommen, P. Salatino, Local and global hydrodynamics in a two-phase internal loop airlift, Chem. Eng. Sci. 62 (2007) 7068–7077.

40. G. Olivieri, A. Marzocchella, P. Salatino, A novel three-phase airlift reactor without circulation of solids, Can. J. Chem. Eng. 88 (2010) 574–578.

41. G. Olivieri, M. E. Russo, M. Simeone, A. Marzocchella, P. Salatino, Effects of viscosity and relaxation time on the hydrodynamics of gas-liquid systems, Chem. Eng. Sci. 66 (2011) 3392–3399.

42. M. C. Ruzicka, M. M. Vecer, S. Orvalho, J. Drahos, Effect of surfactant on homogeneous regime stability in bubble column, Chem. Eng. Sci. 63 (2008) 951–967.

43. J. Y. An, B. W. Kim, Biological desulfurization in an optical-fiber photobioreactor using an automatic sunlight collection system, J. Biotechnol. 80(1) (2000) 35–44.

44. D. J. Bayless, G. Kramer, M. Vis, B. Stuart, L. Shi, E. Cono, J. L. Cuello, Photosynthetic CO_2 mitigation using a novel membrane-based photobioreactor, J. Environ. Eng. Manage. 16 (2006) 209–215.

45. D. Briassoulis, P. Panagakis, M. Chionidis, D. Tzenos, A. Lalos, C. Tsinos, K. Berberidis, A. Jacobsen, An experimental helical-tubular photobioreactor for continuous production of *Nannochloropsis* sp., Bioresour. Technol. 101 (2010) 6768–6777.

46. P. Carlozzi, Hydrodynamic aspects and *Arthrospira* growth in two outdoor tubular undulating row reactors, Appl. Microbiol. Biotechnol. 54 (2000) 14–22.

47. G. Chini Zittelli, R. Pastorelli, M. R. Tredici, A modular flat panel photobioreactor (MFPP) for indoor mass cultivation of *Nannochloropsis* sp. under artificial illumination, J. Appl. Phycol. 12 (2000) 521–526.

48. G. Chini Zittelli, L. Rodolfi, M. R. Tredici, Mass cultivation of *Nannochloropsis* sp. in annular reactors, J. Appl. Phycol. 15 (2003) 107–114.

49. F. García Camacho, A. Contreras Gómez, F. G. Acién Fernández, J. Fernández Sevilla, E. Molina Grima, Use of concentric-tube airlift photobioreactors for microalgal outdoor mass cultures Enzyme. Microb. Technol. 24 (1999) 164–172.

50. J. U. Grobbelaar, N. Kurano, Use of photoacclimation in the design of a novel photobioreactor to achieve high yields in algal mass cultivation, J. Appl. Phycol. 15 (2003) 121–126.

51. D. O. Hall, F. G. Acien Fernandez, E. Canizares Guerrero, K. Krishna Rao, E. Molina Grima, Outdoor helical tubular photobioreactors for microalgal production: modeling of fluid-dynamics and mass transfer and assessment of biomass productivity, Biotechnol. Bioeng. 82 (2003) 62–73.

52. Q. Hu, H. Guterman, A. Richmond, A flat inclined modular photobioreactor for outdoor mass cultivation of photoautotrophs, Biotechnol. Bioeng. 51 (1996) 51–60.

53. M. Javanmardian, B. O. Palsson, High-density photoautotrophic algal cultures: design, construction, and operation of a novel photobioreactor system, Biotechnol. Bioeng. 38 (1991) 1182–1189.

54. B. Ketheesan, N. Nirmalakhandan, Feasibility of microalgal cultivation in a pilot-scale airlift-driven raceway reactor, Bioresour. Technol. 108 (2012) 196–202.

55. C. G. Lee, B. O. Palsson, High-density algal photobioreactors using light-emitting diodes, Biotechnol. Bioeng. 44 (1994) 1161–1167.

56. C. G. Lee, B. O. Palsson, Light emitting diode-based algal photobioreactor with external gas exchange, J. Ferment. Bioeng. 79 (1995) 257–263.

57. J. Masojídek, Š. Papáček, M. Sergejevová, V. Jirka, J. Červený, J. Kunc, J. Korečko, O. Verbovikova1, J. Kopecký, D. Štys, G. Torzillo, A closed solar photobioreactor for cultivation of microalgae under supra-high irradiance: basic design and performance, J. Appl. Phycol. 15 (2003) 239–248.

58. M. Morita, Y. Watanabe, T. Okawa, H. Saiki, Photosynthetic productivity of conical helical tubular photobioreactors incorporating *Chlorella* sp. under various culture medium flow conditions, Biotechnol. Bioeng. 74 (2001) 136–144.

59. M. Morita, Y. Watanabe, H. Saiki, Investigation of photobioreactor design for enhancing the photosynthetic productivity of microalgae, Biotechnol. Bioeng. 69 (2002) 693–698.

60. A. Muller-Feuga, J. Pruvost, R. Le Guédes, L. Le Déan, P. Legentilhomme, J. Legrand, Swirling flow implementation in a photobioreactor for batch and continuous cultures of *Porphyridium cruentum*, Biotechnol. Bioeng. 84 (2003) 544–551.

61. J. C. Ogbonna, H. Yada, H. Masui, H. Tanaka, A novel internally illuminated stirred tank photobioreactor for large-scale cultivation of photosynthetic cells, J. Ferment. Bioeng. 82 (1996) 61–67.

62. J. C. Ogbonna, H. Tanaka, An integrated solar and artificial light system for internal illumination of photobioreactors, J. Biotechnol. 70 (1999) 289–297.

63. B. Pushparaj, E. Pelosi, M. R. Tredici, E. Pinzani, R. Materassi, An integrated culture system for outdoor production of microalgae and cyanobacteria, J. Appl. Phycol. 9 (1999) 113–119.

64. L. Travieso, D. O. Hall, K. K. Rao, F. Benitez, E. Sanchez, R. Borja, A helical tubular photobioreactor producing *Spirulina* in a semicontinuous mode, Int. Biodeter. Biodeg. 47 (2001) 151–155.

65. L. N. Tsoglin, B. V. Gabel, T. N. Fal'kovich, V. E. Semenenko, Closed photobioreactors for microalgal cultivation, Russ. J. Plant Physiol. 43 (1996) 131–136.

66. C. U. Ugwu, J. C. Ogbonna, H. Tanaka, Improvement of mass transfer characteristics and productivities of inclined tubular photobioreactors by installation of internal static mixers, Appl. Microbiol. Biotechnol. 58 (2002) 600–607.

67. C. U. Ugwu, J. C. Ogbonna, H. Tanaka, Characterization of light utilization and biomass yields of *Chlorella sorokiniana* in inclined outdoor tubular photobioreactors equipped with static mixers, Proc. Biochem. 40 (2005) 3406–3411.

68. C. U. Ugwu, J. C. Ogbonna, H. Tanaka, Light/dark cyclic movement of algal culture (*Synechocystis aquatilis*) in outdoor inclined tubular photobioreactor equipped with static mixers for efficient production of biomass, Biotechnol. Lett. 27 (2005) 75–78.

69. Z. Csogör, M. Herrenbauer, I. Perner, K. Schmidt, C. Posten, Design of a photo-bioreactor for modelling purposes, Chem. Eng. Proc. 38 (1999) 517–523.

70. Z. Csgör, M. Herrenbauer, K. Schmidt, C. Posten, Light distribution in a novel photobioreactor—modelling for optimization, J. Appl. Phycol. 13 (2001) 325–333.

71. C. Y. Chen, J. S. Chang, Enhancing phototropic hydrogen production by solid-carrier assisted fermentation and internal optical-fiber illumination, Proc. Biochem. 41 (2006) 2041–2049.

72. J. C. Ogbonna, H. Tanaka, Light requirement and photosynthetic cell cultivation—development of processes for efficient light utilization in photobioreactors, J. Appl. Phycol. 12 (2000) 207–218.

73. K. Mori, Photoautotrophic bioreactor using solar rays condensed by Fresnel lenses, Biotechnol. Bioeng. Symp. 15 (1985) 331–345.

74. D. Feuermann, J. M. Gordon, High-concentration photovoltaic designs based on miniature parabolic dishes, Solar Energy 70 (2001) 423–430.

75. E. Ono, J. L. Cuello, Design parameters of solar concentrating systems for CO_2-mitigating algal photobioreactors, Energy 29 (2004) 1651–1657.

76. C. Y. Chen, G. D. Saratalea, C. M. Lee, P. C. Chen, J. S. Chang, Phototrophic hydrogen production in photobioreactors coupled with solar-energy-excited optical fibers, Int. J. Hydrogen Energy 33 (2008) 6886–6895.

77. J. W. F. Zijffers, M. Janssen, J. Tramper, R. H. Wijffels, Design process of an area-efficient photobioreactor, Mar. Biotechnol. 10 (2008) 404–415.

78. C. L. Guo, X. Zhu, Q. Liao, Y. Z. Wang, R. Chen, D. J. Lee, Enhancement of photo-hydrogen production in a biofilm photobioreactor using optical fiber with additional rough surface, Bioresour. Technol. 102 (2011) 8507–8513.

79. V. Jirka, V. Kuceravy, M. Maly, F. Pech, J. Pokorny, Energy flow in a greenhouse equipped with glass raster lenses, Renew. Ener. 16 (1999) 660–664.

80. J. Masojídek, M. Sergejevová, K. Rottnerová, V. Jirka, J. Korečko, J. Kopecký, I. Zaťková, G. Torzillo, D. Štys, A two-stage solar photobioreactor for cultivation of microalgae based on solar concentrators, J. Appl. Phycol. 21 (2009) 55–63.

81. P. Carlozzi, Dilution of solar radiation through "culture" lamination in photobioreactor rows facing south–north: a way to improve the efficiency of light utilization by cyanobacteria (*Arthrospira platensis*), Biotechnol. Bioeng. 81 (2003) 305–315.

82. J. Pruvost, J. Legrand, P. Legentilhomme, A. Muller-Feuga, Simulation of microalgae growth in limiting light conditions: flow effect, AIChE J. 48 (2002) 1109–1120.

83. A. Lucumi, C. Posten, Establishment of long-term perfusion cultures of recombinant moss in a pilot tubular photobioreactor, Proc. Biochem. 41 (2006) 2180–2187.

84. I. Perner-Nochta, A. Lucumi, C. Posten, Photoautotrophic cell and tissue culture in a tubular photobioreactor, Eng. Life Sci. 7 (2007) 127–135.

85. K. Loubiere, E. Olivo, G. Bougaran, J. Pruvost, R. Robert, J. Legrand, A new photobioreactor for continuous microalgal production in hatcheries based on external-loop airlift and swirling flow, Biotechnol. Bioeng. 102 (2008) 132–147.

86. K. Loubiere, J. Pruvost, F. Aloui, J. Legrand, Investigations in an external-loop airlift photobioreactor with annular light chambers and swirling flow, Chem. Eng. Res. Des. 89 (2011) 164–171.

87. J. Pruvost, J. Legrand, P. Legentilhomme, A. Mueller-Fuega, Effect of inlet type on shear stress and mixing in an annular photobioreactor involving a swirling decaying flow, Can. J. Chem. Eng. 82 (2004) 495–503.

88. Y. Sato, K. Sekoguchi, Liquid velocity distribution in two-phase bubble flow, Int. J. Mult. Flow. 2 (1975) 79–95.

89. M. J. Barbosa, M. J. Hadiyanto, R. H. Wijffels, Overcoming shear stress of microalgae cultures in sparged photobioreactors, Biotechnol. Bioeng. 85 (2004) 78–85.

90. M. J. Barbosa, M. Albrecht, R. H. Wijffels, Hydrodynamic stress and lethal events in sparged microalgae cultures, Biotechnol. Bioeng. 83 (2003) 112–120.

91. J. C. Merchuk, Shear effects on suspended cells, Adv. Biochem. Eng. Biotechnol. 44 (1991) 65–95.

92. K. Issarapayupa, S. Powtongsook, P. Pavasant, Flat panel airlift photobioreactors for cultivation of vegetative cells of microalga *Haematococcus pluvialis*, J. Biotechnol. 142 (2009) 227–232.

93. R. Reyna-Velarde, E. Cristiani-Urbina, D. J. Hernández-Melchor, F. Thalasso, R. O. Canizares-Villanueva, Hydrodynamic and mass transfer characterization of a flat-panel airlift photobioreactor with high light path, Chem. Eng. Proc. 49 (2010) 97–103.

94. O. Pulz, Performance Summary Report: Evaluation of GreenFuel's 3D matrix Algae Growth Engineering Scale Unit, 2007.

95. P. M. Schenk, S. R. Thomas-Hall, E. Stephens, U. C. Marx, J. H. Mussgnug, C. Posten, O. Kruse, B. Hankamer, Second generation biofuels: high-efficiency microalgae for biodiesel production, Bioenergy Res. 1 (2008) 20–43.

96. J. Degen, A. Uebele, A. Retze, U. Schmidt-Staiger, W. Trosch, A novel airlift photo-bioreactor with baffles for improved light utilization through the flashing light effect, J. Biotechnol. 92 (2001) 89–94.

97. L. Christenson, M. Sims, Production and harvesting of microalgae for wastewater treatment, biofuels, and bioproducts, Biotechnol. Adv. 29 (2011) 686–702.

98. M. Šúri, T. A. Huld, E. D. Dunlop, H. A. Ossenbrink, Potential of solar electricity generation in the European Union member states and candidate countries, Solar Ener. 81 (2007) 1295–1305.

99. Y. Chisti, Biodiesel from microalgae, Biotechnol. Adv. 25 (2007) 294–306.

100. L. Rodolfi, G. Chini-Zittelli, N. Bassi, G. Padovani, N. Biondi, G. Bonini, M. R. Tredici, Microalgae for oil: strain selection, induction of lipid synthesis and outdoor mass cultivation in a low-cost photobioreactor, Biotechnol. Bioeng. 102 (2009) 100–112.

101. M. E. Huntley, D. G. Redalje, CO_2 mitigation and renewable oil from photosynthetic microbes: a new appraisal. Mitig. Adapt. Strateg. Glob. Change 12 (2007) 573–608.

102. J. Happel, D. G. Jordan, *Chemical Process Economics*, Dekker, New York, 1995.

103. E. Molina Grima, E.-H. Belarbi, F. G. Acien Fernandez, A. Robles Medina, Y. Chisti, Recovery of microalgal biomass and metabolites: process options and economics, Biotechnol. Adv. 20 (2003) 491–515.

104. M. S. Peters, K. Timmerhaus, R. E. West, *Plant Design and Economics of Chemical Engineers*, McGraw-Hill, New York, 2003.

105. S. Babel, S. Takizawa, H. Ozaki, Factors affecting seasonal variation of membrane filtration resistance caused by *Chlorella algae*, Water Res. 36 (2002) 1193–1202.

106. D. F. Rudd, C. C. Watson, *Strategy of Process Engineering*, Wiley, Hoboken, NJ, 1968.

107. G. Olivieri, A. Marzocchella, R. Andreozzi, G. Pinto, A. Pollio, Biodiesel production from *stichococcus* strains on laboratory scale, J. Chem. Technol. Biotechnol. 86 (2011) 776–783.

CHAPTER 13

Microalgae Biodiesel and Macroalgae Bioethanol: The Solar Conversion Challenge for Industrial Renewable Fuels

NAVID R. MOHEIMANI, MARK P. McHENRY, and POURIA MEHRANI

13.1 INTRODUCTION

Large-scale biofuel production from a range of biomass feedstocks (i.e., oil seeds, tallow, sugar cane, corn, etc.) is technically achievable, yet, a fundamental challenge is reducing the high costs of production. While Malthus [1] popularized the tension between food supply and population, and despite the green revolution and more recent advances in agricultural biotechnology, maintaining food security for the growing global population remains a major concern. While nitrogen (N) fertilizer biofuel inputs can be synthesized from a number of available energy sources, phosphorus (P) is geographically concentrated, is a nonrenewable resource, and world phosphate rock reserves are projected to be depleted in less than 100 years based on the current (predominantly nonbiofuel) agricultural demand and technology [2, 3]. Nonetheless, there is an enormous range of published literature available discussing various aspects of biomass-to-liquid fuel production [4, 5, 6, 7, 8]. Yet, this chapter aims to summarize the current advances and challenges of novel biofuel technologies and feedstocks, their relationship to nonrenewable and renewable inputs, and the wider advantages and disadvantages they may be expected to produce.

According to Sims et al. [9], the global annual primary energy consumption in 2004 was around 464×10^{12} MJ. Only a minority of this primary energy was derived from renewable energy resources, which underlines the challenge for achieving sufficient global energy supply security for the next 100 to 150 years. In relation to liquid fuels, between 2005 and 2008 a barrel of petroleum oil fluctuated between US$13.44 and US$165.3, which at present seems to have stabilized at around US$95 (all in 2013

Natural and Artificial Photosynthesis: Solar Power as an Energy Source, First Edition. Edited by Reza Razeghifard.
© 2013 John Wiley & Sons, Inc. Published 2013 by John Wiley & Sons, Inc.

dollar terms) [10, 11]. Tighter emission regulations and the recent Middle Eastern political tensions add further pressures that increase prices and exacerbate issues of inequitable geographical energy resource distribution. The periodic tendency for temporal and geographic supply chain insecurity of fossil fuel sources, particularly liquid fuels, has been a major driving force in searching for alternative sources since the early 20th century [12, 13].

Apart from fission, fusion, and ocean tides, all other demonstrated energy resources, including fossil fuels, are converted forms of solar energy and can be described as energy carriers. Three factors determine the effectiveness of energy carriers: reliability, power at which the energy can be released, and density at which energy can be stored. These factors determine to a great extent the cost effectiveness for a number of energy applications. Energy carriers used in road and air transport engines require high power-to-mass characteristics. Therefore an energy carrier that provides energy at a reasonably high rate and is able to be transported a suitable range requires high density storage. The only cost-effective highly available energy carriers that meet present requirements at scale are fossil fuels. Solar energy is the origin of these carriers with a low annual density at an average of only ~ 200 W m^{-2}. Direct harvesting of solar energy using modern technology, as opposed to over geological time in the case of fossil fuels, presents a viable solution for displacing high density fossil fuel energy carriers at a reasonable price [14]. Despite the low solar density, it is relatively reliable, predictable, and widely distributed geographically in comparison to energy carriers such as wind. Nonetheless, large collection areas are required to provide sufficient power output. Thus a solar collector cost is a dominant factor in direct harvesting and requires technical and cost considerations for times of low seasonal availability (in some areas fluctuating from around 6 to 25 MJ m^{-2} day^{-1}), or when the solar resource is unavailable. This often necessitates larger collection areas, energy storage, and other complementary approaches to match solar output to demand.

Present technological development in the relatively mature solar–electric industry tends to focus on cost reduction of existing technical applications and designs. In contrast to solar–electric generation, the numerous advantages transforming solar energy to chemical energy (biofuel) using photosynthesis, either natural or artificial, faces several fundamental technological and economic challenges. These limitations include the theoretically maximum photosynthetic efficiency of 11.6%, leading to a maximum conversion of solar to chemical energy of between 1.4 and 2.55 MJ m^{-2} day^{-1} [15].

13.2 BIOFUEL SUPPLY, DEMAND, PRODUCTION, AND NEW FEEDSTOCKS

Biomass has been used widely for millennia as a source of chemical energy [16], yet modern commercial liquid biofuels are bioethanol and biodiesel [17], which are derived from short-rotation crops [15]. Bioethanol is currently produced on very large scales from fermenting sugarcane in Brazil under a national project which started in 1975 [18]. It is to be noted that the principle of fermentation technology used for converting sugar to bioethanol has not changed to a great extent. In contrast, biodiesel production primarily occurs in Europe and the United States, but at a much smaller volume. According to the OECD [19], the global annual production volumes of bioethanol and biodiesel are projected to increase from 93 and 21 billion L in 2010, to 159 and 41 billion L in 2019, respectively. This is a projected annual growth rate of 6.8% and 8% for bioethanol and biodiesel, respectively. For comparison, the 2010 total U.S. petroleum consumption of light vehicles was around 550 billion L [20], and the total global biofuel production in 2010 would only supply around 20% of U.S. demand. From these simple standpoints, it is clear that current biofuel production methods from terrestrial plant and animal sources are unlikely to offset current demand, future growth, or improve global energy or food security to a great extent [21, 22]. Therefore a greater diversity of biofuels grown in systems outside of current agricultural or forested lands will be necessary to avoid several concerns from a large expansion in current biofuel production using existing systems [22, 23, 24].

One option enjoying significant attention is algae (both microalgae and macroalgae) due to the ability to reduce land competition from existing biofuel feedstock production by an order of magnitude for an equivalent biofuel volume output [21, 25, 26]. The Pacific Northwest National Laboratory, part of the U.S. Department of Energy, recently reported that renewable fuel from algae alone could eventually replace 17% of U.S. oil imports [27]. For instance, both microalgal and macroalgal biomass can be used to produce ethanol from fermentation, generate methane through anaerobic digestion, and photobiologically generate hydrogen; additionally, microalgae, due to high oil content, can be used as a source of bio-oil for biodiesel production from transesterification [21, 28]. Macroalgae and microalgae biofuel production systems have a biological advantage in requiring less arable land than conventional agricultural biofuels [21, 25, 26]. Yet, generating efficient commercial microalgae or macroalgae biofuels will require large investment in research and development through the entire production chain [29, 30].

Renewable energy availability is highly dependent on geographical energy and water resource characteristics [30]. Ironically, the best solar resources often occur in the dry arid regions, which fundamentally limit biomass production. For example, Australia is the driest habitable continent, yet has excellent solar resources. The annual daily mean incident solar energy in Australia ranges from 13 to 24 MJ m^{-2}, although 90% of the continent ranges between 18 and 24 MJ m^{-2}, and almost 50% receives between 22 and 24 MJ m^{-2} [30]. Industrial scale microalgae ponds require extremely large volumes of water, and evaporative losses can be sizeable [21]. If seawater or suitable groundwater sources are used for industrial scale pond facilities in dry regions, the security of microalgal production increases, as heavy rainfall can negatively impact pond productivity [28, 31]. However, many groundwaters exhibit various levels of salinity and additional dissolved contaminants, and only some microalgae are able to grow in water with salinity levels higher than seawater ($>35,000$ ppm) [28, 32, 33]. Local groundwater contaminants are also likely to introduce additional pre- and posttreatments. Nevertheless, using fresh water for industrial microalgal open ponds, in most cases, will not be sustainable. It will add new industrial consumers of water and will be a major limiting factor in arid regions [34]. Therefore the selected microalgae for a sustainable large-scale production must be able to tolerate a wide range of salinities, and preferably salinities higher than seawater [34].

13.3 FEASIBILITY OF PHOTOSYNTHETIC FUEL PRODUCTION

Because of the low solar energy density, relatively large areas are required, which comprise a significant capital outlay. To gain an idea of the scale and cost of the required investment, it is instructive to estimate the land use. According to the 2007 Australian census data, the annual liquid fuel consumption in Australia at the time was $30,047 \times 10^6$ L, where 62.8% was gasoline and the remainder diesel. With a higher heating value (HHV) of 34.8 MJ L^{-1} for gasoline and 38.7 MJ L^{-1} for diesel, Australia's total final energy consumption at the time was 1019.489×10^6 GJ yr^{-1}. Assuming a photosynthetic biofuel production rate of 20 g m^{-2} day^{-1} [35], which is equal to around 2–8% photosynthetic efficiency, and an energy content of 21.9 MJ kg^{-1}, the annual production rate of biofuel will be around 160 MJ m^{-2} yr^{-1}. Therefore the estimated productive area required for biofuel production to supply Australian liquid fuel demand is approximately 6377 km^2. For comparison, this is equivalent to a band of 1.545 km width running the 4127 km from Perth to Sydney in Australia, or approximately 15% of the total land area of The Netherlands.

Clearly, while such a high land use in a sporadically populated country such as Australia might be acceptable in many places, it will not be feasible in areas with high density populations and low land availability, such as Europe, which also has significantly less solar resources. Even in Australia, in reality, the maintenance of such a large production site will be a daunting task (if not technically impractical), and evaporation alone can render the production infeasible, especially if any photosynthetic organism except hypersaline microalgae are cultivated. In most cases, places with high sunlight also have a high evaporation rate. For instance, the evaporative loss in Karratha ($20°53'$S, $116°40'$E) in Western Australia is greater than 100 cm yr^{-1}. This will equate to approximately 6.4 TL of evaporation each year for a microalgae production area of 6377 km^2 (for reference, Sydney Harbour holding volume is ~0.56 TL). This signifies the importance of efficiency in resource input and energy conversion efficiency. As the maximum annual solar radiation in Australia is around 24 MJ m^{-2} day^{-1} (or 8760 MJ m^{-2} yr^{-1}), a solar technology with a 10% conversion efficiency (i.e., generic solar photovoltaics) will require 1163.8 km^2 to produce an equivalent amount to Australia's annual combined gasoline and diesel fuel demand. Since living cells have not evolved to only produce fuel, their conversion efficiency is not optimized for this application. Plants use only a small range of specific bands of the solar spectrum, and to prevent damage they also cease conversion at very high levels of irradiation, precisely when a solar technology should be working at its peak rating. These considerations lead to the conclusion that a dedicated and highly controlled photosynthetic reactor will become essential in harvesting solar energy using biological organisms.

Other than solar and water inputs, other limiting factors to consider are plant photosaturation characteristics, and also nutrient requirements [15, 36]. Biofuel production using commercial P fertilizers will be exposed to supply shortages over the long term and may be a major competitor to food production [3, 30]. Therefore a balance between technical, commercial, and security objectives plays a fundamental role in biofuel production. For instance, if algae is the autotrophic target organism, growth media must contain essential nutrients, including nitrogen, phosphorus, potassium, and often silicon, akin to conventional agriculture [21], and also CO_2, which may be provided by the atmosphere, soluble carbonate salts, industrial exhaust gases, and other sources [37]. One advantage of microalgae is that they can alternatively achieve higher photosynthetic efficiencies than current agricultural biofuel crops due to negligible photosaturation and improved nutrient access due to their single-cell nature [15, 25, 28]. Higher photosynthetic efficiencies aside, microalgae still require fertilizer inputs [21, 38], although microalgal growth medium nutrient content can be highly controlled and can often approximate a "closed system." Nonetheless, new laboratory-based trace mineral optimization research is showing promising results that may incrementally enhance microalgae commercial system yield and cost effectiveness [39].

13.4 BIODIESEL PRODUCTION AND FEEDSTOCKS

While the first diesel engine was run on vegetable oils, these oils are too viscous for modern diesel engines as a straight replacement fuel. The four main methods for reducing viscosity of vegetable oils are blending [40], microemulsification [41], pyrolysis [42], and transesterification [43], all of which exhibit various technical and economic advantages and disadvantages [44]. The most common method for converting vegetable and animal oils to biofuel is transesterification. Transesterification (alcoholysis) was originally established in 1853 by Duffy and Patrick for producing glycerol as a substance for soap production with biodiesel as a by-product of the process. In a conventional alkali-catalyzed transesterification method (Scheme 13.1),

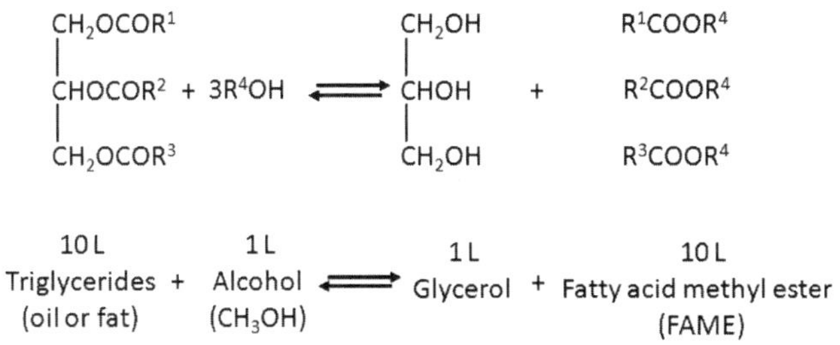

$$
\begin{array}{llll}
CH_2OCOR^1 & CH_2OH & R^1COOR^4 \\
| & | \\
CHOCOR^2 + 3R^4OH \rightleftharpoons CHOH & + & R^2COOR^4 \\
| & | \\
CH_2OCOR^3 & CH_2OH & R^3COOR^4
\end{array}
$$

$$
\begin{array}{llll}
10\,L & 1\,L & 1\,L & 10\,L \\
Triglycerides + & Alcohol \rightleftharpoons & Glycerol & + \; Fatty\ acid\ methyl\ ester \\
(oil\ or\ fat) & (CH_3OH) & & (FAME)
\end{array}
$$

SCHEME 13.1 A summary of transesterification reaction. For more details see review by Meher et al. [6].

the fat is mixed with alcohol (i.e., methanol) and a base (KOH or NaOH) in the absence of water, and at 60 °C for up 120 minutes to produce fatty acid methyl esters (biodiesel) and glycerol [43]. This reaction is reversible and a little excess alcohol is used to shift the equilibrium to the product side and also to reduce reaction time [6]. The excess alcohol is removed by distillation. The process of transesterification is influenced by (1) effect of free fatty acid and moisture [45, 46, 47]; (2) type and concentration of the catalyst [46, 48, 49, 50], (3) type of alcohol and its molar ratio to oil [51, 52, 53], (4) the effect of temperature and reaction time [47, 54], (5) mixing intensity [54], and (6) the effect of using organic cosolvents [49, 55, 56, 57]. Glycerol produced in transesterification (Scheme 13.1) can be sold as a commodity or can be converted to another chemical. The quality of produced biodiesel must meet required standards (i.e., The European Standard EN 14214) and is characterized based on (1) ignition quality, (2) viscosity, (3) density, (4) calorific value, (5) cloud point, cold filter plugging point, and pour point, (6) sulfur content, (7) particulate matter, and (8) the carbon residue.

Biodiesel (fatty acid methyl ester, FAME) is a nontoxic compound and can (1) be blended with diesel from 0% to 100% [58]; (2) provide good engine performance[1]; (3) have significant reduction in particulate matter emission [60]; (4) provide a much higher lubricity [61]; and (5) emit less CO and SO_x emissions [62]. However, some biodiesel can emit slightly more NO_x than diesel [63]. To date, almost 98% of worldwide commercial biodiesel producers use homogeneous alkaline catalyst (either NaOH of KOH) transesterification methods [64], mainly due to the simplicity of this method (summarized in Scheme 13.1). Nevertheless, the effectiveness of this method is low due to (1) the low cost oil feedstocks used (i.e., with high free fatty acid content) [46]; (2) the need for complex purification operations (i.e., the catalyst can be recovered, and high volumes of undesired wastes); and (3) the limited possibilities for continuous processing [65]. As such, there has been significant research on effectively handling low cost feedstocks and also simple purification techniques. In terms of transesterification catalyst options, there are four common types: (1) a homogeneous catalyst [8, 66], (2) a heterogeneous catalyst [8, 66], (3) an enzymatic catalyst [67], and (4) non catalytic methods (i.e., supercritical alcohol) [5]. All of these options have their respective advantages and disadvantages, and there have been several attempts to scale them up, yet none up to the larger scale production levels. After all, the main issue involved with biodiesel production is not the methodology, but the cost and availability of raw material.

At present, biodiesel is primarily produced from edible oil seeds (i.e., canola, soybean), palm oil, used cooking oil, or tallow, the quality of which can determine the quality of biodiesel to a great extent. For instance, canola biodiesel is often very high quality. As such, several regions have developed independent fuel standards to cater for this variability (United States, Europe, etc.). However, a more fundamental global challenge for biodiesel producers is sourcing both low cost and suitable input

---

[1]However, based on an EIA [59], FAME has slightly lower energy content than diesel (1 L of diesel = 1.03 L of 100% biodiesel).

biomass materials in sufficient quantity. The biomass raw material can represent more than 80% of the total cost of biodiesel [68]. Furthermore, the price of canola oil, a major raw material, has increased by almost 500% between 2001 and 2008 and has tended to follow mineral oil trends [11]. These fundamental issues undermine some of the advantages of using biodiesel to displace mineral petroleum diesel, reducing CO_2-e emissions, and may also contribute to food insecurity in addition to higher costs of fuels without government agricultural subsidies or from blending mandates. In practice, it will be impossible to replace even close to 100% of the liquid fuel consumption in most developed countries with first generation biodiesel technologies due to low plant productivity and land availability. To address this problem, researchers have been looking into alternative feedstocks for biodiesel. At present, hopes are pinned on generating feedstocks for the biodiesel industry from microalgae and jatropha. Jatropha has received a lot of attention recently as a new biofuel raw material. Jatropha is native to South America, is well adapted to semiarid regions, and bears seeds with a high oil content (up to 40%), with oil productivities between 100 and 2000 kg ha^{-1} yr^{-1}. However, these higher productivities of jatropha can only be achieved during wetter seasons and optimal horticultural conditions. Microalgae have also been identified as a possible source of raw material for the biodiesel industry and it is mainly due to their fast growth rate and high lipid content. For instance, the high productivities of jatropha are approximately 30 times less than the projected microalgae oil productivities [34, 35, 69]. As such, microalgae are fast becoming the favored biodiesel raw feedstock, and there are more than 25 companies investing billions worldwide trying to establish an economical process for growing microalgae [70].

While most microalgae biofuel activity is in the United States, other companies are active in Australia, Israel, Brazil, China, and some European countries. Akin to any other phototrophic organism, microalgae when grown autotrophically need sunlight, water, and minerals to be able to produce sugar, which can then be converted to lipids. The most important advantages of microalgae over any other oil-containing crops are their capability to grow on saline and hypersaline water, and also the possibility of growing them on nonagricultural lands [34, 35]. This means that microalgae can be cultivated in places which are not suitable for many other crop plants (i.e., coastal regions and even on the surface of the ocean). However, the process of growing algae as a source of raw material for biodiesel production is yet to be commercial. While microalgae can in theory achieve a much higher biomass productivity compared to current oil seed plants, even canola or palm [35], several issues must be resolved. These include (but are not limited to) improving oil productivity, developing a cost-effective method for removing the biomass from the water (dewatering), and also new oil extraction technologies [71]. Nonetheless, a dominant factor for commercial microalgae production will be the relative price of crude oil; a reasonable medium-term price target for microalgal oil for biodiesel production to become cost competitive with petroleum diesel is US$0.48 L^{-1} pretax [30]. It is to be noted that, to date, the lowest cost of microalgae produced commercially (mostly for high-value products) is greater than US$5 kg^{-1} [72]. Considering that no more than 50% of microalgae biomass can be oil [35], this means that the cheapest oil we can produce to date is

TABLE 13.1 Comparisons of Yield, Hydrolyzable Carbohydrate, and Potential Bioethanol Production Between Major Terrestrial Bioethanol Crops, Microalgae, and Macroalgae

	Wheat (grain)	Corn (kernel)	Sugar Beet	Sugarcane	Macroalgae	Microalgae
Average world yield $(kg\ ha^{-1}\ yr^{-1})^a$	2800	4815	47,070	68,260	54,020	73,000
Dry weight of hydrolyzable carbohydrates $(kg\ ha^{-1}\ yr^{-1})$	1560	3100	8825	11,600	$16,206^b$	$19,710^b$
Potential volume of bioethanol $(L\ ha^{-1}\ yr^{-1})^c$	1010	2010	5150	6756	9445	11,487

[a]The yields for wheat, corn, sugar beet, and sugarcane are equal to harvested total weight "off the paddock" and were not adjusted to the dry weight value. In the case of macroalgae (*Gracilaria parvispora*) and microalgae (*Pleurochrysis carterae*), the yield units are g for dry weight m^{-2} (this biomass can alternatively contain up to 15% ash).
[b]Assumption: *Gracilaria parvispora* and *Pleurochrysis carterae* biomass contains 30% hydrolyzable carbohydrate.
[c]Assumption: Carbohydrate to ethanol conversion rate is equal to 64% for wheat and corn, and 58% for sugar beet, sugarcane, macroalgae, and microalgae.
Source: Reproduced from Adams et al. [80] and data given in Glenn et al. [83] and Moheimani and Borowitzha [84] with permission of Springer.

$\sim$US\$10 L^{-1}. The only way to reduce costs is to either increase the biomass productivity of algae, or reduce the total capital expenditure (Capex) and operating expenses (Opex). While there is very little influence we can have on the optimum microalgae biomass productivity (20 g m^{-2} d^{-1}, see Table 13.1), there are opportunities to reduce Capex and Opex.

13.5 MACROALGAE BIOFUEL FEEDSTOCKS AND PRODUCTION

Many feedstocks can be fermented to produce bioethanol, including sucrose-based ones (sweet sorghum, sugarcane, etc.), starch-based ones (wheat, corn, etc.), lignocellulosics (wood, grasses), and both microalgae and macroalgae [66, 73]. However, problems associated with a major expansion of existing bioethanol production are the generally poor energy balances that are close to or less than unity, unless bagasse or other wastes are used in the thermal distillation process [23]. In addition to microalgae, macroalgae are gaining some attention as an alternative renewable source of biomass for the production of bioethanol. However, in contrast to microalgae, macroalgae is already a major global industry [38]. Current macroalgal production is around 16 million wet tonnes ($\sim$1 million tonnes dry) per year and growing, and around 90% of current production is harvested from near-shore mariculture farms

[74]. Macroalgae can also be grown in several aquaculture polycultures to prevent high-nutrient effluent release with no net loss of productivity from primary species [75, 76, 77]. As such, macroalgae are potentially cost-effective biomass feedstocks (with a market price of around US$1.5 to US$7.5 per dry tonne), and may be a new major supply of biofuel feedstocks [75]. Some macroalgae species exhibit greater hydrolyzable carbohydrate contents than current bioethanol feedstocks [78], including *Sargassum, Gracilaria, Prymnesium parvum, Euglena gracilis, Gelidium amansii* [79], and *Laminaria* [80] species. Over 50% (by dry weight) of brown macroalgae such as *Laminaria* spp. can be composed of carbohydrates laminarin and mannitol [80, 81, 82].

The high yield of macroalgae is attributed to little need for supporting structural tissues [78], no internal nutrient transport requirements [79], and their ability to absorb nutrients over their entire surface area [78]. Macroalgae mass productivities of 13.1 kg dry weight m^{-2} over a 7-month growth period compares preferentially against terrestrial plants achieving 0.5–4.4 kg dry weight m^{-2} over an entire year [79, 85, 86]. Macroalgae are historically divided into three major groups based on their photosynthetic pigments: Chlorophyta (green algae), Rhodophyta (red algae), and Phaeophyta (brown algae) [87, 88], as cited by Borines et al. [89]. The photosynthetic product of green macroalgae is starch, and the outer and inner layers of their cell walls are predominantly pectin and cellulose, respectively [90]. Red macroalgae cell walls contain small amounts of cellulose, while the majority is gelatinous or amorphous sulfated galactan polymers [87]. Brown macroalgae cell walls are composed of alginic acid, cellulose, and other polysaccharides, and the food reserves are the carbohydrates laminarin and mannitol, which are particularly suited to ethanol production [82, 91]. Brown macroalgae such as *Laminaria* spp. contain up to 55% (dry weight) of laminarin and mannitol [80, 81], although the location, time of year, and habitat type all influence the production of macroalgae biofuel substrate components [92].

Laminarin is a storage glucan (a linear polysaccharide of glucose) comprised of $\beta(1, 3)$-glucan [93]. Mannitol is a low molecular mass sugar alcohol composed of carbon (C), hydrogen (H), and multiple hydroxyl groups, and forms a component of the laminarin molecule [94]. Various microorganisms can hydrolyze laminarin by glucanases to its glucose monomer, which is a suitable fermentation substrate. However, mannitol requires oxidation to fructose by the enzyme mannitol dehydrogenase to produce the reduced from of nicotanimide adenine dinucleotide (NADH) [81, 95]. Fermentation is usually undertaken using yeasts, although some bacteria can be utilized. Research by Horn et al. [81] demonstrated the possibility of fermenting extracts from *Laminaria hyporbae* (a brown algae) to ethanol using *Pichia angophorae* (a yeast) with a maximum yield of 0.43 g ethanol per g of substrate. Further research by Horn et al. [95] used glucose and mannitol as a mixed substrate in combination with an extract from *L. hyporbea* to determine the efficacy of the bacterium *Zymobacter palmae* to ferment mannitol. The ethanol yields were 0.53 g ethanol per g of mannitol after 11.7 h, and 0.61 g ethanol per g of mannitol after 21.9 h. Further research by Adams et al. [80] explored ethanol production using laminarin from *Saccharina latissima* (a brown macroalga) fermented with *Saccharomyces cerevisiae* (a yeast) achieving ethanol yields of 0.45% (by volume).

Alongside microalgae as a source for biodiesel and jet fuel, macroalgae are noted as one major new bioresource with superior productivities of bioethanol precursor components, as compared to many terrestrial crops [75, 89]. This is primarily due to low energy requirements for supporting mechanisms and nutrient absorption abilities [78, 79]. Furthermore, macroalgae farms need no fertilizer inputs to grow effectively and have a range of postbiofuel production uses [96, 97, 98, 99]. Macroalgae maricul-ture systems are one of the only known commercial farming activities that result in a net capture of useful nutrients from saline water bodies [76], creating a new source of commercial fertilizer, and also can be used for water treatment akin to phytoreme-diation [75]. Therefore macroalgae and microalgae are two major potential biofuel feedstocks, yet, they are in need of basic R&D investment in species selection, harvesting, and preprocessing [35, 74]. However, akin to conventional agricultural crops, any successful algal production technology or method is heavily dependent on marrying the right alga with the right culturing conditions [100].

13.6 CONCLUSION

Volumetric algal biomass productivity needs to be much higher to offset high produc-tion costs in highly controlled biomass growing environments [101, 102]. Such high volumetric production often requires energy-efficient production systems to minimize energy requirements to circulate cultures [73]. Mass transfer limits the practicality of cultivation systems for large-scale production [102], and significant optimization must occur in the initial prototype designs and must continue into commercial sys-tems. Experience to date has shown great difficulty in production system scale-up, and extrapolations from small laboratory operations are often unreliable [103]. This scaling issue forms the premise of much technology development uncertainty for both technology providers and potential investors.

Species-specific characteristics, such as high biofuel precursor productivities and suitable cultivation and harvesting methods, are vital to the success of mass biofuel production facilities [36, 103]. Productivity must be maximized to offset the often very high capital costs of growth systems, harvesters, and dryers [104]. The reduction of these costs is a primary concern for industrial scale production, as it may represent a major component of total fuel production costs [21]. Additionally, the cost of extraction and purification of various commercial bioproducts will be dependent on the species, product, and final use [103], and integrated production systems (such as biorefineries) may enable cost-effective production to reduce total production costs in aggregate across the range of final products, effectively cross-subsidizing biofuels [21, 105, 106]. Nonetheless, fundamental technical and economic limitations remain, including effective solar energy capture, limiting photosaturation, minimizing water requirements and nutrient availability, and their associated commercial and environmental costs [15, 36].

In the end, the question that remains to be answered is: "Can biomass, including microalgae and macroalgae, provide enough sustainable oil and sugar to support economical biofuel production?" It is our view that while biomass, including but not limited to microalgae and macroalgae, can and most likely will play a very important role for our future energy needs, with current technologies and biological limitations, it is very unlikely that these feedstocks can replace 100% of our liquid transport fuel demands. Supplying the growing world demand for liquid fuels will also require more investment for new technologies such as synthetic photosynthesis. As mentioned earlier in this chapter, the highest current annual biomass productivity achievable is no greater than 20 g m^{-2} d^{-1} (from microalgae). The fundamental challenge is the limit of biological photosynthetic efficiency of around 11% [15]. If, on the other hand, various forms of synthetic photosynthesis are developed (possibly using biomimicry techniques), with an efficiency of around 20–30%, it may revolutionize the environmental and commercial credentials of biofuels.

REFERENCES

1. T. R. Malthus, An essay on the principle of population as it affects the future improvement of society with remarks on the speculations of Mr. Gowin, M. Condorcet, and other writers. London, UK, Printed for J. Johnston, in St. Paul's church-yard, 1798.

2. L. Steen, Phosphorus availability in the 21st century: management of a non-renewable resource, Phosphorus Potassium 217 (1998) 25–31.

3. D. Cordell, J.-O. Drangert, S. White, The story of phosphorus: global food security and food for thought, Global Environmental Change 19 (2009) 292–305.

4. A. Demirbas. *Biodiesel: A Realistic Fuel Alternative for Diesel Engines*, Springer, London, 2008.

5. K. de Boer, P. A. Bahri, Supercritical methanol for fatty acid methyl ester production: a review, Biomass and Bioenergy 35(3)(2011) 983–991.

6. L. C. D. Meher, D. Vidyasagar, S. Naik, Technical aspects of biodiesel production by transesterification—a review, Renewable and Sustainable Energy Reviews 10(3) (2006) 248–268.

7. S. Behzadi, M. M. Farid, Review: examining the use of different feedstock for the production of biodiesel, Asia-Pacific J. Chem. Eng. 2(5) (2007) 480–486.

8. K. Narasimharao, A. Lee, K. Wilson, Catalysts in production of biodiesel: a review, J. Biobased Materials Bioenergy 1(1) (2007) 19–30(12).

9. R. E. H. Sims, R. N. Schock, A. Adegbululgbe, J. Fenhann, I. Konstantinaviciute, W. Moomaw, H. B. Nimir, B. Schlamadinger, J. Torres-Martinez, C. Turner, Y. Uchiyama, S. J. V. Vuori, N. Wamukonya, X. Zhang, 2007: Energy supply. In: *Climate Change 2007: Mitigation. Contribution of Working Group III to the Fourth Assessment Report of the Intergovernmental Panel on Climate Change*, Cambridge University Press, Cambridge, UK, 2007.

10. Board of Governors of the Federal Reserve System. Website. http://www.federalreserve.gov/ 2011 (accessed December 7, 2011).

11. S. Kerekes, S. Luda, Climate or rural development policy? Society and Economy 33 (2011) 145–159.

12. M. K. Hubbert, Nuclear energy and the fossil fuels. In: *Drilling and Production Practice*, American Petroleum Institute, Washington DC, 1956, pp. 7–25.

13. M. K. Hubbert, *Energy Resources*, National Academy of Sciences, National Research Council; Washington DC 1962.

14. J. M. Wallace, P. V. Hobbs, *Atmospheric Science: An Introductory Survey*, Academic Press, New York, 1977.

15. P. T. Vasudevan, M. Briggs, Biodiesel production—current state of the art and challenges, J. Ind. Microbiol. Biotechnol. 35 (2008) 421–430.

16. D. Klass, Biomass as an energy resource: concepts and markets. In: D. Klass, (Ed.), *Biomass Renewable Energy, Fuels and Chemicals*, Academic Press, New York, 1998.

17. L. D. Gomez, C. G. Steele-King, S. J. McQueen-Mason, Sustainable liquid biofuels from biomass: the writing's on the walls, New Phytologist 178(3) (2008) 473–485.

18. F. Rosillo-Calle, L. A. B. Cortez, Towards Proalcool—a review of the Brazilian bioethanol programme, Biomass and Bioenergy 14(2) (1998) 115–124.

19. Organisation for Economic Cooperation and Development, *OECD–FAO Agricultural Outlook 2009–2018*, OECD, Paris, 2009.

20. Energy Information Administration, *Annual Energy Outlook*, U.S. Department of Energy, Washington D.C., 2008.

21. Y. Chisti, Biodiesel from microalgae. Biotechnol. Adv. 25 (2007) 294–306.

22. Centre for International Economics. Western Australia biofuels taskforce report: Appendix 4—The economics of biofuels for Western Australia. The Department of Industry and Resources, Canberra, 2007.

23. S. Matsuda, H. Kubota, The feasibility of national fuel-alcohol programs in Southeast Asia, Biomass 4(3) (1984) 161–182.

24. S. I. Mussatto, G. Dragone, P. M. R. Guimaraes, J. P. A. Silva, L. M. Carneiro, I. C. Roberto, Technological trends, global market, and challenges of bio-ethanol production, Biotechnol. Adv. 28(6) (2010) 817–830.

25. J. Sheehan, T. Dunahay, C. E. Hanson, P. Roessler, A look back at the US Department of Energy's Aquatic Species Program—biodiesel from algae, National Renewable Energy Laboratory, Golden, 1998.

26. M. E. Huntley, D. G. Redalje, CO_2 mitigation and renewable oil from photosynthetic microbes: a new appraisal. Mitigation and Adaptation Strategies for Global Change 12 (2006) 573–608.

27. M. S. Wigmosta, A. M. Coleman, R. J. Skaggs, M. H. Huesemann, L. J. Lane, National microalgae biofuel production potential and resource demand. Water Resources Res. 47 (2011) 1–13.

28. S. Amin, Review on biofuel oil and gas production processes from microalgae, Energy Conversion and Management 50 (2009) 1834–1840.

29. U.S. Department of Energy, *National Algal Biofuels Technology Roadmap*, U.S. Department of Energy, Office of Energy Efficiency and Renewable Energy, Biomass Program, College Park, MD, 2010.

30. M. P. McHenry, Microalgal bioenergy, biosequestration, and water use efficiency for remote resource industries in Western Australia. In: A.M. Harris (Ed.), *Clean Energy: Resources, Production and Developments*, Nova Science Publishers, Hauppauge, NY, 2010.

31. N. R. Moheimani, M. Borowitzka, Limits to productivity of the alga *Pleurochrysis carterae* (Haptopphyta) grown in outdoor raceway ponds, Biotechnol. Bioeng. 96(1) (2006) 27–36.

32. M. Hightower, Energy and water issues and challenges—completing the energy sustainability puzzle. In: *REUSE 09*, Brisbane, Australia, Sandia National Laboratories, New Mexico, 2009.

33. L. L. Beer, E. S. Boyd, J. W. Peteres, M. C. Posewitz, Engineering algae for biohydrogen and biofuel production, Curr. Opini. Biotechnol. 20 (2009) 264–271.

34. M. A. Borowitzka, N. R. Moheimani, Sustainable biofuels from algae. Mitigation and Adaptation Strategies for Global Change 2010;doi: 10.1007/s11027-010-9271-9.

35. S. Fon Sing, A. Isdepsky, M. Borowitzka, N. R. Moheimani, Production of biofuels from microalgae. Mitigation and Adaptation Strategies for Global Change 2011;doi: 10.1007/s11027-011-9294-x.

36. M. J. Griffiths, S. T. L. Harrison, Lipid productivity as a key characteristic for choosing algal species for biodiesel production, Appl. Microbiol. Biotechnol. 21 (2009) 493–507.

37. B. Wang, Y. Li, N. Wu, L. C. Q, CO_2 bio-mitigation using microalgae, Appl. Microbiol. Biotechnol. 79 (2008) 707–718.

38. D. Krishnaiah, R. Sarbatly, D. M. R. Prasad, A. Bono, Mineral content of some seaweeds for Sabah's South China Sea, Asian J. Sci. Res. 1(2) (2008) 166–170.

39. W. Xiong, X. Li, J. Xiang, Q. Wu, High-density fermentation of microalga *Chlorella prototothecoides* in bioreactor for microbio-diesel production, Appl. Microbiol. Biotechnol. 78 (2008) 29–36.

40. A. S. Ramadhas, S. Jayaraj, C. Muraleedharan, Use of vegetable oils as I.C. engine fuels—a review, Renewable Energy 29(5) (2004) 727–742.

41. J. Czerwinski, Performance of HD-DI-diesel engine with addition of ethanol and rapeseed oil. SAE Paper No. 940545, SAE International, Warrendale, USA, 1994.

42. P. Srinivasa Rao, K. V. Gopalakrishnan, Esterified oils as fuel in diesel engines. In: 11th National Conference on I.C. Engines, I.I.T. vol., Madras, India, 1983.

43. A. Demirbas, Comparison of transesterification methods for production of biodiesel from vegetable oils and fats, Energy Conversion and Management 49 (1) (2008) 125–130.

44. F. Ma, M. A. Hanna, Biodiesel production: a review, Bioresource Technol. 70(1) (1999) 1–15.

45. M. P. Dorado, E. Ballesteros, J. A. Almeida, C. Schellet, H. P. Lohrlein, R. Krause, An alkali-catalyzed transesterification process for high free fatty acid oils, Trans ASAE 45(3) (2002) 525–529.

46. F. Ma, L. D. Clements, M. A. Hanna, The effect of catalyst, free fatty acids, and water on transesterification of beef tallow, Trans ASAE 41(5) (1998) 1261–1264.

47. B. Freedman, E. H. Pryde, T. L. Mounts, Variables affecting the yield of fatty esters from transesterified vegetable oils, J. Am. Oil Chemists' Soc. 61(10) (1984) 1638–1643.

48. E. G. Shay, Diesel fuel from vegetable oil: status and opportunities, Biomass and Bioenergy 4(4) (1993) 227–242.

49. B. Freedman, R. O. Butterfield, E. H. Pryde, Transesterification kinetics of soybean oil, J. Am. Oil Chemists' Soc. 63(10) (1986) 1375–1380.

50. E. Ahn, M. Mittelbach, R. Marr, A low waste process for the production of biodiesel, Separation Sci. Technol. 30(7-9) (1995) 2021–2033.

51. A. V. Tomasevic, S. S. Marinkovic, Methanolysis of used frying oils, Fuel Processing Technol. 81 (2003) 1–6.

52. J. M. Enciner, J. F. Gonzalez, J. J. Rodriguez, A. Tajedor, Biodiesel fuel from vegetable oils: transesterification of *Cynara cardunculus* L. oils with ethanol, Energy & Fuels 16 (2002) 443–450.

53. W. Zhou, S. K. Konar, D. G. V. Boocock, Ethyl esters from the single-phase base-catalyzed ethanolysis of vegetable oils, J. Am. Oil Chemists' Soc. 80(4) (2003) 367–371.

54. F. Ma, L. D. Clements, M. A. Hanna, The effect of mixing on transesterification of beef tallow, Bioresource Technol. 69 (1999) 289–293.

55. D. G. V. Boocock, S. K. Konar, V. Mao, H. Sidi, Fast one-phase oil-rich processes for the preparation of vegetable oil methyl esters, Biomass and Bioenergy 11 (1996) 43–50.

56. D. G. V. Boocock, S. K. Konar, V. Mao, C. Lee, S. Bulugan, Fast formation of high-purity methyl esters from vegetable oils, J. Am. Oil Chemists' Soc. 75(9) (1998) 1167–1172.

57. D. G. V. Boocock, Single phase processes for production of fatty acid methyl esters from mixtures of triglycerides and fatty acids. In: WO, ed. 2001.

58. S. Romano, Vegetable Oils: a new alternative. In: *Vegetable oils. Fuels-Proceedings of the International Conference on Plant and Vegetable Oils as Fuels*, ASAE Publication, USA, 1982, pp. 106–116.

59. Energy Information Administration. Biofuels in the U.S. transportation sector. U.S. Energy Information Administration Statistics and Analysis, Washington DC, 2007.

60. Department of the Environment, Water, Heritage and the Arts, Orbital Australia Pty Ltd. Effects of fuel composition and engine load on emissions from heavy duty engines. `http://www.environment.gov.au/atmosphere/fuelquality/publ ications/heavy-duty-engine-emissions.html` 2010 (accessed June 1, 2010).

61. G. Knothe, K. R. Steidley, Lubricity of components of biodiesel and petrodiesel. The origin of biodiesel lubricity, Energy & Fuels 19(3) (2005) 1192–1200.

62. H. A. Fukuda, A. Kondo, H. Noda, Biodiesel fuel production by transesterification of oils, J. Biosci. Bioeng. 92(5) (2001) 405–416.

63. C. S. Cheung, C. Cheng, T. L. Chan, S. C. Lee, C. Yao, K. S. Tsang, Emissions characteristics of a diesel engine fueled with biodiesel and fumigation methanol, Energy & Fuels 22(2) (2008) 906–914.

64. M. Mittelbach, C. Remschmidt, *Biodiesel: The Comprehensive Handbook*, Martin Mittelbach, Vienna, Austria, 2006.

65. E. Lotero, Y. Liu, D. E. Lopez, K. Suwannakarn, D. A. Bruce, J. G. Goodwin, Synthesis of biodiesel via acid catalysis, Ind. Eng. Chem. Res. 44(14) (2005) 5353–5363.

66. A. Demirbas, Progress and recent trends in biofuels, Prog. Energy Combustion Sci. 33 (2007) 1–18.

67. M. Iso, B. Chen, M. Eguchi, T. Kudo, S. Shrestha, Production of biodiesel from triglycerides and alcohol using mobilized lipase, J. Mol. Catalysis Enzymatic 16 (2001) 53–58.

68. Energy Information Administration. World crude oil prices. `http://www.eia.doe/ gov/dnav/pet/pet_pri_wco_k_w.htm` 2011 (accessed November 1, 2011).

69. W. M. Achten, J. E. Mathijs, L. V. Verchot, V. P. Singh, R. Aerts, B. Muys, Jatropha biodiesel fueling sustainability? Biofuels, Bioproducts, and Biorefining 1 (4) (2007) 283–291.

70. M. Borowitzka, N. R. Moheimani, M. P. McHenry, *Qantas sustainable aviation fuel competition review*, Murdoch University, Perth, 2010.

71. N. R. Moheimani, D. Lewis, M. A. Borowitzka, S. Pahl, Harvesting, thickening and dewatering microalgae. In: J. O. B. Carioca (Ed.), *International Microalgae and Biofuels Workshop*, Fortaleza, Brasil, 2011, p. 227.

72. M. A. Borowitzka, Economic evaluation of microalgal processes and products. In: Z. Cohen, (Ed.), *Chemicals from Microalgae*, Taylor & Francis, London, 1999, pp. 387–409.

73. B. Hankamer, F. Lehr, J. Rupprecht, J. H. Mussgnug, C. Posten, O. Kruse, Photosynthetic biomass and H_2 production by green algae: from bioengineering to bioreactor scale-up, Physiologia Plantarum 131 (2007) 10–21.

74. G. Roesijadi, A. E. Copping, M. H. Huesemann, J. Forster, J. R. Benemann, Techno-economic feasibility analysis of offshore seaweed farming for bioenergy and biobased products, Battelle and Pacific Northwest Division, Richland, WA, 2008.

75. M. G. Borines, M. P. McHenry, R. L. de Leon, Integrated macroalgae production for sustainable bioethanol, aquaculture and agriculture in Pacific island nations, Biofuels, Bioproducts, and Biorefining 5(6) (2011) 599–608.

76. H. Mai, R. Fotedar, J. Fewtrell, Evaluation of *Sargassum sp.* as a nutrient-sink in an integrated seaweed-prawn (ISP) culture system, Aquaculture 2010;doi:10.1016/j.aquaculture.2010.09.010.

77. R. R. Stickney, J. P. McVey, Responsible marine aquaculture. In: W. A. Society, (Ed.), CABI Publishing, New York, 2002.

78. R. P. John, G. S. Anisha, K. M. Nampoothiri, A. Pandey, Micro and macroalgal biomass: a renewable source for bioethanol, Bioresource Technol. 2010;doi:10.1016/j.biortech.2010.06.139.

79. S. G. Wi, H. J. Kim, S. A. Mahadevan, D. Yang, H. Bae, The potential value of the seaweed Ceylon moss (*Gelidium amansii*) as an alternative bioenergy resource, Bioresource Technol. 100 (2009) 6658–6660.

80. J. Adams, J. Gallagher, I. Donnison, Fermentation study on *Saccharina latissima* for bioethanol production considering variable pre-treatments, J. Appl. Phycology 21 (2009) 569–574.

81. S. J. Horn, I. M. Aasen, K. Ostgaard, Ethanol production from seaweed extract, J. Ind. Microbiol. Biotechnol. 25 (2000) 249–254.

82. E. Percival, R. H. McDowell, *Chemistry and Enzymology of Marine Algal Polysaccharides*, Academic Press, New York, 1967.

83. E. P. Glenn, D. Moore, J. Brown, M. Akutigawa, S. Napolean, K. Fitzsimmons, Sustainable culture of Gracilaria Parvispora using sporelings, reef outplantings and floating cages in Hawaii, Aquaculture 165 (1998) 221–232.

84. N. R. Moheimani, M. Borowitzka, The long-term culture of the coccolithophore *Pleurochrysis carterae* (Haptophyta) in outdoor raceway ponds, J. Appl. Phycology 18 (2006) 703–712.

85. M. Aizawa, K. Asaoka, M. Atsumi, T. Sakou, Seaweed bioethanol production in Japan—the Ocean Sunrise Project. In: *Oceans 2007*, Vancouver, British Colombia, Canada, 2007.

86. C. Goh, K. Lee, A visionary and conceptual macroalgae-based third generation bioethanol (TGB) biorefinery in Sabah, Malaysia as an underlay for renewable and sustainable development. Renewable and Sustainable Energy Rev. 14(2) (2010) 842–848.

87. E. Ganzon-Fortes, Characteristics and economic importance of seaweeds. In: *Proceedings of the Seaweed Research Training and Workshop for Project Leaders*, Marine Science Institute, University of the Philippines, Diliman, Quezon City, Philippines, Philippine Council for Aquatic and Marine Research and Development, Department of Science and Technology, 1991.

88. U.S. Department of Energy. *National Algal Biofuels Technology Roadmap*, U.S. Department of Energy, College Park, MD, 2010.

89. M. G. Borines, R. L. De Leon, M. P. McHenry, Bioethanol production from farming non-food macroalgae in Pacific island nations—chemical constitutents, bioethanol yields, and prospective species in the Philippines, Renewable and Sustainable Energy Rev. 15(9) (2011) 4432–4435.

90. J. G. C. Trono, E. Ganzon-Fortes, *Philippine Seaweeds*, National Bookstore, Manila, Philippines, 1988.

91. T. Davis, B. Volesky, A. Mucci, A review of the biochemistry of heavy metal biosorption by brown algae, Water Res. 37(18) (2003) 4311–4330.

92. R. Lewin, *Physiology and Biochemistry of Algae*, Academic Press, New York, 1962.

93. N. Miyanishi, Y. Inaba, H. Okuma, C. Imada, E. Watanabe, Amperometric determination of laminarin using immobilized B-1,3 glucanase, Biosensors and Bioelectronics 19 (2004) 557–562.

94. R. H. Reed, I. R. Davison, J. A. Chudek, R. Foster, The osmotic role of mannitol in the Phaeophyta: an appraisal, Phycologia 24(1) (1985) 35–47.

95. S. J. Horn, I. M. Aasen, K. Ostgaard, Production of ethanol from mannitol by *Zymobacter palmae*, J. Ind. Microbiol. Biotechnol. 24 (2000) 51–57.

96. P. Avino, P. L. Carconi, L. Lepore, A. Moauro, Nutritional and environmental properties of algal products used in healthy diet by INAA and ICP-AES, J. of Radioanal. and Nuclear Chem. 244(1) (2000) 247–252.

97. Y. Chisti, Response to Reijnders: do biofuels from microalgae beat biofuels from terrestrial plants? Trends Biotechnol. 26(7) (2008) 351–352.

98. P. Matanjun, S. Mohamed, N. M. Mustapha, K. Muhammad, Nutrient content of tropical edible seaweeds, *Eucheuma cottonii*, *Caulerpa lentillifera* and *Sargassum polycystum*, J. Appl. Phycology 21 (2009) 75–80.

99. J. I. Nirmal Kumar, R. N. Kumar, K. Patel, S. Viyol, R. Bhoi, Nutrient composition and calorific value of some seaweeds, Our Nature 7(1) (2009) 18–25.

100. R. Raja, S. Hemaiswarya, N. Ashok Kumar, S. Sridar, R. Rengasamy, A perspective on the biotechnological potential of microalgae, Crit. Rev. Microbiol. 34 (2008) 77–88.

101. Y.-K. Lee, Microalgal mass culture systems and methods: their limitation and potential, J. Appl. Phycology 13 (2001) 307–315.

102. C. U. Ugwu, H. Aoyagi, H. Uchiyama, Photobioreactors for mass cultivation of algae, Bioresource Technol. 99 (2008) 4021–4028.

103. M. Borowitzka, Algal biotechnology products and processes—matching science and economics, J. Appl. Phycology 4 (1992) 267–279.

104. M. Borowitzka, Commercial production of microalgae: ponds, tanks, tubes and fermenters, J. Biotechnol. 70 (1999) 314–321.

105. C. E. Wyman, B. J. Goodman, Biotechnology for production of fuels, chemicals, and materials from biomass, Appl. Biochem. Biotechnol. 39(40) (1993) 41–59.

106. A. Arundel, D. Sawaya, *The Bioeconomy to 2030. Designing a Policy Agenda*, Organisation for Economic Cooperation and Development, Paris, 2009.

CHAPTER 14

Technoeconomic Assessment of Large-Scale Production of Bioethanol from Microalgal Biomass

RAZIF HARUN, HASSAN J, LI J. S. SHU, LUCY A. ARTHUR, and MICHAEL K. DANQUAH

14.1 INTRODUCTION

As the scarcity of fossil fuels becomes more apparent, researchers are compelled to reduce the dependency on these limited resources with alternative fuels that are sustainable, environmentally friendly, and cost effective. Bioethanol, a viable option to the aforementioned problems, has been reported to burn cleanly especially when the amount of gasoline with which it is blended decreases [1]. It also provides a higher octane rating, hence eliminating the expense of engine modification [2].

Bioethanol is a combustible fuel produced from renewable biomass. The biomass for bioethanol production can be categorized into three groups; sucrose based (sugarcane, sugar beets, fruits, and sweet sorghum), starch based (corn, wheat, rice, and barley), and lignocellulosic biomass (wood, straw, grasses). However, various social, economic, and environmental issues hinder the utilization of these materials. Sucrose- and starch-based materials are direct sources of food. As a result, the increase in the price of food on the global market and the cost associated with growing this feedstock have rendered them less viable. Similar issues are encountered with lignocellulosic-type biomass. Although they are cheaper and in abundance, higher production costs eliminate their suitability for bioethanol production [2].

These concerns have highlighted the need for an alternative biomass and have consequently resulted in a growing interest in microalgae as a feedstock for bioethanol production. Microalgal biomass has been found to have a potential to develop into an important fermentation feedstock due to its many advantages [3, 4, 5]. Microalgae provide a faster growth rate and shorter harvesting time compared to the crop biomass [6]. The structural composition of most microalgae species displays a high

Natural and Artificial Photosynthesis: Solar Power as an Energy Source, First Edition. Edited by Reza Razeghifard.
© 2013 John Wiley & Sons, Inc. Published 2013 by John Wiley & Sons, Inc.

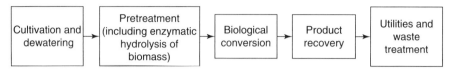

FIGURE 14.1 Schematic representation of biomass to bioethanol conversion process [1].

carbohydrate content, which is beneficial for the production of bioethanol [2]. In relation to the environment, microalgae have the ability to capture carbon dioxide and other greenhouse gases during photosynthesis, thus reducing the release of carbon emissions into atmosphere [2, 7]. Some strains of microalgae are capable of growing in seawater and wastewater streams, and able to treat contaminated water systems, hence diminishing the reliance on freshwater sources [7]. The overall process to convert microalgal biomass to bioethanol is presented in Fig. 14.1.

To ensure that the production of bioethanol from microalgal biomass is a viable process, the economic aspect of the process needs to be evaluated. This chapter discusses the technoeconomic aspects of bioethanol production from microalgal biomass. This involves the conceptual design of the overall process, the technical assessment and design process, covering the microalgae to bioethanol conversion process, and the economic analysis to assess the feasibility of the proposed project.

14.2 TECHNOLOGY SELECTION AND PROCESS DESIGN

14.2.1 Design Basis

The bioethanol production plant is designed based on 50,000 tonnes of dry biomass per year. The life span of the plant is set at 10 years with 330 operating days in a year. A total of 2 weeks of plant shutdown is scheduled for cleaning and maintenance of the facilities and equipment. As the total completion time of biomass production is estimated around 10 days, the production will be distributed into 33 batches a year. Therefore cultivation and dewatering stages are assumed to generate 1515 tonnes of dry biomass for downstream processes in a batch production.

14.2.2 Strain Selection

The culture strain that is adopted in this study is marine algae, *Chlorococcum humicola*. This algae strain has carbohydrate and starch content of 32.52 (% w/w) and 11.32 (% w/w), respectively, with the composition of monosaccharides shown in Table 14.1.

14.2.3 Technology Selection

In selecting the most suitable method at each stage, it is important to keep in mind the efficiency of the technologies in the design process configurations, which later contributes to the total production cost of the produced bioethanol.

TABLE 14.1   Composition of *Chlorococcum* sp. [7]

Component	Composition (% w/w)
Total carbohydrates	32.52
Xylose	9.54
Mannose	4.87
Glucose	15.22
Galactose	2.89
Starch	11.32
Others	56.16

The reliability of the cultivation is observed and evaluated in open pond (raceway pond) and tubular photobioreactors (horizontal and helical coiled (external loop)). Harun et al. [8] has proved that dual stage dewatering with flocculation reduces the cost burden of a single dewatering system. Chitosan is chosen as a flocculant in the flocculation process due to its reliability and working capacity. Solar drying is selected to dry the biomass since it provides lower capital cost and minimizes cell degradation.

Considering the large volume of dry biomass to be processed, the application of mechanical-type treatment will not be feasible due to its high energy consumption. Also, biological treatment is not suitable to pretreat the biomass due to its slow process; hence this method is not advisable for industrial scale production. On the other hand, chemical methods have been recognized as the most preferred pretreatment method for a wide range of biomass. Study by Schell et al. [9] explains that the sugar yields obtained from corn stover using dilute sulfuric acid pretreatment were considerably high, at about 90% cellulose yield conversion. Furthermore, Harun and Danquah [10] conducted an in-depth investigation of *Chlorococcum* sp. using two different chemical methods—alkaline and dilute acid. The highest ethanol yield in 0.75% NaOH concentration for 30 minutes was 26.1% g ethanol/g algae [7], while 0.5% (v/v) of sulfuric acid treatment for 15 min produced 52% g ethanol/g algae [10]. The study concludes that bioethanol production from dilute acid pretreatment provides higher glucose yield compared to alkaline treatment. Therefore dilute acid can be considered as the most appropriate method for microalgal biomass pretreatment and thus will be applied in this chapter.

The proposed technology for glucose to ethanol conversion is conducted in two different methods: *separate hydrolysis and fermentation* (SHF) and *simultaneous saccharification and fermentation* (SSF). Both methods involve saccharification by amylase and cellulase enzymes to release the fermentable sugars. *Saccharomyces cerevisiae* is used as a biocatalyst to ferment glucose to ethanol.

Dilute ethanol is concentrated in two stages of distillation; namely, beer column and rectifying column distillation. The molecular sieve recovers most of the water content, while CO_2 is absorbed in the scrubber.

14.2.4 Process Design

The following section illustrates the process design of bioethanol production from microalgae strain in large-scale production. In general, the information required for design purposes was based on the literature; further calculations have been carried out to estimate the design processes of the entire bioethanol production.

14.2.4.1 Biomass Production The study of biomass production from microalgae strain has previously been conducted by Harun et al. [8]. The overall biomass production configurations are presented in Fig. 14.2. In brief, strains are pumped from the inoculum storage tank (RP T-2) to the raceway pond and growth media such as nutrients including nitrate, phosphate, silicate, vitamins, and other trace metals are added to enhance culture growth. A mixture of flue gas and air is injected to provide sufficient gas exchange within the culture pond. Similar mechanisms are adopted for both photobioreactors (horizontal and external loop). From the closed chamber, the strains are pumped to a solar array, before being recycled to the column. It is assumed that only 80% of the biomass is harvested from both cultivation methods.

Dewatering is done in a dual stage process. The dewatering process is illustrated in Fig. 14.3. The culture is pumped to open tank flocculants (DW E-1) with 40 mg/L of chitosan flocculants loading at up to 90% efficiency. Following flocculation, biomass slurry is then transferred to a disk stack centrifuge (DW E-2). The power consumed during centrifugation is estimated at 1.5 MW for a horizontal PBR, 1.6 MW for an external loop PBR, and 115 MW for raceway ponds. The next step is the drying stage, where the process is carried out using solar drying.

14.2.4.2 Pretreatment Pretreatment of microalgal biomass with dilute sulfuric acid at high temperatures will decompose the crystalline structure of microalgal cell walls and solubilize cellulose compositions. The hydrolysate liquid and solid after-treatment are flashed to atmospheric temperature to vent out most of the water through steam. The remaining hydrolysate is separated from the solid biomass by allowing settling using a centrifugation technique. To provide an appropriate condition for the enzymatic hydrolysis process, the liquid has to be neutralized by raising the pH to 5. Dilution water containing soluble polymeric sugar is then transferred to the SHF or SSF system. The process configurations of the pretreatment stage are shown in Fig. 14.4.

On the basis of 33 batches of production per year, around 1515×10^3 kg of dried biomass per batch will be pretreated and each batch is distributed into 24 cycles, where only 63,100 kg of dry biomass per cycle is transported through the belt conveyor (PT BC-1) to the pretreatment tank PT E-1. Dry biomass is treated with 0.5% v/v of sulfuric acid in 10 g/L biomass loading for 15 min at a temperature of 160 °C and 6 atm. The conditions are implemented based on the highest ethanol yield using dilute sulfuric acid pretreatment as discussed by Harun et al. [10]. Apparently, the 3% v/v of sulfuric acid adopted by Harun et al. [10] is relatively high for large-scale production. This concern arises as sulfuric acid is expensive and incurs high operational cost. On the other hand, Gírio et al. [11] proposed that 0.5–1.5% of sulfuric acid treatment at

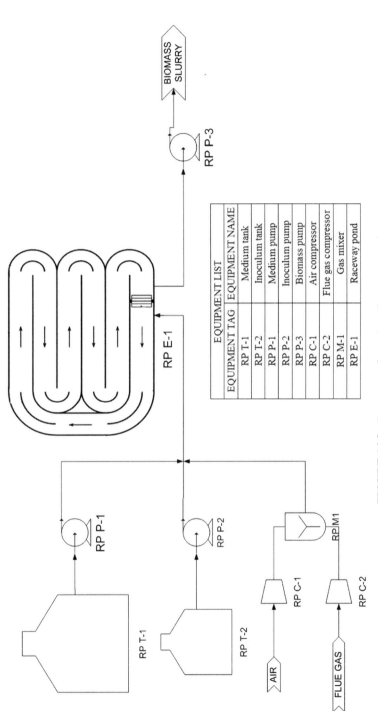

EQUIPMENT LIST	
EQUIPMENT TAG	EQUIPMENT NAME
RP T-1	Medium tank
RP T-2	Inoculum tank
RP P-1	Medium pump
RP P-2	Inoculum pump
RP P-3	Biomass pump
RP C-1	Air compressor
RP C-2	Flue gas compressor
RP M-1	Gas mixer
RP E-1	Raceway pond

FIGURE 14.2 Process configuration of raceway pond.

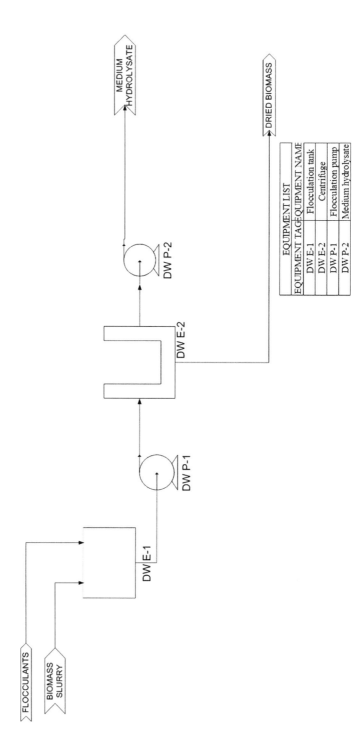

FIGURE 14.3 Process configuration for dewatering.

EQUIPMENT LIST	
EQUIPMENT TAG	EQUIPMENT NAME
DW E-1	Flocculation tank
DW E-2	Centrifuge
DW P-1	Flocculation pump
DW P-2	Medium hydrolysate

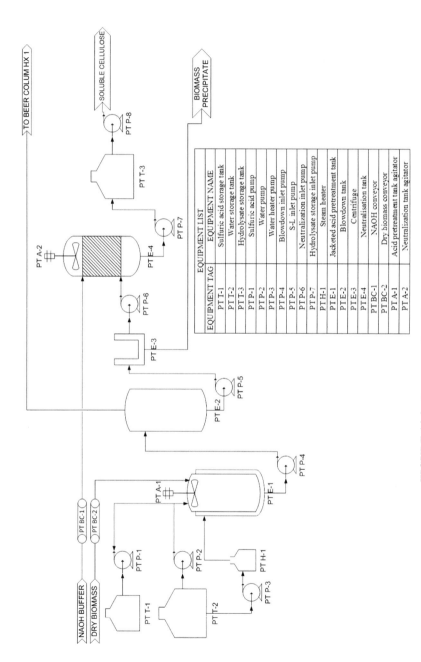

EQUIPMENT LIST	
EQUIPMENT TAG	EQUIPMENT NAME
PT T-1	Sulfuric acid storage tank
PT T-2	Water storage tank
PT T-3	Hydrolysate storage tank
PT P-1	Sulfuric acid pump
PT P-2	Water pump
PT P-3	Water heater pump
PT P-4	Blowdown inlet pump
PT P-5	S-L inlet pump
PT P-6	Neutralization inlet pump
PT P-7	Hydrolysate storage inlet pump
PT H-1	Steam heater
PT E-1	Jacketed acid pretreatment tank
PT E-2	Blowdown tank
PT E-3	Centrifuge
PT E-4	Neutralisation tank
PT BC-1	NAOH conveyor
PT BC-2	Dry biomass conveyor
PT A-1	Acid pretreatment tank agitator
PT A-2	Neutralization tank agitator

FIGURE 14.4 Process configuration in pretreatment stage.

TABLE 14.2   Monomeric Sugars Converted
from Cellulose per Batch

Component	Composition (kg)
Total carbohydrate	850,910
Xylose	144,546
Mannose	73,788
Glucose	230,606
Galactose	43,748
Starch	171,515

121–160 °C is the most favorable condition for industrial application. Furthermore, water from water storage tank PT T-2 is pumped to the tank to attain biomass loading. As the sulfuric acid treatment requires a high temperature environment, preheated saturated steam of 162 °C and 6.03 atm is injected directly into the jacketed tank. A total of six 1270 m³ pretreatment tanks (PT E-1) are required to accommodate the entire single batch of biomass. In addition, it is assumed that 100% of the carbohydrate (32.52% w/w) and starch (11.32% w/w) content is digested by sulfuric acid activity. The amount of monomeric sugar and starch that is converted from biomass cellulose is presented in Table 14.2.

From the pretreatment tank, the liquid hydrolysate and solid move to tank PT E-2 to be flash cooled to 1 atm with residence time of 10 min. About 26.2% w/w of 100 °C steam is vapored out and used to preheat the beer column feed on PR HX-1. The condensed steam is later sent to the wastewater treatment. The remaining solid and hydrolysate are then pumped for solid–liquid separation a using solid bowl centrifuge (PT E-3) with 3.28 kg/s of solid capacity and 750 kW of energy requirement [12]. By allowing a settling time of 30 min, it is expected that 95% of the solid material will be removed, leaving 111×10^3 m³ of hydrolysate per batch containing soluble cellulose. The remaining biomass residue is collected and stored for further commercial use.

Following the solid–liquid separation, the hydrolysate with pH of 3.61 has to be neutralized by NaOH. The amount of NaOH is calculated according to Eq. 14.1:

$$H_2SO_4 + 2NaOH \rightarrow Na_2SO_4 + 2H_2O \tag{14.1}$$

The addition of a total of 4.45 kg of NaOH pellet for every batch will bring up the hydrolysate pH to 5 [13]. The reaction that occurs forms a sodium sulfate that acts as a buffer to maintain pH condition, and hence further NaOH addition in enzyme hydrolysis will be unnecessary.

Four neutralization tanks (PT E-4) with capacity of 1275 m³ will accommodate the entire acidic hydrolysate per batch. The residence time for neutralizing is around 1 hour [14]. An agitator (PT A-2) with approximate power requirement of 98.5 W/m³ is introduced to each tank to aid homogenization.

After the pretreatment stage is completed, the neutralized hydrolysate is stored in a 18,600 m³ hydrolysate storage tank (PT T-3), which holds half of the total volume of hydrolysate in a single batch, while the other half is transferred to the enzyme hydrolysis stage (SHF E-3) or saccharification tanks (SSF E-2).

14.2.4.3 *Separate Hydrolysis and Fermentation (SHF)*   The overall SHF configuration is presented in Fig. 14.5. The received hydrolysate from the pretreatment stage is transferred to hydrolysis tank SHF E-3 through the centrifugal tank. The capacity of a single tank is approximately 3060 m^3 with a total of 20 tanks. Cellulase and amylase are loaded to SHF BC-1 and SHF BC-2, respectively. The amounts of cellulase and amylase loading at the hydrolyzing tank are maintained at 0.02 and 0.01 g enzyme/g substrate, respectively [15]. The tank conditions are maintained at 40 °C and the pH is around 5 for 72 h [15]. Mixing (SHF A-3) is available in every tank with a power requirement of 19.4 W/m^3. During enzymatic hydrolysis, approximately 90% of the complex sugar in carbohydrate and starch is hydrolyzed to hexose sugars. Upon completion, the hydrolyzed liquid is cooled to 37 °C before being transferred to the fermentation tank.

Saccharomyces cerevisiae is cultivated in a yeast jacketed fermenter vessel SHF E-1 before entering the fermentation tank. The loading of the yeast is kept at 5 g/L, while 3% v/v of Luria broth and supporting media such as ammonium chloride, ammonium sulfate, and magnesium sulfate supplied at concentrations of 2 g/L, 1 g/L, and 0.8 g/L are added to the tank. Yeast is cultured aerobically at 30 °C for 24 h. Oxygen exchange is supplied by vigorous mixing (SHF A-1), with estimated agitation power of 14 watts. Moreover, tank sterilization is necessary since yeast is very susceptible to contamination. This can be achieved by steam injection to the vessel jacket at 160 °C for 15 min.

Saccharomyces cerevisiae is used as a biological catalyst to convert the fermentable sugar into ethanol and carbon dioxide according to the Eq. 14.2:

$$C_6H_{12}O_6 \rightarrow 2C_2H_5OH + 2CO_2 \qquad (14.2)$$

Fermentation takes place in twenty 3060 m^3 vessel jacketed fermentation tanks SHF E-2 and split into two batches. As the fermentation is carried out in anaerobic conditions, nitrogen is purged to vent out the air inside the vessel followed by similar sterilization as mentioned earlier. Finally, the cultured yeast, hydrolyzed hydrolysate, and nutrients are fed to the vessel at 37 °C and the fermentation is accomplished in 48 h [13]. The 54 kW of agitation is applied to each tank to mix the fermentation broth. The addition of sodium hydroxide is necessary at this stage ($\sim$10 gram) to maintain the bioactivity of yeast at pH 7. This is done as the CO_2 produced during conversion gives an acidic effect to solution.

It is assumed that the biocatalyst activity of *Saccharomyces cerevisiae* can only be achieved at 90% conversion of hexose sugar. Pentose sugar xylose is not converted due to the limitation of the selected yeast. After both batches of the fermentation are completed, approximately 209 × 10^3 kg of dilute ethanol (40% w/w) produced is stored at beer well SHF T-1 and 220 × 10^3 kg of gaseous carbon dioxide is released per batch. The CO_2 off-gas is condensed and sent to scrubber PR E-5. The ethanol carried in the CO_2 is scrubbed and returned to the beer well.

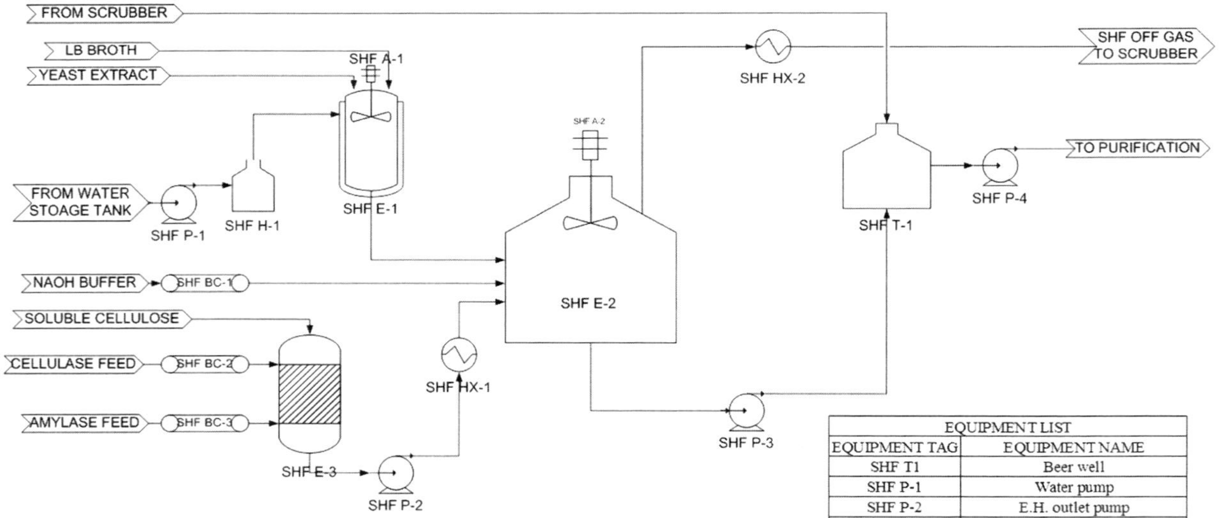

EQUIPMENT LIST	
EQUIPMENT TAG	EQUIPMENT NAME
SHF T1	Beer well
SHF P-1	Water pump
SHF P-2	E. H. outlet pump
SHF P-3	Fermentation tank outlet pump
SHF P-4	Beer pump
SHF E-1	Jacketed yeast fermentation tank
SHF E-2	Enzyme hydrolysis tank
SHF E-3	Jacketed fermentation tank
SHF H-1	Steam heater
SHF BC-1	NAOH conveyor
SHF BC-2	Cellulase conveyor
SHF BC-3	Amylase conveyor
SHF HX-1	Hydrolysate cooler
SHF HX-2	CO_2 condenser
SHF A-1	Yeast fermentation tank agitator
SHF A-2	Fermentation tank agitator

FIGURE 14.5 SSF process configuration.

14.2.4.4 *Simultaneous Saccharification and Fermentation (SSF)* Figure 14.6 illustrates the overall process configuration for SSF. SSF is carried out in 20 jacketed fermentation tanks SSF E-2; each has a capacity of 3060 m^3. This process is divided into two batches from a single batch biomass. Similar to SHF, the SSF vessel involves nitrogen and steam injection into the vessel to meet sterile and anaerobic requirements. The yeast is cultivated in yeast fermenter tank SSF E-1. After the addition of the enzymes, yeast, nutrients, and hydrolysate, the temperature is kept at 37 °C and the pH at 5. The agitation power for SSF (SSF A-2) is assumed to be 54 kW. The total estimated residence time for one fermentation batch is 72 h and the conversion of glucose to bioethanol is assumed to be 90%. This results in a total of 160 × 10^3 kg of ethanol and 153 × 10^3 kg of CO_2 per batch. In addition, approximately 23% w/w of dilute ethanol is stored in the beer well SSF T-1 prior to purification. CO_2 off-gas (99.8% w/w) is captured and condensed in SSF HX-1, followed by CO_2 absorption in PR E-5. Recovered ethanol in CO_2 off-gas from scrubbing is collected back to the beer well SSF T-1.

14.2.4.5 *Purification* The broth after fermentation is transferred to the distillation and molecular sieve sections to recover up to 90% of raw dilute ethanol to concentrated ethanol. The first column removes most of dissolved CO_2, water, and soluble material and the second column, the rectifying column, produces the concentrated ethanol. The molecular sieve dehydrates the remaining water vapor from the rectifying column overhead, resulting in pure ethanol as the final product, while some of the regenerated water–ethanol mixture is recycled to the distillation column. The scrubber is applied to recover ethanol carried in the off-gas from the fermentation tank and beer column overhead. The product recovery process is not discussed in this section. The estimated design for industrial scale bioethanol purification is carried out based on the study by Aden et al. [16].

Before entering the beer column, the dilute raw ethanol is heated to 95 °C using flash steam from the blowdown tank PT E-2 in PR HX-1, followed by a further heating to 100 °C with reboiled stream from the bottom beer column (Fig. 14.7). The bottom of the column contains most of the solid residue and soluble material (Fig. 14.7). The beer column (PR E-1), which operates at 2 atm, consists of 32 actual trays with 48% efficiency and reflux ratio of 3:1. The column diameter is 4.4 meters with a tray spacing of 0.61 meter. The overhead of the beer column contains about 88% w/w of CO_2. The 12% w/w of ethanol dissolved in a mixture is absorbed through scrubber PR E-5 and the recovered ethanol is stored in the beer well. Overall, 99% of ethanol from the beer feed is recovered and recycled to the rectifying column PR E-2, while the other 0.5% is lost at the beer bottom (Fig. 14.7).

The designed scrubber in this work consists of 7.6 meters of plastic packing with four theoretical stages. The water from a storage tank is fed at the top scrubber; the CO_2 is absorbed and vented at the top scrubber. The water scrubber is fed with 97% w/w of CO_2 from the fermentation tank off-gas and 88% w/w of CO_2 from the beer overhead. Approximately 99% of the CO_2 is collected and stored in the CO_2 storage tank as a coproduct.

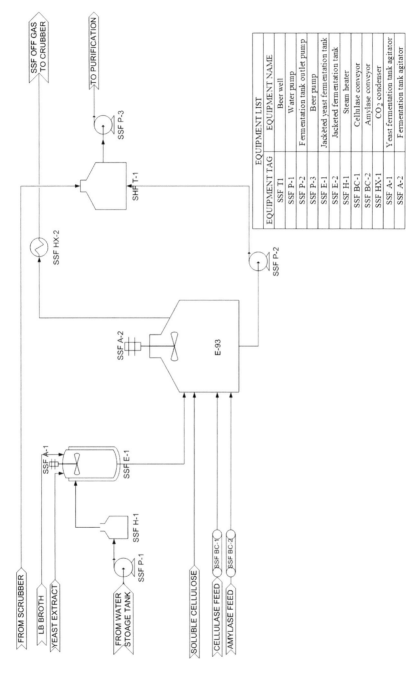

FIGURE 14.6 Overall process configuration for SSF.

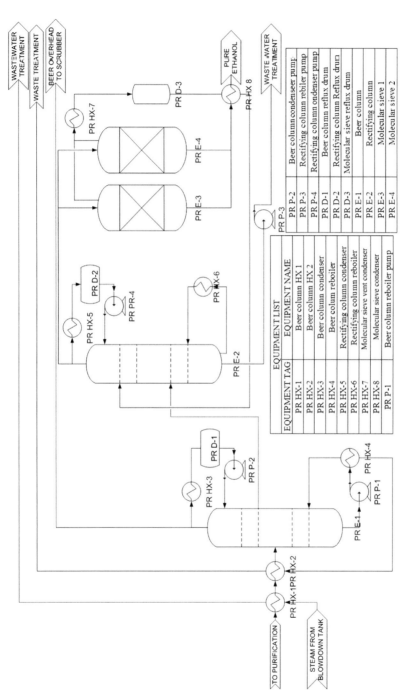

FIGURE 14.7 Process purification of bioethanol.

EQUIPMENT LIST	
EQUIPMENT TAG	EQUIPMENT NAME
PR HX-1	Beer column HX 1
PR HX-2	Beer column HX 2
PR HX-3	Beer column condenser
PR HX-4	Beer colum reboiler
PR HX-5	Rectifying column condenser
PR HX-6	Rectifying column reboiler
PR HX-7	Molecular sieve vent condenser
PR HX-8	Molecular sieve condenser
PR P-1	Beer column reboiler pump
PR P-2	Beer column condenser pump
PR P-3	Rectifying column rebiler pump
PR P-4	Rectifying column ondenser pump
PR D-1	Beer column reflux drum
PR D-2	Rectifying column Reflux drun
PR D-3	Molecular sieve reflux drum
PR E-1	Beer column
PR E-2	Rectifying column
PR E-3	Molecular sieve 1
PR E-4	Molecular sieve 2

Rectifying column PR E-2 (Fig. 14.7) operates at 2 atm, has 60 trays with 57% efficiency, and has a reflux ratio of 3:2:1. About 98.4% w/w of ethanol in the water mixture from the beer column enters the rectifying column after dehydration. The ethanol vapor from the beer column is fed at tray 16 and the recovered mixture from the molecular sieve is fed at tray 41. The column diameter above tray 44 is around 1.2 meters and below tray 44 is 3.5 meters. About 99% of the total ethanol fed to the column flows to the overhead in 0.3% w/w of the water mixture. About 93% water content from the rectifying feed is collected to the rectifying bottom and recycled to water storage.

The ethanol vapor in the overhead continues to molecular sieves PR E-4 and E-5 for water removal. Two molecular sieve chambers (PR E-3 and E-4) work correspondently every 8 h for bed regeneration. About 95% of the carried water will be removed by dehydration, which is recycled to the rectifying column feed. At the overhead, 80% of purified and recovered ethanol vapor is about 99% pure and the product is stored after condensation.

14.2.4.6 *Product and Coproduct Recovery*

From the entire process of the bioethanol production from microalgae, there are four components that can be considered as major outputs: bioethanol, as the main product, CO_2 and biomass residue as the coproduct, and water and solid material as waste products. As the pure bioethanol is sold to the fuel market, CO_2 is also considered as a saleable product to be marketed and/or recycled for microalgae cultivation. Biomass residue can be sold as animal feed or utilized as a feedstock of biodiesel and methane production, which can be added to plant revenues. Water output is treated and recycled back to the water well, while solid waste material is disposed for waste treatment. Table 14.3 highlights the major output produced in both SHF and SHF. SHF performed better over SSF in terms of bioethanol yield. From Fig. 14.8A, about 14% of dry biomass can be converted into bioethanol by implementing the SHF approach. This yield drops by 4% when fermentation technology is alternated to the SSF approach. The difference between SHF and SSF bioethanol yield may not be substantial, but this will greatly influence the total production cost per liter or bioethanol produced. Comparing the actual and theoretical yields, the estimated bioethanol yield is slightly lower by 1% from the maximum conversion biomass to bioethanol, regardless of the fermentation method

TABLE 14.3   Summary of Products, Coproducts, and Waste Products in Ethanol Production Using SHF and SSF

Material	SSF (kg/ batch production)	SHF (kg/ batch production)
99% Bioethanol	158,125	213,352
CO_2	152,309	224,194
Biomass residue	850,909	850,909
Water	9,981	9,981
Solid material	425,780	295,590

(A)

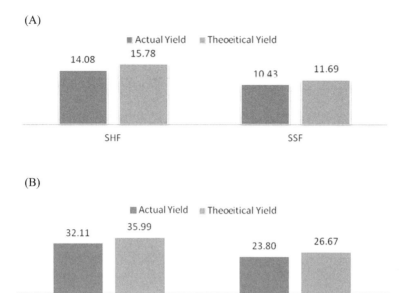

(B)

FIGURE 14.8 (A) Annual bioethanol yield in % weight per weight of biomass. (B) Annual bioethanol yield in % weight per weight of reducing sugar.

applied. Also, annual bioethanol converted per gram of fermentable sugar released is higher in SHF than SSF (Fig. 14.8B). This is due to the appropriate saccharification condition in SHF, which gives a higher conversion of polymeric to monomeric sugars. On the other hand, carbon dioxide (CO_2) yield is around 4% lower in SSF implementation (Fig. 14.9). This can be a good prospect for SSF if minimum emission is the main consideration. However, higher CO_2 capture increases production revenue if CO_2 can be sold or audited to carbon credit.

14.3 ECONOMIC ANALYSIS

The economic analysis is carried out to further evaluate the economic viability of the production of bioethanol from microalgal biomass. The production costs of bioethanol

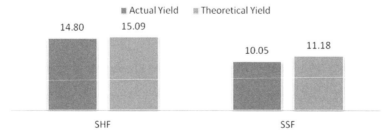

FIGURE 14.9 Annual CO_2 produced in % weight per weight of biomass.

production from microalgal biomass must be cheaper or lower than existing feed-stocks, in particular, lignocellulosic-based materials. This evaluation includes capital cost estimation, operating cost estimation, cash flow, and sensitivity analysis.

14.3.1   Capital Cost Estimates

The developed process design discussed in the earlier section is used to estimate the costs of each unit operation in accordance with the capacity of the proposed project. Since there is no precedent assessment of bioethanol production from microalgae on a large scale, the evaluation is based on a factored estimate with probable accuracy up to ±30% [12]. The total capital investment is classified into fixed capital and working capital expenditures, which are discussed in a later section.

14.3.2   Major Equipment Cost (MEC)

Specialized equipment costs used under cultivation and dewatering stages such as raceway ponds, photobioreactors, conveyor belts, centrifuge, and filters are based on a previous study by Harun et al. [8].

In relation to the stages involved in the pretreatment and fermentation process, the major equipment costs such as jacketed vessels, pumps, storage tanks, and heat exchangers are based on the data provided by Peters et al.[12].

The overall costs in the purification stage are based on a technical report completed by Aden et al. [16]. These involved four major equipment types including the beer column, rectification column, molecular sieve dehydrator, and scrubber.

All major equipment costs were estimated using a cost index to determine the equivalent installed cost for the project in 2010/2011.

14.3.3   Fixed Capital Investments and Working Capital

Fixed capital comprises all paid costs to the contractors at the initial stage of the project, including electrical systems, piping installations, land and buildings, and any costs associated with auxiliary facilities, utilities, and administration. The mentioned components are listed under direct costs and estimated as a proportion of the major equipment costs, based on the data reported by Molina Grima et al. [14]. The indirect costs, on the other hand, are proportions of the total direct costs.

In the case of the required land for cultivation, each system is modeled based on the cost of two large agricultural properties situated in Gippsland, Victoria. The location of the plant is chosen to ensure a free supply of carbon dioxide for microalgal cultivation [8].

Figure 14.10 shows the capital cost involved in all stages of the project. It was found that cultivation predominates the overall cost by almost 50% due to the large capacity of biomass production and complexity of equipment in photobioreactor systems. Therefore we can assume that the technologies involved in biomass production (cultivation and dewatering) play a major role in contributing to the overall cost of the bioethanol plant.

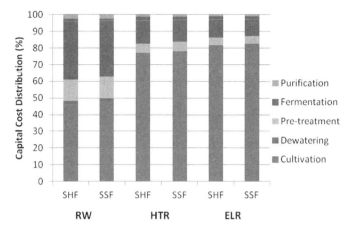

FIGURE 14.10   Breakdown of capital cost based on section of plant.

Table 14.4 summarizes the capital investment for all systems. A raceway pond with SSF system provides the lowest cost of about AU\$1788 million, whereas the external loop reactor with the same fermentation approach shows a total cost of AU\$5158 million.

14.3.4   Operating Cost Estimates

*14.3.4.1 **Fixed Operating Cost***   Fixed operating costs considered in this calculation include maintenance of labor and materials, operating supplies, plant overheads, insurance, property taxes, book depreciation, and royalties. These costs are estimated based on the fixed capital cost and overall wages and are generally independent of the production rate. A significant increase in fixed operating costs is attributed

TABLE 14.4   Capital Cost Comparison for Different Design Process Configuration

	Horizontal					
Plant Area	Raceway Pond (AU\$ Million)		Tubular (AU\$ Million)		External Loop (AU\$ Million)	
Cultivation	734		2721		3617	
Dewatering	23		4.8		5.4	
Pretreatment	198		198		198	
SHF	574		574		574	
SSF	526		526		526	
Purification	38		38		38	
Fixed capital investment	1567.4	1519.7	3535.9	3488.2	4432.5	4384.8
Working capital	277	268	623	615	782	774
Total capital investment	1844	1788	4160	4103	5214	5158

to book depreciation, where 10% of FCI is used and is based on the assumption of a 10-year plant straight-line depreciation.

14.3.4.2 *Variable Operating Cost*   Variable operating costs are reliant on the production rate of the plant. Electricity consumption and raw materials constitute the major costs in production operation and the costs involved are calculated based on the approach adopted by Harun et al. [8].

The electricity consumption throughout the bioethanol production operation was estimated based on the power requirements from HYSYS simulation (HYSYS, Australia). However, energy requirements in biomass production (cultivation and dewatering) were estimated based on the data reported by Harun et al. [8]. Their study took into consideration the energy requirements for mixing provided by the airlift pump for photobioreactors and paddle wheels for raceway ponds. In the use of a dual-stage dewatering process, chitosan is used as a flocculant with cost estimated at US$ 11/kg [17]. Industrial sulfuric acid with 99.5% purity is used and priced at AU$ 300/m^3 (Alibaba express). The costs of cellulose and amylase used in hydrolysis calculations are based on 14–20 and 2–4 cents per gallon bioethanol (US$) [2]. Yeast from *Saccharomyces cerevisiae* is priced at US$86/kg (Sigma Aldrich, Australia). It is assumed that 20% of the annual broth throughput required treatment associated with the cost of wastewater treatment [8].

Moreover, labor cost accounts for a total of 12 shift operators, 4 engineers, and 2 supervisors employed at standard Australian pay rates. Corresponding pay rates are also applied for 2 nonprocess laborers working during normal operation and 4 security guards working on a two-shift swing.

In this economic model, several key assumptions have been made with regard to water use, funding, and exchange rate. Since *Chlorococcum* sp. is a marine algae, seawater is used as a culture broth for microalgae and it is free of charge throughout the production line with the exception of fresh water used during yeast fermentation. All funding was also assumed to be acquired through venture capital; therefore debt service has been eliminated in cost estimation. As most costs and pricing were given in U.S.dollars (US$), an exchange rate of AU$0.95 to US$1 is used whenever necessary.

14.3.5   Cost of Ethanol Production

14.3.5.1 *Cultivation*   According to Harun et al. [8], the raceway pond was found to be the cheapest biomass production system (AU$3/kg) followed by the horizontal tubular reactor at AU$10/kg and the external loop reactor at AU$13/kg. At the given production capacity, the cost required to construct a complex photobioreactor (HTR or ELR) is definitely higher as compared to the cost required in constructing a raceway pond. A breakdown of the annual operating cost incurred by each system can be found in Harun et al. [8].

Wastewater treatment and culture medium are the major cost contributors in a raceway pond [8]. Its lower volumetric productivity has resulted in greater culture medium requirement and larger fluid volume for processing in wastewater treatment. On the other hand, photobioreactors have higher electricity and maintenance costs.

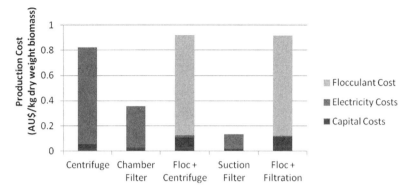

FIGURE 14.11 Biomass dewatering costs—raceway pond.

The energy consumption in both photobioreactors is greater as compared to the energy used to operate the paddle wheel in raceway pond configurations.

14.3.5.2 Dewatering The cost of the dewatering process for raceway ponds is approximately 15 times higher than photobioreactors [8]. This is mainly due to large processing volumes in raceway ponds, which requires longer harvesting period and consequently greater energy costs [8]. As can be seen from Fig. 14.11, the capital cost of dewatering is less significant compared to its operating cost. Based on the findings reported by Harun et al. [8], the reliability of standalone centrifuge options or centrifuge coupled with flocculation has resulted in its suitability as a raceway pond dewatering alternative. The study also suggested that a single-stage centrifugal dewatering process is introduced for photobioreactor cultivation systems. However, the above-mentioned study has failed to make comparisons with dual-stage flocculation and filtration. In the case of the raceway pond, upon conducting similar calculations based on the data provided by Harun et al. [8], the costs of standalone centrifugation and both dual-stage dewatering options (flocculation + centrifuge, flocculation + filtration) are comparable, with the latter lower in energy cost.

Referring to Fig. 14.11, the cost of standalone filtration is significantly lower in comparison to other dewatering alternatives. However, this option raises a concern regarding the membrane clogging and formation of compressible filter cakes when applied in a large scale, resulting in higher operating costs.

Figure 14.11 also shows that the dual-stage dewatering process incurs higher cost due to the high cost of flocculants. However, process development in this technology is believed to achieve greater impact in cost reduction, thus improving the bioethanol production process [18]. In addition, the preceding flocculation process has led to a significant decrease in energy consumption in centrifugation by approximately 98%. In the case of dewatering methods used in photobioreactors, centrifugation clearly resulted in the lowest additional annual costs at AU$0.01–0.02/kg. Therefore dual-stage dewatering unit (flocculation + filtration) and single-stage centrifugation

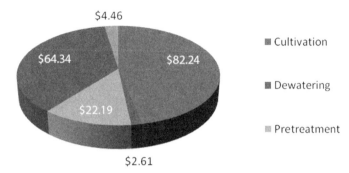

FIGURE 14.12   Cost (AU$ /L ethanol) of bioethanol production at each individual stage—SHF.

would be the most suitable technologies for harvesting microalgae in raceway ponds and photobioreactors, respectively.

14.3.5.3 *Pretreatment*   Pretreatment accounts for as high as 13% of the total production cost in a raceway pond configuration given AU$27 per liter of ethanol is produced (on the basis of capital investment) as shown in Harun et al. [8]. Note that the costs of other pretreatment alternatives such as physical methods are not evaluated in this section, despite the viability of its greater energy requirements [19]. Research has shown that one of the options to reduce the production of bioethanol is by coupling pretreatment with enzymatic hydrolysis to increase ethanol yield [19]. Therefore this section examines different fermentation approaches including *simultaneous saccharification* and *fermentation* and *separate hydrolysis and fermentation*.

By comparing the fermentation methods and assuming the costs are constant for all systems, SHF has a higher production cost at AU$570 million as compared to SSF at AU$530 million due to greater equipment costs. However, due to higher ethanol yield in SHF, it resulted in lower a ethanol cost of approximately AU$22/liter bioethanol for the SHF production stage. A cost comparison between the different systems can be seen in Figs. 14.12 and 14.13, where ethanol cost for SSF production is approximately AU$30/L bioethanol.

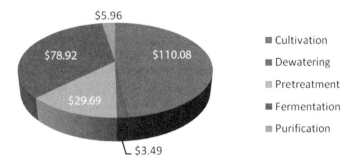

FIGURE 14.13   Cost (AU$ /L ethanol) of bioethanol production at each individual stage—SSF.

SSF remains the preferred option for fermentation due to its ability to prevent inhibitive reactions. Unfortunately, this study does not discuss the effect of enzymatic inhibitions and unwanted products formation from SHF; however, its impact on cost can only be deduced to incur greater loss. Ultimately, in view of the cost effectiveness of the two alternatives, taking into consideration the possible contamination, SSF would be the better option and with little difference in terms of ethanol cost, improvements on ethanol yield will undoubtedly lower its ethanol cost.

14.3.5.4 *Purification* Product recovery in the NREL/Chem Systems studies is based on distillation technology [16]. Given that distillation is a mature technology, it has a lower impact on cost and energy consumption in the production of bioethanol, contributing less than 10% of the total capital cost.

14.3.6 Overall Production Cost

Figures 14.12 and 14.13 illustrate the ethanol cost at each stage. Each system is represented in terms of the overall cost per liter of bioethanol produced.

The cost of ethanol per liter ranges from AU$56 to AU$131 depending on the system used, as shown in Table 14.5. As mentioned in the earlier section, dual-stage dewatering is recommended for the raceway pond to accommodate the large volume of medium circulating throughout the system. Centrifugation, on the other hand, is more cost effective for dewatering culture from photobioreactors. SSF is a cheaper method; however, when looking at an entire process configuration, SHF has a total production cost lower than SSF.

14.3.7 Profitability

14.3.7.1 *Revenue* As mentioned in the previous chapter, coproducts such as carbon credit, electricity, and residue from pretreatment (to be processed and used to produce biodiesel or used as a livestock feed) can be very important to enhancing revenues. However, considering bioethanol as the only saleable product for this project at a selling price of AU$1.05 per liter (based on lignocellulosic materials) [20], the lowest obtainable net production cost of bioethanol per liter is AU$55.

TABLE 14.5   Summary of Design Process Configuration

System	Technology					
Cultivation	Raceway pond		Horizontal tubular		External loop	
Dewatering	Flocculation + centrifugation		Centrifugation		Centrifugation	
Pretreatment	Acid hydrolysis		Acid Hydrolysis		Acid Hydrolysis	
Fermentation	SHF	SSF	SHF	SSF	SHF	SSF
Product recovery	Purification distillation		Purification distillation		Purification distillation	
AU$/liter ethanol	56	60	75	106	129	131

14.3.7.2 Net Present Value (Cash Flow Analysis) The total project invest-ments as well as total operating costs (fixed and variable) are determined first to facilitate the cost estimation for ethanol production. These costs are the key compo-nents to determine the discounted cash flow rate of return when the net present value of the project is zero.

Cash flow analysis predicts the profitability of the project. The following analyses are based on the project plant life of 10 years (not including years required for design and construction), and scheduled as below:

1. First and second years ($t = 1, 2$): Design and procurement is 30% (15% each year) of fixed capital cost.

2. Third and fourth years ($t = 3, 4$): Construction and Commissioning is 70% (35% each year) of fixed capital cost.

3. Fifth year ($t = 5$): Start of production. Targeted 50% of normal operating production, 100% fixed operating cost + 50% variable operating cost + 100% working capital.

4. From sixth year on ($t = 6$–14): 100% normal operating production, 100% fixed operating cost + 100% variable operating cost.

From Fig. 14.14, there is a noticeable decline in the curve indicating the revenue from bioethanol is insufficient to generate profit for the project. The total capital investment is not repaid until the end of the project plant life, thus reiterating that the minimum viable selling price of bioethanol has to be near the production cost of approximately AU\$55–60.

14.3.7.3 Sensitivity Analysis Sensitivity analysis is performed to investigate the impact of changes in key performance parameters on process economics. This section examines the sensitivity of the projected costs to revenue from coproducts,

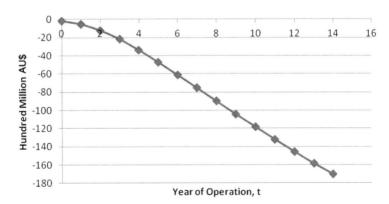

FIGURE 14.14 Cumulative net present value—raceway pond,dual stage (flocculation + centrifugation)—SHF.

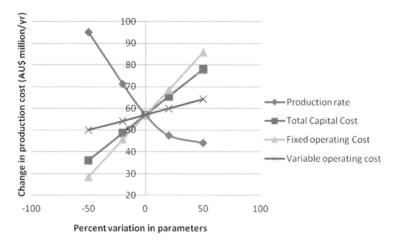

FIGURE 14.15   Sensitivity analysis based on raceway pond system using SHF option.

capital investment, fixed and variable operating cost, and production rate. The sensitivity analysis of the raceway pond system is shown in Fig. 14.15. According to Fig. 14.15, the production cost is mostly affected by the production rate. The production rate is the only parameter that is inversely proportional to the production cost. On the contrary, the capital, and the fixed operating and variable operating costs are directly proportional to the ethanol production cost with the latter parameter having the least impact on the production cost. The percentage difference from the base production cost is only ±5% at ±20% variation. As mentioned earlier, bioethanol is considered the only saleable product with an approximate net production cost of AU$56 per liter bioethanol. However, when residue from the process is allowed to be sold and carbon credits allowed to be received, the cost can only be reduced to AU$55.

14.4   REDUCTION OF OVERALL PRODUCTION COST

Based on the economic analysis, it is clear that bioethanol production from microalgae remains incomparable to those produced from starch-based and lignocellulosic-based materials. The analyses undertaken in this chapter have included six major production stages from cultivation to product recovery, each containing similar economic models for an easy comparison. The stages involved in biomass production were found to be the major contributor to the overall costs as discussed in the previous section.

Although the process designs and economic evaluations described in this chapter provide useful estimates of the cost of production, these are highly dependent on the selection of specific process designs. It is possible to produce different results due to the complexity of each process. Changes in the study approach should be made to further improve the economic viability of the overall process.

First, the carbohydrate content of the microalgae strain used in this study, *Chlorococcum* sp. is approximately 44%, of which 34% is assumed to have the ability to

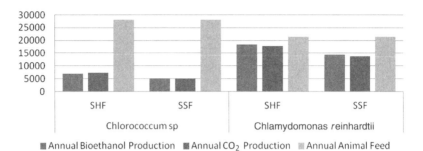

■ Annual Bioethanol Production ■ Annual CO_2 Production ▨ Annual Animal Feed

FIGURE 14.16 Comparison of annual production of product and coproducts between *Chlorococcum* sp. and *Chlamydomonas reinhardtii.*

be reduced to further produce bioethanol. Using a different microalgae strain with different compositions might influence the overall cost, by increasing the revenue either from coproduct or from the main product itself. Figure 14.16 shows a comparison of annual production of product and coproducts between *Chlorococcum* sp. and *Chlamydomonas reinhardtii.*

Second, the technologies chosen in the dewatering stage can be researched further. This is due to several ambiguities in terms of its reliability and cost effectiveness. The cost of filtration is undoubtedly lower than the other dewatering alternatives; however, studies have also shown that it is most suitably used in laboratories with small-scale capacity. The actual performance and the frequency for filter change when coupled with flocculation are not examined in this chapter as well; therefore it remains comparable to dual-stage flocculation–centrifugation.

The sensitivity analysis discussed in the previous section has shown that the production rate has the greatest impact on the production cost. However, this study did not include the use of xylase in SHF or SSF, which can potentially increase the ethanol yield, as xylase is responsible for pentose fermentation.

Finally, the integration of methane production with bioethanol production plants also provides a great potential to reduce the overall production costs by converting methane into electricity for plant usage [21].

14.5 CONCLUSION

The production of bioethanol from microalgal biomass is technologically viable and has vast potential for continued advancements and large-scale benefits. Among the three cultivation options studied, the raceway pond is the most cost-effective method requiring simple construction and little maintenance. The production cost for a raceway pond is significantly lower as compared to photobioreactors, thus reinforcing its feasibility. The analysis on harvesting of microalgae shows that dual-stage dewatering with flocculation followed by centrifugation is a more attractive method as the former technique greatly reduces the energy and cost requirements for centrifugation.

Extensive studies have been conducted on the dilute acid pretreatment approach for its effectiveness to improve the enzymatic digestibility of cellulose. Improving the yield of sugar to be fermented will ultimately increase the ethanol yield or ethanol production rate, thereby reducing the overall cost of bioethanol production. In relation to enzymatic hydrolysis and fermentation, SHF gives rise to an overall ethanol production cost of AU$57, whereas the SSF yields AU$60 per liter ethanol. The findings from the sensitivity analysis show the production rate as a vital parameter in the analysis; hence the cost estimations have directly reflected the performance of both SHF and SSF with the former producing a higher ethanol yield. Finally, as a mature technology for product purification, distillation contributes less than AU$5 per liter ethanol produced in the overall production cost.

The overall findings from this study indicate that bioethanol production from microalgae at a scale of 50,000 tonnes per year is not economically feasible, albeit viable technologically. The extensively high cost of production of AU$57/liter bioethanol has led to a negative net present value throughout the 10-year plant life. The recommended improvements described in Section 14.4 may help to further reduce the cost of bioethanol so it is competitive with other feedstocks and current fossil fuels even without tax incentives. The implementation of carbon credit and carbon tax regulations will promote interest in bioethanol, creating a good competition between bioethanol and petroleum fuel.

REFERENCES

1. L. R. Lynd, Overview and evaluation of fuel ethanol from cellulosic biomass: technology, economics, the environment, and policy, Annu. Rev. Energy Environ. 21 (1996) 403–465.

2. R. Harun, W. S. Y. Jason, T. Cherrington, M. K. Danquah, Microalgal biomass as a cellulosic fermentation feedstock for bioethanol production, Renew Sustainable Energy Revs. 2010. doi:10.1016/j.rser.2010.07.071.

3. R. Davis, A. Aden, P. T. Pienkos, Techno-economic analysis of autotrophic microalgae for fuel production, Appl. Energy 88 (2011) 3524–3531.

4. A. Singh, S. I. Olsen, A critical review of biochemical conversion, sustainability and life cycle assessment of algal biofuels, Appl. Energy 88 (2011) 3548–3555.

5. H. Tang, N. Abunasser, M. E. D. Garcia, M. Chen, K. Y. Simon Ng, S. O. Salley, Potential of microalgae oil from *Dunaliella tertiolecta* as a feedstock for biodiesel, Appl. Energy. 88 (2011) 3324–3330.

6. Y. Chisti, Biodiesel from microalgae, Biotechnology Adv. 25 (2007) 294–306.

7. R. Harun, W. S. Y. Jason, T. Cherrington, M. K. Danquah, Exploring alkaline pre-treatment of microalgal biomass for bioethanol production, Appl. Energy 10 (2011) 3464–3467.

8. R. Harun, M. Doyle, R. Gopiraj, M. Davidson, G. M. Forde, M. K. Danquah, Process economics and greenhouse gas audit for microalgal biodiesel production. In: *Advanced Biofuels and Bioproducts*, Springer Science & Business Media.

9. D. J. Schell, J. Farmer, M. Newman, J. D. McMillan, Dilute-sulfuric acid pretreatment of corn stover in pilot-scale reactor: investigation of yields, kinetics, and enzymatic digestibilities of solids, Appl. Biochem. Biotechnol. 105-108 (2003) 69–85.

10. R. Harun, M. K. Danquah, Influence of acid pre-treatment on microalgal biomass for bioethanol production, Process Biochem. 46 (2010) 304–309.

11. F. M. Gírio, C. Fonseca, F. Carvalheiro, L. C. Duarte, S. Marques, R. Bogel-Łukasik, Hemicelluloses for fuel ethanol: a review, Bioresource Technol. 101 (2010) 4775–4800.

12. M. S. Peters, K. D. Timmerhause, R. E. West, *Plant Design and Economics for Chemical Engineers*, McGraw Hill, New York, 2003.

13. R. Harun, M. K. Danquah, G. M. Forde, Microalgal biomass as a fermentation feedstock for bioethanol production, J. Chem. Technol. Biotechnol. 85 (2010) 199–203.

14. E. Molina Grima, E. H. Belarbi, F. G. Acién Fernández, A. Robles Medina, Y. Chisti, Recovery of microalgal biomass and metabolites: process options and economics, Biotechnol. Adv. 20 (2003) 491–515.

15. R. Harun, M. Singh, G. M. Forde, M. K. Danquah, Bioprocess engineering of microalgae to produce a variety of consumer products, Renewable and Sustainable Energy Rev. 14 (2010) 1037–1047.

16. A. Aden, M. Ruth, K. Ibsen, J. Jechura, K. Neeves, J. Sheehan, B. Wallace, L. Montague, A. Slayton, J. Lukas, Lignocellulosic biomass to ethanol process design and economics utilizing co-current dilute acid prehydrolysis and enzymatic hydrolysis for corn stover, Technical rept. accession Number: ADA436469.

17. Department of Planning (DOP) (2009) Model Project on Chitosan Preparation. Department of Planning, Uttar Pradesh, http://planning.up.nic.in/innovations/inno3/fi/Chitosan.htm (accessed April 14, 2009).

18. N. Uduman, M. K. Danquah, A. F. A. Hoadley, Marine microalgae flocculation and focused beam reflectance measurement, Chem. Eng. J. 162 (2010) 935–940.

19. C. E. Wyman, Ethanol from lignocellulosic biomass: technology, economics, and opportunities, Bioresource Technol. 50(1) (1994) 3–15.

20. E. Gnansounou, A. Dauriat, Techno-economic analysis of lignocellulosic ethanol: a review, Bioresource Technol. 101 (2010) 4980–4991.

21. R. Harun, M. Doyle, R. Gopiraj, M. Davidson, G. M. Forde, M. K. Danquah, Technoeconomic analysis of an integrated microalgae photobioreactor, biodiesel and biogas production facility, Biomass and Bioenergy 35 (2011) 741–747.

CHAPTER 15

Microalgae-Derived Chemicals: Opportunity for an Integrated Chemical Plant

AZADEH KERMANSHAHI-POUR, JULIE B. ZIMMERMAN, and PAUL T. ANASTAS

15.1 INTRODUCTION

Microalgae, a diverse group of photosynthetic microorganisms, with certain species being capable of heterotrophic and mixotrophic growth, are typically referred to eukaryotic unicellular and prokaryotic cyanobacteria (blue-green algae) [1]. This group has been investigated as a source of a broad spectrum of chemicals including hydrocarbons, triacyglycerols (TAGs), proteins, pigments, polysaccharides, omega-3/6 fatty acids, and bioactive compounds with applications ranging from biofuels to pharmaceuticals [2, 3, 4, 5, 6, 7]. The emerging field of microalgal biotechnology is driven by intrinsic characteristics of microalgae including species diversity, unique biochemical composition, higher growth rate compared to other autotrophs, and their simple growth medium requirement (e.g., CO_2, light, and minerals) [8]. Furthermore, recognition of microalgae as renewable feedstocks to substitute for the depleting fossil fuel and reduce its environmental impacts has also contributed to advances in the biotechnology of microalgae [9].

Today, the most recognized product of microalgae with respect to production volume and economical value is its biomass, which is used as nutritional food [10]. Species of cyanobacteria including *Aphanizomenon* and *Spirulina*, both of which are currently produced at commercial scale, have been in use as food for thousands of years [11]. Producing microalgae as a source of protein was among the primary research efforts in algal mass cultivation [12, 13] and the first commercial plant of microalgae was built for production of *Chlorella* as a health food in the early 1960s in Japan [14]. Later, *Spirulina* as a health food and *Dunaliella salina* as a source of *B-carotene* (provitamin A carotenoid) were commercialized in the 1970s and 1980s,

Natural and Artificial Photosynthesis: Solar Power as an Energy Source, First Edition. Edited by Reza Razeghifard.
© 2013 John Wiley & Sons, Inc. Published 2013 by John Wiley & Sons, Inc.

respectively [11, 14]. Advancements in production of fine chemicals were preceded by commercial cultivation of *Haematococcus pluvialis* for astaxanthin (food colorant in aquaculture industries) and heterotrophic cultivation of *Crypthecodinium cohnii* and *Schizochytrium* sp. for docosahexaenoic acid (DHA, functional food additive) productions [11, 15]. Producing microalgae for animal feed is currently an established technology [10]. Species of *Spirulina, Nannochloropsis*, and heterotrophically grown *Chlorella, Nitzschia*, and *Tetraselmis* have been used in aquaculture industries [15, 16, 17, 18, 19]. Microalgae as sources of fuels were first explored for methane production over 50 years ago and later for hydrocarbons and biodiesel [3, 20, 21]. However, a commercial plant for algal biofuel production does not exist yet.

Despite the thousands of algal species identified [3, 22] and several thousands of chemicals isolated from microalgae [23], commercial application of microalgae is confined to a few species and products [5]. A major challenge in commercialization is low biomass and product yield in cultivation systems and lack of a cost-effective technology [14].

Due to low biomass yield, any strategy that allows maximizing the use of biomass can potentially enhance the production economics and broaden the applied field of microalgal biotechnology. Deriving multiple chemicals from microalgae also offers an opportunity for an integrated chemical plant, within which the use of feedstocks is optimized and environmental impact is minimized in accord with the principle of green engineering: systems must include integration and interconnectivity with available energy and material flows [24].

This chapter highlights the strategy of process integration as a partial solution to advance the economics and sustainability of algal biotechnology. Examples of valuable chemicals derived from lipid, carbohydrate, and protein fractions of microalgae are presented and their relevant biotechnological processes and production systems are discussed. Prior to this main scope, a brief overview of the cultivation systems, employed for the production of fine chemicals either in applied research or on commercial scale, is introduced. The chapter concludes with examples of chemicals that appear promising for simultaneous production within integrated bioprocesses.

15.2 MICROALGAE CULTIVATION SYSTEMS

Microalgae cultivation is practiced in open and closed photobioreactors under autotrophic conditions or in fermenter type reactors under mixotrophic and heterotrophic conditions [14]. The main technical consideration in photobioreactors is providing light and carbon dioxide efficiently to support optimum growth [25]. A large surface area to volume ratio in open ponds favors the transfer of available light energy and carbon dioxide to the culture media for photosynthesis activity. On the other hand, loss of CO_2 to the atmosphere, water evaporation, and the high chance of bacterial and protozoan contamination lead to low biomass yield and unpredictable productivity [25, 26]. Due to technical constraints, applications of open ponds are limited to a few microalgal species and specific geographical locations [14]. Closed systems are designed to broaden microalgal biotechnology and overcome the limitations of the open pond through reactor design approaches and controlled culture conditions [27]. Enclosed photobioreactors enable axenic microalgae cultivation, particularly important for production of compounds of pharmaceutical interest, and provide efficient control of the processing parameters such as temperature, mixing pattern, gaseous transfer (e.g., CO_2 transfer to microalgae and oxygen removal), and light regime, all of which influence system performance [27, 28]. On the other hand, closed photobioreactors are difficult to scale up due to their specific geometry and require expensive cooling devices to prevent overheating as a result of light absorption [29]. In both open and closed outdoor systems, photoinhibition of microalgae that occurs at high irradiance impacts the ability of the cells to harvest the light [25, 30, 31]. Photosynthesis activity is directly proportional to light intensity up to a certain point, above which photoinhibition occurs [25].

Comparison of open ponds and enclosed photobioreactors as well as design parameters that influence the photobioreactors performance have been discussed in several studies [26, 28, 31, 32, 33, 34].

15.2.1 Outdoor Open Systems

Open ponds include natural waters, artificial ponds, and containers [14]. Outdoor open ponds, available in two configurations of unmixed (extensive cultivation) and mixed ponds (intensive cultivation), are currently the most commonly used systems for commercial cultivation of microalgae [25]. Extensive cultivation is conducted in very large shallow ponds and relies on wind power for mixing and absorption of CO_2, whereas paddle-wheel or rotating mixers are employed in intensive cultivation with the supply of CO_2 to support microalgal growth and also to control pH of the media [14]. Unmixed open ponds are less expensive to build and maintain compared to raceways but the cost of the downstream harvesting process is higher in the former case due to lower biomass yield [14]. Both of these cultivation approaches are used commercially, with raceways being the most popular open systems [25]. The schematic of a raceway pond is shown in Fig. 15.1A.

The high possibility of bacterial and protozoan contamination and uncontrolled culture conditions have limited the application of open ponds to species that can tolerate extreme environmental conditions. A number of microalgal species including *Chlorella* in a nutrient-rich environment, *Spirulina platensis* in a high pH medium, and *Dunaliella salina* at high salinity are cultivated in open ponds [14, 27, 35].

15.2.2 Outdoor Enclosed Systems

Enclosed bioreactors are available in three main configurations: tubular, flat plate, and fermenters. Schematics of vertical, horizontal, and helical arrangements of tubular photobioreactors are presented in Fig. 15.1B–E. Tubular reactors are more common for outdoor cultivation than other configurations [36]. The main design consideration in a photobioreactor is to provide a configuration with an appropriate angle toward the sun and a high surface area to volume ratio in order to optimize the light availability to

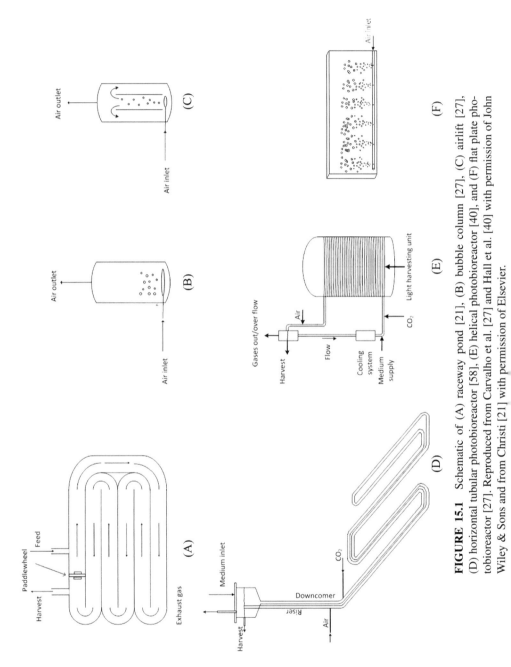

FIGURE 15.1 Schematic of (A) raceway pond [21], (B) bubble column [27], (C) airlift [27], (D) horizontal tubular photobioreactor [58], (E) helical photobioreactor [40], and (F) flat plate photobioreactor [27]. Reproduced from Carvalho et al. [27] and Hall et al. [40] with permission of John Wiley & Sons and from Christi [21] with permission of Elsevier.

microalgae [27]. A gas exchange device is often envisioned in reactor design, where CO_2 is supplied and O_2 removal is enhanced [31]. Accumulation of photosynthetically generated oxygen in the medium, above the air saturation concentration, results in reversing the photosynthetic reaction and reducing biomass productivity [27,31,32]. Additionally, oxygen accumulation causes photooxidative cell damage and adversely affects biomass productivity [27]. Mixing is another important design parameter that facilitates O_2 removal and enhances CO_2 and nutrient mass transfer to the cells [37].

15.2.2.1 *Vertical Tubular Photobioreactors (VTRs)* These are available in conventional configurations of bubble column (Fig. 15.1B) and airlift (Fig. 15.1C), in which CO_2 is sparged at the bottom of the reactor, resulting in an efficient mixing, enhanced CO_2 mass transfer, and oxygen removal. In airlift reactors, rising air creates a liquid flow pattern that reduces shear stress on cells and enhances the gas–liquid mass transfer. A high level of gas hold up (fraction of gas in liquid medium), which is an indication of poor mass transfer and mixing properties [26], is a drawback of these types of reactors [28]. In these types of reactors, light availability is limited due to their large angles toward the sun [27].

15.2.2.2 *Horizontal Tubular Photobioreactors (HTRs)* These (Fig. 15.1D) consist of small-diameter tubes, arranged horizontally or with an angle relative to the horizontal plane [16]. The high area to volume ratio and their angle toward the sun maintain a high photosynthetic efficiency but on the contrary results in elevated culture temperature and oxygen oversaturation [26]. Shading, spraying with water, overlapping the tubes, and submerging in a water bath are some of the techniques that are used to control the temperature [38]. Cooling systems can be expensive with technical barriers when implementing on a large scale [26]. A gas exchange unit is often connected to the light-harvesting device to overcome oxygen oversaturation in HTRs [26]. These reactors are in commercial use for astaxanthin production, for which water spraying is employed to decrease the temperature if needed [38].

15.2.2.3 *Helical Tubular Photobioreactors* These are made of circular tubes coiled horizontally around a vertical cylindrical or conical framework made of polyethylene tubes [27]. This arrangement with a high surface to volume ratio provides an efficient light-harvesting unit when combined with heat and gas exchangers. A centrifuge or airlift pump circulates the culture media, with the latter imposing less shear stress on cells. A schematic of a helical reactor with a cylindrical framework, commonly known as "Biocoil" [39], combined with an airlift is shown in Fig. 15.1E. Cultivation of several microalgal species including *Spirulina platensis, Monodus subterraneus* and *Phaeodactylum tricornutum* have been examined in outdoor and indoor helical reactors [30,39,40].

15.2.2.4 *Flat Plate Photobioreactors* These are available in different configurations including V shaped and alveolar panel or in a simple structure of rectangular shape, with air bubbled at the bottom as shown in Fig. 15.1F [27]. Their narrow

light path leads to efficient penetration of light. The light path is usually optimized to enhance biomass and product concentrations [41, 42].

15.2.3 Fermenter-Type Reactors

These are used for heterotrophic or mixotrophic cultivation of microalgae [27]. Design parameters of conventional fermenters are better studied compared to photobioreactors owing to their widespread applications in biotechnology industries [27]. These reactors are commercially employed for heterotrophic cultivation of *Chlorella* spp., which are used as nutritional food [14], and *Crypthecodinium cohnii* for production of DHA [43]. The stirred fermenter is more common than the airlift fermenter but the latter is more efficient in reducing shear stress and more appropriate for sensitive species [44].

15.3 LIPIDS

Cellular lipids of microalgae have diverse structures and functionality and may include polar lipid, hydrocarbons, sterols, wax esters, and carotenoids [3]. The lipid content of some microalgae species may reach as high as 75% [21]. The lipid fraction of algae serves as a source of several valuable chemicals and as such has been of particular interest with respect to their cellular functionality, properties, and applications [3].

The major polar membrane lipids, glycosylglycerides in chloroplast and phosphoglycerides in plasma membrane and endoplasmic membrane, are often enriched in polyunsaturated fatty acids (PUFAs) [3, 45]. Several chemicals in the class of PUFAs are identified for their nutraceutical and pharmaceutical importance [15, 19].

Neutral lipids, mainly triacylglycerols (TAGs), are often constituted of saturated and monounsaturated fatty acids. In some species, TAGs can also be rich in PUFAs [44, 46]. TAGs, present in the lipid body of cytoplasm, function as storage lipid and have drawn attention for their potential application as biofuel feedstocks [3]. In this section, some common valuable chemicals derived from the lipid fraction of microalgae and their production processes are discussed.

15.3.1 Polyunsaturated Fatty Acids

Long chain ω-3 and ω-6 PUFAs with more than 18 carbons including eicosapentaenoic acid (EPA, 20:5ω3), docosahexaenoic acid (DHA, 22:6ω3), and arachidonic acid (ARA, 20:4ω6) (Fig. 15.2A–C) have received increasing attention in human nutrition [47, 48] due to their widely reported therapeutic effects on the cardiovascular system, inflammatory disease, and blood pressure [49, 50]. Nutritional values of these fatty acids also warranted their application as a dietary source in poultry and aquaculture feed [19, 51].

PUFAs mostly exist as constituents of polar membrane lipids of microalgae, imparting fluidity and selective permeability to cell membranes [19]. Currently, PUFAs

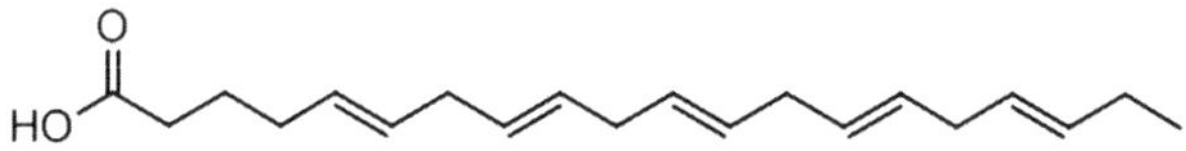

(A)

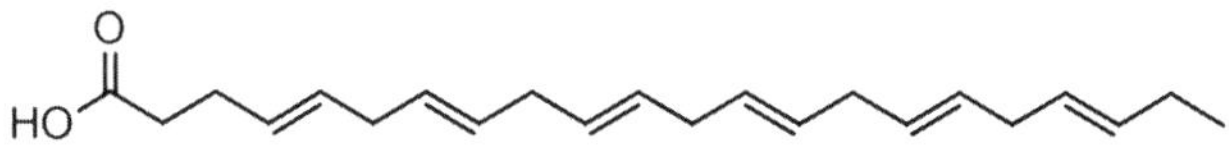

(B)

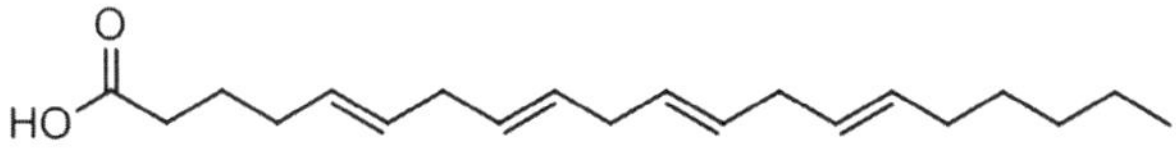

(C)

FIGURE 15.2 Chemical structures of (A) eicosapentaenoic acid, (B) docosahexaenoic acid, and (C) arachidonic acid.

such as EPA and DHA are obtained mainly from fish oil [47]. However, microalgae have been considered as alternative sources of production due to the increase in market demand, possible contamination of fish fat tissue, and declining fish population [43, 47, 52, 53]. Additionally, microalgae are capable of producing the desired fatty acid with high purity under controlled culture conditions, which is particularly important for sensitive applications of PUFAs such as DHA as an ingredient in infant formula [19, 43].

15.3.1.1 *Eicosapentaenoic Acid (EPA)* EPA mainly accumulates as fatty acid fractions of galactolipids (e.g., monogalactosyldiacylglycerol, digalactosyldia-cylglycerol) and phospholipids of autotrophic microalgae *Nannochloropsis* sp. (Eustigmatophyte), *Monodus subterraneus* (Eustigmatophyte) and *Porphyridium cruentum* (Rhodophyta) or as a constituent of TAGs of the heterotrophic diatom *Nitzschia* (Bacillariophyceae) under favorable conditions [44, 54].

Environmental and nutritional conditions—for example, temperature, pH, light/dark cycle, light intensity, salinity, and concentration of carbon, nitrogen, and phosphorus as well as cell density, culture age, and mode of cultivation—have been shown in numerous investigations to influence EPA biosynthesis, and fatty acid content and composition of microalgae [54, 55, 56, 57, 58, 59]. Factors that affect autotrophic and heterotrophic production and the biosynthesis pathway of EPA have been reviewed [52, 60].

Effect of culture conditions on EPA cellular content varies with microalgae species [61, 62]. For instance, in the case of *Porphyridium cruentum*, culture conditions (e.g., temperature, pH, light intensity, and salinity) that led to maximum EPA content were consistent with the optimum growth condition [55], whereas in the marine eustigmato-phyte *Nannochloropsis* sp., maximum EPA content was observed at growth-limiting light intensity [63]. In the freshwater eustigmatophyte *Monodus subterraneus*, maximum EPA content was observed toward the end of the exponential phase, at which nutrients were limited and growth rate was reduced [45]. The EPA content of the diatom *Phaeodactylum tricornutum* was highest at the later stage of the exponential phase [57] and overproduced under optimum growth conditions [61].

Culture age also has a profound effect on fatty acid content and composition [3]. Accumulation of EPA, as a constituent of polar membrane lipids, has often been observed during the exponential growth phase (Table 15.1). When cells reach the stationary growth phase, saturated and monounsaturated fatty acids, mainly present in TAGs, become predominant [3, 59]. Therefore this transition should be identified to determine the appropriate harvesting time to recover the highest EPA content in batch operational mode [64].

In semicontinuous operation, dilution rate, which determines the cell density, influences fatty acid composition and content. Cells are usually kept at the exponential phase for autotrophic EPA production in semicontinous mode and optimum cell density was often attempted to be identified by varying dilution rate (Table 15.1). EPA production by *Monodus subterraneus* was enhanced in an outdoor flat plate photobioreactor with increasing the cell density up to an optimum value of 4 gL^{-1} and then decreased with further increase of biomass concentration [65] (Table 15.1). A relatively close value of 3.4 gL^{-1} for the optimum cell density in an indoor helical tubular photobioreactor was reported [39] (Table 15.1). Increase in EPA content in response to increase in cell density was attributed to shade adaptation by increasing the cellular membrane content and surface area to absorb more light [65, 66].

Interaction of influential parameters was also demonstrated to affect EPA cellular content and productivity [16, 39]. For instance, the relationship of cell density and EPA content was different when cells were exposed to high irradiance compared to low irradiance, which was attributed to higher exposure of cells to light at low cell density [39].

Variables such as light/dark cycle, temperature, and light fluctuations add to the complexity of outdoor systems [58] and may be responsible for the inconsistencies reported on the effect of environmental parameters on productivity in indoor and outdoor cultivation systems [16].

Several investigations have focused on outdoor cultivation of microalgae for EPA production. Some of these findings including the yields and productivities obtained in these systems are summarized in Table 15.1. Both yields and productivities of biomass and product impact the economics of a process and should be considered as an objective function in optimization studies. Furthermore, selection of the objective function depends on the end-user application [65]. For animal feed application, in which dry biomass is the final product, high cellular content is generally desirable, whereas for cases where the product is extracted, such as pharmaceutical applications, optimizing the product yield may be more important [65].

As expected, concentration and productivity of EPA in outdoor open pond cultivation of the diatom *Thalassiosira weissflogii* was significantly lower than those in closed photobioreactors (Table 15.1). Although outdoor open ponds require less

TABLE 15.1 EPA Production in Autotrophic and Heterotrophic Microalgae Cultivation Systems

Species	Culture System/Volume	Growth Mode/Culture Mode	Biomass Concentration (g L⁻¹)	Biomass Productivity (g L⁻¹ day⁻¹)	EPA Concentration (mg L⁻¹)	EPA Productivity (mg L⁻¹ day⁻¹)	EPA Content of the Cell (% dry weight)	Reference	Notes
Thalassiosira weissflogii	Outdoor mixed open pond/100,000 L	Autotrophic/semicontinuous	—	—	10–15	2–3	—	[53]	a
Phaeodactylum tricornutum	Outdoor tubular photobioreactor/50 L	Autotrophic/continuous	7.12 ± 0.11	2.57 ± 0.04	135	47.8	1.86 ± 0.09	[58]	b
Nannochloropsis sp.	Outdoor tubular photobioreactor/10.2 L	Autotrophic/semicontinuous	5	0.69 ± 0.30	200	32	4	[16]	c
Monodus subterraneus	Outdoor flat plate photobioreactor/14 L	Autotrophic/semicontinuous	4	—	152	58.9	3.8	[65]	d
Monodus subterraneus	Indoor helical tubular photobioreactor/4.5 L	Autotrophic/semicontinuous	2.6 / 3.4	1.7 / 1.1	104 / 146	56 / 46	4 / 4.3	[39]	e
Nitzschia alba	Fermenter/30 L	Heterotrophic/fed batch	45–48	16.8–18	675–1200	253–450	1.5–2.5	[44]	f
Nitzschia laevis	Fermenter/2.2 L	Heterotrophic/perfusion	27	2.09	700	55.5	2.59	[71]	g
Nitzschia laevis	Fermenter/2.2 L	Heterotrophic/perfusion-bleeding	9.5	6.75	250	175	2.59		
Nitzschia laevis	Fermenter /2 L	Heterotrophic/fed batch	22.1	1.5	695.2	49.7	3.14	[72]	h

[a]Light intensity was controlled at 40,000–50,000 lux by covering the pond with netting.

[b]Photobioreactor was equipped with an airlift system. Incident solar irradiance was 56.25 ± 2.45 × 10⁻¹⁷ quanta day⁻¹ cm⁻¹. Dilution rate was 0.400 h⁻¹. EPA concentration was estimated based on the biomass concentration and EPA cellular content.

[c]Near horizontal tubular photobioreactor (NHTR) was operated for 7 months. Air was injected into the tubes by means of a perforated pipe to provide mixing and degassing. Semicontinuous mode was maintained by daily harvesting and addition of fresh medium. Data in the table correspond to the month of June and the mean solar irradiance in this month was 18.9 ± 4.4 MJ m⁻² day⁻¹. EPA concentration was estimated based on biomass concentration and EPA cellular content.

[d]Photobioreactor was equipped with an airlift system. Semicontinuous mode was maintained by daily harvesting and addition of fresh medium. EPA concentration was estimated based on the biomass concentration and EPA cellular content.

[e]Photobioreactor was equipped with airlift system. Reactor was illuminated continuously by cool white fluorescence lamps at 82 μmol m⁻² s⁻¹. Semicontinuous mode was maintained by daily harvesting and addition of fresh medium. EPA concentration was estimated based on the biomass concentration and EPA cellular content.

[f]Glucose and sodium nitrate were used as carbon and nitrogen sources. EPA cellular content was calculated based on fatty acid content of the cells and EPA content of fatty acids estimated to be 50% and 3–5%, respectively. EPA concentration was estimated based on biomass concentration and EPA cellular content. EPA productivity was estimated based on biomass productivity and EPA cellular content.

[g]Glucose and sodium nitrate were used as carbon and nitrogen sources, respectively.

[h]Glucose and sodium nitrate were used as carbon and nitrogen sources, respectively. After glucose was depleted, a continuous feeding was employed and the feeding rate was adjusted based on the glucose concentration in the medium.

capital and operating costs than the enclosed systems, they generate a large volume of waste and need an energy-intensive harvesting process, which could potentially make open ponds less favorable. Both tubular and flat plate photobioreactors have been examined for EPA production in several autotrophic species (Table 15.1). As can be seen in Table 15.1, product yields and productivities appear to be comparable.

Relatively low yield and productivity, obtained in autotrophic systems, has been attributed to light limitation and low photosynthetic activity of microalgae [67]. To overcome these limitations, heterotrophic growth in the absence of light, using organic carbon sources such as glucose, has been explored [52]. Diatoms have been particularly screened for heterotrophic EPA production and *Nitzschia laevis* was identified as a promising EPA producer [68]. Higher product yield and productivity in heterotrophic cultivation are achieved due to higher biomass concentrations and growth rate [67] (Table 15.1).

EPA accumulates as a fraction of TAGs in heterotrophic microalgae of *Nitzschia* species and since TAGs are overproduced under nitrogen-limiting conditions, the highest EPA concentration also occurs at the stationary phase of growth when nitrogen is limited [44]. Carbon to nitrogen ratio and silicate and dissolved oxygen concentrations are often among the important optimization parameters to enhance EPA cellular content in diatoms [44, 69, 70]. Maximizing glucose consumption efficiency by adjusting glucose concentration in the feed has also been studied as an optimization parameter [70].

EPA heterotrophic production involves a two-stage process. At the first stage biomass is grown on an organic carbon source such as glucose. The second stage, namely, the oleogenic phase, starts when biomass growth has ceased and EPA starts to accumulate under the nitrogen limitation condition [44]. Fed batch mode, which is often employed for heterotrophic EPA production, allows better control of the carbon to nitrogen ratio by incremental addition of carbon and nitrogen sources and also avoids substrate inhibition, usually encountered in batch operations [44, 69].

Another studied culture mode for EPA production is the *perfusion-bleeding* strategy [52]. In this regime, a portion of biomass suspension was continuously transferred to a settling tank, where algal cells were settled and recycled to the fermenter while the spent medium was discarded (perfusion). Perfusion permits partial removal of toxic by-products that might inhibit cell growth. Fresh medium was continuously fed to the fermenter and a flow of cell suspension was continuously removed to allow a continuous culture harvesting (bleeding). This system resulted in a high EPA productivity (Table 15.1). However, yield was lower compared to only "perfusion" [71] and fed batch [72] due to continuous product removal [71] (Table 15.1).

Although higher biomass concentration and EPA yield were achieved in all cases of heterotrophic compared to autotrophic growth, EPA cellular contents of heterotrophic and autotrophic species were comparable (Table 15.1). A higher EPA content of heterotrophic microalgae was expected given that TAGs constitute a larger portion of the dry weight (20–50%) compared to polar membrane lipids (5–20%) and EPA

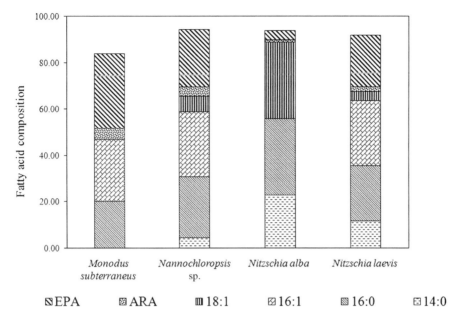

FIGURE 15.3 Fatty acid composition in autotrophically grown *Monodus subterraneus* [65], *Nannochloropsis* sp. [16], and heterotrophically grown *Nitzschia alba* [44] and *Nitzschia laevis* [71].

accumulates in TAGs of *Nitzschia* species [3, 53]. Fatty acid composition of the biomass generated in the processes presented in Table 15.1 shows that saturated or monounsaturated fatty acids are the major constituents of fatty acids of both autotrophic and heterotrophic EPA producers (Fig. 15.3). This profile may imply that screening of more EPA producers in order to find species that can accumulate EPAs as building blocks of TAGs is a worthwhile attempt. In addition to determining EPA cellular content as an optimization parameter, fatty acid composition has also been a point of interest [45, 59] since the presence of fatty acids with various chain lengths and degree of unsaturations complicates the purification process [73]. Furthermore, fatty acid distribution should be determined to achieve a proper balance of ω-3 and ω-6 in nutraceutical and pharmaceutical products and aquaculture feed due to the adverse health effects of a high ratio of ω-6 to ω-3 [74, 75]. Absence of a variety of fatty acids, particularly other PUFAs such as ARA, is more desirable and simplifies EPA recovery [45, 76].

15.3.1.2 Docosahexaenoic Acid (DHA) DHA is most known for its positive effect on brain and eye development in infants and on the cardiovascular system and has been recommended by numerous health agencies to be included in infant's formula [19]. DHA is also an essential nutrient for fish in aquaculture industries [19]. *Crypthecodinium cohnii* and *Schizochytrium limacinum*, both heterotrophic marine microalgae, are commercial sources of DHA. Docosahexaenoic acid accumulates as a constituent of TAGs of the above species [43, 77].

TAGs are generally overproduced under adverse environmental and nutritional conditions [3] and thus production of DHA as a fatty acid fraction of TAGs is also stimulated under unfavorable growth conditions (e.g., nitrogen-deficient condition). Similar to heterotrophic EPA production, a two-stage process is employed to produce biomass under favorable growth conditions in the first stage and then shift to adverse environmental conditions to stimulate lipid production in the second stage [43, 46]. For instance, a temperature shift from the optimum growth temperature to a lower temperature enhanced DHA content of *C. cohnii* [78] and a dissolved oxygen shift from a higher to a lower level showed improvement in DHA content of *S. limacinum* [46, 48].

Commercial DHA production in *C. cohnii* is reported in fed batch fermenters, at which carbon and nitrogen sources are added incrementally [43] (Table 15.2). Fed batch mode permits adjusting the carbon to nitrogen ratio in the medium to support optimum growth at the early stage of fermentation and to provide a nitrogen-limited condition to induce lipid production at the later stage of fermentation [43] (Table 15.2). A maximum biomass concentration of 40 gL^{-1} was reported in this patented process [43].

In addition to the carbon to nitrogen ratio, the nature of the carbon source, temperature, salinity, and dissolved oxygen concentration influence the growth rate and DHA content [43, 79]. DHA accumulation in *C. cohnii* is also reported under photoautotrophic conditions [80] at a much lower level relative to heterotrophic growth (Table 15.2). In addition to glucose, alternative carbon sources such as ethanol and acetate are used for growth of *C. cohnii* [81, 82]. DHA in *C. cohnii* reached a high concentration, when acetic acid was used as a substrate [82] (Table 15.2) and DHA productivity was highest when ethanol was the substrate, compared to glucose or acetic acid [81] (Table 15.2).

A low level of EPA content [43] and other PUFAs [83] reported in *C. cohnii* [43] is desirable since EPA impurity in DHA, which is used as an ingredient of infants' formula, should be below a certain level due to the anticoagulant properties of EPA [43].

Schizochytrium limacinum is another commercial producer of DHA [48], with a simple PUFA profile, which is beneficial in product isolation and purification processes [84]. Similar to *C. cohnii*, a high carbon to nitrogen ratio supports an increase in total fatty acids and DHA content in *S. limacinum* [84]. A low level of oxygen is reported to be essential for lipid production since a high oxygen concentration suppresses lipid accumulation [48]. Commercial production employs a two-stage process with a higher dissolved oxygen concentration to support biomass growth in the first stage and with reduced dissolved oxygen in the second stage to induce lipid production [46]. Gradual addition of corn syrup as the carbon source, in a fed batch mode, maintains the nitrogen-limiting condition at the lipid production stage [46]. Under controlled pH, dissolved oxygen, and sugar concentration in the medium, a high biomass concentration of up to 220 gL^{-1} was achieved with DHA concentration

TABLE 15.2 DHA Production in Microalgae Cultivation Systems

Species	Culture System/Volume	Growth Mode/Culture Mode	Biomass Concentration (g L^{-1})	Biomass Productivity (g L^{-1} day^{-1})	DHA Concentration (mg L^{-1})	DHA Productivity (mg L^{-1} day^{-1})	DHA Content of the Cell (% dry weight)	Reference	Notes
Cryptecodinium cohnii	Fermenter/30 L	Heterotrophic/fed batch	25	7.9	2625	829	10.5	[43]	a
Cryptecodinium cohnii	Fermenter/1 L	Heterotrophic/fed batch	109	6.54	19,000	1104 ± 48	17.4	[82]	b
Cryptecodinium cohnii	Fermenter/2 L	Heterotrophic/fed batch	83	8.7	11,700	1270	14	[81]	c
Cryptecodinium cohnii	Shake flask/150 mL	Autotrophic/batch	1.510	0.15	19.5	47.1	1.3	[80]	d
Schizochytrium sp.	Fermenter/53,000 L	Heterotrophic/fed batch	190.4	48.6	36,747	9380	19.3	[46]	e
Schizochytrium limacinum	Two-stage: fermenter/5 L; shake flask/50 mL	Heterotrophic/fed batch	37.9	3.79	6560	656	17.3	[48]	f
Schizochytrium limacinum	Fermenter/4.5 L	Heterotrophic/continuous	11.78 ± 0.76	3.48 ± 0.20	1740 ± 100	520 ± 30	14.8	[64]	g

^{a}Glucose and yeast extract were used as carbon and nitrogen sources and added incrementally during the course of cultivation. DHA cellular content was estimated based on approximate oil content of biomass and DHA content of oil, indicated to be 30% and 35%, respectively. DHA concentration and productivity were calculated based on the corresponding biomass data and DHA cellular content.

^{b}Acetic acid and yeast extract were carbon and nitrogen sources, respectively.

^{c}Ethanol and yeast extract were used as carbon and nitrogen sources, respectively.

^{d}Culture was continuously illuminated at 60 μE m^{-2} s^{-1}.

^{e}Corn syrup was used as the carbon source and added gradually. Dissolved oxygen was kept at 8% during the first 24 hours and was gradually reduced to 0.5% by the end of fermentation. Biomass productivity was estimated based on DHA cellular content and productivity.

^{f}Glycerol was used as the carbon source. Biomass was grown in the fermenter with controlled dissolved oxygen at 50%. Then 50 mL of medium was harvested and placed in fresh medium in a shake flask for the lipid production stage. Biomass and DHA productivity were estimated based on the increase in their concentrations during the 10 days of culture (the time of the whole two-stage process).

^{g}Crude glycerol and corn steep were used as the carbon and nitrogen sources, respectively. Crude glycerol from a biodiesel refinery was pretreated through pH adjustment. Productivity data in the table corresponds to the maximum value, which was achieved at the dilution rate of 0.3 day^{-1}.

and productivity of $40 \, \text{gL}^{-1}$ and $10 \, \text{gL}^{-1} \, \text{day}^{-1}$, respectively [46]. Fed batch is a common strategy in DHA production similar to most biotechnology processes, because it provides a more controlled culture condition as well as reducing the inhibitory effect of substrate on growth rate [46, 81].

Corn syrup [46], sorghum juice [85], and pure glycerol [84] were used to grow *S. limacinum*. Crude glycerol, the major by-product of biodiesel refineries, also supported growth of *S. limacinum* [86]. Medium composition and temperature for an optimum DHA production in this process were determined [86]. Elimination of methanol and soap impurities present in crude glycerol enhanced both biomass concentration and DHA cellular content [87]. Given the sensitive application of DHA as food or feed additive, DHA enriched biomass, grown on crude glycerol, requires careful analysis to verify the nutritional levels and presence of impurities [87].

Schizochytrium limacinum achieved a higher biomass concentration and productivity, implying a higher growth rate compared to *C. cohnii,* in the processes described in Table 15.2. As a result, higher DHA concentration and productivity was achieved in *S. limacinum* [46] (Table 15.2).

15.3.1.3   *Arachidonic Acid (ARA)*   ARA is a main fatty acid in the brain and central nervous system [88] and is recognized as a nutrient of high importance for brain development of infants and in particular premature infants [19, 89, 90]. The current commercial microbial source of ARA is the fungus *Mortierella alpina* [91]. Red microalga *Porphyridium cruentum* and green microalga *Parietochloris incisa* have been identified as promising sources of ARA [45, 92, 93].

ARA in *Porphyridium cruentum* deposits in both TAGs and membrane polar lipids, mostly galactolipids. ARA production in *Porphyridium cruentum* has been explored under various nutritional and environmental conditions and mode of cultivation [55, 94, 95]. Suboptimal growth conditions stimulated ARA production [55]. Conditions that lower the growth rate, including increased cell concentration and suboptimal temperature, pH, and salinity, resulted in ARA overproduction at the expense of EPA [55, 94]. Increasing dilution rate (ranging from 0.048 to 0.3 day^{-1}) in a chemostat photobioreactor led to an increase in both ARA and EPA cellular content, with EPA being the predominant PUFA [96]. Increasing renewal rate in a semicontinous mode also resulted in an increase of ARA and EPA fractions of fatty acids, with ARA being the predominant PUFA [97]. Outdoor cultivation of *Porphyridium cruentum* in an open pond and a closed photobioreactor was successfully practiced [95, 98] (Table 15.3). Biomass and ARA productivities were significantly enhanced in the closed photobioreactor (Table 15.3).

ARA in *Parietochloris incisa* (Trebuxiophyceae, Chlorophyta) deposits as a fraction of TAGs up to more than 60% of total fatty acid content and over 20% of cellular content under appropriate conditions [99]. High cellular content of ARA is due to its deposition in TAGs, since fatty acids of TAGs constitute the major part of the total fatty acids [92]. Accumulation of PUFAs in TAGs of autotrophic microalgae is an uncommon phenomenon [93]. As can be seen in Table 15.3, higher ARA cellular content was obtained in *Parietochloris incisa* relative to *Porphyridium cruentum.*

TABLE 15.3 ARA Production in Autotrophic Microalgae Cultivation Systems

Species	Culture System/Volume	Growth Mode/ Culture Mode	Biomass		ARA			Reference	Notes
			Concentration $(g\,L^{-1})$	Productivity $(g\,L^{-1}\,day^{-1})$	Concentration $(mg\,L^{-1})$	Productivity $(mg\,L^{-1}\,day^{-1})$	ARA Content of the Cell (% dry weight)		
Porphyridium cruentum	Outdoor mixed open pond/130 L	Autotrophic/ semicontinuous	—	0.17	—	1.5	1.5	[95]	a
Porphyridium cruentum	Outdoor horizontal tubular photobioreactor/220 L	Autotrophic/ continuous	3.5	1.76	70	35.2	2	[98]	b
Parietochloris incisa comb. nov.	Indoor flat plate photobioreactor/1.3 L	Autotrophic/batch	21 ± 0.3	1.23	2667 ± 13	70.2	12.7	[42]	c
Parietochloris incisa comb. nov.	Outdoor flat plate photobioreactor/26 L	Autotrophic/batch	4.2 ± 0.1	0.28	403 ± 8	12.7 ± 5	6.9 ± 9	[42]	d

[a]Surface area and depth of the pond were 1 m² and 13 cm, respectively. Biomass and ARA productivities correspond to the mean values within 48 days of operation. ARA cellular content is the maximum value reported during the outdoor operation.

[b]The photobioreactor was equipped with an airlift pump. Values of concentration and productivity correspond to the average of three days. External irradiance on the surface changed between 1600 and 2000 $\mu E\,m^{-2}\,s^{-1}$.

[c]Light path length of the reactor was 1 cm. Reactor was continuously illumination at 2000 $\mu E\,m^{-2}\,s^{-1}$. Biomass reached stationary phase in 17 days. After 17 days, cell mass precipitated and the supernatant was removed and replaced by nitrogen-free growth medium.

[d]Light path length of the reactor was 10 cm. Air enriched with 2% CO_2 was supplied through perforated plastic tube placed in the bottom of the reactor. Temperature was controlled by water spray. Cultivation was conducted in full growth medium for 15 days, followed by 17 days in nitrogen-free medium.

Increase in cell density and nitrogen starvation were shown to enhance fatty acid content [99]. Increasing the light irradiance from 250 to 2000 $\mu E\ m^{-2}\ s^{-1}$ decreased ARA cellular content but increased the volumetric and areal biomass concentration in a flat plate bioreactor [42]. Cultivation of *P. incisa* in an outdoor flat plate system (during winter) resulted in a lower biomass concentration and productivity and ARA cellular content compared to the indoor system [42] (Table 15.3). In both reactors, after reaching a constant cell density, biomass was separated from the supernatant and nitrogen-free medium was added to induce lipid synthesis [42].

15.3.2 Carotenoids

Carotenoids are a class of tetraterpene lipids [100] with an important role in the photosynthetic pathway (e.g., light-absorbing properties, quenching chlorophyll fluorescence) that protects the cells against light and oxygen [101, 102], owing to their polyene chain structure [103]. The major chemical groups of carotenoids are carotene with a nonoxygenated aromatic ring (e.g., β-carotene, Fig. 15.4A) and xanthophylls with an oxygenated aromatic ring (e.g., astaxanthin and lutein, Fig. 15.4B, C) attached to their polyene chain [103]. Carotenoids are currently used as colorants and vitamins in food products and as feed additives in poultry and aquaculture industries [103]. Some carotenoids including β-carotene and lutein have been shown to

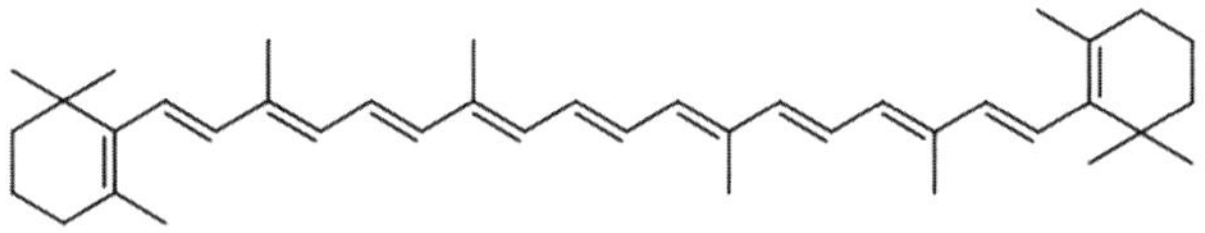

(A)

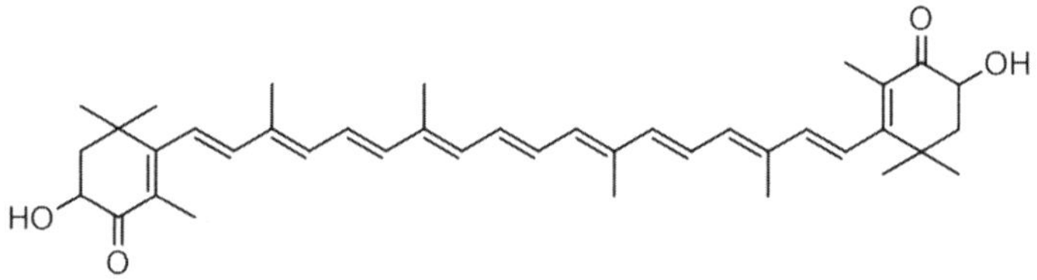

(B)

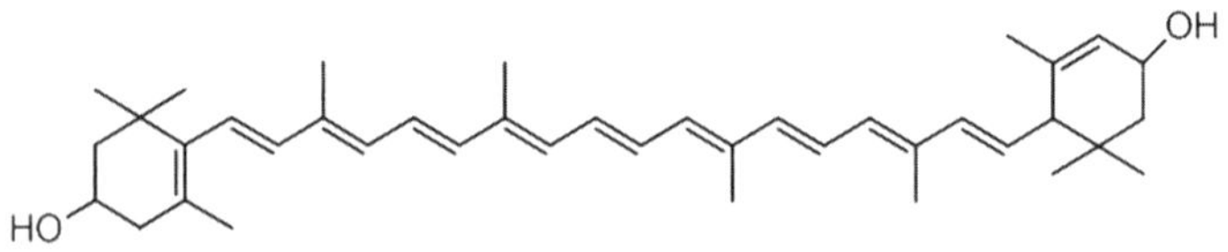

(C)

FIGURE 15.4   Chemical structures of (A) β-carotene, (B) astaxanthin, and (C) lutein.

possess antioxidant properties [104, 105], introducing new applications in the emerging nutraceutical market [103]. *Dunaliella salina* (Chlorophyceae, Chlorophyta) and *Haematococcus pluvialis* (Chlorophyceae, Chlorophyta) are currently used for commercial production of β-carotene and astaxanthin, respectively [11]. Carotenoid production in outdoor microalgae cultivation systems is reviewed by Del Campo et al. [103].

15.3.2.1 *β-Carotene* β-Carotene has applications as a food coloring agent and provitamin A in health food and animal feed [103]. Production of β-carotene by *D. salina*, a marine halophilic microalga [35], was commercialized in the mid-1980s [11]. β-Carotene accumulates in oil globules in the interthylakoid space of the chloroplast of *D. salina* [103] and is overproduced by exposure to environmental and nutritional stresses such as hypersalinity, elevated temperature, high light intensity, and nitrate starvation [103, 106].

Commercial β-carotene production plants using *D. salina* (reviewed by Ben-Amotz [35]) are currently limited to unmixed open ponds (extensive) or raceways (intensive) in areas where a high solar intensity and saline water are available [35]. For extensive ponds, which occupy a large land area (up to 250 ha), biomass concentration of $0.1–0.5$ gL^{-1} [14] and productivity of 0.1 mg L^{-1} day^{-1} were reported [35]. Productivity was further improved in a two-stage process, wherein *D. salina* was grown in open raceways and when the desired cell density was achieved, the culture was transferred to the production raceways and diluted with nitrate-deficient medium to induce carotenogenesis [106] (Table 15.4).

β-Carotene production in an outdoor enclosed photobioreactor, equipped with a sunshade screen, has shown the feasibility of enhanced biomass productivity through a controlled culture condition (e.g., temperature, irradiation, and flow velocity) [107]. Removing the sunshade screen further enhanced the productivity and β-carotene content of the cell to 42 mg L^{-1} and 10%, respectively (Table 15.4).

Heterotrophic growth of *D. salina* did not succeed at the commercial scale due to the high light requirement [35]. Mixotrophic growth on malonate on a small laboratory scale showed a comparable β-carotene concentration and productivity to those of photoautotrophic condition [108] (Table 15.4).

Milking microalgae, which allows simultaneous production and extraction, was applied for β-carotene production at the laboratory scale [109]. *Dunaliella salina* was grown in the bubble column reactor at low light intensity and then cells were transferred to a two-phase bioreactor to simultaneously induce carotenogenesis at high light intensity and extract β-carotene in dodecane, a biocompatible solvent [109]. β-Carotene cellular content and concentration were higher compared to other systems (Table 15.4). The two-phase bioreactor system not only eliminates the harvesting and dewatering process in favor of production economics but also expands its production sites to broader locations, independent of climate and environmental conditions.

15.3.2.2 *Astaxanthin* Astaxanthin, a ketocarotenoid, is used as a pigment and feed supplement in the aquaculture, nutraceutical, cosmetics, and food industries [103, 110]. The majority of commercially available astaxanthin is either synthetically

TABLE 15.4   β-Carotene Production in *Dunaliella salina*

Culture System/Volume	Growth Mode/ Culture Mode	Biomass		β-Carotene		Product Content of the Cell (% dry weight) (pg cell⁻¹)*	Reference	Notes
		Concentration (g L⁻¹) (cells L⁻¹)*	Productivity (g L⁻¹ d⁻¹) (cells L⁻¹ day⁻¹)*	Concentration (mg L⁻¹)	Productivity (mg L⁻¹ d⁻¹) (mg m⁻² day⁻¹)*			
Raceway nursery pond, 0.3–1000 m² raceway production pond/3000–4000 m²	Autotrophic/ batch	$(0.5 \pm 0.1 \times 10^9)$*	—	15	$1.5\ (300 \pm 50)$*	(30)*	[106]	a
Open pond/2000–2600 L	Autotrophic/ semicontinuous	—	0.01	—	0.38–0.5	4%	[203]	b
Outdoor tubular photobioreactor/ 55 L (2.2 m²)	Autotrophic/ semicontinuous	$0.42(4.9 \times 10^9)$*	0.063	42	6.3	10%	[107]	c
Photobioreactor/400 mL	Mixotrophic/ batch	(1.35×10^9)*	(0.28×10^9)*	18	3.8	(13.3)*	[108]	d
Two-stage: bubble column/450 mL – two-phase bioreactor/ 875 mL	Autotrophic/ batch – fed batch	$(0.8–1.6 \times 10^9)$*	0.02×10^9	115	2.7	(51)*	[109]	e

[a]Culture depth was 20 cm in production ponds. Biomass and product concentrations and productivities correspond to the production stage. β-Carotene cellular content was estimated by division of β-carotene concentration to cell density.

[b]Areal biomass and β-carotene productivities were 1.3 and 50 g m⁻² day⁻¹ and volumetric biomass (g L⁻¹ day⁻¹) and β-carotene (mg L⁻¹ day⁻¹) productivities were estimated based on the varying depth of the pond (varied between 10 and 13 cm) and the surface area of 20 m². Productivities correspond to the mean value throughout the year.

[c]Horizontal tubular photobioreactor was equipped with an airlift system. The solar harvesting unit was submerged in a thermostatic pond of water to control the temperature. Photosynthetic active irradiation was 805 ± 100 µmol m⁻² s⁻¹. Culture was diluted during a 4-day interval.

[d]Cells were grown in a photobioreactor and then transferred to a shake flask at continuous illumination of 400 µE/m² s for carotenogenesis. Then 45 mM malonate was used as an organic carbon source. Productivities were estimated based on the increase in biomass or product during 114 hours of cultivation.

[e]Cells were grown in a bubble column at a light intensity of 250 µmol m⁻² s⁻¹ and were transferred to a flat-panel photobioreactor with a light path of 2.5 cm for simultaneous β-carotene production and extraction at light intensity of 1200 µmol m⁻² s⁻¹. The reactor contained 80% aqueous phase and 20% organic phase. A total of 101 mg β-carotene was produced in the system during 46 days, with approximately 60 mg in the organic phase and 41 mg inside the cells.

produced or extracted from crustaceans [103, 110]. Microalgae including *Haematococcus* and *Chlorella* have demonstrated accumulation of astaxanthin [111]. The freshwater biflagellate *Haematococcus pluvialis* (Chlorophyceae, Chlorophyta) is currently a commercial source of astaxanthin [103]. Astaxanthin is an extrachloroplastic lipid, which mostly deposits in the form of esters of fatty acids around the nucleus of *H. pluvialis* and expands throughout the entire protoplast in the transition of green vegetative cells into red aplanospores under stressed culture conditions [110, 112, 113]. Astaxanthin accumulation was mostly reported during the formation of nonmotile resting cyst cells called aplanospores [114] and was also observed in flagellated growing vegetative cells [114, 115].

Haematococcus pluvialis is capable of accumulating astaxanthin under autotrophic, heterotrophic (in the absence of light), and mixotrophic conditions [116, 117]. In autotrophic growth, high irradiance, elevated temperature, high salt concentration, and nitrogen or phosphorus limitation stimulated carotenogenesis [113, 115, 118, 119]. Irradiance was among the most important factors that influence carotenogenesis [114, 119, 120] and thus its effect on carotenoid production was extensively investigated both independently and in combination with other parameters including temperature and salt concentration [120].

Stress conditions that induce carotenogenesis also tend to inhibit biomass growth and increase the cell mortality [119]. This has led to adaptation of a two-stage commercial process that separates biomass growth and production [38]. In the first stage, vegetative cells of *Haematococcus pluvialis* are grown at optimal growth conditions in an enclosed photobioreactor, and in the second stage, the axenic cell culture is transferred to an open pond to induce carotenogenesis and accumulate astaxanthin under stressed conditions caused by nutrient limitation, elevated temperature, and high irradiance [38]. Both of these reactors are operated in batch mode.

Feasibility of continuous cultivation of *H. pluvialis* for astaxanthin production was also demonstrated [115, 121]. Dilution rate and nitrate concentration in the feed medium were reported to affect the astaxanthin accumulation [115, 121]. At constant irradiance, ranging from 160 to 430 μmol m^{-2}s^{-1}, lower dilution rate led to higher astaxanthin cellular content [121]. Moderate nitrate limitation showed enhancement of astaxanthin cellular content and productivity [115] (Table 15.5). It was also observed that astaxanthin accumulates in vegetative cells, which were actively growing and dividing. Vegetative cells, which lack the cyst's characteristic cell wall, can potentially result in an enhanced product extraction versus cyst cells in downstream product recovery [115].

In heterotrophic and mixotrophic growth of *H. pluvialis*, the optimum concentration of organic carbon source (e.g., glucose, acetate) should be determined, since above a certain level, growth rate decreases due to the substrate inhibitory effect [116, 122]. Heterotrophic and mixotrophic growth often result in a higher biomass and astaxanthin concentration (Table 15.5). However, astaxanthin accumulation is lowest in the heterotrophic condition [116, 123]. Mixotrophic cultivation of *Haematococcus pulvialis* on acetate as the carbon source led to a higher cell density and astaxanthin concentration than the heterotrophic condition (in dark) and photoautotrophic cultivation [116]. Higher cell density was attributed to the simultaneous

TABLE 15.5 Astaxanthin Production in Photoautotrophic and Mixotrophic *Haematococcus pluvialis*

Culture System/Volume	Growth Mode/Culture Mode	Biomass		Astaxanthin			Reference	Notes
		Concentration ($g\,L^{-1}$)	Productivity ($g\,L^{-1}\,day^{-1}$)	Concentration ($mg\,L^{-1}$)	Productivity ($mg\,L^{-1}\,day^{-1}$)	Product Content of the Cell (% dry weight)		
Two-stage: outdoor closed photobioreactor/ 25,000 L mixed open pond	Autotrophic/batch	0.2–0.36	0.036–0.054	5.6–10.8	—	2.8–3	[38]	a
Bubble column/1.8 L	Autotrophic/continuous	1	0.9	6.1	5.6	0.6	[115]	b
Fermenter/2.5 L	Mixotrophic/fed batch	2.74	0.14	64.36	3.2	2.3	[117]	c
Fermenter/2.5 L	Mixotrophic/batch	2.65	0.13	53.43	2.7	2		
Two-stage: fermenter/2.5 L – photobioreactor/900 mL	Heterotrophic/fed batch – photoautotrophic/ batch	6	0.23	114	4.4	1.9	[122]	d

[a] Biomass concentration and productivity are the average of 9-month operation. Astaxanthin concentration was estimated based on biomass concentration and cellular astaxanthin content.
[b] Reactor was illuminated at 1220 $\mu E\,m^{2}\,s^{-1}$. Nitrate concentration in feed medium was 2.7 mM.
[c] Sodium acetate was used as the carbon source. The reactor was illuminated stepwise 45 to 180 $\mu mol\,m^{-2}\,s^{-1}$.
[d] Sodium acetate was used as the organic carbon source in the heterotrophic stage. The photoautotrophic stage was started when most cells changed from vegetative to cyst. The reactor was continuously illuminated at 950 $\mu mol\,m^{2}\,s^{-1}$ at the second photoautotrophic stage.

photosynthetic metabolism of CO_2 and oxidative metabolism of acetate [116]. It was also demonstrated that growth rate is enhanced in the mixotrophic condition to the extent of the summation of specific growth rates in autotrophic and heterotrophic conditions [116]. A fed batch mode is often preferred to reduce the substrate inhibitory effect [122].

In a sequential heterotrophic–photoautotrophic system, a high cell density was achieved in fed batch heterotrophic cultivation using sodium acetate as the carbon source [122]. Astaxanthin concentration, reported in this process, was high compared to available data in the literature [122] (Table 15.5).

In a two-stage growth-production process, cultivation conditions (e.g., nature of nitrogen source), culture age, and physiological state of the cells in the first stage affect astaxanthin accumulation and cell mortality under the stressed condition of the second stage. In a sequential heterotrophic–autotrophic system, when vegetative cells were switched from heterotrophic to autotrophic conditions, a higher cell death rate was reported compared to the condition that cyst cells underwent during the same transition [122]. Older cell cultures also showed more resistance to sodium chloride, added to induce carotenogenesis [124].

15.3.2.3   *Lutein*

Lutein is mainly used as a feed additive in poultry farming and aquaculture and also as a natural colorant of foods, drugs, and cosmetics [100, 103]. It is also recognized for its potential therapeutic and preventative effects on age-related macular degeneration, cataracts, and atherosclerosis [125].

Lutein is currently obtained from the petals of marigold in the form of diesters, which is then isolated as free lutein by a saponification reaction [100, 103, 126]. Currently, there is no commercial plant for production of lutein from microalgae [103]. However, a horizontal tubular photobioreactor pilot scale of 28,000 L has been operating to investigate the potential of industrial production of lutein from *Scenedesmus almeriensis* [100, 127]. Promising microalgal species for lutein production as well as advances in outdoor production of lutein have been reviewed by Del Campo et al. [103] and Fernández-Sevilla et al. [100].

Several chlorophycean microalgae have been shown to be rich in lutein and among these species, *Scenedesmus almeriensis* and *Muriellopsis* sp. have been of particular interest due to their high lutein content and growth rate [100, 103, 128]. Species of *Chlorella* including *Chlorella pyrenoidosa* and *Chlorella protothecoides* have also been grown heterotrophically for lutein production [129, 130].

Muriellopsis sp. is a moderate halotolerant that can grow in the restrictive environmental condition of high pH and salinity [131]. Lutein was reported to reach its maximum cellular content in *Muriellopsis* sp. at an early stationary phase under conditions relatively close to those for optimum growth (e.g., sufficient nitrogen) [111, 128].

In *Muriellopsis* sp., increasing irradiance to 460 μmol photon m^{-2} s^{-1} (below the photoinhibition level) in an indoor laboratory scale photobioreactor resulted in an increase in lutein content [128, 132]. A further increase in irradiance led to a decrease in lutein cellular content [132]. Whereas in an outdoor tubular photobioreactor, operated in continuous mode, the highest lutein content in *Muriellopsis* sp was observed at an irradiance of 1900 µmol photon $m^{-2} s^{-1}$ around noon [132]. Interrelationships between temperature and irradiance and also solar cycle in outdoor conditions were suggested to be possible reasons for the differences in outdoor and indoor results [127]. The cell density and the hydrodynamics of the system also influence light penetration [28], which can be responsible for inconsistencies in different cultivation systems.

Scenedesmus almeriensis, a mesophilic chlorophycean, was capable of overproducing lutein as high as 0.6% of dry weight, among the highest reported (Table 15.6). This species is reported to produce lutein of up to 70% total carotenoid, simplifying further the purification steps [127]. In *Scenedesmus almeriensis*, both biomass and lutein concentrations and productivities were enhanced by increasing the irradiance to 1625 µmol photon $m^{-2} s^{-1}$ in an indoor bubble column photobioreactor, operated in continuous mode [133].

Lutein production has been attempted in open ponds, in tubular photobioreactors, and in fermenters (Table 15.6). Some of these efforts, which were mainly focused on optimizing lutein cell content and productivity, are summarized in Table 15.6.

Lutein-rich *Muriellopsis* sp. was successfully cultivated in an outdoor open pond [131] (Table 15.6). The requirement of a restrictive environment such as high pH and salinity gives *Muriellopsis* sp. a chance to overgrow the outdoor contamination, which makes it particularly interesting for economical cultivation of algae in an open pond [131]. Lutein content of the biomass in the open pond was comparable to that of an outdoor tubular photobioreactor with the productivity being much lower (Table 15.6). Lutein cellular content was higher in *Scenedesmus almeriensis* in an outdoor tubular reactor and productivity was comparable with that of *Muriellopsis* sp. in the same photobioreactor configuration (Table 15.6).

Heterotrophic production of lutein was explored with several *Chlorella* species [129, 134], using various nitrogen sources such as nitrate, ammonium, and urea [129]. As expected, a much higher cell density was obtained in comparison with autotrophic conditions. However, lutein content of biomass dry weight was comparable in both conditions (Table 15.6). Fed batch mode resulted in a biomass concentration of 48 gL^{-1} that is significantly higher than 19.6 gL^{-1} in batch mode [129]. Furthermore, a three-stage process, which involved a batch cultivation with sufficient nitrogen source followed by feeding the reactor a nitrogen-limited medium at a late exponential phase and finally increasing the temperature to 32 °C led to enhanced lutein yield. The last step, a temperature stress, led to a decrease in biomass concentration but due to a larger effect on lutein cellular content, lutein yield was enhanced [129] (Table 15.6).

15.4 CARBOHYDRATES

Algae have a diverse carbohydrate fraction, which varies among the species [135]. Commercially available polysaccharides including agar, carrageenans, alginates, fucoidan, and laminaran are mainly obtained from red and brown macroalgae with applications in the food, textile, and biotechnology industries [10, 136, 137]. Carbohydrates of microalgae are also of interest as feedstocks in fermentation industries

TABLE 15.6 Lutein Production in Photoautotrophic and Heterotrophic Microalgae Cultivation Systems

Species	Culture System/Volume	Growth Mode/Culture Mode	Biomass		Lutein			Reference	Notes
			Concentration (g L⁻¹)	Productivity (g L⁻¹ day⁻¹)* (g m⁻² day⁻¹)*	Concentration (mg L⁻¹)	Productivity (mg L⁻¹ day⁻¹)* (mg m⁻² day⁻¹)*	Product Content of the Cell (% dry weight)		
Muriellopsis sp.	Mixed pen pond/100 L	Autotrophic/ semicontinuous	0.4–1	0.14 (14)*	1.6–6	0.64 (64)*	0.4–0.6	[131]	a
Muriellopsis sp.	Outdoor tubular photobioreactor/55 L	Autotrophic/ continuous	1.6–2	0.92–1.6 (23–40)*	6.4–9.5	3.6–7.2 (90–180)*	0.4–0.45	[132]	b
Scenedesmus almeriensis	Outdoor tubular photobioreactor/4000 L	Autotrophic/ continuous	—	0.6	—	3.6 (360)*	0.6	[100]	c
Chlorella pyrenoidosa	Fermenter/19 L	Heterotrophic/ fed-batch	70	13.4	178	28.5	0.25	[130]	d
Chlorella protothecoides	Fermenter/20 L	Heterotrophic/ fed-batch	45.8	4.6	209.08	19.18	0.53	[129]	e

[a]Surface area and depth of the pond were 1 m² and 10 cm, respectively. A part of cell suspension was removed and replaced by fresh medium. Productivities correspond to the average of the 9-month operation.

[b]Surface area of the horizontal tubular photobioreactor was 2.2 m². Biomass and lutein concentrations correspond to the data, collected during a day in May. The lowest and highest productivities correspond to the data, collected in November and July, respectively.

[c]Values of concentration and productivity correspond to the yearly average.

[d]Medium containing glucose as the carbon source was fed into the fermenter.

[e]Carbon and nitrogen sources were glucose and urea, respectively. A three-step cultivation system was employed as follows: algal cultivation with sufficient carbon and nitrogen sources, followed by feeding with nitrogen-deficient medium, and finally a temperature stress was induced. Productivities correspond to the average values. Biomass and lutein concentrations correspond to the maximum values.

including bioethanol production. Sulfated polysaccharides of red microalgae, which are gaining attention for their therapeutic effects, are discussed below.

15.4.1 Polysaccharides

Polysaccharides are present in the cell wall of several algae groups including red, brown, and green microalgae [135]. Cellulose, xylan, and mannan are among the sugar components commonly present in the green and red algae species [135]. Sulfated polysaccharides are high molecular weight heteropolymers of industrial interest due to their rheological and thermal properties [10, 138]. Relative to the polysaccharides of macroalgae, those of microalgae are unexplored both in terms of chemical characterization and industrial production [138]. However, interest is growing for the large-scale production of polysaccharides, particularly from red microalgae, due to accumulating evidence of their antiviral, antioxidant, and anti-inflammatory activities [138, 139, 140, 141]. Characterization, bioactivity properties, and mass production of red microalgae for sulfated polysaccharides have been reviewed [138, 142, 143].

Polysaccharides of red microalgae exist as soluble or cell-bound fractions on the cell wall and inside the cell [144]. Soluble polysaccharides are released from the cell wall and reach their highest concentration at the stationary phase, resulting in an increase in the viscosity of the medium [144].

The effects of light quality and intensity, light path length (in flat plate photobioreactors), and concentration of nitrogen and phosphorus on polysaccharide production have been the focus of several investigations in both outdoor and indoor laboratory cultivation systems [41, 145, 146, 147]. Light intensity is an important optimization parameter to achieve high biomass and polysaccharide yields. Similar to other microalgal species, in *Porphyridium*, increasing the light intensity above a certain level will decrease the cell density and polysaccharide production to the extent of complete cell deactivation [147]. Concentrations of CO_2, nitrate, and sulfate also influence the solubilization of polysaccharide as well as its chemical composition [148, 149].

Porphyridium sp. and *Porphyridium cruentum* were cultivated in an outdoor open pond [95, 144], and horizontal tubular [150] and flat plate photobioreactors (Table 15.7). A low cell density is often reported in the open pond (Table 15.7). In enclosed bioreactors a higher cell density could be achieved (Table 15.7). In an indoor bioreactor for which light and temperature were controlled, product concentration and productivity were highest among the case studies presented in Table 15.7.

15.5 PROTEIN

Protein fraction of microalgae is mostly recognized for its nutritional value for health food or animal feed applications [151]. *Chlorella* sp. (Chlorophyceae, Chlorophyta), *Scenedesmus obliquus* (Chlorophyceae, Chlorophyta), and the cyanobacteria *Spirulina* sp. and *Athrospira* sp. are produced in large scale as sources of protein [151]. Phycobiliproteins, food colorants, and a fluorescence probe, which are discussed below, are also obtained from the protein fraction of some microalgal species.

TABLE 15.7 Polysaccharide Production in *Porphyridium* Species

Species	Culture System/Volume	Growth Mode/ Culture Mode	Biomass		Product		Reference	Notes
			Concentration (cells L^{-1})	Productivity (cells L^{-1} day^{-1})	Concentration ($g L^{-1}$)	Productivity ($g L^{-1}$ day^{-1}) ($g m^{-2}$ day^{-1})*		
Porphyridium sp.	Outdoor mixed open pond/1 m^2	Autotrophic/batch	1.1–2.6×10^7	—	—	—	[144]	a
Porphyridium sp.	Column/ 1 L	Autotrophic/batch	5×10^{10}	0.8×10^{10}	0.47	0.03	[145]	b
		Autotrophic/ semicontinuous	5×10^{10}	0.8×10^{10}	0.43	0.028		
Porphyridium sp.	Outdoor flat plate photobioreactor/ 96 L (0.482 m^2)	Autotrophic/batch	3×10^{10}	—	0.547	$0.028\ (2.8)^*$	[41]	c
		Autotrophic/ semicontinuous	3×10^{10}	—	0.542	$0.034\ (3.36)^*$		
Porphyridium cruentum	Indoor bioreactor	Photoautotrophic/ batch	4.44×10^9	0.23×10^9	4.63	0.24	[147]	d

[a]Highest cell density was observed in the summer and lowest in the winter.
[b]The reactor was continuously illuminated at 150 $\mu E/m^2 s$.
[c]Light path length was 20 cm. A 2% CO_2-enriched air was supplied through a perforated tube. Temperature of the reactor was controlled by water spray. In semicontinuous operation 75% of the reactor content was replaced with fresh medium every 12 days. Product concentration and productivity of both batch and semicontinuous correspond to the maximum values in the month of May to July.
[d]Cells were grown in a controlled bioreactor under blue light (400–500 nm) at an intensity of 40 $\mu E/m^2 s$. Productivity was calculated based on the maximum concentration of cells produced during 19 days of growth.

15.5.1 Phycobiliproteins

Phycobiliproteins (PBPs) are water-soluble oligomeric proteins, consisting of phycobilin chromophores linked to polypeptides [152]. These colored proteins function as accessory pigments in photosynthetic organisms including cyanobactreia (blue-green alga, prokaryotic) and rhodophytes (red algae, eukaryotic) by absorbing light energy and transferring the excitation energy to the chlorophyll–protein complexes present in the thylakoid membrane for conversion to chemical energy [152, 153]. PBPs are commonly classified into three main groups according to their spectroscopic properties: phycocyanins (PCs) (ranging from purple to deep blue), allophycocyanins (blue), and phycoerythrins (PEs) (red) [152], with commercial applications as fluorescence probes in immunodiagnostics [154] and natural colorants in the food and cosmetics industries [152]. PBPs have also recently been recognized for their therapeutic value [155]. The status of applied research and commercialization of phycobiliproteins as well as their current and prospective applications are reviewed [152].

Rhodophyta *Phorphyridium* and the cyanobacterium *Spirulina* are the major industrial producers of PEs and PCs, respectively [152]. Other species of red microalgae, *Rhodosorus marinus* [156], and the cyanobacteria *Nostoc* sp. [153] and *Anabaena* sp. [157] have also been studied in large-scale indoor or outdoor cultivation systems.

The effects of environmental parameters (e.g., temperature, light intensity), media composition (e.g., nature of carbon and nitrogen sources, and carbon dioxide concentration), and mode of cultivation on PBPs biosynthesis have been demonstrated in small lab scale systems as well as outdoor photobioreactors [96, 97, 150, 158, 159, 160, 161].

Being light-harvesting pigments, the PBPs content is greatly influenced by light quality and intensity [150, 162], and increases with a decrease in irradiance [97, 146, 156, 163]. Increasing light intensity from 4 to 9 klx resulted in a decrease in PC content but an increase in yield due to the higher biomass concentration with the highest yield achieved at the upper limit of light intensity, 9 klx [164]. There has also been a report on the increase of PC content of mixotrophically grown *Spirulina platensis* when increasing light intensity from 0.5 to 4 klx [165]. Modification of PE and PC cellular contents in response to changes in light quality and intensity, observed in some strains of cyanobacteria and rhodophytes, has been attributed to their chromatic adaptation capability [162, 163], a phenomenon that can be employed to induce the desired PBPs.

Nitrogen concentration in the media also affects the PBP content. A lower PBP content, reported in nitrogen-deficient environments, was attributed to the function of PBPs as nitrogen reserves [166, 167, 168]. The hypothesis was supported by continued cell growth during nitrogen starvation and increase in proteolytic activity [167].

Compared to a batch operation, semicontinuous cultivation of *Porphyridium cruentum* gave higher PE concentration, which was attributed to removal of toxic metabolites and addition of micronutrients in the semicontinuous mode [97]. The yield was further enhanced with increasing renewal rate to an optimum value [97].

Mixotrophic growth of *Spirulina platensis* using either glucose or acetate resulted in higher biomass and PC yields compared to autotrophic conditions (Table 15.8). In a mixotrophic fed batch cultivation strategy, in which the light intensity was increased stepwise from 80 to 160 $\mu E\ m^{-2}\ s^{-1}$, a high PC concentration of 795 mg L^{-1} was achived [169] (Table 15.8).

The recovery and purification of PBPs is identified as a costly process needing multiple units for chemical or physical disruption of cells and protein extraction in buffer solution or distilled water [152, 153, 160]. Among the most commonly applied cell disruption techniques are ultrasound, lyophilization, repeated freezing and thawing, and enzymatic cell wall disintegration [152]. Supernatant or crude extract, containing PBPs, can be separated by centrifugation and precipitated by ammonium sulfate [153,160,170,171]. Expanded bed chromatography is also examined to recover PBPs from the crude extract [171]. Rivonal treatment prior to ammonium sulfate addition was employed to precipitate and eliminate cellular polysaccharide from crude extract of *Porphyridium cruentum* that may interfere with further purification of PBPs [170]. Ion exchange chromatography is often practiced for the final purification step [172].

15.6   PROCESS INTEGRATION

Renewable energy is positioned in the spotlight of research efforts, a response to concerns over fossil fuels depletion and their environmental consequences including oil spillage and greenhouse gas emissions. Microalgae have received growing attention as a nonfood feedstock for biofuel production due to their carbon dioxide sequestration capability and also high growth rate and lipid content compared to crop plants [173]. The major challenge, however, that has yet to be overcome is finding an efficient algal cultivation system [14], which is not energy intensive and subsequently cost effective in terms of the harvesting/dewatering process. Life cycle assessment predicts a poor energy balance [174, 175, 176] with the major capital cost claimed to be associated with the algae cultivation step [177]. Thus considerable advances in the area of photobioreactor design and also screening and development of new microalgal species with increased photosynthetic efficiency are required to make microalgal biofuel technology commercially viable [21, 178].

Alternative approaches have also been considered to improve the energy balance such as production economics including in situ lipid extraction from living microalgae using a biocompatible solvent in a two-phase bioreactor [178, 179]. This approach, referred to as "milking microalgae," combines cultivation, harvesting, and extraction steps and has successfully been practiced on the laboratory scale for *Dunaliella salina* [109] and *Botryococcus braunii* [180] for β-carotene and hydrocarbon productions, respectively.

Biofuel production in conjunction with wastewater treatment has also been proposed to overcome the cost of algal growth [181]. Algae are currently used for nitrogen and phosphorus removal in wastewater treatment processes [11] and biomass, generated in these processes, is proposed to be the feedstock for biofuel production [182]. However, such integration is debatable as to whether the low biomass yield

TABLE 15.8 Phycoerythrin and Phycocyanin Production in *Porphyridium cruentum* and *Spirulina platensis*, Respectively

Species	Culture System/Volume	Growth Mode/Culture Mode	Biomass Concentration ($g\,L^{-1}$)	Biomass Productivity ($g\,L^{-1}\,day^{-1}$)	PBPs Concentration ($mg\,L^{-1}$)	PBPs Productivity ($mg\,L^{-1}\,day^{-1}$)	Cellular Content of Product (%)	Reference	Notes
Porphyridium cruentum	Outdoor horizontal tubular photobioreactor/200 L	Autotrophic/continuous	—	1.2	—	20	—	[160]	a
Porphyridium cruentum	Glass tube/80 mL	Autotrophic/semicontinuous	—	—	—	18.3	—	[97]	b
Porphyridium cruentum	Glass tube/97 mL	Autotrophic/continuous	2.3	0.69	84.4	25.4	—	[96]	c
Spirulina platensis	Fermenter/1 L	Autotrophic/batch	1.38	0.13	169.7	16.9	12.3	[164]	d
Spirulina platensis	Shake flask/100 mL	Mixotrophic/batch	2.66	1.6	322	200	13	[165]	e
Spirulina platensis	Fermenter/1 L	Mixotrophic/batch	2.14	0.2	257	25.7	12	[164]	f
Spirulina platensis	Fermenter/2.5 L	Mixotrophic/fed batch	7.8	0.8	795	76	10.2	[169]	g

[a]Photobioreactor was equipped with an airlift to circulate the culture medium. Temperature and pH were controlled.

[b]Light intensity was 152 μmol quanta $m^{-2}\,s^{-1}$ with light/dark cycle of 12:12. The data in the table correspond to maximum productivity, which was reported at a renewal rate of 30%.

[c]The data in the table correspond to the highest PE productivity at the dilution rate of 0.3 day^{-1} and temperature of 25 °C.

[d]Bicarbonate was the inorganic carbon source. Reactor was continuously illuminated at light intensity of 11.39 × $10^{-3}\,kJ\,cm^{-2}\,h^{-1}$.

[e]Glucose was used as the carbon source. Reactor was continuously illuminated at 4 klx. Maximum productivity, presented in the table, was achieved at 2.5 g/L glucose concentration.

[f]Glucose was used as the carbon source. Reactor was continuously illuminated at a light intensity of 11.39 × $10^{-3}\,kJ\,cm^{-2}\,h^{-1}$ (9 klx).

[g]Glucose was used as the carbon source, which was continuously fed to the reactor. Reactor was continuously illuminated by fluorescent lamp with irradiance intensity increasing from 80 to 160 $\mu E\,m^{-2}\,s^{-1}$ at 30 °C.

414

generated in wastewater treatment plants could support large volume production of biofuel [183].

Converting the waste generated in the algal–biofuel process to energy and nutrients through anaerobic digestion is another alternative, proposed to have a positive effect on energy balance according to life cycle assessment [176, 184]. This approach also offers a sustainable solution for solid waste management. Anaerobic digestion was introduced as a key contributor to the sustainability of a biorefinary [184], requiring further research to develop an understanding of the process conditions (e.g., retention time, organic loading rate) and to evaluate the feasibility of nutrient recycling to the photobioreactor [177, 185].

Coproduction of biofuel in combination with high-value (e.g., carotenoids) and moderate-value chemicals (e.g., polymers, surfactants) has been evaluated to be promising and necessary for the economical improvement of large-scale biofuel production [177, 186]. High-value-added chemicals such as astaxanthin and β-carotene are presently derived from microalgae at low volume; therefore production of biofuel as a by-product of these processes will not likely be profitable due to the low production volume and high cost [183]. However, since production of several value-added chemicals from microalgae is a well-established process, such integration provides an opportunity to advance the science of microalgal biofuel and improve our knowledge of the economics of a large-scale production plant [187], given the level of uncertainty involved in capital and operating cost estimations [177].

Primary consideration in designing an integrated process is screening microalgal species, capable of efficient coproduction of the desired chemicals under relatively similar nutritional and environmental conditions. Efficiency and cost of the product isolation and purification should also be analyzed carefully when recovery of multiple products is attempted. In this section a number of examples of microalgal species and chemicals that appear to be promising when employed in the practice of an integrated chemical plant are presented. The possibility of coproduction of fuel and valuable chemicals is discussed, followed by an examination of the available literature on simultaneous production of nonfuel chemicals from microalgae.

Microalgal TAGs, consisting of saturated and monounsaturated fatty acids of chain lengths ranging from C_{14} to C_{18}, are the primary feedstock for biodiesel production [3]. Maximum content of these fatty acids is generally observed at the stationary growth phase under nutrient-deprived conditions that often coincide with a decrease in polyunsaturated fatty acids, which are constituents of polar membrane lipids [3, 54, 98, 188]. Similar to TAGs, accumulation of carotenoids such as astaxanthin is stimulated under nutrient-deficient conditions [119]. As a result, integration of triacylglycerols and astaxanthin production combined with anaerobic digestion, illustrated in a conceptual flowsheet in Fig. 15.5, appears to be a compelling option to explore. *Chlorella vulgaris* and *Chlorella zofingiensis* have shown accumulation of astaxanthin [111, 189, 190] and lipids (unsaturated and monounsaturated) [191, 192, 193] under appropriate conditions and therefore can be considered as potential feedstocks for simultaneous production of both chemicals.

Commercial astaxanthin production in microalgae involves a two-stage process using *Haematococcus pluvialis*, known as a hybrid production system. In the first

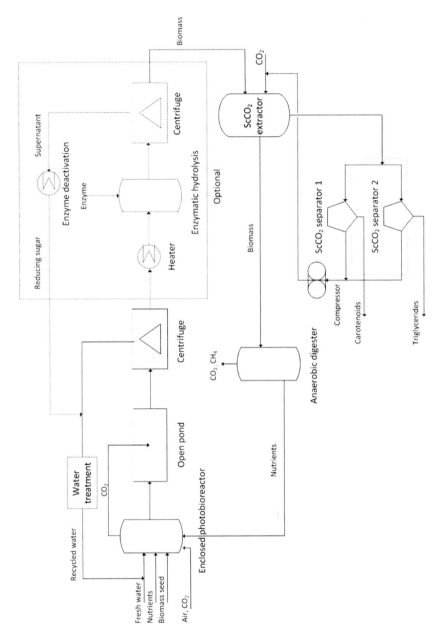

FIGURE 15.5 Conceptual flowsheet of coproduction of carotenoid and triacylglycerols.

stage, *H. pluvialis* is autotrophically grown in an enclosed photobioreactor to obtain a high cell density under axenic conditions. Biomass is then transferred to an open pond to induce carotenogenesis under environmental and nutritional stressed conditions [38, 173]. To generate a higher cell density, relative to the autotrophic growth conditions, heterotrophic or mixotrophic cultivations are occasionally adopted for astaxanthin [190] and lipid production [192]. A sequential mixotrophic–autotrophic culture may be employed, wherein microalgae (e.g., *Chlorella vulgaris, Chlorella zofingiensis*) are mixotrophically grown to a high cell density and subsequently transferred to an open pond to induce carotenogenesis under photoautotrophic and adverse environmental conditions (Fig. 15.5).

Downstream processes including harvesting and product recovery are usually energy intensive, which need to be optimized in order to achieve a sustainable integrated process. Industrial harvesting of *Haematococcus pluvialis* involves a passive settling, followed by centrifugation [38]. Centrifugation as a method of harvesting is particularly costly and energy intensive but efficient and is shown to be feasible for high-value products applications [172]. Occasionally, a combination of several harvesting methods such as flocculation and centrifugation is employed to reduce the cost and improve the efficiency [172]. Although effective in reducing costs and in saving energy, the toxicity and environmental impacts of flocculants must be fully addressed, particularly for those cases in which biomass is intended to be used as aquaculture and animal feed [194]. Organic solvent extraction of the chemicals is a cost-effective method that is widely used in industry [172]. However, the dehydration step prior to solvent extraction that enhances the extraction yield [73, 195] is significantly energy intensive [175]. Another separation method is supercritical carbon dioxide ($scCO_2$) extraction, which is superior to solvent extraction, notably for compounds with nutraceutical and pharmaceutical importance due to eliminating the residual organic solvent in the extracted products and spent biomass [196, 197]. Supercritical carbon dioxide also allows selective extraction of fatty acids and carotenoids under different temperatures and pressures [198, 199].

Cell disruption pretreatment prior to extraction was demonstrated to enhance both organic solvent and $scCO_2$ extraction yields of lipids and carotenoids [172, 196, 197, 199]. Disruption of the cellulosic cell wall of *Chlorella* sp. using cellulase enzyme was reported to simultaneously enhance the lipid recovery and generate reducing sugar by hydrolysis of polysaccharides [200]. Therefore the enzyme treatment may be employed as a cell disruption method prior to $scCO_2$ extraction. Reducing sugar, generated in this process, can be transferred to the photobioreactor to be used as an organic carbon source to support mixotrophic growth of microalgal as shown in Fig. 15.5. The potential application of reducing sugar to support the mixotrophic growth of microalgae or to grow any other microorganisms has yet to be fully assessed. Lipid-extracted microalgae debris was previously used as substrate to support growth of the yeast *Saccharomyces bayanus* to produce bioethanol [201]. However, more information is required on carbohydrate composition of microalgae in order to evaluate their potential applications in fermentation processes [177].

At the final stage of the process, lipid-extracted microalgal cell debris may be processed in an anaerobic digester to produce methane, carbon dioxide, and nutrients (Fig. 15.5). Although anaerobic digestion is a well-established technology in wastewater treatment, the process needs to be developed for microalgal debris with regard to the nature and kinetics of reactions [177].

Another species that can be considered for simultaneous production of biofuel and valuable chemicals is *Schizochytrium limacinum*, which is cultivated heterotrophically as an industrial source of DHA [195]. *Schizochytrium limacinum* is also rich in saturated fatty acids [195]. Unlike most photosynthetic microalgae, where the saturated and monounsatured fatty acids are present in TAGs while PUFAs are deposited in the membrane structural lipid [3], in *S. limacinum* both PUFAs and saturated fatty acids are stored in TAGs and thus are overproduced under similar nitrogen-deficient conditions [84]. This important feature makes this microalga a promising candidate for both DHA and biodiesel production. Furthermore, this microalga is shown capable of consuming biodiesel-derived glycerol as a substrate for DHA production, adding to the economical benefits of a potential biorefinery [86].

In the context of employing microalgae as sources of nonfuel chemicals, the red microalga *Porphyridium cruentum* is identified as a promising species for production of EPA, ARA, sulfated polysaccharides, and PE [143]. Several studies have investigated the pattern of simultaneous accumulation of these chemicals in response to various cultivation conditions on the laboratory scale [94, 96, 97], in an outdoor pond [95], and in outdoor photobioreactors [150]. Such efforts to monitor the evolution of multiple coproducts in microalgae are essential to determine the optimum culture strategies that support simultaneous production of valuable chemicals for the benefit of the economical feasibility of an integrated process.

In *Porphyridium cruentum*, EPA and ARA are not overproduced under similar culture conditions [55]. EPA production is stimulated under optimum growth conditions, in contrast to ARA, which is overproduced at suboptimal growth conditions [55]. Therefore a cultivation strategy can accordingly be adjusted to produce either of these products [97]. For instance, ARA was the predominant PUFA in *Porphyridium cruentum*, cultivated in a semicontinuous operation [97], whereas in another study, in which a chemostat cultivation system was employed, EPA was the predominant PUFA [96].

Figures 15.6 and 15.7 illustrate the effect of renewal and dilution rates on accumulation of the chemicals in *Porphyridium cruentum* in semicontinuous and chemostat modes, respectively. As shown in Fig. 15.6, in semicontinuous operation, the concentration of PE and polysaccharide reached a maximum value at a renewal rate of 10%. At dilution rates of 10% and 20%, ARA as a portion of total fatty acid fraction reached its highest value [97]. ARA cellular content and concentration were not reported in this study. Therefore it is not known whether the ARA highest concentration also occurred at dilution rates of 10–20% and coincided with that of polysaccharides and PE.

In the chemostat system, the effect of dilution rate on polysaccharide concentration is more pronounced than EPA and PE production, as can be seen in Fig. 15.7. At a dilution rate of 0.154 day^{-1} at which both PE and EPA concentrations were at their

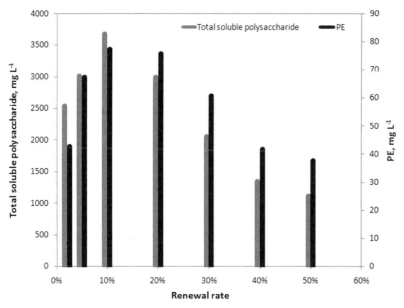

FIGURE 15.6 Effect of renewal rate on accumulation of soluble polysaccharide and PE in *P. cruentum* in a semicontinous mode. Numerical data taken from Fábregas et al. [97] to produce this graph.

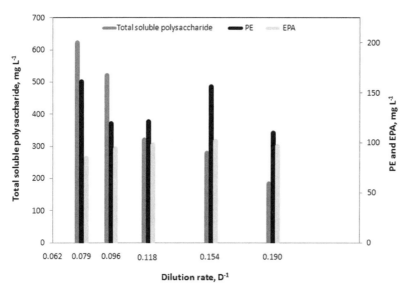

FIGURE 15.7 Effect of dilution rate on accumulation of soluble polysaccharide, EPA, and PE in *P. cruentum* in a continuous cultivation mode. Numerical data taken from Lee and Tan [96] to produce this graph.

maximum, polysaccharide concentration was not optimum and reached its maximum at lowest dilution rate.

As shown in Figs. 15.6 and 15.7, conditions in the semicontinuous mode led to a higher concentration of polysaccharide [97], while those in the chemostat system resulted in a higher yield of EPA and PE [96]. Further systematic research has to be done to rationalize these findings. Multivariable data analysis (principal component analysis), which has been applied to analyze the effect of influential environmental and operational parameters on the nutrition composition of *P. cruentum* in outdoor photobioreactors [150], can determine the conditions that target optimization of several products.

Figure 15.8 illustrates extraction and purification of PE according to a process developed by Bermejo Roman et al. [160] and the possibility of simultaneous recovery of EPA (or ARA) and PE from *P. cruentum*. Combination of autoflocculation and centrifugation may be applied to harvest the cells. Autoflocculation of *P. cruentum* by increasing pH is shown to be a feasible harvesting method on a large scale [95]. Cells are disrupted by sonication and PE is extracted from the disrupted cells by a buffer solution and precipitated by ammonium sulfate. Product is further purified by dialysis and ion chromatography [160]. The waste streams of this process should be analyzed to examine the recovery of intracellular polysaccharides since they can constitute up to 49% of the microalgae [96]. Precipitation of cellular polysaccharide by rivanol prior to PBP precipitation by ammonium sulfate was already reported [170]. EPA or ARA from biomass waste may be extracted by organic solvent or supercritical carbon dioxide and purified by means of column chromatography [73]. In case of $scCO_2$ extraction of PUFAs from seaweed and microalgae, pressure and temperature can be adjusted toward selection of PUFA versus saturated and monounsaturated fatty acids [168, 202].

15.7 CONCLUSION

The potential of microalgae as renewable sources of a versatile class of chemicals is widely accepted. There are, however, major challenges in both microalgae cultivation and downstream processes that have to be overcome in order to make microalgae-derived chemicals economically competitive. Screening and development of new microalgae with high photosynthetic efficiency, designing innovative cultivation systems, and achieving efficient downstream processes are the top research priorities. The significance of the coproduction of chemicals in microalgae as an alternative strategy to improve the economics and sustainability is also emphasized by several studies. An integrated chemical plant approach requires multidisciplinary research on identification and development of microalgal species that are capable of producing multiple chemicals, designing culture strategies to simultaneously overproduce these chemicals, and developing downstream technolgies for efficient separation. Furthermore, the practice of deriving multiple chemicals from microalgae within an integrated demonstration plant permits a careful economical analysis to evaluate the feasibility of a full-scale plant.

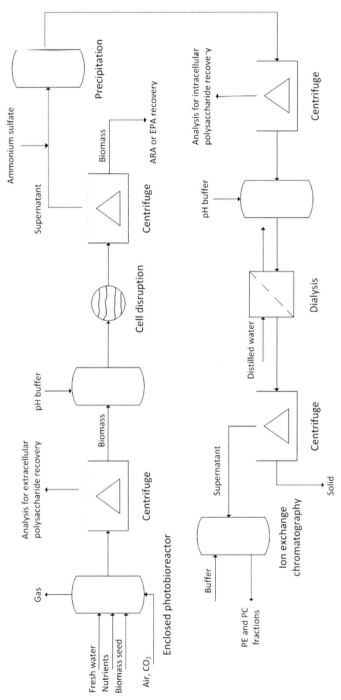

FIGURE 15.8 Conceptual flowsheet of polysaccharide and ARA (or EPA) recovery from PE purification process. Process of PE purification was adapted from Bermejo Román et al. [160] with permission of Elsevier.

REFERENCES

1. L. Tomaselli, The microalgal cell. In: A. Richmond (Ed.), *Handbook of Microalgal Culture: Biotechnology and Applied Phycology*, Blackwell Science Ltd., Oxford, 2004, pp. 3–19.

2. A. Banerjee, R. Sharma, Y. Chisti, U. C. Banerjee, *Botryococcus braunii*: a renewable source of hydrocarbons and other chemicals, Crit. Rev. Biotechnol. 22 (2002) 245–279.

3. Q. Hu, M. Sommerfeld, E. Jarvis, M. Ghirardi, M. Posewitz, M. Seibert, A. Darzins, Microalgal triacylglycerols as feedstocks for biofuel production: perspectives and advances, The Plant Journal 54 (2008) 621–639.

4. A. C. Guedes, H. M. Amaro, F. X. Malcata, Microalgae as sources of high added-value compounds—a brief review of recent work, Biotechnol. Prog. 27 (2011) 597–613.

5. R. Raja, S. Hemaiswarya, N. A. Kumar, S. Sridhar, R. Rengasamy, A perspective on the biotechnological potential of microalgae, Crit. Rev. Microbiol. 34 (2008) 77–88.

6. M. A. Borowitzka, Microalgae as sources of pharmaceuticals and other biologically active compounds, J. Appl. Phycology 7 (1995) 3–15.

7. S. Singh, B. N. Kate, U. C. Banerjee, Bioactive compounds from cyanobacteria and microalgae: an overview, Crit. Rev. Biotechnol. 25 (2005) 73–95.

8. J. C. Goldman, Outdoor algal mass cultures—I. Applications, Water Res. 13 (1979) 1–19.

9. P. T. Pienkos, A. Darzins, The promise and challenges of microalgal-derived biofuels, Biofuels, Bioproducts and Biorefining 3 (2009) 431–440.

10. O. Pulz, W. Gross, Valuable products from biotechnology of microalgae, Appl. Microbiol. Biotechnol. 65 (2004) 635–648.

11. P. Spolaore, C. Joannis-Cassan, E. Duran, A. Isambert, Commercial applications of microalgae, J. Biosci. Bioeng. 101 (2006) 87–96.

12. E. A. Davis, J. Dedrick, C. S. French, H. W. Milner, J. Myers, J. H. C. Smith, and H. A. Spoehr, Growth of algae in mass culture. In: J. S. Burlew (Ed.), *Algal Culture from Laboratory to Pilot Plant*, Carnegie Institution of Washington Publication, Washington DC, 1953, pp. 1–357.

13. D. Chaumont, Biotechnology of algal biomass production: a review of systems for outdoor mass culture, J. Appl. Phycology 5 (1993) 593–604.

14. M. A. Borowitzka, Commercial production of microalgae: ponds, tanks, and fermenters, Prog. Ind. Microbiol. 35 (1999) 313–321.

15. C. C. Becker, D. J. Kyle, Developing functional foods containing algal docosahexaenoic acid, Food Technol. 52 (1998) 68–71.

16. G. Chini Zittelli, F. Lavista, A. Bastianini, L. Rodolfi, M. Vincenzini, M. R. Tredici, Production of eicosapentaenoic acid by *Nannochloropsis* sp. cultures in outdoor tubular photobioreactors, J. Biotechnol. 70 (1999) 299–312.

17. J. D. Day, A. P. Edwards, G. A. Rodgers, Development of an industrial-scale process for the heterotrophic production of a micro-algal mollusc feed, Bioresource Technol. 38 (1991) 245–249.

18. R. M. Gladue, J. E. Maxey, Microalgal feeds for aquaculture, J. Appl. Phycology 6 (1994) 131–141.

19. O. P. Ward, A. Singh, Omega-3/6 fatty acids: alternative sources of production, Process Biochem. 40 (2005) 3627–3652.

20. O. Pulz, K. Scheibenbogen, Photobioreactors: design and performance with respect to light energy input, Adv. Biochem. Eng./Biotechnol. 59 (1998) 123–152.

21. Y. Chisti, Biodiesel from microalgae, Biotechnol. Adv. 25 (2007) 294–306.

22. T. A. Norton, M. Melkonian, R. A. Andersen, Algal biodiversity, Phycologia 35 (1996) 308–326.

23. J. W. Blunt, B. R. Copp, R. A. Keyzers, M. H. G. Munro, M. R. Prinsep, Marine natural products, Natural Product Reports 29 (2012) 144–222.

24. P. T. Anastas, J. B. Zimmerman, Peer reviewed: design through the 12 principles of green engineering, Environ. Sci. Technol. 37 (2003) 94A–101A.

25. O. Pulz, Photobioreactors: production systems for phototrophic microorganisms, Appl. Microbiol. Biotechnol. 57 (2001) 287–293.

26. C. U. Ugwu, H. Aoyagi, H. Uchiyama, Photobioreactors for mass cultivation of algae, Bioresource Technol. 99 (2008) 4021–4028.

27. A. P. Carvalho, L. A. Meireles, F. X. Malcata, Microalgal reactors: a review of enclosed system designs and performances, Biotechnol. Prog. 22 (2006) 1490–1506.

28. A. Sánchez Mirón, A. Contreras Gómez, F. García Camacho, E. Molina Grima, Y. Chisti, Comparative evaluation of compact photobioreactors for large-scale mono-culture of microalgae, J. Biotechnol. 70 (1999) 249–270.

29. J. C. Weissman, R. P. Goebel, J. R. Benemann, Photobioreactor design: mixing, carbon utilization, and oxygen accumulation, Biotechnol. Bioeng. 31 (1988) 336–344.

30. Y. Watanabe, D. O. Hall, Photosynthetic production of the filamentous cyanobacterium *Spirulina platensis* in a cone-shaped helical tubular photobioreactor, Appl. Microbiol. Biotechnol. 44 (1996) 693–698.

31. E. Molina, J. Fernández, F. G. Acién, Y. Chisti, Tubular photobioreactor design for algal cultures, J. Biotechnol. 92 (2001) 113–131.

32. M. Tredici, R. Materassi, From open ponds to vertical alveolar panels: the Italian experience in the development of reactors for the mass cultivation of phototrophic microorganisms, J. Appl. Phycology 4 (1992) 221–231.

33. A. Richmond, Z. Cheng-Wu, Optimization of a flat plate glass reactor for mass production of *Nannochloropsis* sp. outdoors, J. Biotechnol. 85 (2001) 259–269.

34. C. Posten, Design principles of photo-bioreactors for cultivation of microalgae, Eng. Life Sci. 9 (2009) 165–177.

35. A. Ben-Amotz, Industrial production of microalgal: cell-mass and secondary products–major industrial species. In: A. Richmond (Ed.), *Handbook of Microalgal Culture: Biotechnology and Applied Phycology*, Blackwell Science Ltd., Oxford, 2004, pp. 273–280.

36. R. Harun, M. Singh, G. M. Forde, M. K. Danquah, Bioprocess engineering of microalgae to produce a variety of consumer products, Renewable and Sustainable Energy Rev. 14 (2010) 1037–1047.

37. M. Javanmardian, B. Palsson, High-density photoautotrophic algal cultures: design, construction, and operation of a novel photobioreactor system, Biotechnol. Bioeng. 38 (1991) 1182–1189.

38. M. Olaizola, Commercial production of astaxanthin from *Haematococcus pluvialis* using 25,000-liter outdoor photobioreactors, J. Appl. Phycology 12 (2000) 499–506.

39. C. Lu, K. Rao, D. Hall, A. Vonshak, Production of eicosapentaenoic acid (EPA) in *Monodus subterraneus* grown in a helical tubular photobioreactor as affected by cell density and light intensity, J. Appl. Phycology 13 (2001) 517–522.

40. D. O. Hall, F. G. Acién Fernández, E. C. Guerrero, K. K. Rao, E. M. Grima, Outdoor helical tubular photobioreactors for microalgal production: modeling of fluid-dynamics and mass transfer and assessment of biomass productivity, Biotechnol. Bioeng. 82 (2003) 62–73.

41. S. Singh, S. M. Arad, A. Richmond, Extracellular polysaccharide production in outdoor mass cultures of *Porphyridium* sp. in flat plate glass reactors, J. Appl. Phycology 12 (2000) 269–275.

42. Z. Cheng-Wu, Z. Cohen, I. Khozin-Goldberg, A. Richmond, Characterization of growth and arachidonic acid production of *Parietochloris incisa* comb. nov (Trebouxiophyceae, Chlorophyta), J. Appl. Phycology 14 (2002) 453–460.

43. D. J. Kyle, S. E. Reeb, V. J. Sicotte, Docosahexaenoic acid, methods for its production and compounds containing the same, Patent WO 1991/011918.

44. D. J. Kyle, R. Gladue, Eicosapentaenoic acid-containing oil and methods for its production, U.S. Patent 5244921 (1993).

45. Z. Cohen, Production potential of eicosapentaenoic acid by *Monodus subterraneus*, J. Am. Oil Chemists' Soc. 71 (1994) 941–945.

46. R. B. Bailey, D. DiMasi, J. M. Hansen, P. J. Mirrasoul, C. M. Ruecker, G. I. Veeder, T. Kaneko, W. R. Barclay, Enhanced production of lipids containing polyenoic fatty acid by very high density cultures of eukaryotic microbes in fermentors, U.S. Patent 6,607,900 (2003).

47. W. R. Barclay, K. M. Meager, J. R. Abril, Heterotrophic production of long chain omega-3 fatty acids utilizing algae and algae-like microorganisms, J. Appl. Phycology 6 (1994) 123–129.

48. Z. Chi, Y. Liu, C. Frear, S. Chen, Study of a two-stage growth of DHA-producing marine algae *Schizochytrium limacinum* SR21 with shifting dissolved oxygen level, Appl. Microbiol. Biotechnol. 81 (2009) 1141–1148.

49. A. P. Simopoulos, Omega-3 fatty acids in health and disease and in growth and development, Am. J. Clin. Nutrition 54 (1991) 438–463.

50. A. P. Simopoulos, Omega-3 fatty acids in inflammation and autoimmune diseases, J. Am. College of Nutrition 21 (2002) 495–505.

51. V. Patil, T. Källqvist, E. Olsen, G. Vogt, H. Gislerød, Fatty acid composition of 12 microalgae for possible use in aquaculture feed, Aquaculture Int. 15 (2007) 1–9.

52. Z. Y. Wen, F. Chen, Heterotrophic production of eicosapentaenoic acid by microalgae, Biotechnol. Adv. 21 (2003) 273–294.

53. S. S. Thomas, S. Kumaravel, Photoautotrophic growth of microalgae for omega-3 fatty acid production, U.S. 2008/ 0166779 A1.

54. I. Khozin-Goldberg, Z. Cohen, The effect of phosphate starvation on the lipid and fatty acid composition of the fresh water eustigmatophyte *Monodus subterraneus*, Phytochemistry 67 (2006) 696–701.

55. Z. Cohen, A. Vonshak, A. Richmond, Effect of environmnetal conditions on fatty acid composition of the red alga *Porphyridium cruentum*: correlation to growth rate, J. Phycology 24 (1988) 328–332.

56. V. Veloso, A. Reis, L. Gouveia, H. L. Fernandes, J. A. Empis, J. M. Novais, Lipid production by *Phaeodactylum tricornutum*, Bioresource Technol. 38 (1991) 115–119.

57. W. Yongmanltchal, O. Ward, Growth and eicosapentaenoic acid production by *Phaeodactylum tricornutum* in batch and continuous culture systems, J. Am. Oil Chemists' Soc. 69 (1992) 584–590.

58. E. Molina Grima, F. García Camacho, J. A. Sánchez Pérez, J. Urda Cardona, F. G. Acién Fernández, J. M. Fernández Sevilla, Outdoor chemostat culture of *Phaeodactylum tricornutum* UTEX 640 in a tubular photobioreactor for the production of eicosapentaenoic acid, Biotechnol. Appl. Biochem. 20 (1994) 279–290.

59. R. Huerlimann, R. de Nys, K. Heimann, Growth, lipid content, productivity, and fatty acid composition of tropical microalgae for scale-up production, Biotechnol. Bioeng. 107 (2010) 245–257.

60. Z. Cohen, *Chemicals from Microalgae*, Taylor & Francis Ltd., Padstow, UK, 1999.

61. W. Yongmanltchal, O. Ward, Growth of and omega-3 fatty acid production by *Phaeodactylum tricornutum* under different culture conditions, Appl. Environ. Microbiol. 57 (1991) 419–425.

62. Z. Cohen, *Porphyridium cruentum*. In: Z. Cohen (Ed.), *Chemicals from Microalgae*, Taylor & Francis Ltd. Padstow, UK, 1999, p. 11.

63. A. Sukenik, Y. Carmeli, B. Tamar, Regulation of fatty acid composition by irradiance level in the eustigmatophyte *Nannochloropsis* sp., J. Phycology 25 (1989) 686–692.

64. S. Ethier, K. Woisard, D. Vaughan, Z. Wen, Continuous culture of the microalgae *Schizochytrium limacinum* on biodiesel-derived crude glycerol for producing docosahexaenoic acid, Bioresource Technol. 102 (2011) 88–93.

65. H. Qiang, H. Zheungu, Z. Cohen, A. Richond, Enhancement of eicosapentaenoic acid (EPA) and γ-linolenic acid (GLA) production by manipulating algal density of outdoor cultures of *Monodus subterraneus* (Eustigmatophyta) and *Spirulina platensis* (cyanobacteria), Eur. J. Phycology 32 (1997) 81–86.

66. A. Sukenik, P. G. Falkowski, J. Bennett, Potential enhancement of photosynthetic energy conversion in algal mass culture, Biotechnol. Bioeng. 30 (1987) 970–977.

67. F. Chen, High cell density culture of microalgae in heterotrophic growth, Trends Biotechnol. 14 (1996) 421–426.

68. C. K. Tan, M. R. Johns, Screening of diatoms for heterotrophic eicosapentaenoic acid production, J. Appl. Phycology 8 (1996) 59–64.

69. Z. Y. Wen, F. Chen, Continuous cultivation of the diatom *Nitzschia laevis* for eicosapentaenoic acid production: physiological study and process optimization, Biotechnol. Prog. 18 (2002) 21–28.

70. Z. Y. Wen, F. Chen, Heterotrophic production of eicosapentaenoid acid by the diatom *Nitzschia laevis*: effects of silicate and glucose, J. Ind. Microbiol. Biotechnol. 25 (2000) 218–224.

71. Z. Y. Wen, F. Chen, A perfusion-cell bleeding culture strategy for enhancing the productivity of eicosapentaenoic acid by *Nitzschia laevis*, Appl. Microbiol. Biotechnol. 57 (2001) 316–322.

72. Z. Y. Wen, Y. Jiang, F. Chen, High cell density culture of the diatom *Nitzschia laevis* for eicosapentaenoic acid production: fed-batch development, Process Biochem. 37 (2002) 1447–1453.

73. E. H. Belarbi, E. Molina, Y. Chisti, A process for high yield and scaleable recovery of high purity eicosapentaenoic acid esters from microalgae and fish oil, Enzyme Microb. Technol. 26 (2000) 516–529.

74. K. W. Lee, G. Y. H. Lip, The role of omega-3 fatty acids in the secondary prevention of cardiovascular disease, QJM 96 (2003) 465–480.

75. W. E. M. Lands, Dietary fat and health: the evidence and the politics of prevention: careful use of dietary fats can improve life and prevent disease, Ann. N.Y. Acad. Sci. 1055 (2005) 179–192.

76. W. Yongmanitchai, O. P. Ward, Screening of algae for potential alternative sources of eicosapentaenoic acid, Phytochemistry 30 (1991) 2963–2967.

77. E. Morita, Y. Kumon, T. Nakahara, S. Kagiwada, T. Noguchi, Docosahexaenoic acid production and lipid-body formation in *Schizochytrium limacinum* SR21, Marine Biotechnol. 8 (2006) 319–327.

78. Y. Jiang, F. Chen, Effects of temperature and temperature shift on docosahexaenoic acid production by the marine microalge *Crypthecodinium cohnii*, J. Am. Oil Chemists' Soc. 77 (2000) 613–617.

79. M. E. de Swaaf, T. C. de Rijk, G. Eggink, L. Sijtsma, Optimisation of docosahexaenoic acid production in batch cultivations by *Crypthecodinium cohnii*, Prog. Ind. Microbiol. 35 (1999) 185–192.

80. R. Vazhappilly, F. Chen, Eicosapentaenoic acid and docosahexaenoic acid production potential of microalgae and their heterotrophic growth, J. Am. Oil Chemists' Soc. 75 (1998) 393–397.

81. M. E. de Swaaf, J. T. Pronk, L. Sijtsma, Fed-batch cultivation of the docosahexaenoic-acid-producing marine alga *Crypthecodinium cohnii* on ethanol, Appl. Microbiol. Biotechnol. 61 (2003) 40–43.

82. M. E. de Swaaf, L. Sijtsma, J. T. Pronk, High-cell-density fed-batch cultivation of the do-cosahexaenoic acid producing marine alga *Crypthecodinium cohnii*, Biotechnol. Bioeng. 81 (2003) 666–672.

83. G. Harrington, G. Holz, The monoenoic and docosahexaenoic fatty acids of a heterotrophic dinoflagellate, Biochim. Biophys. Acta 164 (1968) 137–139.

84. T. Yokochi, D. Honda, T. Higashihara, T. Nakahara, Optimization of docosahexaenoic acid production by *Schizochytrium limacinum* SR21, Appl. Microbiol. Biotechnol. 49 (1998) 72–76.

85. Y. Liang, N. Sarkany, Y. Cui, J. Yesuf, J. Trushenski, J. W. Blackburn, Use of sweet sorghum juice for lipid production by *Schizochytrium limacinum* SR21, Bioresource Technol. 101 (2010) 3623–3627.

86. Z. Chi, D. Pyle, Z. Wen, C. Frear, S. Chen, A laboratory study of producing docosahex-aenoic acid from biodiesel-waste glycerol by microalgal fermentation, Process Biochem. 42 (2007) 1537–1545.

87. D. J. Pyle, R. A. Garcia, Z. Wen, Producing docosahexaenoic acid (DHA)-rich algae from biodiesel-derived crude gycerol: effects of impurities on DHA production and algal biomass composition, J. Agric. Food Chem. 56 (2008) 3933–3939.

88. S. M. Innis, Essential fatty acids in growth and development, Prog. Lipid Res. 30 (1991) 39–103.

89. B. Koletzko, E. Schmidt, H. J. Bremer, M. Haug, G. Harzer, Effects of dietary long-chain polyunsaturated fatty acids on the essential fatty acid status of premature infants, Eur. J. Pediatrics 148 (1989) 669–675.

90. M. Crawford, K. Costeloe, K. Ghebremeskel, A. Phylactos, L. Skirvin, F. Stacey, Are deficits of arachidonic and docosahexaenoic acids responsible for the neural and vascular complications of preterm babies? Am. J. Clin. Nutrition 66 (1997) 1032S–1041S.

91. D. J. Kyle, Arachidonic acid and methods for the production and use thereof, U.S. Patent 5,658,767 (1997).

92. C. Bigogno, I. Khozin-Goldberg, S. Boussiba, A. Vonshak, Z. Cohen, Lipid and fatty acid composition of the green oleaginous alga *Parietochloris incisa*, the richest plant source of arachidonic acid, Phytochemistry 60 (2002) 497–503.

93. C. Bigogno, I. Khozin-Goldberg, Z. Cohen, Accumulation of arachidonic acid-rich tri-acylglycerols in the microalga *Parietochloris incisa* (Trebuxiophyceae, Chlorophyta), Phytochemistry 60 (2002) 135–143.

94. T. J. Ahern, S. Katoh, E. Sada, Arachidonic acid production by the red alga *Porphyridium cruentum*, Biotechnol. Bioeng. 25 (1983) 1057–1070.

95. A. Vonshak, Z. Cohen, A. Richmond, The feasibility of mass cultivation of *Porphyridium*, Biomass 8 (1985) 13–25.

96. Y. K. Lee, H. M. Tan, Effect of temperature, light intensity and dilution rate on the cellular composition of red alga *Porphyridium cruentum* in light-limited chemostat cultures, World J. Microbiol. Biotechnol. 4 (1988) 231–237.

97. J. Fábregas, D. García, E. Morales, A. Domínguez, A. Otero, Renewal rate of semi-continuous cultures of the microalga *Porphyridium cruentum* modifies phycoerythrin, exopolysaccharide and fatty acid productivity, J. Fermentation Bioeng. 86 (1998) 477–481.

98. M. M. Rebolloso Fuentes, J. L. Garcia Sánchez, J. M. Fernández Sevilla, F. G. Acién Fernández, J. A. Sánchez Pérez, E. Molina Grima, Outdoor continuous culture of *Porphyridium cruentum* in a tubular photobioreactor: quantitative analysis of the daily cyclic variation of culture parameters, Prog. Ind. Microbiol. 35 (1999) 271–288.

99. I. Khozin-Goldberg, C. Bigogno, P. Shrestha, Z. Cohen, Nitrogen starvation induces the accumulation of arachidonic acid in the fresh water green alga *Parietochloris incisa* (Trebuxiophyceae), J. Phycology 38 (2002) 991–994.

100. J. Fernández-Sevilla, F. Acién Fernández, E. Molina Grima, Biotechnological production of lutein and its applications, Appl. Microbiol. Biotechnol. 86 (2010) 27–40.

101. D. Siefermann-Harms, The light-harvesting and protective functions of carotenoids in photosynthetic membranes, Physiologia Plantarum 69 (1987) 561–568.

102. K. K. Niyogi, O. Björkman, A. R. Grossman, The roles of specific xanthophylls in photoprotection, Proc. Nat. Acad. Sci. U.S.A. 94 (1997) 14162–14167.

103. J. Del Campo, M. García-González, M. Guerrero, Outdoor cultivation of microalgae for carotenoid production: current state and perspectives, Appl. Microbiol. Biotechnol. 74 (2007) 1163–1174.

104. R. Edge, D. J. McGarvey, T. G. Truscott, The carotenoids as anti-oxidants—a review, J. Photochem. Photobiol. B: Biology 41 (1997) 189–200.

105. A. Bendich, J. Olson, Biological actions of carotenoids, FASEB J. 3 (1989) 1927–1932.

106. A. Ben-Amotz, New mode of *Dunaliella* biotechnology: two-phase growth for β-carotene production, J. Appl. Phycology 7 (1995) 65–68.

107. M. García-González, J. Moreno, J. C. Manzano, F. J. Florencio, M. G. Guerrero, Production of *Dunaliella salina* biomass rich in 9-cis-β-carotene and lutein in a closed tubular photobioreactor, J. Biotechnol. 115 (2005) 81–90.

108. M. Mojaat, J. Pruvost, A. Foucault, J. Legrand, Effect of organic carbon sources and Fe^{2+} ions on growth and β-carotene accumulation by *Dunaliella salina*, Biochem. Eng. J. 39 (2008) 177–184.

109. M. A. Hejazi, E. Holwerda, R. H. Wijffels, Milking microalga *Dunaliella salina* for β-carotene production in two-phase bioreactors, Biotechnol. Bioeng. 85 (2004) 475–481.

110. P. Bubrick, Production of astaxanthin from *Haematococcus*, Bioresource Technol. 38 (1991) 237–239.

111. J. A. Del Campo, H. Rodríguez, J. Moreno, M. Á. Vargas, J. Rivas, M. G. Guerrero, Accumulation of astaxanthin and lutein in *Chlorella zofingiensis* (Chlorophyta), Appl. Microbiol. Biotechnol. 64 (2004) 848–854.

112. N. J. Lang, Electron microscopic studies of extraplastidic astaxanthin in *Haematococcus*, J. Phycology 4 (1968) 12–19.

113. M. Harker, A. J. Tsavalos, A. J. Young, Autotrophic growth and carotenoid production of *Haematococcus pluvialis* in a 30 liter air-lift photobioreactor, J. Fermentation Bioeng. 82 (1996) 113–118.

114. P. Z. Margalith, Production of ketocarotenoids by microalgae, Appl. Microbiol. Biotechnol. 51 (1999) 431–438.

115. E. D. Río, F. G. Acién, M. C. García-Malea, J. Rivas, E. Molina-Grima, M. G. Guerrero, Efficient one-step production of astaxanthin by the microalga *Haematococcus pluvialis* in continuous culture, Biotechnol. Bioeng. 91 (2005) 808–815.

116. M. Kobayashi, T. Kakizono, K. Yamaguchi, N. Nishio, S. Nagai, Growth and astaxanthin formation of *Haematococcus pluvialis* in heterotrophic and mixotrophic conditions, J. Fermentation Bioeng. 74 (1992) 17–20.

117. X. W. Zhang, X. D. Gong, F. Chen, Kinetic models for astaxanthin production by high cell density mixotrophic culture of the microalga *Haematococcus pluvialis*, J. Ind. Microbiol. Biotechnol. 23 (1999) 691–696.

118. L. Fan, A. Vonshak, S. Boussiba, Effect of temperature and irradiance on growth of *Haematococcus pluvialis* (Chlorophyceae), J. Phycology 30 (1994) 829–833.

119. M. Harker, A. J. Tsavalos, A. J. Young, Factors responsible for astaxanthin formation in the Chlorophyte *Haematococcus pluvialis*, Bioresource Technol. 55 (1996) 207–214.

120. M. Harker, A. Tsavalos, A. Young, Use of response surface methodology to optimise carotenogenesis in the microalga *Haematococcus pluvialis*, J. Appl. Phycology 7 (1995) 399–406.

121. Y.-K. Lee, C.-W. Soh, Accumulation of astaxanthin in *Haematococcus lacustris* (Chlorophyta), J. Phycology 27 (1991) 575–577.

122. N. Hata, J. C. Ogbonna, Y. Hasegawa, H. Taroda, H. Tanaka, Production of astaxanthin by *Haematococcus pluvialis* in a sequential heterotrophic–photoautotrophic culture, J. Appl. Phycology 13 (2001) 395–402.

123. M. Kobayashi, Y. Kurimura, Y. Tsuji, Light-independent, astaxanthin production by the green microalga *Haematococcus pluvialis* under salt stress, Biotechnol. Lett. 19 (1997) 507–509.

124. R. Sarada, U. Tripathi, G. A. Ravishankar, Influence of stress on astaxanthin production in *Haematococcus pluvialis* grown under different culture conditions, Process Biochem. 37 (2002) 623–627.

125. F. Granado, B. Olmedilla, I. Blanco, Nutritional and clinical relevance of lutein in human health, Br. J. Nutrition 90 (2003) 487–502.

126. F. Khachik, Process for extraction and purification of lutein, zeaxanthin and rare carotenoids from marigold flowers and plants, U.S. Patent 6,262,284.

127. J. M. Fernández-Sevilla, F. G. Acien Fernandez, J. Perez-Parra, J. J. Magán Cañadas, F. Granado-Lorencio, B. Olmedilla, Large-scale production of high-content lutein extracts from *S. almeriensis*. In: *Proceedings of the 11th International Conference on Applied Phycology*, Galway, Ireland, 2008.

128. J. A. Del Campo, J. Moreno, H. Rodrıguez, M. Angeles Vargas, J. Rivas, M. G. Guerrero, Carotenoid content of chlorophycean microalgae: factors determining lutein accumulation in *Muriellopsis* sp. (Chlorophyta), J. Biotechnol. 76 (2000) 51–59.

129. X. M. Shi, Y. Jiang, F. Chen, High-yield production of lutein by the green microalga *Chlorella protothecoides* in heterotrophic fed-batch culture, Biotechnol. Prog. 18 (2002) 723–727.

130. Z. Wu, S. Wu, X. Shi, Supercritical fluid extraction and determination of lutein in heterotrophically cultivates *Chlorella pyrenoidosa*, J. Food Process Eng. 30 (2007) 174–185.

131. A. Blanco, J. Moreno, J. A. Del Campo, J. Rivas, M. Guerrero, Outdoor cultivation of lutein-rich cells of *Muriellopsis* sp. in open ponds, Appl. Microbiol. Biotechnol. 73 (2007) 1259–1266.

132. J. A. Del Campo, H. Rodrıguez, J. Moreno, M. A. Vargas, J. Rivas, M. G. Guerrero, Lutein production by *Muriellopsis* sp. in an outdoor tubular photobioreactor, J. Biotechnol. 85 (2001) 289–295.

133. J. F. Sánchez, J. M. Fernández-Sevilla, F. G. Acién, M. C. Cerón, J. Pérez-Parra, E. Molina-Grima, Biomass and lutein productivity of *Scenedesmus almeriensis*: influence of irradiance, dilution rate and temperature, Appl. Microbiol. Biotechnol. 79 (2008) 719–729.

134. X. M. Shi, F. Chen, J. P. Yuan, H. Chen, Heterotrophic production of lutein by selected *Chlorella* strains, J. Appl. Phycology 9 (1997) 445–450.

135. Z. A. Popper, M. G. Tuohy, Beyond the green: understanding the evolutionary puzzle of plant and algal cell walls, Plant Physiol. 153 (2010) 373–383.

136. E. L. McCandless, Sulfated polysaccharides in red and brown algae, Annu. Rev. Plant Physiol. 30 (1979) 41–53.

137. R. J. P. Cannell, Algal biotechnology, Appl. Biochem. Biotechnol. 26 (1990) 85–105.

138. S. Arad, O. Levy-Ontman, Red microalgal cell wall polysaccharides: biotechnological aspects, Curr. Opin. Biotechnol. 21 (2010) 358–364.

139. M. Matsui, N. Muizzuddin, S. Arad, K. Marenus, Sulfated polysaccharides from red microalgae have antiinflammatory properties in vitro and in vivo, Appl. Biochem. Biotechnol. 104 (2003) 13–22.

140. T. Tannin-Spitz, M. Bergman, D. van-Moppes, S. Grossman, S. Arad, Antioxidant activity of the polysaccharide of the red microalga *Porphyridium* sp. J. Appl. Phycology. 17 (2005) 215–222.

141. B. Chen, W. You, J. Huang, Y. Yu, W. Chen, Isolation and antioxidant property of the extracellular polysaccharide from *Rhodella reticulata*, World J. Microbiol. Biotechnol. 26 (2010) 833–840.

142. S. Arad, Polysaccharides of red microlgae. In: Z. Cohen (Ed.), *Chemicals from Microalgae*, Taylor & Francis Ltd. Wallington, UK, 1999, pp. 282–291.

143. S. Arad, A. Richmond, Industrial production of microalgal cell-mass secondary products-species of high potential, *Porphyridium* sp. In: A. Richmond (Ed.), *Handbook of Microalgal Culture: Biotechnology and Applied Phycology*, Blackwell Science Ltd, Oxford, 2004, pp. 289–298.

144. S. Arad, M. Adda, E. Cohen, The potential of production of sulfated polysaccharides from *Porphyridium*, Plant and Soil 89 (1985) 117–127.

145. S. Arad, O. Friedman, A. Rotem, Effect of nitrogen on polysaccharide production in a *Porphyridium* sp., Appl. Environ. Microbiol. 54 (1988) 2411–2414.

146. O. Friedman, Z. Dubinsky, S. Arad, Effect of light intensity on growth and polysaccharide production in red and blue-green *rhodophyta unicells*, Bioresource Technol. 38 (1991) 105–110.

147. T. You, S. M. Barnett, Effect of light quality on production of extracellular polysaccharides and growth rate of *Porphyridium cruentum*, Biochem. Eng. J. 19 (2004) 251–258.

148. M. Adda, J. C. Merchuk, S. Arad, Effect of nitrate on growth and production of cell-wall polysaccharide by the unicellular red alga *Porphyridium*, Biomass 10 (1986) 131–140.

149. S. Y. Li, J. P. Lellouche, Y. Shabtai, S. Arad, Fixed carbon partitioning in the red microalgae *porphyridium* sp. (Rhodophyta), J. Phycology 37 (2001) 289–297.

150. M. M. Rebolloso Fuentes, G. G. Acién Fernández, J. A. Sánchez Pérez, J. L. Guil Guerrero, Biomass nutrient profiles of the microalga *Porphyridium cruentum*, Food Chem. 70 (2000) 345–353.

151. E. W. Becker, Micro-algae as a source of protein, Biotechnol. Adv. 25 (2007) 207–210.

152. S. Sekar, M. Chandramohan, Phycobiliproteins as a commodity: trends in applied research, patents and commercialization, J. Appl. Phycology 20 (2008) 113–136.

153. A. Reis, A. Mendes, H. Lobo-Fernandes, J. A. Empis, J. M. Novais, Production, extraction and purification of phycobiliproteins from *Nostoc* sp., Bioresource Technol. 66 (1998) 181–187.

154. A. N. Glazer, Phycobiliproteins—a family of valuable, widely used fluorophores, J. Appl. Phycology 6 (1994) 105–112.

155. C. Romay, R. Gonzalez, N. Ledon, D. Remirez, V. Rimbau, C-Phycocyanin: a biliprotein with antioxidant, anti-inflammatory and neuroprotective effects, Curr. Protein Peptide Sci. 4 (2003) 207–216.

156. C. Dupré, J. C. Guary, D. Grizeau, Effect of photon fluence rate, nitrogen limitation and nitrogen recovery on the level of phycoerythrin in the unicellular alga, *Rhodosorus marinus* (Rhodophyceae), Physiologia Plantarum 92 (1994) 521–527.

157. J. Moreno, H. Rodríguez, M. Vargas, J. Rivas, M. G. Guerrero, Nitrogen-fixing cyanobacteria as source of phycobiliprotein pigments. Composition and growth performance of ten filamentous heterocystous strains, J. Appl. Phycology 7 (1995) 17–23.

158. J. H. Eley, Effect of carbon dioxide concentration on pigmentation in the blue-green alga *Anacystis nidulans*, Plant Cell Physiol. 12 (1971) 311–316.

159. H. J. Silva, T. I. Cortiñas, R. J. Ertola, Effect of nutritional factors on the culture of *Nostoc* sp. as a source of phycobiliproteins, Appl. Microbiol. Biotechnol. 31 (1989) 293–297.

160. R. Bermejo Román, J. M. Alvárez-Pez, F. G. Acién Fernández, E. Molina Grima, Recovery of pure B-phycoerythrin from the microalga *Porphyridium cruentum*, J. Biotechnol. 93 (2002) 73–85.

161. S. Kathiresan, R. Sarada, S. Bhattacharya, G. A. Ravishankar, Culture media optimization for growth and phycoerythrin production from *Porphyridium purpureum*, Biotechnol. Bioeng. 96 (2007) 456–463.

162. N. Tandeau de Marsac, Occurrence and nature of chromatic adaptation in cyanobacteria, J. Bacteriol. 130 (1977) 82–91.

163. W. Jahn, J. Steinbiss, K. Zetsche, Light intensity adaptation of the phycobiliprotein content of the red alga *Porphyridium*, Planta 161 (1984) 536–539.

164. F. J. Marquez, N. Nishio, S. Nagai, K. Sasaki, Enhancement of biomass and pigment production during growth of *Spirulina platensis* in mixotrophic culture, J. Chem. Technol. Biotechnol. 62 (1995) 159–164.

165. F. Chen, Y. Zhang, S. Guo, Growth and phycocyanin formation of *Spirulina platensis* in photoheterotrophic culture, Biotechnol. Lett. 18 (1996) 603–608.

166. M. M. Allen, A. J. Smith, Nitrogen chlorosis in blue-green algae, Arch. Microbiol. 69 (1969) 114–120.

167. S. Boussiba, A. E. Richmond, C-phycocyanin as a storage protein in the blue-green alga *Spirulina platensis*, Arch. Microbiol. 125 (1980) 143–147.

168. G. Andrich, U. Nesti, F. Venturi, A. Zinnai, R. Fiorentini, Supercritical fluid extraction of bioactive lipids from the microalga *Nannochloropsis* sp., Eur. J. Lipid Sci. Technol. 107 (2005) 381–386.

169. F. Chen, Y. Zhang, High cell density mixotrophic culture of *Spirulina platensis* on glucose for phycocyanin production using a fed-batch system, Enzyme Microb. Technol. 20 (1997) 221–224.

170. A. A. Tcheruov, K. M. Minkova, D. I. Georgiev, N. B. Houbavenska, Method for B-phycoerythrin purification from *Porphyridium cruentum*, Biotechnol. Techniques 7 (1993) 853–858.

171. R. Bermejo, F. Gabriel Acién, M. José Ibáñez, J. M. Fernández, E. Molina, J. M. Alvarez-Pez, Preparative purification of B-phycoerythrin from the microalga *Porphyridium cruentum* by expanded-bed adsorption chromatography, J. Chromatography B 790 (2003) 317–325.

172. E. Molina Grima, E. H. Belarbi, F. G. Acién Fernández, A. Robles Medina, Y. Chisti, Recovery of microalgal biomass and metabolites: process options and economics, Biotechnol. Adv. 20 (2003) 491–515.

173. L. Brennan, P. Owende, Biofuels from microalgae—a review of technologies for production, processing, and extractions of biofuels and co-products, Renewable and Sustainable Energy Rev. 14 (2010) 557–577.

174. L. F. Razon, R. R. Tan, Net energy analysis of the production of biodiesel and biogas from the microalgae: *Haematococcus pluvialis* and *Nannochloropsis*, Appl. Energy 88 (2011) 3507–3514.

175. L. Lardon, A. Hélias, B. Sialve, J.-P. Steyer, O. Bernard, Life-cycle assessment of biodiesel production from microalgae, Environ. Sci. Technol. 43 (2009) 6475–6481.

176. L. B. Brentner, M. J. Eckelman, J. B. Zimmerman, Combinatorial life cycle assessment to inform process design of industrial production of algal biodiesel, Environ. Sci. Technol. 45 (2011) 7060–7067.

177. P. J. L. B. Williams, L. M. L. Laurens, Microalgae as biodiesel & biomass feedstocks: review & analysis of the biochemistry, energetics & economics, Energy Environ. Sci. 3 (2010) 554–590.

178. P. M. Schenk, S. R. Thomas-Hall, E. Stephens, U. C. Marx, J. H. Mussgnug, C. Posten, O. Kruse, B. Hankamer, Second generation biofuels: high-efficiency microalgae for biodiesel production, BioEnergy Res. 1 (2008) 20–43.

179. E. Stephens, I. L. Ross, J. H. Mussgnug, L. D. Wagner, M. A. Borowitzka, C. Posten, O. Kruse, B. Hankamer, Future prospects of microalgal biofuel production systems, Trends Plant Sci. 15 (2010) 554–564.

180. J. Frenz, C. Largeau, E. Casadevall, Hydrocarbon recovery by extraction with a biocompatible solvent from free and immobilized cultures of *Botryococcus braunii*, Enzyme Microb. Technol. 11 (1989) 717–724.

181. J. K. Pittman, A. P. Dean, O. Osundeko, The potential of sustainable algal biofuel production using wastewater resources, Bioresource Technol. 102 (2011) 17–25.

182. T. van Harmelen, H. Oonk, Microalgae biofixation processes: applications and potential contributions to green house gas mitigation options. In: *TNO Built Environment and Geosciences*, 2006.

183. J. B. van Beilen, Why microalgal biofuels won't save the internal combustion machine, Biofuels, Bioproducts and Biorefining, 4 (2010) 41–52.

184. B. Sialve, N. Bernet, O. Bernard, Anaerobic digestion of microalgae as a necessary step to make microalgal biodiesel sustainable, Biotechnol. Adv. 27 (2009) 409–416.

185. M. Ras, L. Lardon, S. Bruno, N. Bernet, J. P. Steyer, Experimental study on a coupled process of production and anaerobic digestion of *Chlorella vulgaris*, Bioresource Technol 102 (2011) 200–206.

186. P. M. Foley, E. S. Beach, J. B. Zimmerman, Algae as a source of renewable chemicals: opportunities and challenges, Green Chem. 13 (2011) 1399–1405.

187. B. Yeh, Commercializing algae—challenges and opportunities. Inform 22 (2011) 473–536.

188. P. G. Roessler, Environmental control of glycerolipid metabolism in microalgae: commercial implications and future research directions, J. Phycology 26 (1990) 393–399.

189. L. Gouveia, V. Veloso, A. Reis, H. Fernandes, J. Novais, J. Empis, Evolution of pigment composition in *Chlorella vulgaris*, Bioresource Technol. 57 (1996) 157–163.

190. P. F. Ip, K. H. Wong, F. Chen, Enhanced production of astaxanthin by the green microalga *Chlorella zofingiensis* in mixotrophic culture, Process Biochem. 39 (2004) 1761–1766.

191. Z. Y. Liu, G. C. Wang, B. C. Zhou, Effect of iron on growth and lipid accumulation in *Chlorella vulgaris*, Bioresource Technol. 99 (2008) 4717–4722.

192. Y. Liang, N. Sarkany, Y. Cui, Biomass and lipid productivities of *Chlorella vulgaris* under autotrophic, heterotrophic and mixotrophic growth conditions, Biotechnol. Lett. 31 (2009) 1043–1049.

193. P. Feng, Z. Deng, Z. Hu, L. Fan, Lipid accumulation and growth of *Chlorella zofingiensis* in flat plate photobioreactors outdoors, Bioresource Technol. 102 (2011) 10577–10584.

194. M. Heasman, J. Diemar, W. O'Connor, T. Sushames, L. Foulkes, Development of extended shelf-life microalgae concentrate diets harvested by centrifugation for bivalve molluscs—a summary, Aquaculture Res. 31 (2000) 637–659.

195. M. B. Johnson, Z. Wen, Production of biodiesel fuel from the microalga *Schizochytrium limacinum* by direct transesterification of algal biomass, Energy & Fuels 23 (2009) 5179–5183.

196. R. L. Mendes, H. L. Fernandes, J. P. Coelho, E. C. Reis, J. M. S. Cabral, J. M. Novais, A. F. Palavra, Supercritical CO_2 extraction of carotenoids and other lipids from *Chlorella vulgaris*, Food Chem. 53 (1995) 99–103.

197. R. L. Mendes, B. P. Nobre, M. T. Cardoso, A. P. Pereira, A. F. Palavra, Supercritical carbon dioxide extraction of compounds with pharmaceutical importance from microalgae, Inorg. Chim. Acta 356 (2003) 328–334.

198. A. P. R. F. Canela, P. T. V. Rosa, M. O. M. Marques, M. A. A. Meireles, Supercritical fluid extraction of fatty acids and carotenoids from the microalgae *Spirulina maxima*, Ind. Eng. Chem. Res. 41 (2002) 3012–3018.

199. L. Soh, J. B. Zimmerman, Biodiesel production: the potential of algal lipids extracted with supercritical carbon dioxide, Green Chem. 13 (2011) 1422–1429.

200. C. C. Fu, T. C. Hung, J. Y. Chen, C. H. Su, W. T. Wu, Hydrolysis of microalgae cell walls for production of reducing sugar and lipid extraction, Bioresource Technol. 101 (2010) 8750–8754.

201. R. Harun, M. K. Danquah, G. M. Forde, Microalgal biomass as a fermentation feedstock for bioethanol production, J. Chem. Technol. Biotechnology, 85 (2010) 199–203.

202. P. C. K. Cheung, A. Y. H. Leung, P. O. Ang, Comparison of supercritical carbon dioxide and Soxhlet extraction of lipids from a brown seaweed, *Sargassum hemiphyllum* (Turn.) C. Ag., J. Agric. Food Chem. 46 (1998) 4228–4232.

203. M. García-González, J. Moreno, J. P. Cañavate, V. Anguis, A. Prieto, C. Manzano, F. J. Florencio, M. G. Guerrero, Conditions for open-air outdoor culture of *Dunaliella salina* in southern Spain, J. Appl. Phycology 15 (2003) 177–184.

CHAPTER 16

Fuels and Chemicals from Lignocellulosic Biomass

IAN M. O'HARA, ZHANYING ZHANG, PHILIP A. HOBSON, MARK D. HARRISON, SAGADEVAN G. MUNDREE, and WILLIAM O. S. DOHERTY

16.1 INTRODUCTION

Lignocellulosic biomass from plants is the most abundant renewable resource on earth. The amount of biomass from the growth of plants is estimated to be in excess of 117 billion t/y with 62% of this resource in tropical and other forests [1]. Biomass currently provides the highest proportion of renewable energy, primarily in cooking and heating applications [2].

Carbon dioxide is fixed by plants during photosynthesis and is converted into biomass. Depending on the type of plant, the carbon may be released through natural degradation processes on short carbon cycles such as with annual plants and crops or in the case of long-lived woody plants on longer cycles which may last hundreds or thousands of years.

The utilization of biomass, particularly waste residues from agriculture and forestry industries, green wastes, and energy crops, for the production of fuels and chemicals provides significant opportunities to reduce global carbon dioxide emissions and reduce reliance on high emission and fossil-based resources.

Fossil-based crude oil is the major feedstock for the production of liquid fuels such as petrol, diesel, jet, and marine fuels. Crude oil is also the major feedstock for the chemical industry where, through catalytic processes, platform chemicals such as ethylene and propylene are converted into the domestic and industrial chemicals, such as plastics and resins, upon which modern society depends. Crude oil, however, is a nonrenewable feedstock and prices are increasing with increasing global demand and declining production from existing oil fields [3].

Lignocellulosic biomass can be converted into fuels and chemicals through biochemical, or thermochemical processes. Biochemical processes typically focus on

Natural and Artificial Photosynthesis: Solar Power as an Energy Source, First Edition. Edited by Reza Razeghifard.
© 2013 John Wiley & Sons, Inc. Published 2013 by John Wiley & Sons, Inc.

breaking down the complex carbohydrates in biomass into simple sugars via saccharification. Modern industrial biotechnology then converts the sugars into products using enzymes, often with much lower energy consumption in production and at higher product specificity than can be achieved through chemical processing technologies.

Thermochemical processes such as gasification, pyrolysis, and liquefaction utilize higher temperature catalysis to convert the biomass to a higher value final or intermediate product. Intermediate products can be gases which can be converted into fuels and chemicals via Fischer–Tropsch (FT) processes or liquid bio-oils that can be upgraded into chemicals.

The economics of fuel and chemical production from biomass is primarily influenced by the cost of the feedstock and the capital cost of the industrial plant required for conversion. The magnitude of the available biomass resource and the costs of harvesting, collection, and transport to a processing facility are key factors in determining the economic viability of these technologies and can differ markedly between feedstocks and regions.

Chemicals and fuels can also be derived from the simple sugars in starch and sugar producing crops such as corn, sugarcane, cassava, and grains (wheat, sorghum, rice) and from edible oil-producing plants such as soy and canola. However, the use of these edible components of crops raises concerns with the use of food crops for fuels production. Economic factors including the high price of many starch and sugar-based feedstocks also limit the viability of these crops as fuel and chemical feedstocks. The utilization of the waste lignocellulosic residues, however, does not displace food production and generally offers a lower price feedstock for use in industrial processes. Therefore the use of these lignocellulosic feedstocks for fuel and chemical production is a key focus of global research and of this chapter.

16.2   THE NATURE OF LIGNOCELLULOSIC BIOMASS

Lignocellulose is the primary building block of plant cell walls and hence plant biomass. Plant biomass is mainly composed of cellulose, hemicellulose, and lignin, along with smaller amounts of pectin, protein, extractives (soluble nonstructural materials such as sugars, nitrogenous compounds, chlorophyll, and waxes), and ash [4]. The composition of lignocellulosic materials is not uniform and varies between plant species. Within a single plant, the ratios between these constituents vary with the tissue, age, stage of growth, and other conditions [4]. Table 16.1 shows typical values for cellulose, hemicellulose, and lignin composition for several lignocellulosic biomasses. Physically, woody biomasses are typically larger, structurally stronger, and denser than agricultural biomasses. Chemically, woody biomasses typically have higher lignin contents than nonwoody biomasses [5].

16.2.1   Cellulose

In plant biomass, cellulose is ordered into fibrils, which are surrounded, and chemically bound, by lignin and hemicellulose [7]. Most cellulose components exist in

TABLE 16.1   Cellulose, Hemicellulose, and Lignin Content of Lignocellulosic Materials [6]

Type of Biomass	Composition (% w/w)		
	Cellulose	Hemicellulose	Lignin
Sugarcane bagasse	34–47	24–29	16–28
Corn stover	34–39	22–29	11–21
Corn cob	39–45	31–37	12–17
Wheat straw	38–45	22–32	14–21
Rice straw	35–41	16–24	14–24
Rapeseed straw	38	18	28
Switchgrass	32–45	12–31	12–27
Spruce	58–60	5–7	28–30
Douglasfir	60–64	4–8	28–31
Poplar	42–52	12–21	28–29
Oak	49	20	24

crystalline forms, although there is also a small portion of amorphous cellulose present in lignocellulosic fiber. Native cellulose structure and morphology can be considered at various levels including the molecular scale, the supramolecular level, and the fiber scale [8].

At the molecular level, cellulose is an unbranched homopolysaccharide polymer consisting of D-anhydroglucose repeating units joined by 1,4-β-D-glycosidic linkages at the C1 and C4 positions with hydroxy groups at C-2, C-3, and C-6 [8, 9]. The hydroxy group at the C-1 end of the glucose chain has reducing properties and the hydroxy group at C-6 is nonreducing [8]. The β-linkage produces linear glycan chains that can be stacked, forming cellulose fibers. The structure is stabilized by hydrogen bonds formed between chains through the hydroxyl groups at C-2, C-3, and C-6 positions residing on the outside of the chains and van der Waals forces [4].

At the supramolecular level, the cellulose is aggregated into highly ordered structures through an extensive network of hydrogen bonds. The cellulose supramolecules contain regions of both high crystalline order and amorphous regions of lower crystallinity [8, 10]. The morphology of cellulose is characterized by a well-ordered aggregation of microfibrils into macrofibrils. The macrofibrils contain pores, capillaries, voids, and interstices that increase the surface area of the cellulose fibrils [8].

The structure of cellulose, along with the intermolecular hydrogen bonds, gives cellulose high tensile strength, makes it insoluble in most solvents, and is partly responsible for the resistance of cellulose to microbial degradation [11]. The supposedly hydrophobic surface of cellulose results in formation of a dense layer of water that may hinder diffusion of enzymes and degradation products near the cellulose surface [11].

16.2.2 Hemicellulose

Hemicellulose is not a form of cellulose and the name is something of a misnomer [10]. Hemicellulose serves as a connection between the lignin and the cellulose fibers and gives the cellulose–hemicellulose–lignin network rigidity [13]. Hemicellulose is heterogeneous, consisting of C5 sugars (xylose, arabinose), C6 sugars (mannose, glucose and galactose), and uronic acids [10, 14, 15]. The proportion of each component in hemicellulose varies among different types of biomass. For example, softwood hemicellulose has a higher proportion of mannose and glucose than hemicellulose from hardwood and broad-acre monocot crops, which usually contains a higher proportion of xylose [16, 17].

Hemicellulose is typically a branched heteropolysaccharide with much lower molar mass than cellulose (typically DP 80–300) [10, 18]. Hemicellulose exhibits a considerable degree of chain branching containing pendant side groups giving rise to its noncrystalline nature [10]. In cell walls, hemicelluloses hydrogen bond to cellulose microfibrils. They act to coat the microfibrils, restricting enzyme access to the cellulose, and also span the microfibrils to link them together into macrofibrils [13]. The properties of the hemicellulose components differ. For example, xylan from hemicellulose can be extracted readily in an acidic or alkaline environment, while glucomannan can hardly be extracted in an acidic environment and needs a stronger alkaline environment than xylan for extraction [16].

16.2.3 Lignin

Lignin is a large, complex molecule containing crosslinked polymers of phenolic monomers with both aliphatic and aromatic constituents [10] that imparts structural support, impermeability, and resistance to microbial attack to the plant cell wall [4]. Lignin is totally amorphous and hydrophobic [16]. Hydroxyl, methoxyl, and carbonyl groups have been identified. Lignin has been found to contain five hydroxyl and five methoxyl groups per building unit [10]. It is believed that the structural units of lignin molecule are derivatives of 4-hydroxy-3-methoxy phenylpropane. There are three major phenylpropane units in lignin, as shown in Fig. 16.1, p-hydroxyphenyl (H), guaiacyl (G), and syringyl (S), which differ in the methoxyl substitution of the aromatic ring. The main difficulty in lignin chemistry is that no method has been established by which it is possible to isolate lignin in its native state from the fiber. Therefore the structure of lignin that has been isolated from biomass is dependent on both the plant and the process used for delignification [19, 20].

In lignocellulosic biomass, lignin is covalently linked to both cellulose and hemicellulose through ester and ether linkages, and the structure is further stabilized by combinations of hydrogen bonds, ionic interactions, and van der Waals interactions, which strongly affect enzymatic digestibility of the carbohydrates [13]. Lignin normally starts to dissolve into water around 180 °C under neutral conditions [21]. The solubility of lignin in acidic, neutral, or alkaline environments depends, however, on the proportions of monomeric units in the lignin [22].

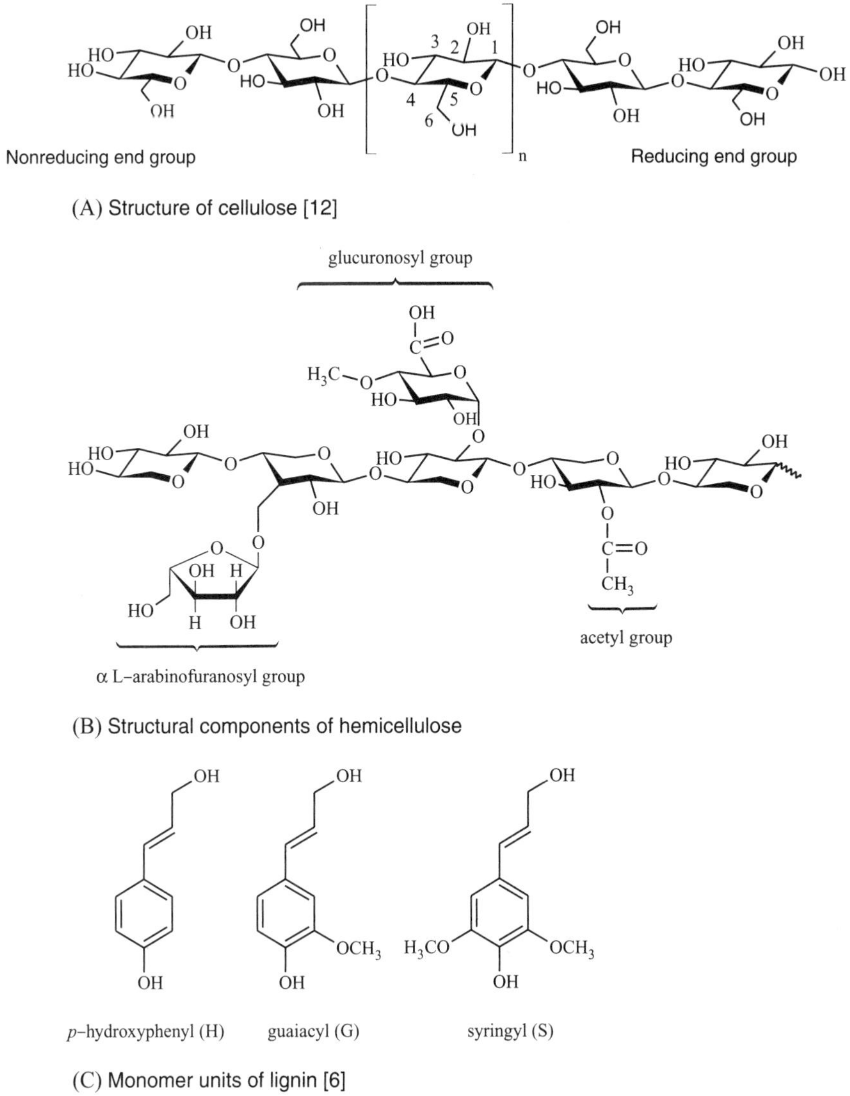

(A) Structure of cellulose [12]

(B) Structural components of hemicellulose

p–hydroxyphenyl (H)   guaiacyl (G)   syringyl (S)

(C) Monomer units of lignin [6]

FIGURE 16.1   Chemical structures of (A) cellulose, (B) hemicellulose, and (C) lignin.

16.3 FEEDSTOCKS FOR BIOMASS PROCESSING

Renewable biomass can be obtained from a wide range of primary sources. Bioprocesses can potentially convert these materials into a range of products such as fuels
and chemicals that are ecologically and increasingly economically competitive with
their nonrenewable counterparts [23]. To facilitate this transition, there is a need to
prioritize the processing of biomass by identifying local comparative advantages and

assist with the vertical integration of research, development, and commercialization of technologies. The production of fuels and chemicals from renewable biomass feedstocks will require large quantities of biomass to reach economies of scale. To keep transport costs down, processing facilities must be located regionally. Thus, a successful renewable fuels and chemicals sector will be built on rural and regional foundations.

In general, prospective lignocellulosic materials for fuels and chemicals production can be divided into five main groups: (a) agricultural residues (sugarcane and energy cane bagasse, corn stover, wheat straw, rice straw, rice hulls, barley straw, sweet sorghum bagasse, olive stones, and palm fronds), (b) forest residues including hardwood (eucalyptus, poplar, etc.) and softwood (pine, spruce, etc.), (c) cellulose wastes (newsprint, waste office paper, recycled paper sludge), (d) herbaceous biomass (alfalfa hay, switchgrass, reed canary grass, coastal Napier grass), and (e) municipal solid wastes [23]. Some of these major potential feedstocks are discussed in the following sections.

16.3.1 Agricultural Residues

16.3.1.1 Sugarcane Sugarcane is perhaps placed most advantageously to become the leading biomass feedstock for second generation fuels and chemicals production. Sugarcane is supported by significant research and development programs across the globe. It has a natural advantage due to its large existing network of sugarcane producers, sugar factories, and other infrastructure, established sugar and ethanol markets, and established, economically-viable electricity cogeneration technologies. Sugarcane offers the fuels and chemicals sector three opportunities. The first opportunity is through first generation technologies where fuels and chemicals are made from the sucrose produced from crushing sugarcane, which is both a food and a feedstock for production of chemicals and fuels. The second opportunity is in the use of bagasse, the lignocellulosic material left behind in the sugar extraction process. The third opportunity is through the processing of sugarcane trash (i.e. tops and leaves). Green harvesting of sugarcane with trash recovery could create large quantities of biomass for the production of fuels and chemicals.

16.3.1.2 Energy Cane Energy cane and sugarcane are from the same genus, *Saccharum*, with the major difference between them being that energy cane is bred for high fiber content while sugarcane is bred for low fiber content but high sugar content. Energy cane has the capacity to reduce the average unit cost of fiber and provide additional benefits in the form of season length extension if grown on marginal land in low rainfall areas, currently not under sugarcane production. The germplasm resource that supports the breeding program is diverse and includes material sourced from several regions across the world that may be useful in breeding varieties suited to a range of climates, including temperate climates. Energy cane grows taller than regular sugarcane and there is evidence that energy cane may become a potentially useful bioenergy crop on unmanaged mineral soils in various regions. The success of energy cane as a source of biomass for fuels and chemicals will be predicated on the development of new varieties with higher biomass and longer-lived stands, and more efficient management practices, particularly fertilizer nutrient inputs, along with the lignocellulosic ethanol conversion process to lower per liter processing costs.

16.3.1.3 Sweet Sorghum   While the origins of sweet sorghum are in the tropics, it is also well adapted to temperate climates. There is keen interest in developing sweet sorghum as a fermentation feedstock in the United States, Europe, India, China, and Australia. Its advantages as an energy crop are its high water use efficiency, requirements for lower levels of fertilizer, herbicides, and pesticides, and its association with lower levels of soil erosion [24]. In addition, sweet sorghum has faster growth cycles (up to three crops a year) than most crops.

Sweet sorghum matures in approximately one-third of the time of sugarcane and, in fact, thrives under drier and warmer. It has some tolerance to salinity and good drought tolerance with best results in semiarid areas with 500–800 mm rainfall. It can yield up to 45 tonnes of biomass per hectare in 110 days with fermentable solids in the stalks from 2.5 to 5 tonnes per hectare. As with energy cane, sweet sorghum may provide an alternative cropping system for biomass feedstocks that would support the fuels and chemicals industries.

16.3.2 Forest Residues

Most of the world's plant biomass is in the form of woody forest materials. Forest species store energy for relatively long periods of time. There is considerable interest in the commercial opportunities and potential environmental benefits of producing fuels and chemicals from Eucalyptus biomass. Eucalyptus is a forest genus that meets most of the desired features for low-cost delivered biomass. Eucalyptus is indigenous to Australia, Indonesia, and Papua New Guinea and it is the most frequently planted fast-growing hardwood in the world. In addition, it is the main hardwood raw material supplied to the pulp and paper industry in Brazil, Portugal, South Africa, Uruguay, and other countries [25]. Shorter rotation lengths, development of more freeze-tolerant seedlings, higher stand tree density together with other silviculture practices are being developed to improve plantation productivity. These outcomes indicate that Eucalyptus is a promising biomass for bioenergy production. Trees may be grown in belts and harvested every three to four years in short rotation coppicing. Replanting is not required after harvest because the trees are allowed to regrow from the stump. Agricultural production can continue to be practiced on the land in between belts of trees.

16.4   PRODUCTION OF FERMENTABLE SUGARS FROM BIOMASS

The cellulose and hemicellulose in plant cell walls (biomass) represent an abundant, renewable source of fermentable sugars that have the potential to underpin global economic production of biofuels and other products from bacterial, fungal, and algal fermentation. However, the plant cell wall is a resilient, structurally heterogeneous barrier that requires either pretreatment or enzymatic hydrolysis to liberate the fermentable sugars. The following section focuses on the production of fermentable sugars from biomass pretreatment and enzymatic hydrolysis processes.

16.4.1 Pretreatment of Biomass

The effect of biomass pretreatment on subsequent enzymatic hydrolysis has been long recognized [26]. Pretreatment is required to improve the accessibility of hydrolytic enzymes to their substrates and this can be achieved through the removal of lignin, removal of hemicellulose, reduction in cellulose crystallinity, physical deconstruction of the fiber, or combinations thereof. An effective pretreatment must not only improve the accessibility of hydrolytic enzymes to their substrate but must also (1) avoid carbohydrate degradation or loss, (2) avoid formation of by-products that are inhibitory to enzymatic hydrolysis or fermentation, and (3) be cost-effective. Chemical, physical, and biological processes have been used successfully for pretreatment of biomass and some of the more commonly used processes are described.

The most commonly used process for chemical pretreatment of biomass is mild acid pretreatment. In this process, the hemicellulose chains are hydrolyzed resulting in hemicellulose depolymerization and solubilization [27]. Some amorphous cellulose is also solubilized. Carbohydrate degradation products including furfural, 5-hydroxymethylfurfural, levulinic acid, acetic acid, and formic acid are also formed by chemical pretreatment under acidic conditions. These degradation products are inhibitory to many fermentation organisms, and hydrolysate detoxification may be necessary to minimize their impact on downstream processes [28, 29].

Alkaline processes such as sodium hydroxide pretreatments act by solubilizing a portion of the hemicellulose from the biomass and breaking lignin–carbohydrate bonds resulting in a more open biomass structure [27]. Aqueous ammonia processes have been shown to result in lignin solubilization and swelling of the fiber, increasing enzyme accessibility [30].

Solvent-based pretreatments include organosolv processes using alcohols (e.g., methanol, ethanol, glycerol) or organic acids (e.g., formic, maleic, acetic). In such processes, lignin and/or hemicellulose is removed, thereby increasing pore volumes and enzyme accessibility [31]. Significant work is also being undertaken with green solvents (in particular, ionic liquids) that offer greater selectivity in biomass dissolution or pretreatment and more benign processing conditions [32], although these solvents are currently considered too expensive for commercial processes.

Physicochemical pretreatments of biomass include steam explosion, ammonia explosion, and CO_2 explosion pretreatments. Explosion-based pretreatments operate by heating the biomass under pressure to maintain the solvent in the liquid phase and then rapidly depressurizing the biomass to atmospheric pressure. Depending on the solvent used, hydrolysis of the hemicellulose and/or solubilization of the lignin occur during the high pressure reaction, which is followed by explosive depressurization, disrupting biomass structure. Fiber explosion processes can be very effective in combination with chemical pretreatments.

16.4.2 Enzymatic Hydrolysis of Cellulose

Cellulose is hydrolyzed by highly specific enzymes called cellulases [33]. Cellulolytic enzymes are produced naturally by a wide range of bacteria and fungi [34]. It has long been known that bacteria from the genus *Clostridium, Cellulomonas, Bacillus, Thermomonospora, Ruminococcus, Bacteroides, Erwinia, Acetovibrio, Microbispora,* and *Streptomyces* produce cellulases [35]. A recent analysis of ~1500 complete bacterial genomes has revealed that ~40% encoded at least one cellulase gene [36]. While most of the bacteria were soil and marine saprophytes, cellulase genes are present in nonsaprophytic bacteria, such as *Mycobacterium tuberculosis, Legionella pneumophila, Yersinia pestis,* and even *Escherichia coli,* suggesting that they are involved in cellulose synthesis or in intra-amebal persistence and so need to degrade amebal cellulose [36]. Many cellulolytic bacteria, particularly anaerobes such as *Clostridium thermocellum* and *Bacteroides cellulosolvans* produce cellulases with high specific activity and do not produce high enzyme titers [37]. Given that these bacteria grow relatively slowly, the majority of research for commercial cellulase production has focused on fungi [37].

Brown-, white-, and soft-rot fungi are used to degrade biomass [38] and engineered strains of these organisms can secrete proteins to titers greater than 100 g L^{-1} [39]. Brown-rot fungi mainly degrade cellulose, while white-rot and soft-rot fungi degrade both cellulose and lignin. White-rot fungi are the most effective basidiomycetes for degradation of biomass [38]. Fungi from the genus *Sclerotium, Phanerochaete, Trichoderma, Aspergillus, Shizophyllum,* and *Penicillium* have been reported to produce cellulases [37, 38, 40, 41]. Of these fungal genera, *Trichoderma* has been the most extensively studied for cellulase production [40] and *Trichoderma* cellulases are the best characterized. It should be noted that fungi and most bacteria use uncomplexed, secreted cellulases, while some bacteria aggregate their enzymes to large scaffolds in complexes called cellulosomes [42].

The enzymes required for efficient and complete degradation of cellulose to glucose are derived from three functional classes: (1) endoglucanases (endo-1,4-β-D-glucanases, EC 3.2.1.4), (2) exoglucanases or cellobiohydrolases (exo-1,4-β-D-glucanases), and (3) β-glucosidases (EC 3.2.1.21) [43]. Endoglucanases hydrolyze the internal $\beta(1\rightarrow4)$ bonds at random in the amorphous regions of cellulose. The free chain ends generated are the sites of action for the processive, unidirectional cellobiohydrolases. These enzymes act on either the nonreducing (EC 3.2.1.91) or reducing (EC 3.2.1.176) ends of cellulose polysaccharide chains, liberating cellobiose. Cellobiose is hydrolyzed by β-glucosidase producing glucose, which prevents feedback inhibition of cellobiohydrolases [44]. Most cellulases have a modular structure consisting of a catalytic domain and a cellulose-binding domain, connected by a flexible linker region [45].

The cellulase enzymes required for efficient degradation of cellulose biomass to glucose currently represent 20–40% of the cost "at-pump" of lignocellulosic ethanol [46]. Reducing the cost of these enzymes, through reduction of enzyme loading and increased enzyme activity, and reducing the cost of producing the enzymes are major strategic goals in making conversion of lignocellulosic biomass to fermentable sugars a cost-competitive process to other alternatives. Production of recombinant cellulases *in planta* provides an opportunity to substantially reduce the saccharification costs during the production of lignocellulosic ethanol [46] and we have previously demonstrated expression of multiple recombinant cellulase enzymes in transgenic sugarcane [47].

16.4.3   Enzymatic Hydrolysis of Hemicellulose

There are a range of enzymes responsible for the degradation of hemicellulose, a heteropolysaccharide. The degradation of the xylan component requires the coordinated activity of endo-1,4-β-xylanase, β-xylosidase, α-glucuronidase, α-L-arabinofuranosidase, and acetylxylan esterase. Glucomannan degradation also requires β-mannanase and β-mannosidase to cleave the polymer chain.

16.4.4   Enzymatic Hydrolysis of Pretreated Biomass by Industrial Cellulase Mixtures

Industrial cellulase mixtures are produced in fungal culture and are a complex mixture of secreted enzymes. For example, *Trichoderma reesei* secretes a mixture of at least eight endoglucanases, two cellobiohydrolases, seven β-glucosidases, four xylanases, β-xylosidase, two acetyl xylan esterases, two arabinofuranosidases, β-mannase, four α-galactosidases, β-galactosidase, swollenin, and two proteins of unknown function [48]. Originally isolated from native fungi, current commercial cellulase mixtures are produced by fungi that have been genetically modified to hypersecrete modified ratios of individual enzymes and/or to express proteins to enhance the performance of the cellulase mixture. For example, while the β-glucosidase content in fungal exudates is sufficient to meet the needs of the native organism as it grows on biomass, it is not suffcient to maintain cellobiose concentration below the level that inhibits the function of cellobiohydrolases in a commerical biomass hydrolysis reaction and so increased relative abundance of β-glucosidase was the earliest target for genetic modification of fungi producing commercial cellulase mixtures.

In order for the cellulases in commercial mixtures to efficiently hydrolyze cellulose in pretreated biomass, they must be able to access the cellulose chains. This is facilitated by hemicellulose and/or lignin removal during pretreatment, and by the presence of hemicellulase-degrading enzymes in the cellulase mixture. Assuming that pretreatment is effective, enzymatic hydrolysis of the crystalline cellulose residue is initially characterized by a process called *amorphogenesis* [49]. During amorphogenesis, crystalline cellulose swells or delaminates, resulting in reduction in the crystallinity and the generation of a larger accessible surface for cellulases to act upon. Proteins that have been proposed to participate in amorphogenesis include the carbohydrate-binding modules of cellulases and expansin-like proteins, such as swollenin [49].

TABLE 16.2   Operating Conditions in Liquefaction and Pyrolysis Processes

Process	Temperature (°C)	Pressure (bar)	Drying	Solvent	Product
Liquefaction	250–330	1–240	Unnecessary	Yes	Liquid
Pyrolysis	280–630	1–5	Necessary	No	Liquid and gas
Gasification	800–1000	1–20	Usually necessary	No	Solids and gas

Source: The data were taken from Goyal et al. [51] with permission of Elsevier, from Balat [52] with permission of John Wiley & Sons, and from Behrendt et al. [53] with permission of Taylor & Francis.

16.5   THERMOCHEMICAL CONVERSION OF BIOMASS TO FUELS AND CHEMICALS

Thermochemical processes are a means of upgrading biomass to higher value fuels or commodity chemicals using elevated reaction temperatures ($> 200\,°C$), usually in the absence of air or in a low oxygen environment. Thermochemical processes are potentially highly efficient as most organic components in biomass are amenable to conversion at these elevated temperatures. In thermal processing, lignin can be broken down or depolymerized to potentially form high value, aromatic chemicals [50]. Thermochemical processes generally require shorter reaction times (from seconds to minutes) compared with biochemical conversion processes (from days to weeks) with the same requirement for water removal from the end product which generally results in lower process energy inputs [50].

Based on reaction conditions and products, thermochemical processes can broadly be divided into four different conversion processes, these being combustion, gasification, pyrolysis, and direct liquefaction (hereafter called liquefaction) [51]. As combustion is used almost exclusively for heat generation, it will not be considered in the following review of thermochemical processes. The operating conditions for gasification, pyrolysis, and liquefaction processes are summarized in Table 16.2.

Depending on the conversion method, intermediate products of thermochemical processes can be upgraded further to fuels or chemicals (secondary products). Some examples of secondary products include hydrocarbon transport fuels, oxygenated transport fuels, and bulk chemicals including hydrogen and ammonia. Oxygenated fuel additives such as γ-valerolactone and furans (e.g., dimethylfuran and methyltetrahydrofuran) can readily be blended with petroleum products to create cleaner-burning fuels [54, 55].

16.5.1   Gasification

The main (intermediate) product targeted by gasification processes is syngas, a mixture of gases consisting primarily of CO, CO_2, H_2, and (depending on the gasification technology) N_2. Syngas can be combusted directly as a fuel (it typically has half the heating value of natural gas) or used in the FT process for the production of fuels.

Gasification and FT synthesis typically produce yields of between 140 L [56] and 250 L [57] of hydrocarbons (with a range of C5$^+$ alkanes similar to those found in diesel) or alternatively 330 L of ethanol per tonne of dry ash-free biomass [58].

Gasification and FT production of hydrocarbons and fuel alcohol from fossil fuel derived feedstocks are well-established technologies. The South African-based company SASOL currently runs large-scale, commercial operations utilizing coal as feedstock for gasification and FT synthesis to produce gasoline, diesel, waxes, hydrocarbon lubricants, natural gas, phenol, ammonia, and detergents [59]. Once the syngas from biomass gasification has been adjusted (usually by a steam supplied water gas shift reactor) to produce the H$_2$/CO ratios required for fuels methanol production, the FT process becomes identical to that required for fossil fuel-derived syngas. Research toward the production of FT fuels and chemicals from biomass has therefore focused on biomass gasification rather than the syngas reforming process.

Biomass gasification research has been reviewed extensively in the open literature (e.g., Bulushev and Ross [60] and Alauddin et al. [61]) with the bulk of the reported research being directed toward issues impacting on carbon conversion efficiency, cold gas efficiency, and minimizing the prevalence of biomass derived mineral (ash) contaminants in syngas. There is little evidence in the literature of major technical barriers to the implementation of gasification and FT production of biofuels. One of the major reasons for the lack of transfer of this technology to a commercially-viable process for biomass conversion is the dispersed nature of the feedstock, which makes it difficult to achieve the economies of scale required to offset the high associated capital costs.

Studies on novel supply chain options where biomass undergoes energy densification (by torrefaction) at local collection points and is transported to a large centralized FT plant have been undertaken as a proposed means of improving the economies of scale [62,63]. These studies have shown that FT fuel production costs can be reduced significantly by large-scale centralized production, although with associated minimum production costs of AU\$1.6/L (US\$6/gallon) of FT diesel [63], costs remain too high to compete directly with oil-derived hydrocarbon fuels.

16.5.2   Pyrolysis

With some exceptions, pyrolysis is the general term given to the thermal degradation of dry biomass at moderate temperatures without catalysts in the absence of oxygen [64]. Pyrolysis processes and technologies are broadly characterized as slow, fast, and flash pyrolysis according to the key determining reaction conditions of temperature, particle size, residence time, and heating rate (Table 16.3).

The products of pyrolysis include char, condensable organic liquids (bio-oil), and noncondensable gases. Bio-oil is generally the target product and contains acids, esters, alcohols, ketones, aldehydes, and a range of aromatic compounds formed from the depolymerization of the lignin component of the biomass. The incondensable gas typically contains CO, CO$_2$, CH$_4$, as well as H$_2$ and smaller quantities of hydrocarbon gases. These gases are usually used to provide energy for the biomass drying and pyrolysis process. The majority of ash in the biomass feedstock is retained in the char

TABLE 16.3   Range of Operating Parameters for Pyrolysis

Parameter	Slow Pyrolysis	Fast Pyrolysis	Flash Pyrolysis
Temperature (°C)	130–680	580–980	780–1030
Particle size (mm)	5–50	< 1	< 0.2
Residence time (s)	450–550	0.5–10	< 0.5
Heating rate (°C/s)	0.1–1	10–200	> 1000
High yield product	Char	Liquid (up to 17 wt%) and gas	Liquid (up to 70 wt%) and gas

Source: Data from Balat [52] with permission of John Wiley & Sons.

together with unreacted lignin and is used to supplement the energy requirements of the process or is used as a soil enhancer. Bio-oil can be substituted for coal, fuel oil, or diesel in some static applications including boilers and furnaces [65].

Bio-oil properties that prevent it being a drop-in liquid fuel substitute include incompatibility (immiscibility) with conventional fuels, solids (char and ash) content, high viscosity, and chemical instability. Significant work has been carried out in the area of chemical and physical upgrading of bio-oil; this has been thoroughly reviewed by Diebold [66]. Hot-gas filtration using ceramic or sintered steel filters (rather than simple cyclone separators) can reduce the ash content of the oil to less than 0.01% and the alkali content to less than 10 ppm.

Chemical/catalytic upgrading processes to produce hydrocarbon fuels that can be processed conventionally are more complex and expensive. Full deoxygenation to high-grade products such as transportation fuels can be accomplished by two main routes, namely, hydrotreating and catalytic vapor cracking of pyrolysis products over zeolites [67,68]. None of these chemical upgrading processes have been proved at the commercial- or demonstration-scale, although Dynamotive Energy Systems claims to have the capability to produce jet fuel, naphtha, and diesel at costs between US$1.82 and US$3.25 per U.S. gallon for biomass feedstock costs of US$30 and US$130 per tonne, respectively [69].

Low-temperature flash pyrolysis has been investigated as a means of fractionating biomass to produce phenol from the lignin component [70]; the production of furfural from the hemicellulose component of biomass has been studied and reported extensively (e.g. see Lau et al. [71]). Unlike the more "conventional" pyrolysis process for bio-oil production, phenol production was enhanced in an oxygen (air) environment and elevated levels of initial biomass moisture; a catalyst ($ZnCl_2$) was required to obtain significant furfural yields.

16.5.3   Liquefaction

During liquefaction of biomass, which is undertaken at relatively low reaction temperatures (Table 16.2), the macromolecules undergo degradation in a solvent (typically water) environment into highly unstable and reactive monomers that readily repolymerize into oil-like compounds (referred to hereafter as bio-crude to differentiate it from bio-oil produced by pyrolysis). Other products such as acetic acid and acetone remain dissolved in the solvent phase. The bio-crude products from conventional liquefaction are generally less oxygenated and less reactive when compared to pyrolysis products and hence are considered more suitable as a fuel. The reduction in oxygen is achieved during liquefaction by a combination of dehydration and decarboxylation reactions whereby oxygen is removed in the form of H_2O or CO_2, respectively. If oxygen is removed as H_2O, the ratio of O/C is decreased; the removal of oxygen as CO_2 results in an oil product with a higher H/C ratio [72].

In addition to the reduced oxygen content of the bio-crude, a further attraction of liquefaction is that the biomass feedstock does not require predrying, thereby permitting the utilization of high-moisture (also high ash), low-value lignocellulosic feedstocks. A disadvantage of the process is the higher pressures and longer residence times compared to gasification and pyrolysis processes.

Liquefaction yield and bio-crude composition depend on different physical and operating parameters, such as the lignin content of the feedstock, the solvents and choice of catalyst, residence time, and processing temperature. An extensive review of the literature pertaining to the impact of these factors on the liquefaction process is presented by Rackemann et al. [73] and is summarized here.

Biomass feedstocks with a higher proportion of lignin result in a larger amount of solid residue formed during liquefaction. The thermal decomposition of lignin during liquefaction leads to the formation of phenoxyl radicals which form solid residues through repolymerization [53, 74, 75]. The use of catalysts (such as K_2CO_3) has the greatest effect on reducing repolymerization, thereby increasing oil yield for biomass feedstocks with a high lignin content [75].

For reasons of cost, the solvent most commonly used in biomass liquefaction is water. Alternative organic solvents including glycerine, ethylene glycol, and acetone have been trialed. The yield of oil from these solvents (and water) was found to be limited at any given pressure by the corresponding boiling temperature of the solvents. In all cases the oil yield increased with increasing catalyst (KOH) concentration [76].

Heterogeneous catalysts have been shown to have low activity in biomass liquefaction compared with homogeneous catalysts such as organic and inorganic acids, NaOH, and salts. NaOH showed higher catalytic activity compared with salts such as phosphates, sulfates, chlorides, and acetates [53].

Oil yield is dependent on residence time and temperature. An initial increase in both residence time and temperature facilitates the depolymerization and fragmentation of cellulose, hemicellulose, and lignin into the oil-rich phase. Further increase in residence time and temperature beyond a critical point results in repolymerization and a reduction in oil yield [75, 77].

Liquefaction technology has yet to reach the commercial demonstration stage. The HydroThermal Upgrading (HTU) technology developed initially at the Shell Research Laboratory in the 1980s has made the greatest advances toward commercialization with the extended operation of a 100 kg/h dry feedstock rate pilot plant based at Apeldoorn in The Netherlands. This plant depolymerizes biomass in water at (subcritical) temperatures of 300–360 °C and pressures between 12 and 18 MPa. The reaction times are between 5 and 20 minutes during which a large fraction of the oxygen is removed as CO_2 and H_2O to produce bio-crude with oxygen contents as low as 10–18 wt% and lower heating values (LHVs) of 30–35 MJ/kg (approximately double that of the original dry feedstock). The fuel value of the bio-crude product as a percentage of the fuel value of the feedstocks and external fuel, associated with the HTU® process, is given as 74.9%.

16.6 FUELS AND CHEMICALS FROM BIOMASS

Most organic chemicals currently produced from fossil-based feedstocks can also be produced directly from biomass [78]. Over 300 potential value-added chemicals and products from biomass have been identified [78, 79]. These include chemicals that can readily be utilized as fuels in internal combustion engines such as ethanol and butanol. Chemicals can be classified as commodity, polymers, and plastics, or specialty and fine chemicals based on the interrelationship between market volume and price [78]. Building block chemicals are molecules that have multiple functional groups and can readily be transformed into new families of useful molecules. These can be further converted to secondary chemicals and intermediates and ultimately to industrial and consumer products [79].

The challenge for manufacturers of chemicals and fuels from biomass, however, is cost-competitive production. The economics of biofuels production can be enhanced through the coproduction of value-added chemicals and by-products in integrated biorefineries.

The fermentable sugars stream in a biorefinery may contain a complex mixture of glucose, fructose, galactose, xylose, mannose, and arabinose. In addition, depending on the pretreatment and hydrolysis conditions, carbohydrate degradation products including 5-hydroxymethyl furfural, furfural, formic acid, levulinic acid, and acetic acid, and phenolic compounds from lignin degradation may be present as well as lignin itself. Many of these degradation products are toxic or inhibitory to fermentation and may limit the fermentability of the biorefinery carbohydrates streams. In this case, hydrolysate detoxification strategies may need to be employed, or alternatively there may be a need to develop more resilient fermentation organisms through genetic engineering or adaptation strategies.

One of the main issues with thermochemical conversion of biomass to fuels and chemicals by pyrolysis and liquefaction is the multitude of products formed during biomass decomposition. A detailed understanding of the reaction pathways associated with the depolymerization and chemical kinetics of intermediate compounds will provide guidance in solving this issue.

A summary of assessments of the screening of potential candidates for the production of value-added chemicals from sugars, syngas, lignin, and biomass are shown in Table 16.4.

16.7 CONCLUSION

Lignocellulosic biomass offers great potential as a feedstock for the production of fuels and chemicals to replace fossil-based products. Significant advances have been made over the past decade in understanding the opportunities available from

TABLE 16.4 Summary of the Leading Potential Candidates for the Production of Value-Added Chemicals from Biomass

Chemicals from Sugars [79]	Chemicals from Syngas [80]	Chemicals from Lignin [81]	Products from Biomass [78]
1,4-Succinic, fumaric, and malic acids	Hydrogen	Macromolecules	1,2-Propanediol
	Ammonia	Carbon fiber	Epichlorohydrin
2,5-Furan dicarboxylic acid	Methanol and derivatives	Polymer modifiers	Lactic acid
3-Hydroxy propionic acid	dimethyl ether (DME)	Thermoset resins	Diesel
	Acetic acid	Aromatic chemicals	Gasoline
Aspartic acid	Formaldehyde	BTX (benzene,	Kerosine
Glucaric acid	Methyl *tert*-butyl	toluene, xylene)	Ethanol
Glutamic acid	ether (MTBE)	derivatives	Methanol
Itaconic acid	Methanol to olefins	Phenol	DME
Levulinic acid	Methanol to gasoline	Lignin monomers	Char
3-Hydroxybutyrolactone	Ethanol	Propylphenol	Wood pellets
Alcohols (e.g., glycerol,	Mixed higher	Eugenol	Animal feed
sorbitol, xylitol/	alcohols	Syringol	1,3-Propanediol
arabinitol)	Oxosynthesis	Oxidized lignin	Carbon dioxide
	products ($C_3 - C_{15}$	monomers	
	aldehydes)	Syringaldehyde	
	Isosynthesis products	Vanillan	
	(isobutene,	Vanillic acid	
	isobutane)		

waste residues and dedicated energy crops, and in developing cost-effective processing technologies. Many pilot-scale and demonstration-scale facilities are currently in operation globally, resulting in rapid advancement toward commercialization. Given the scale of the fuel and chemical markets globally, significant opportunities exist to produce products from these renewable and sustainable feedstocks and processes.

REFERENCES

1. O. Bobleter, Hydrothermal degradation and fractionation of saccharides and polysaccharides. In: S. Dimitriu (Ed.), *Polysaccharides: Structural Diversity and Functional Versatility*, Marcel Dekkar, New York, 2005, pp. 893–936.
2. A. Bauen, G. Berndes, M. Junginger, M. Londo, F. Vuille, Bioenergy—a sustainable and reliable energy source. A review of status and prospects, IEA Bioenergy, 2009.
3. IEA, *World Energy Outlook*, OECD/IEA, Paris, France, 2011.
4. P. Kumar, D. M. Barrett, M. J. Delwiche, P. Stroeve, Methods for pretreatment of lignocellulosic biomass for efficient hydrolysis and biofuel production, Ind. Eng. Chem. Res. 48 (2009) 3713–3729.

5. J. Y. Zhu, X. J. Pan, Woody biomass pretreatment for cellulosic ethanol production: technology and energy consumption evaluation, Bioresource Technol. 101 (2010) 4992–5002.

6. I. M. O'Hara, Z. Zhang, W O. S. Doherty, C. M. Fellows, Lignocellulosics as a renewable feedstock for chemical indutstry: chemcial hdyrolysis and pretreatment processes. In: R. Sanghi, V. Singh (Eds.), *Green Chemistry for Environmental Remediation*, Scrivener Salem, 2011, pp. 505–560.

7. J. X. Sun, X. F. Sun, H. Zhao, R. C. Sun, Isolation and characterization of cellulose from sugarcane bagasse, Polymer Degradation and Stability 84 (2004) 331–339.

8. D. Klemm, *Comprehensive Cellulose Chemistry*, Wiley-VCH, Weinheim, 1998.

9. C. E. Carraher, *Introduction to Polymer Chemistry*, CRC Press, Taylor & Francis Group, Boca Raton, FL, 2007.

10. M. J. John, S. Thomas, Biofibres and biocomposites, Carbohydrate Polymers 71 (2008) 343–364.

11. H. Jorgensen, J. B. Kristensen, C. Felby, Enzymatic conversion of lignocellulose into fermentable sugars: challenges and opportunities, Biofuels Bioprod. Biorefining 1 (2007) 119–134.

12. D. Klemm, B. Heublein, H. P. Fink, A. Bohn, Cellulose: fascinating biopolymer and sustainable raw material, Angew. Chem. Int. 44 (2005) 3358–3393.

13. L. Laureano-Perez, F. Teymouri, H. Alizadeh, B. Dale, Understanding factors that limit enzymatic hydrolysis of biomass, Appl. Biochem. Biotechnol. 124 (2005) 1081–1099.

14. B. C. Saha, Hemicellulose bioconversion, J. Ind. Microbiol. Biotechnol. 30 (2003) 279–291.

15. N. M. L. Hansen, D. Plackett, Sustainable films and coatings from hemicelluloses: a review, Biomacromolecules 9 (2008) 1493–1505.

16. A. Hendriks, G. Zeeman, Pretreatments to enhance the digestibility of lignocellulosic biomass, Bioresource Technol. 100 (2009) 10–18.

17. E. Palmqvist, B. Hahn-Hägerdal, Fermentation of lignocellulosic hydrolysates. II: Inhibitors and mechanisms of inhibition, Bioresource Technol. 74 (2000) 25–33.

18. X.-F. Sun, R.-C. Sun, L. Zhao, J.-X. Sun, Acetylation of sugarcane bagasse hemicelluloses under mild reaction conditions by using NBS as a catalyst, J. Appl. Polym. Sci. 92 (2004) 53–61.

19. W. Doherty, P. Halley, L. Edye, D. Rogers, F. Cardona, Y. Park, T. Woo, Studies on polymers and composites from lignin and fiber derived from sugar cane, Polym. Adv. Technol. 18 (2007) 673–678.

20. C. Bonini, M. D'Auria, New materials from lignin. In: M. D. Brenes (Ed.), *Biomass and Bioenergy : New Research*, Nova Science Publishers, New York, 2006, pp. 141–168.

21. O. Bobleter, Hydrothermal degradation of polymers derived from plants, Prog. Polym. Sci. 19 (1994) 797–841.

22. J. H. Grabber, How do lignin composition, structure, and cross-linking affect degradability? A review of cell wall model studies, Crop Sci. 45 (2005) 820–831.

23. O. J. Sanchez, C. A. Cardona, Trends in biotechnological production of fuel ethanol from different feedstocks, Bioresource Technol. 99 (2008) 5270–5295.

24. I. Dogaris, O. Gkounta, D. Mamma, D. Kekos, Bioconversion of dilute-acid pretreated sorghum bagasse to ethanol by *Neurospora crassa*, Appl. Microbiol. Biotechnol. 95 (2012) 541–550.

25. S. Gonzalez-Garcia, M. A. Moreira, G. Feijoo, Environmental aspects of eucalyptus based ethanol production and use, Sci. Total Environ. 438C (2012) 1–8.

26. J. D. McMillan, Pretreatment of lignocellulosic biomass. In: M. E. Himmel, J. O. Baker, R. P. Overend (Eds.), *Enzymatic Conversion of Biomass for Fuels Production*, Americal Chemical Society, Washington DC, 1994, pp. 292–324.

27. D. J. Fox, P. P. Gray, N. W. Dunn, W. L. Marsden, Factors affecting the enzymic susceptibility of alkali and acid pretreated sugar-cane bagasse, J. Chem. Technol. Biotechnol. 40 (1987) 117–132.

28. C. Martin, O. Almazan, M. Marcet, L. J. Jonsson, A study of three strategies for improving the fermentability of sugarcane bagasse hydrolysates for fuel ethanol production, Int. Sugar J. 109 (2007) 33–39.

29. C. Martin, B. Alriksson, A. Sjöde, N.-O. Nilvebrant, L. Jönsson, Dilute sulfuric acid pretreatment of agricultural and agro-industrial residues for ethanol production, Appl. Biochem. Biotechnol. 137–140 (2007) 339–352.

30. T. H. Kim, F. Taylor, K. B. Hicks, Bioethanol production from barley hull using SAA (soaking in aqueous ammonia) pretreatment, Bioresource Technol. 99 (2008) 5694–5702.

31. X. B. Zhao, K. K. Cheng, D. H. Liu, Organosolv pretreatment of lignocellulosic biomass for enzymatic hydrolysis, Appl. Microbiol. Biotechnol. 82 (2009) 815–827.

32. C. L. Li, B. Knierim, C. Manisseri, R. Arora, H. V. Scheller, M. Auer, K. P. Vogel, B. A. Simmons, S. Singh, Comparison of dilute acid and ionic liquid pretreatment of switchgrass: biomass recalcitrance, delignification and enzymatic saccharification, Bioresource Technol. 101 (2010) 4900–4906.

33. P. Beguin, J. P. Aubert, The biological degradation of cellulose, FEMS Microbiol. Rev. 13 (1994) 25–58.

34. Y. Sun, J. Y. Cheng, Hydrolysis of lignocellulosic materials for ethanol production: a review, Bioresource Technol. 83 (2002) 1–11.

35. V. S. Bisaria, Bioprocessing of agro-residues to glucose and chemicals. In: A. M. Martin (Ed.), *Bioconversion of Waste Materials to Industrial Products*, Elsevier, London, 1991, pp. 210–213.

36. F. M. Medie, G. J. Davies, M. Drancourt, B. Henrissat, Genome analyses highlight the different biological roles of cellulases, Nat. Rev. Microbiol. 10 (2012) 227–U.

37. S. J. B. Duff, W. D. Murray, Bioconversion of forest products industry waste cellulosics to fuel ethanol; a review, Bioresource Technol. 55 (1996) 1–33.

38. L. T. Fan, M. M. Gharpuray, Y.-H. Lee, *Cellulose Hydrolysis Biotechnology Monographs*, Springer, Berlin, 1987, p. 57.

39. D. B. Wilson, Cellulases and biofuels, Curr. Opin. Biotechnol. 20 (2009) 295–299.

40. D. Sternberg, Production of cellulase by *Trichoderma*, Biotechnol. Bioeng. Symp. (1976) 35–53.

41. P. Baldrian, V. Valaskova, Degradation of cellulose by basidiomycetous fungi, FEMS Microbiol. Rev. 32 (2008) 501–521.

42. E. A. Bayer, J. P. Belaich, Y. Shoham, R. Lamed, The cellulosomes: multienzyme machines for degradation of plant cell wall polysaccharides, Ann. Rev. Microbiol. 58 (2004) 521–554.

43. M. P. Coughlan, L. G. Ljungdahl, Comparitive biochemistry of fungl and bacterial cellulolytic system. In: J.-P. Aubert, P. Beguin, J. Millet (Eds.), *Biochemistry and Genetics of Cellulose Degradation*, Academic Press, New York, 1988, pp. 11–30.

44. M. E. Himmel, S. Y. Ding, D. K. Johnson, W. S. Adney, M. R. Nimlos, J. W. Brady, T. D. Foust, Biomass recalcitrance: engineering plants and enzymes for biofuels production, Science 315 (2007) 804–807.

45. I. Levy, O. Shoseyov, Cellulose-binding domains biotechnological applications, Biotechnol. Adv. 20 (2002) 191–213.

46. M. B. Sainz, Commercial cellulosic ethanol: the role of plant-expressed enzymes, In Vitro Cell Dev. Pl. 45 (2009) 314–329.

47. M. D. Harrison, J. Geijskes, H. D. Coleman, K. Shand, M. Kinkema, A. Palupe, R. Hassall, M. Sainz, R. Lloyd, S. Miles, J. L. Dale, Accumulation of recombinant cellobiohydrolase and endoglucanase in the leaves of mature transgenic sugar cane, Plant Biotechnol. J. 9 (2011) 884–896.

48. P. K. Foreman, D. Brown, L. Dankmeyer, R. Dean, S. Diener, N. S. Dunn-Coleman, F. Goedegebuur, T. D. Houfek, G. J. England, A. S. Kelley, H. J. Meerman, T. Mitchell, C. Mitchinson, H. A. Olivares, P. J. M. Teunissen, J. Yao, M. Ward, Transcriptional regulation of biomass-degrading enzymes in the filamentous fungus *Trichoderma reesei*, J. Biol. Chem. 278 (2003) 31988–31997.

49. V. Arantes, J. N. Saddler, Access to cellulose limits the efficiency of enzymatic hydrolysis: the role of amorphogenesis, Biotechnol. Biofuels 3 (2010) 4.

50. L. Zhang, C. Xu, P. Champagne, Overview of recent advances in thermo-chemical conversion of biomass, Energy Convers. Manage. 51 (2010) 969–982.

51. H. B. Goyal, D. Seal, R. C. Saxena, Bio-fuels from thermochemical conversion of renewable resources: a review, Renewable Sustainable Energy Rev. 12 (2008) 504–517.

52. M. Balat, Mechanisms of thermochemical biomass conversion processes. Part 1: Reactions of pyrolysis, Energy Sources Part A–Recovery Util. Environ. Eff. 30 (2008) 620–635.

53. F. Behrendt, Y. Neubauer, M. Oevermann, B. Wilmes, N. Zobel, Direct liquefaction of biomass, Chem. Eng. Technol. 31 (2008) 667–677.

54. J. J. Bozell, L. Moens, D. C. Elliott, Y. Wang, G. G. Neuenscwander, S. W. Fitzpatrick, R. J. Bilski, J. L. Jarnefeld, Production of levulinic acid and use as a platform chemical for derived products, Resources, Conservation and Recycling 28 (2000) 227–239.

55. D. J. Hayes, S. Fitzpatrick, M. H. B. Hayes, J. R. H. Ross, *The Biofine Process— Production of Levulinic Acid, Furfural, and Formic Acid from Lignocellulosic Feedstocks, Biorefineries-Industrial Processes and Products*, Wiley-VCH Verlag GmbH, Weinheim, 2008, pp. 139–164.

56. J. L. Kuester, *Diesel fuel from biomass, Energy from Biomass and Wastes VIII Symposium*, January 30—February 3, Lake Buena Vista, Florida, 1984.

57. H. Boerrigter, Economy of biomass-to-liquids (BTL) plants, Energy Research Centre of The Netherlands (ECN) Report ECN-C-06–019, 2006.

58. S. Philips, A. Aden, J. Jechura, D. C. Dayton, T. Eggeman, Thermochemical ethanol via indirect gasification and mixed alcohol synthesis of lignocellulosic biomass, National Renewable Energy Laboratory Technical Report, 2007.

59. P. L. Spath, D. C. Dayton, Preliminary screening—Technical and economic assessment of synthesis gas to fuels and chemicals with emphasis on the potential for biomass-derived syngas. Report prepared for U.S. Department of Energy, National Renewable Energy Laboratory. Report number NREL/TP-510–51034929: http://www.nrel.gov/docs/fy04osti/34929.pdf, 2003.

60. D. A. Bulushev, J. R. H. Ross, Catalysis for conversion of biomass to fuels via pyrolysis and gasification: a review, Catal. Today 171 (2011) 1–13.

61. Z. A. B. Z. Alauddin, P. Lahijani, M. Mohammadi, A. R. Mohamed, Gasification of lignocellulosic biomass in fluidized beds for renewable energy development: a review, Renewable Sustainable Energy Rev. 14 (2010) 2852–2862.

62. P. A. Hobson, Torrefaction and gasification for high efficiency second-generation biofuel production, In: *Proceedings of the Australian Society of Sugar Cane Technologists*, vol. 31, 2009, pp. 389–399.

63. J. C. Meyer, P. A. Hobson, F. Schultmann, The potential for centralised second generation hydrocarbons and ethanol production in the Australian sugar industry. In: *Proceedings of the Australian Society of Sugar Cane Technologists* (e-copy), 2012.

64. Q. Zhang, J. Chang, T. Wang, Y. Xu, Review of biomass pyrolysis oil properties and upgrading research, Energy Conv. Manag. 48 (2007) 87–92.

65. R. Sturzl, The commercial co-firing of RTP® bio-oil at the Manitowoc public utilities power generation station, vol. 2010. http://www.ensyn.com/docs/manitowoc.htm, 1997.

66. J. P. Diebold, *A Review of the Chemical and Physical Mechanisms of the Storage Stability of Fast Pyrolysis Bio-oils*, CPL Press, Berks, UK, 2002.

67. A. V. Bridgwater, Catalysis in thermal biomass conversion, Appl. Catal. A: General 116 (1994) 5–47.

68. A. V. Bridgwater, Production of high grade fuels and chemicals from catalytic pyrolysis of biomass, Catal. Today 29 (1996) 285–295.

69. Dynamotive, Modern clean burning fuels. http://www.dynamotive.com/fuels, 2012.

70. D. Butt, Formation of phenols from the low-temperature fast pyrolysis of radiata pine (*Pinus radiata*): Part II. Interaction of molecular oxygen and substrate water, J. Anal. Appl. Pyrolysis 76 (2006) 48–54.

71. G. J. Lau, N. Godin, H. Maachi, C. S. Lo, S. J. Wu, J. X. Zhu, M. L. Brezniceanu, I. Chenier, J. Fragasso-Marquis, J. B. Lattouf, J. Ethier, J. G. Filep, J. R. Ingelfinger, V. Nair, M. Kretzler, C. D. Cohen, S. L. Zhang, J. S. Chan, Bcl-2-modifying factor induces renal proximal tubular cell apoptosis in diabetic mice, Diabetes 61 (2012) 474–484.

72. M. Balat, Mechanisms of thermochemical biomass conversion processes. Part 3: Reactions of liquefaction, Energy Sources Part A–Recovery Util. Environ. Eff. 30 (2008) 649–659.

73. D. W. Rackemann, L. Moghaddam, T. J. Rainey, C. F. Fellows, P. A. Hobson, W. O. S. Doherty, *Hydrothermal Technologies for the Production of Fuels and Chemicals from Biomass, Green Chemistry for Environmental Remediation*, John Wiley & Sons, Hoboken, NJ, 2011, pp. 291–342.

74. A. Demirbas, Mechanisms of liquefaction and pyrolysis reactions of biomass, Energy Conv. Manag. 41 (2000) 633–646.

75. C. Zhong, X. Wei, A comparative experimental study on the liquefaction of wood, Energy 29 (2004) 1731–1741.

76. V. R. Rustamov, K. M. Abdullayev, E. A. Samedov, Biomass conversion to liquid fuel by two-stage thermochemical cycle, Energy Conv. Manag. 39 (1998) 869–875.

77. Y. Qu, X. Wei, C. Zhong, Experimental study on the direct liquefaction of *Cunninghamia lanceolata* in water, Energy 28 (2003) 597–606.

78. A. V. Bridgwater, R. Chinthapalli, P. W. Smith, Identification and market analysis of most promising added-value products to be co-produced with the fuels, vol. Deliverable 2 Total, Bioref-Integ, 2010.

79. T. Wcrpy, G. Petersen, A. Aden, J. Bozell, J. Holladay, J. White, A. Manheim, Top value added chemicals from biomass. Volume 1: Results of screening for potential candidates from sugars and synthesis gas, PNNL/NREL, 2004.

80. P. Spath, D. Dayton, Preliminary screening—Technical and economic assessment of synthesis gas to fuels and chemicals with emphasis on the potential for biomass-derived syngas, vol. NREL/TP-510-34929, NREL, 2003.

81. J. E. Holladay, J. J. Bozell, J. F. White, D. Johnson, Top value added chemicals from biomass. Volume II: Results of screening for potential candidates from biorefinery lignin, PNNL/NREL, 2007.

Natural and Artificial Photosynthesis: Solar Power as an Energy Source, First Edition. Edited by Reza Razeghifard.
© 2013 John Wiley & Sons, Inc. Published 2013 by John Wiley & Sons, Inc.